Microwave-Assisted Synthesis
Innovations and Applications in Materials and Chemistry

Edited by

Inamuddin[1], Tariq Altalhi[2], Arwa Alrooqi[3], Maha Khan[1]

[1]Advanced Functional Materials Laboratory, Department of Applied Chemistry, Zakir Husain College of Engineering and Technology, Faculty of Engineering and Technology, Aligarh Muslim University, India

[2]Department of Chemistry, College of Science, Taif University, 21944 Taif, Kingdom of Saudi Arabia

[3]Department of Chemistry, Faculty of Science, Al-Baha University, Al Baha, Saudi Arabia

Published by **Materials Research Forum LLC**
Millersville, PA 17551, USA

Published as part of the book series
Materials Research Foundations
Volume 189 (2026)
ISSN 2471-8890 (Print)
ISSN 2471-8904 (Online)

Print ISBN 978-1-64490-402-2
eBook ISBN 978-1-64490-403-9

Distributed worldwide by

Materials Research Forum LLC
105 Springdale Lane
Millersville, PA 17551
USA
https://mrforum.com

Manufactured in the United States of America
10 9 8 7 6 5 4 3 2 1

Table of Contents

Preface

With the alarming rise in pollution levels and the growing threat of global warming, there is an urgent need for innovative scientific and technological advancements that can help mitigate these global challenges. Sustainable and environmentally responsible approaches in chemistry and materials science are no longer optional but essential. In this context, microwave chemistry has emerged over the past decade as a powerful and transformative tool, offering numerous advantages such as rapid synthesis, simplified experimental setups, higher reaction yields, enhanced reaction rates, and cost-effectiveness. Importantly, it provides a greener alternative for the synthesis of a wide range of organic molecules and materials.

Microwave-Assisted Synthesis: Innovations and Applications in Materials and Chemistry aims to serve as a comprehensive resource for scientists, engineers, researchers, and industrial practitioners interested in understanding the fundamentals and exploring the diverse applications of microwave-assisted synthesis. The book brings together foundational principles, operational advantages, and cutting-edge developments in the microwave-assisted preparation of organic and inorganic compounds, advanced composite materials, polymers, and nanostructures.

The volume begins with an in-depth discussion of the fundamental concepts and working principles underlying microwave-assisted synthesis. It then progresses to detailed accounts of real-time and practical applications, including microwave synthesis of heterocycles, polymer synthesis, the preparation of inorganic compounds e.g. zeolites, synthesis of ionic liquids and radiopharmaceuticals, edible film and coating material, advanced composite materials, and nanostructures. Furthermore, specialized and emerging applications—such as targeted drug delivery systems, heterogeneous catalysis, corrosion inhibition, and future research directions—are thoroughly examined.

This book is designed to be both instructional and forward-looking. It serves as a valuable guide for readers seeking a clear understanding of the principles, advantages, and practical potential of microwave-assisted synthesis. The content is structured to benefit beginners entering the field as well as advanced researchers seeking deeper insights and innovative perspectives. The primary audience for this book includes scientists, organic chemists, polymer chemists, spectroscopic researchers, and industrial professionals who are keen to adopt greener methodologies for synthesizing organic compounds, polymers, and advanced materials. By minimizing or eliminating reliance on toxic and energy-intensive conventional methods, microwave-assisted synthesis offers a sustainable pathway toward modern chemical manufacturing. We hope that this volume will inspire researchers and industrialists alike to embrace environmentally responsible strategies for producing large-chain polymers, zeolites, hybrid composites, drug delivery systems, and other advanced functional materials. We sincerely hope that this book contributes meaningfully to the advancement of sustainable chemistry and materials science, and serves as a catalyst for further innovation in microwave-assisted technologies. The book consists of fourteen chapters which are summarised below:

Chapter 1 examines microwave-assisted synthesis as a rapid and energy-efficient alternative to conventional heating methods. A bibliometric analysis of publications from the last decade highlights key contributors, major applications, and the growing interest in microwave-based techniques. Performance comparisons between microwave-assisted and conventional methods are presented, along with sustainability and green chemistry benefits. Major emphasis is placed on the advantages, limitations, and future potential of microwave technology across academic and industrial contexts.

Chapter 2 provides an in-depth overview of microwave-assisted synthesis of bioactive heterocycles and its significance in medicinal chemistry. Various classes of heterocycles, including pyrimidines, quinazolines, indoles, thiazoles, oxazoles, and imidazoles, are discussed along with their microwave-enabled synthetic strategies. The chapter highlights how microwave irradiation enhances reaction rates, yields, and selectivity while supporting greener, solvent-free, and multicomponent approaches. Particular emphasis is placed on pharmacological applications, including anticancer, antimicrobial, antiviral, anti-inflammatory, antimalarial, and central nervous system activities. Advantages, limitations, and future potential are critically discussed, underscoring the role of microwave-assisted heterocycle synthesis as a transformative and sustainable tool in modern drug discovery.

Chapter 3 focuses on the synthesis of nitrogen-containing heterocycles, which are highly prevalent in pharmaceuticals due to their bioactive properties. Microwave-assisted synthesis of these nitrogen heterocycles is presented across five-, six-, and seven-membered ring systems. In five-membered systems, compounds containing one, two, three, and four nitrogen atoms—such as pyrroles, pyrazoles, imidazoles, triazoles, tetrazoles, and their derivatives—are discussed. The synthesis of six- and seven-membered heterocycles containing one or two nitrogen atoms is outlined. Applications of these compounds in medicinal chemistry are also discussed.

Chapter 4 describes the microwave-assisted synthesis of polypyrrole (PPy) and highlights how this method accelerates polymerization, improves product purity, and enables better control over morphology compared with conventional polymerization approaches. The chapter outlines the fundamentals of microwave heating and presents examples of PPy nanoparticles, thin films, and nanocomposites prepared using microwave irradiation. Potential applications and enhanced performance of microwave-synthesized PPy in sensors, supercapacitors, energy devices, biomedicine, and antimicrobial materials are also discussed.

Chapter 5 highlights different approaches for the synthesis of spiro-heterocyclic compounds via microwave-assisted techniques for pharmacological applications such as anticancer, antimicrobial, and anti-inflammatory therapies, as well as for use in optoelectronic device development.

Chapter 6 provides a comprehensive overview of zeolites, including their significance, advancements in synthesis techniques, and role in modern materials science. Structural characteristics, framework diversity, and classification are discussed in detail, with distinctions between natural and synthetic zeolites and an overview of the zeolitization process. Primary emphasis is placed on microwave-assisted zeolite synthesis and its advantages over other synthesis approaches. The chapter also covers commercial-scale manufacturing, adsorption processes, post-synthetic modification, pH effects on adsorption performance, and applications such as ion exchange, molecular sieving, hydrogen purification, drying, gas removal, pollutant adsorption, dye removal, heavy metal adsorption, and catalytic reduction.

Chapter 7 describes the structural features of zeolites, including their natural and synthetic forms, along with synthesis methods such as hydrothermal, sol–gel, mechanochemical, and microwave-assisted techniques. Special emphasis is placed on microwave heating, which accelerates nucleation and crystallization, reduces reaction time, and improves material uniformity compared with traditional processes. The chapter also discusses the roles of zeolites in catalysis, adsorption, and environmental purification, as well as emerging biomedical applications including drug delivery, antimicrobial systems, tissue engineering scaffolds, and diagnostic biosensors. Challenges related to cost-effective production, structural control, environmental impact, and toxicity assessment are addressed, while highlighting the growing potential of zeolites and zeolitic imidazolate frameworks (ZIFs) in sustainable technologies and biomedical applications.

Chapter 8 discusses the current research status, reaction mechanisms, existing challenges, optimization strategies, and industrial application cases of microwave-assisted synthesis of ionic liquids (ILs), and explores future development directions. The structure, physicochemical properties, and applications of ionic liquids are also described.

Chapter 9 describes microwave-assisted synthesis (MAS) as a powerful technique for the production of radiopharmaceuticals. It explores mechanistic foundations, technological innovations, regulatory considerations, and clinical applications of MAS in radiochemistry. The chapter demonstrates how MAS is reshaping radiopharmaceutical development by offering a scalable and efficient approach aligned with the demands of modern nuclear medicine.

Chapter 10 provides a detailed discussion of microwave-assisted synthesis of metal ferrites for targeted drug delivery. It examines synthesis methods including sol–gel, hydrothermal, precipitation, solvothermal, and microwave-assisted approaches, with particular emphasis on the mechanisms, principles, and advantages of microwave-assisted synthesis. Applications in targeted drug delivery are highlighted, focusing on the role of metal ferrites as drug carriers, contrast agents, imaging tracers, and components in immunoassays.

Chapter 11 presents an in-depth discussion of microwave-assisted transition metal-catalyzed reactions and their importance in accelerating organic transformations and enhancing reaction efficiency. The chapter covers both noble metals (Ru, Rh, Pd) and non-noble metals (Fe, Co, Ni, Cu, Zn) involved in coupling reactions such as Suzuki, Heck, Stille, Sonogashira, and Negishi reactions, as well as other organic transformations. Mechanistic examples demonstrate how microwave irradiation improves yield, selectivity, and sustainability by reducing reaction time and by-product formation. Although scalability remains a challenge, the chapter concludes that integrating microwave technology with advanced catalytic systems offers a powerful tool for green, efficient, and rapid synthetic methodologies.

Chapter 12 discusses biomass valorization through microwave-assisted pyrolysis (MAP). Although this technology is still under development toward full commercialization, MAP offers improved heat distribution and product customization compared with conventional pyrolysis. Current MAP pyrolyzer designs are being refined to address technical complexity, high capital costs, scalability, and localized heating issues. Integration of MAP with existing biorefinery processes is highlighted as a promising approach to improving the sustainability of biomass conversion into value-added products.

Chapter 13 describes the use of microwave processing to develop edible films and coatings from polysaccharides, proteins, composites, and agro-industrial by-products. Emphasis is placed on rapid drying, structural modification, and improvements in barrier, mechanical, optical, and antimicrobial properties. The chapter also discusses microwave-assisted extraction, combined treatments, and challenges related to scalability and non-uniform heating.

Chapter 14 provides a comprehensive assessment of microwave-assisted synthesis of photocatalysts, emphasizing their effectiveness in enhancing hydrogen generation efficiency. It explains the fundamental microwave heating mechanisms and their influence on the reaction dynamics involved in photocatalyst formation. Various microwave-assisted strategies for optimizing the physicochemical properties of photocatalysts are discussed. The chapter concludes with a critical evaluation of the potential of microwave-engineered photocatalysts for high-efficiency and large-scale hydrogen production.

Materials Research Foundations 189 (2026) 1-44 https://doi.org/10.21741/9781644904039-1

Chapter 1

Microwave-Assisted vs Conventional Synthesis: An Overview

H. Chenni[1], H. Kerras[1], N. Outili[1*]

[1]Laboratory of environmental process engineering (LIPE). Process engineering faculty. Constantine3 Salah Boubnider. Algeria

nawel.outili@univ-constantine3.dz

Abstract

Microwave assisted synthesis offers a modern alternative to conventional heating by directly coupling electromagnetic energy with reactants, resulting in rapid heating. This approach accelerates reaction rates, improves selectivity, and reduces energy consumption. Compared to conventional methods, based on long heat transfer times. However, microwave synthesis enhances the process efficiency and supports green chemistry. To provide an overview of this challenge on microwave technology, a bibliometric study of scientific publications between 2014-2024 was conducted, developing significant growth and increasing the interest in microwave methods through various applications such as organic, nano-materials, and pharmaceuticals. This study highlights the transformative potential of microwave synthesis for enhancing sustainability in both academic and industrial scales.

Keywords

Microwave, Conventional Heating, Green Chemistry, Sustainable Processes, Selective Heating, Process Intensification

Abbreviations Word

Abbreviation	Meaning
RF	Radio Frequency
HL	Hysteresis Losses
ECL	Eddy Current Losses
MRL / RL	Magnetic Resonance / Residual Losses
FM	ferro-magnetic
PI	Process Intensification
E	Environmental
Word	Abbreviations
MW	Microwave
MWs	Microwaves
EM	Electromagnetic
Ir	Irradiation
WCO	Waste cooking oil

Materials Research Foundations 189 (2026) 1-44 https://doi.org/10.21741/9781644904039-1

UV	Ultraviolet
CC	Centrale composite

Contents

Materials Research Foundations 189 (2026) 1-44 https://doi.org/10.21741/9781644904039-1

1. Introduction

In recent years, reaction processes via microwave irradiation (MW-Ir) technology have been a progressively interesting research topic in the modern scientific world. This technique has been used in chemical reaction modification since the 1950s. MW- assisted, it obtained understanding acceptance after research papers presented by Richard Gedye and George Majctich in 1986 [1,2], where it was used in domestic kitchen MW equipment to perform amide hydrolysis, transformation of ester to carboxylic acids, conversion of carboxylic acids by esterification reaction using alcohol products, producing aromatic acids by oxidation reaction from alkyl benzenes, and transformation of ethers using alkyl halides, and Diels-Alder organic reactions [3-5]. Taking into account, the difficulty of controlling the reaction process: pressure and boiling temperature of used reagents [6]. During the 1990s, the development to solve reaction media using microwaves was widely studied, in order to avoid boiling and explosion risk problems [7]. Since the year 2000, several commercial microwave synthesizers have been appeared, for chemical synthesis purposes [8]. The fundamental basis of dielectric MW-Ir heating is covered in a few review studies, they are divided into two primary mechanisms: ionic conduction and dipolar polarization [9,10]. In general, MW heats all solid or liquid substances that contain electrically moving charges, i.e., polarized molecules or conductive ions [11-16]. However, irradiation ovens as a heat source in chemistry laboratory research have made it possible to achieve many changes; it can be employed to minimize reaction duration [17], with high energy efficiency [18], enhanced yields, selectivity, and purity [19]. Based on the 12 green chemistry principles, two significant challenges for MW-assisted chemistry are sustainable solvents and green energy [20,21]. Generally, conventional thermal heating synthesis is based on conduction/convection heating transfer, e.g., via a directly contact between two substances or by indirect contact using gas/liquid flow to heat substances, takes place from the external phase to the internal phase in the presence of an interface between the heat donor and the heat absorber, according to the heat mode [22,23]. Compared to microwave-assisted, where it is based on radiation heating transfer. Radiative mode does not involve a direct contact with the material; it can be transmitted through space in the form of electromagnetic infrared radiations, which are absorbed by the substance [24]. MWs take a trend in the intensification methods [25]. The efficiency of MWs for controlled heating is typically viewed in many operations, while MWs need lower energy consumption compared to conventional heating [20]. While MW-assisted process occurs in seconds/minutes, conventional heating takes hours.

Kerras et al. [26] have applied microwave radiation in waste cooking oil (WCO) pretreatment to increases biodiesel conversion, to enhance WCO purity by reducing acid value parameter. Furthermore, microwaves have a wide range of applications. Among them, organic chemistry [27], catalysis synthesis, nanomaterials, and polymers [28,29].

This chapter highlights the difference between specific effects of MW and conventional heating in chemical synthesis, fundamental principles of MW heating, microwave as process intensification, and their role in green chemistry. In addition, part of a doctoral dissertation conducted in Algeria, defended in February 2024 was included in this chapter as an example of microwave-assisted synthesis.

In order to better situate the evolution and research dynamics related to the use of microwaves in chemical synthesis, a bibliometric analysis was conducted. This approach makes it possible to identify scientific trends, priority application areas and key contributors to the field [30]. Interest in this technology has experienced sustained growth, stimulated by the search for more efficient, selective and environmentally friendly processes. For this study, data were extracted from the Lens database [31,32] using a targeted query on keywords related to microwave-assisted synthesis, focusing on the last twenty years. The analysis was carried out using the Lens analysis [31], and VOSviewer tools [33], to map the co-occurrences of terms, the relationships between authors, institutions and temporal trends.

2. Fundamental principles of microwave-assisted synthesis

Microwave and conventional synthesis represent two different methods for heating and driving chemical reactions. In contrast, conventional heating is based on external heating energy. As an example: oil baths, hot plates, ovens, etc. In order to transfer energy from the power source to the reaction mixture through conduction and convection phenomena. On the other hand, the principle of MW-assisted synthesis is that it uses EM-radiation to directly interact with polar molecules and ions to generate heat energy through thermal and non-thermal mechanisms.

2.1 Electromagnetic and fields frequency

Dielectric heating is generally defined as the process of generating heat within a specific material through the use of high-frequency EM fields. Knowing that frequencies above 300 MHz correspond specifically to the MW range. As shown in the EM spectrum Fig.1, it is important to note that MW, radio-frequency (RF), infra-red (IR), and ultraviolet (UV) frequency are classified as non-ionizing radiation. This means they have insufficient energy to remove ions, electrons, molecules, or atoms from substances. Nevertheless, microwave chemistry offers numerous advantages, such as greener processing, higher yields, and improved selectivity [34].

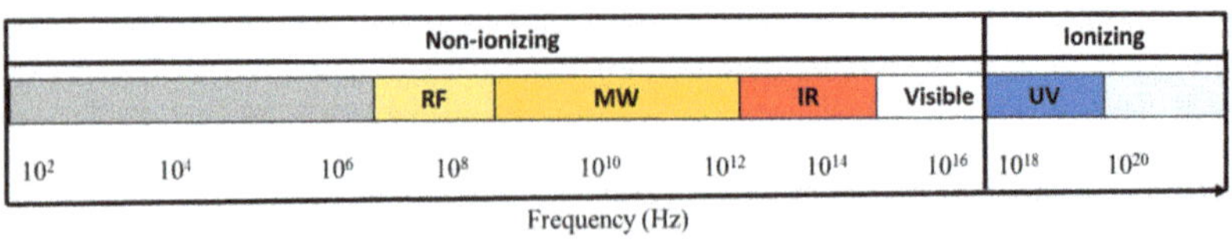

Figure 1. EM-Spectrum. (reproduced by the authors)

2.2 Mechanisms of microwave interaction with matter

Microwaves interact with matter through different fundamental mechanisms, resulting in the absorption, reflection, or transmission of EM energy.

2.2.1 Thermal and non-thermal effects of microwave interaction with matter

Thermal and specific thermal effects come from the transformation of EM microwave energy into heat within the material, which dominate most MW substance interactions. Noting that the principal thermal mechanisms are: dipolar/dipole polarization, ionic conduction, and other dielectric losses. Taking into account the specific phenomena that rarely occur under conventional heating, such as selective, super-heating, and localized hot-spots/temperature gradients.

Moreover, non-thermal microwave effects refer to energy transfer processes arising from the direct interaction between the EM microwave field and polar molecule/intramolecular polar groups, which potentially lead to structural modifications, electron transfer, or other physicochemical changes, without being associated with a measurable increase in bulk temperature [35]. Fig.2 showing a comprehensive view on these effects.

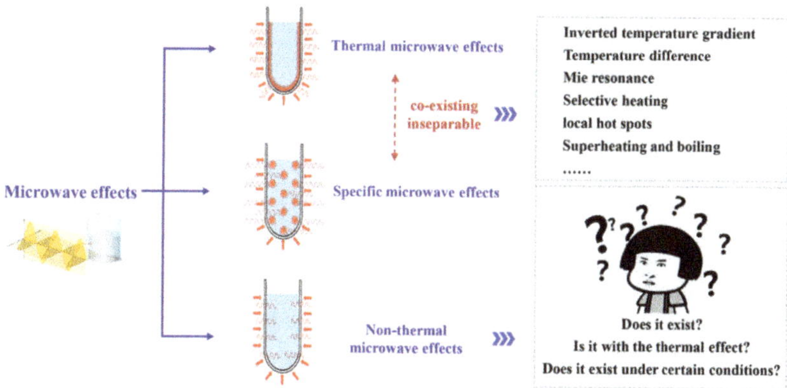

Figure 2. MW effects categorizes (1) thermal, (2) specific, (3) and non-thermal effects, with the latter involving direct electromagnetic interactions that may induce molecular changes beyond simple heat generation [35].

Jacob et al.[36] reviewed studies on both thermal and non-thermal effects, particularly in the context of organic chemistry. When experimental conditions such as temperature (T °C) and pressure (P) are controlled, volumetric heating is observed [34]. This results in a reduced reaction time due to rapid heat generation. On the other hand, the non-thermal interaction of MWs, also referred to as "specific MW interaction", was discussed by Jacob, who highlighted its principal effects in enhancing or reducing reaction times under microwave irradiation [36]. In terms of:

a. Intermolecular zones, which are responsible for the overheating of certain MW heating substances, named hot spots.

b. Molecular agitation of polar substances results in collision phenomena in the reaction medium between reactants.

c. Diffusion phenomena, responsible for reducing reaction rate compared to the classic heating, by increasing the transport properties of the molecule substances.

- Dipolar/ Dipole polarization mechanism

Dipolar compounds such as water, ethanol, and methanol exhibit high sensitivity to external electric fields owing to polarization effects, leading them to reorient in alignment with the EM field [37,38]. Under high-frequency conditions, dipoles interacting with the electric absorption field experience mutual collisions, leading to energy dissipation as heat, since they lack sufficient time to align with the oscillating field [38,39]. As an example H_2O molecule of water ; is known as a polar substance, due to their high specific electronegativity of the negative oxygen atom charge, Fig.3. While oxygen ions are capable. Subsequently, electrons create a negative charge around oxygen and a positive charge around hydrogen, separated and displaced depending on the EM field [40].

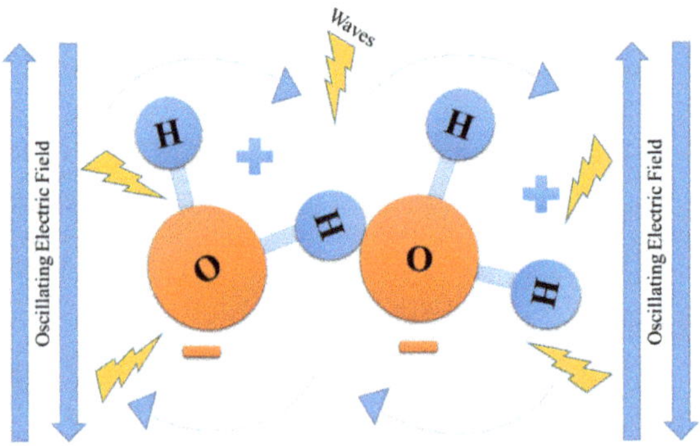

Figure 3. Dipolar polarization mechanism representation (water molecule). Produced by the authors

- Conduction mechanism

In addition to the dipole polarization mechanism matter containing mobile ions, negative and positive charge carriers, as shown in Fig.4. Under microwave EM field effect, the charges can move back and forth, to starting oscillate. Which constitutes an electric current. Heat is generated in the matter as a result of electrical resistance, when the charged species collide with adjacent molecules or atoms [38]. However, in chemical systems, the conduction mechanism is considered a fundamental MW induced mechanism. Which dominates the aqueous electrolytes, ionic solutions and metallic elements [41].

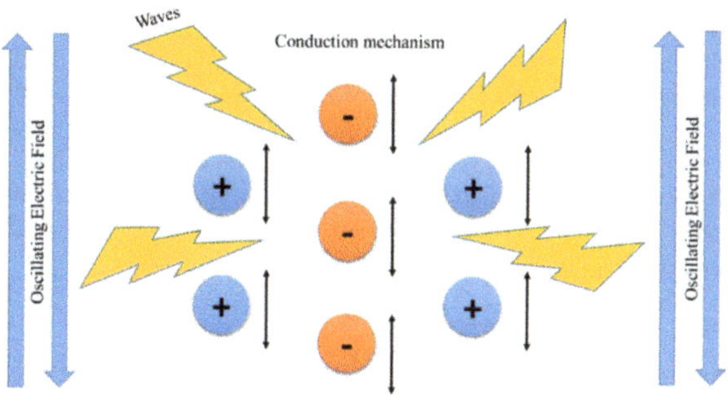

Figure 4. Ionic conduction mechanism representation. Produced by the authors

Anwar et al. [42] research has examined the interactions of MWs with dielectric aqueous systems. The effect of ionic concentration was successfully highlighted on the conversion of EM into thermal energy. Various inorganic ions from aqueous solutions of anhydrous chlorides of lithium (LiCl), sodium (NaCl), potassium (KCl), cesium (CsCl), magnesium ($MgCl_2$), calcium ($CaCl_2$), strontium ($SrCl_2$), barium ($BaCl_2$), nickel ($NiCl_2$), copper, cobalt, urea (CH_4N_2O), and sugar ($C_{12}H_{22}O_{11}$) were exposed to microwaves of 2.45 GHz frequency. As a results, a significant observation on decrease in the medium temperature. With the increase of ion concentration, compared with deionized water.

In certain chemistry MW heating situations, both MW mechanisms can be present together, especially in reaction processes where conducting substances are contained in other non-conducting substances, such as electrolyte mediums. Horikoshi et al. [43] It was proved that under the common MW conditions (at 2.45 GHz), the heat rate of the dissolved salts such as NaBr, NaCl, etc., in water is faster than ultrapure water (H_2O). Taking into consideration the effect of conductive materials in enhancing the MW heating efficiency of the reaction medium. Unlike aqueous electrolyte solutions, solid dielectric materials with mobile free charged particles have the ability to move in certain regions of the material. For example, in carbon materials, π-electrons can interact in phase with the EM-MW field. Nevertheless, π-electrons cannot move rapidly, when the EM-field changes direction at 2.45 GHz. According to Maxwell-Wagner Polarization, shown in Fig.5, these charges accumulate in the interfaces to generate MW heat energy [44].

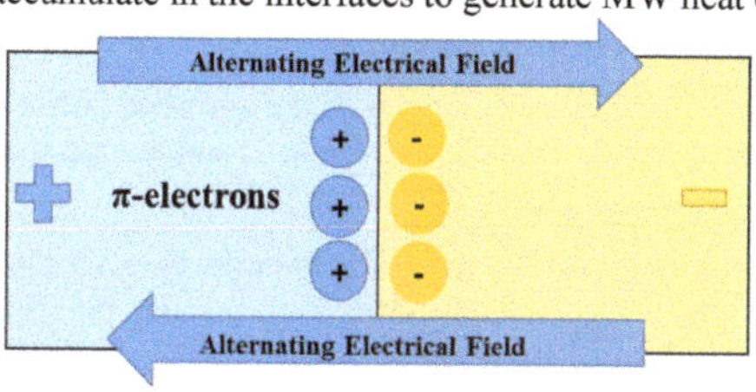

Figure 5. Maxwell-Wagner Polarization (Interfacial). Produced by the authors

Materials Research Foundations 189 (2026) 1-44 https://doi.org/10.21741/9781644904039-1

Basically, the EM theories in matter differ between disciplines mainly in convention and notation terms. In chemical science, due to the presence of complex chemical reaction introduces a significant scientific challenge. Among them are the microscopic competing reaction pathways, in addition to the complexities that complicate understanding at macroscopic levels [45]. Furthermore, previous studies have developed different theoretical approaches based on EM fundamentation to study the MW phenomena, using Maxwell's equations. Which are based on the distribution of EM-waves in terms of interaction, with the objective to estimate the absorbed, transmitted/ reflected energy by dielectric absorbed MW substances, with the necessary knowledge of ($\boldsymbol{\varepsilon}$) and permeability ($\boldsymbol{\mu}$) properties.

2.2.2 Magnetic field effects

MW phenomena have spotlighted the attention on the magnetic field (H). Especially on ferrites and conductive powder materials. Three principal mechanisms account for energy losses in the microwave magnetic field: circulation of eddy currents losses, dissipation linked to magnetic hysteresis, and energy absorption related to magnetic resonance effects [46-49]. As a definition :

Eddy current losses (ECL) : are the circulating currents induced in conductive microwave materials by the changing magnetic field (H), as described by Faraday's rule[50]. The dissipated heat energy is formed through the perpendicular closed loops to the magnetic field (H) in terms of ($P = I^2R$), which represents the undesired loss [51].

Hysteresis losses (HL): due to the oscillation of magnetic field (H), magnetization phenomenon forms an area loop, which is called hysteresis loop. In this case, the magnetic field (H) exerted on the ferro-magnetic (FM) substances, a hysteresis heat loss energy was produced [52,53].

Magnetic resonance / Residual losses (MRL/ RL): this phenomenon produced principally from the magnetic domain wall field and the spain electron resonances. While, the oscillations within the magnitic domains filed (H) and the electrons spain absorbed by the MW energy for heat objective [54-56].

As general results of the previous section, ECL is the main contributor to MW-heating, in different applications in conductive and semiconductive chemical materials [57-59]. In addition, HL it occurs in FM materials [60,61]. Also MRL /RL is involved in induction heating in magnetic materials under oscillating magnetic fields [62-64]. On the other hand, all these losses ECL, HL, and MRL can act at the same time, to promote ferrites and the conductive materials to generate MW-heating [38].

2.3 General equations in microwave heating

As an overview of equation systems used to estimate the absorbed power density, and penetration depth of materials in MW systems. Table 1 and Table 2, summarize some theoretical equations and nomenclature suggestions from previous research [65-68].

Eq.1, estimates the power density of MW absorbing material at (x,y,z) position, depending on the effective dielectric loss factor ε''_{eff}, which takes into account the dipolar polarization, ionic conduction, and the interfacial polarization interactions.

Materials Research Foundations 189 (2026) 1-44 https://doi.org/10.21741/9781644904039-1

Eq.2, estimates the power density of magnetic MW heating at the 3-dimensional space position on W/m^3, depending on magnetic intensity $H_{rms}(A/m)$. This factor takes into account the dominant influencing phenomena, including HL, ECL, and RL.

Eq.3, estimates the power loss density owing, to MW heating, expresses the energy dissipated (per volume unit) in the material, at (x,y,z) position, which integrates both fields at once.

Eq.4, estimates the general penetration depth of MWs heating. Where the heat efficiency may be limited for large samples despite high dielectric loss, due to low penetration depth, which serves as a key factor for evaluating heating performance and uniformity.

Table 1. General equation presented in previous research.

Equations		**Ref**
Power density of dielectric MW heating		
$P\left(\frac{W}{m^3}\right) = \omega . \varepsilon''_{eff} . \varepsilon_0 . E^2_{rms}$ (1)	$\omega\ (\mathrm{Hz}) = 2.\pi.f$	[69,70,38]
	$\varepsilon_0\ (F/m) = 8.854 \times 10^{-12}$	
	$\varepsilon''_{eff} = \varepsilon''_{polarization} + \varepsilon''_{conduction}$ $\varepsilon''_{eff} = \varepsilon''_{Dipolar} + \varepsilon''_{interfacial} + \sigma/\omega.\varepsilon_0$	
Power density of the magnetic MW heating		
$P(W/m^3) = \omega . \mu''_{eff} . \mu_0 . \mathrm{H}^2_{rms}$ (2)	$\mu''_{eff} = \mu''_{hyst} + \mu''_{eddy} + \mu''_{res}$	[65,71]
	$\mu_0\ (H/m) = 4\pi \times 10^{-7}$	
Power density dissipated owing to MW heating		
$P_{d(x,y,z)}(W/m^3) = \omega\mu''_{eff} . \mu_0 . H^2_{rms} + \omega . \varepsilon''_{eff} . \varepsilon_0 . E^2_{rms}$ (3)		[71]
Penetration Depth of Microwaves		
$d = 2D_p = \frac{1}{\alpha} = \frac{\lambda_0}{\sqrt{2\pi}}\left\{\varepsilon''_r\mu''_r - \varepsilon'_r\mu'_r + \left[(\varepsilon'_r\mu'_r)^2 + (\varepsilon''_r\mu''_r)^2 + (\varepsilon'_r\mu''_r)^2 + (\varepsilon''_r\mu'_r)^2\right]^{\frac{1}{2}}\right\}^{\left(-\frac{1}{2}\right)}$ (4)		[73-75]

Materials Research Foundations 189 (2026) 1-44 https://doi.org/10.21741/9781644904039-1

Table 2. General nomenclature suggestion of MW parameters.

Parametre	Definition
P	Power dissipation density.
ω	Angular frequency.
ε_0	Corresponding to the free space permittivity.
ε''_{eff}	Effective dielectric dissipation/ loss factor. Taking into consideration the mechanism type (Accounting for conductivity losses as well as polarization losses).
μ''_{eff}	Imaginary part of the relative effective magnetic permeability.
H_{rms}	Represents the value in terms of A/m, at the 3-dimensional space of magnetic intensity fields.
E_{rms}	Electric field corresponding to the local position (x,y,z).
f	MW frequency.
μ_0	Vacuum magnetic permeability.
λ_0	MW wavelength in free space (m)
D_p	Penetration depth (m).
ε''_r	Imaginary part of complex relative permittivity.
ε'_r	Real part of complex relative permittivity.
μ''_r	Real part of complex relative permeability.
μ'_r	Imaginary part of complex relative permeability.
D_p	Power penetration depth.
d	Penetration depth.

3. Process intensification through microwave technology

A general overview of the intensification process (PI) to show the classification of MW irradiation. In this context, especially in chemical engineering, PI is known as an innovative approach, which focused on transforming conventional processes into smaller size, reduced space, efficient, and eco-friendly systems [76]. According to Stankiewicz et al. [77] the creation of innovative equipment and techniques, providing substantial improvements in production and processing. However, to serve both reactive and non-reactive operations different types of intensified equipment have been developed, for operations involving chemical reactions and non-chemical reactions. Table 3 summarizes various intensified equipment.

Table 3. Process intensification equipment for green synthesis [76-83].

Containing chemical reactions (reactors)	Non chemical reactions
Spinning disk	Centrifugal adsorbers
Static mixer with zero maintenance	Static mixers
Static mixing catalysts (katapaks)	Compact heat exchangers
Monolithic reactor for high selectivity	Microchannel heat exchangers
Micro-reactors with safe operation	Rotor or stator mixers
Supersonic gas or liquid	-
Jet-impingement	-

On the other hand, as technique method used in process intensification. In order to enhancing mass transfer, heat transfer, reducing energy consumption, reducing chemical usage, and minimizing equipment footprint. Table 4 is showing various examples of intensification methods that have been developed in previous research.

Table 4. Process intensification methods for green synthesis [76], 84-88].

Multifunctional reactors and hybrid separations	**Alternative thermal energy sources**
Reverse-flow and dynamic process	Centrifugal fields and electric fields
Reactive distillation, extraction, and crystallization	***Microwaves irradiation***
reactive extrusion, comminution, and fuel cells	Plasma technology
Membrane absorption/ distillation, chromatographic, membrane reactors, and adsorption-distillation	Ultrasound irradiation (cleaning bath, probe, and continuous flow reactor)
Supercritical fluids (H_2O and CO_2)	Solar energy

This section focuses on PI through MW technology, which has emerged as a promising method. Especially, due to its ability to provide rapid, uniform, and selective heating. This makes it particularly suitable for enhancing reaction rates and improving overall process efficiency, highlighting the key contributions to PI, advantages, and comparison with other intensification strategies.

MW chemistry route, has taken its position as a powerful method offering various advantages in chemical synthesis and PI. MWs support automation. Especially when associated with robotic systems, supports combinatorial chemistry with user-friendly tools, and efficient operation.

In green chemistry, MW methods enable free solvent, non-catalytic reactions, relaxed reaction conditions, and improved atom economy. These aspects contribute to lower environmental (E) E-factors, reduced CO_2 emissions, enhanced life cycle performance, and significant energy savings. In terms of chemical reactivity, MW chemistry promotes high stereoselectivity, expeditious synthesis with high crystallinity. The technology also enables the investigation of non-thermal effects, efficient thermal control, and frequency-specific interactions, including discharge lamp-driven or faster reaction catalysts. Moreover, specific heating under MW irradiation offers a new dimension in materials processing, allowing rapid and localized heating, which supports industrial applications such as internal and selective heating, ultimately contributing to improved reaction rates and material performance. Fig.6 summarizes the advantages of different MW in chemical synthesis [89].

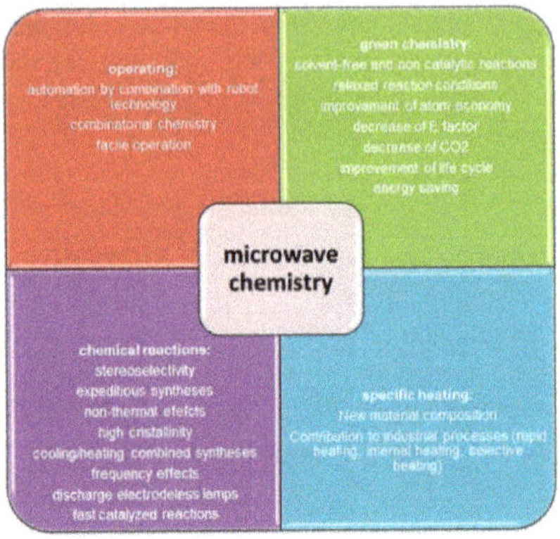

Figure 6. Microwave rules [89].

3.1 Key contributions of process intensification

3.1.1 Selective heating

Microwave-assisted processes MWs selectively couple with the component materials of higher dielectric loss, such as polar solvents, solid catalysts, and polymer composites [90-92]. Antonio et al. [93] reviewed the study of biphasic system (water/chloroform), in the context of Hoffmann elimination. The chemical reaction conducted in H_2O at 105 °C caused polymerization of the resulting product. However, due to the different dielectric properties of polar water (reaches to 110°C) and apolar chloroform (reaches to 50°C). The microwave prevents overheating and polymerization, which is difficult to achieve with conventional heating processes. In addition, the selective absorption heating of MW energy leads to the formation of localized temperature gradients within the hot spots [93,94].

In heterogeneous systems, particularly solid-fluid systems. Ramirez et al. [95] developed a new method for measuring gas-solid temperatures in a MW-heated catalytic quartz reactor in real time, equipped with Ge or ZnSe windows, allowing direct imagination of catalyst under irradiation. Which makes it possible to film the phenomena occurring inside the reactor with a thermal camera, during the reaction process. They applied this technique on ethylene epoxidation reaction catalyzed by a (Ag-CuO) deposited on a monolithic cordierite support. Where, they discovered that the catalyst heats more than the surrounding gas, with a difference of up to 50°C and a solid temperature of up to 200°C. On the other hand, MWs enable selective heating of solid catalyst, which can improve the reactivity but, and the modification of reaction mechanism. Their findings confirm the formation of localized hot spots and highlight the ability of MW-heating to generate the non-uniform thermal environments, that cannot be readily achieved with conventional heating, providing valuable information for the design and optimization of catalytic MW-processes.

3.1.2 Volumetric heating and energy efficiency

MWs tend to penetrate close to the core of components and heat them internally, thereby heating the entire volume simultaneously [96]. This key contribution leads to significant advantages. Demircan et al. [97] investigated a comparative study on different drying methods for bergamot peel revealed that microwave drying effectively dried the peels, by converting them into powder form and identifying the bioactive compounds. Nonetheless, MW had the highest concentration of 5-hydroxymethylfurfural, reaching 73.06 mg/100g. Sickert et al. [98] investigated the impact of EM-field modulation in MW-assisted freeze-drying using solid state microwave generators. They focused on the basis of frequency control strategies, supported by EM simulations, to improve energy efficiency and heating uniformity. Their results, Fig. 7 demonstrated that targeted frequency modulation resulted in an enhancement of the energy distribution, and reduced the drying times. This highlights the potential of frequency-controlled, as a promising technique for improving drying performance and reducing energy consumption in the freeze-drying of sensitive bioactive materials.

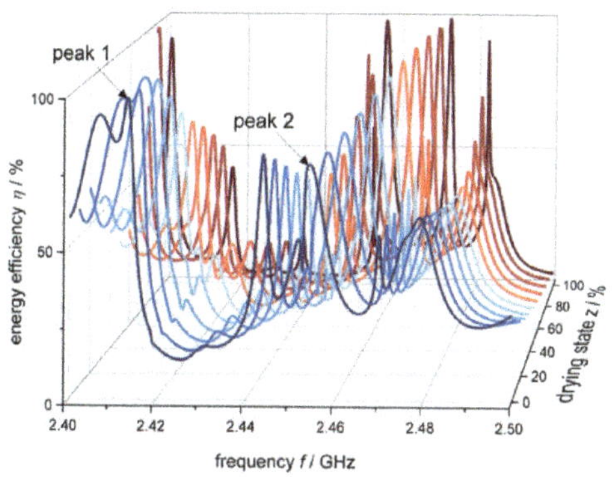

Figure 7. The energy efficiency obtained results across the 2.4–2.5 GHz range frequency, derived from EM models at drying at the range states between 0% to 100%, revealed two prominent energy-efficient peaks, identified as Peak 1 and Peak 2, throughout the drying process, as reported by Sickert et al [98].

4. Bibliometric Study of Scientific Publications (2014-2024)

With the aim of gaining a thorough understanding of the scientific evolution of the use of microwaves in chemical synthesis, a bibliometric study is of strategic interest. This type of analysis makes it possible to trace the temporal evolution of the field, to identify the key phases of its development (emergence, growth, maturity) and to detect dominant or emerging research themes [30]. By focusing on the last twenty years, the study aims to offer a current and relevant view by filtering the results according to the publication date and the disciplinary field of the documents [99]. To ensure the relevance of the data, the search query is restricted to titles and abstracts, which makes it possible to target publications directly dealing with the application of microwaves to chemical synthesis. The Lens.org database is used for the extraction and analysis of publications

[32], combined with the VOSviewer tool Fig.11, for the cartographic visualization [33], of relationships between keywords, thus enriching the analysis of the Lens database tool.

The search query equation is formulated to include the basic concepts (***microwave*** and ***reaction***) of our objective by using the Boolean connectors OR: for synonyms and terms representing the same concept and AND for those representing key concepts and * to take into consideration the singular and plural of the same term. This search equation is therefore formulated as follows:

("microwave assisted " **OR** "microwave-assisted " **OR** "microwave ") **AND** (reaction* **OR** synthesis **OR** method* **OR** experiment*)

4.1 Evolution of microwave use in chemical synthesis over time

The initial results, obtained without date filters, generated approximately 259.587 publications. As shown in Fig. 8a, the first contributions in this field date back several decades, but scientific interest has increased significantly since the 2000s. The last twenty years (51352 with field filter) thus concentrate the majority of research activity Fig.8b, marking a phase of significant growth in the use of microwaves in chemical synthesis.

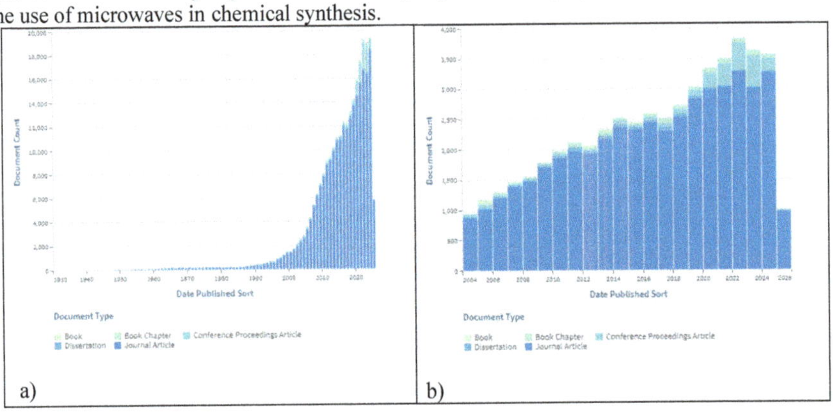

Figure 8. Scientific publications on microwave synthesis over time. (Produced by the author using the Lens.org database)

The use of microwaves in chemical synthesis has undergone significant evolution over time, evolving from a laboratory curiosity to a mature technology with numerous applications.
The main phases and characteristics of this evolution can be summarized as follows:

a. The beginnings (1950s): The use of microwave heating in chemistry dates back to the 1950s, with limited and experimental applications [100,101].

b. The 1980s and the first observations: A significant milestone was reached in 1986, when works of researchers like Giguere reported accelerated reactions in organic chemistry using domestic microwave ovens, and sometimes even improved yields [102-106].

c. The era of dedicated reactors and improved control (1990s): The availability of dedicated microwave reactors increased in the 1990s. This technological development enabled real-time control of crucial parameters such as temperature, pressure, power, and agitation [107,108]. Temperature control was achieved using infrared or fiber optic probes [109-112]. This began to overcome the safety and reproducibility issues encountered with domestic ovens.

d. More mature technology and diversified applications (2000s): Microwave chemistry became an increasingly mature technology. Applications became highly diverse, ranging from organic and inorganic synthesis to the synthesis of nanomaterials and enzymatic reactions. Beyond speed, the advantages highlighted include improved yields, enhanced selectivities, and the ability to work without solvents [113-116]. A major development of this phase is the integration of microwave chemistry with continuous-flow processes "*MW-Assisted Continuous Flow Organic Synthesis, MACFOS*" [117]. The gradual transition from batch to continuous mode is considered essential for the industrial exploitation of microwave energy. New reactors offer better control of operating parameters. Industrial interest in microwaves has increased, although large-scale adoption has been delayed by the demand for safer equipment and facilities.

4.2 Main Scientific Contributors: Authors, Institutions, and Countries

This section aims to identify the main scientific contributors to the field, which provides insight into the expertise, collaborative dynamics, and knowledge production centers in the use of microwaves in chemical synthesis.

The analysis of the main scientific contributors in the field of microwaves in chemical synthesis reveals a concentration of research among twenty authors, institutions, and countries. The most prolific researchers include Ping Liang recognized for his work on microwave ablation [118-120], and Wei Li recognized for microwave activated carbon preparation [121-122]. On the other hand, Chinese academy of sciences [123,124], and Russian academy of sciences [125,126], appear as the major centers of innovation and scientific production in this field. Geographically, research is dominated by China, followed by the United States and India, reflecting both national priorities in green chemistry and investments in emerging technologies. This data makes it possible to identify centers of excellence, better target partners for collaborations, and understand the overall dynamics of the field, Fig.9.

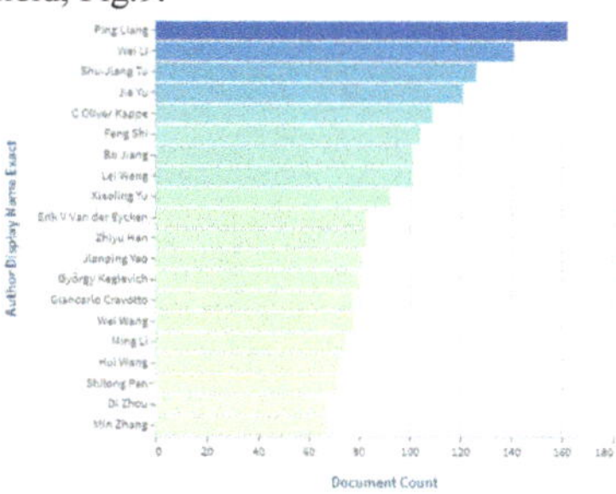

a)

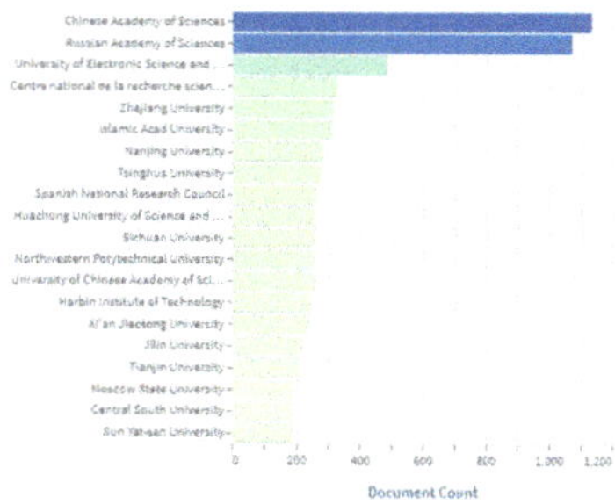

b)

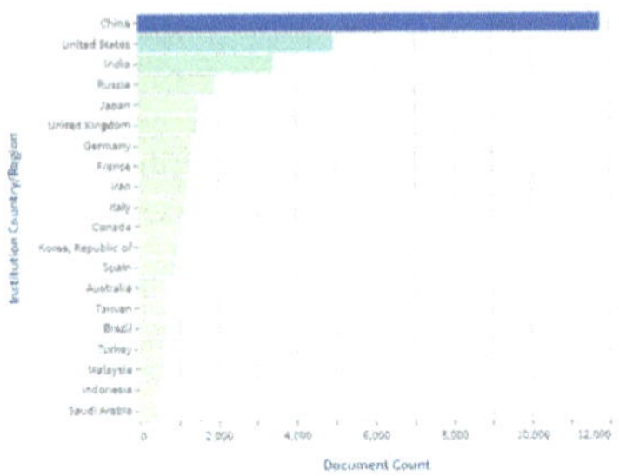

c)

Figure 9. Main scientific contributors a) authors, b) institutions c) countries. (Produced by the author using the Lens.org database)

4.3 Research Areas, Key Journals, and Thematic Trends

This section aims to analyze the dominant thematic areas, targeted applications, key journals of publication, as well as recurring terms and their temporal evolution. These elements help to understand the strategic direction of the field, its recent developments, and emerging trends.

The analysis of scientific fields and publication sources highlights the structuring axes of research on the use of microwaves in chemical synthesis. The main disciplines concerned are chemistry, physics, organic chemistry, analytical chemistry, catalysis, engineering, and materials science, nanotechnology, medicine, biology, reflecting the interdisciplinary nature of the field as presented in Fig. 10a. The most active journals include ChemInform, an academic journal that publishes majorly in the area(s): Catalysis & Aryl [127], confirming the topic's anchoring in process intensification and green chemistry approaches Fig.10b.

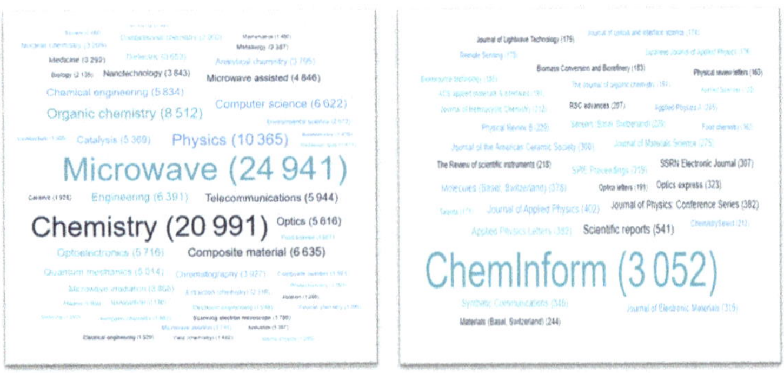

a) *b)*

Figure 10. Research Areas, Key Journals. (Produced by the author using the Lens.org database)

Materials Research Foundations 189 (2026) 1-44 https://doi.org/10.21741/9781644904039-1

The co-occurrence analysis of key terms identified three distinct thematic clusters, reflecting the main directions of research on the use of microwaves in chemical synthesis. The first cluster brings together terms related to organic and medicinal chemistry (organic synthesis, greener, biological activity, medicinal chemistry), highlighting the importance of sustainable approaches aimed at obtaining high yields with reduced environmental impact. The second cluster focuses on the procedural characteristics of the synthesis (short reaction time, rapid, novel, aromatic aldehyde), illustrating the interest in efficiency and innovation in reaction conditions. Finally, the third cluster focuses on the characterization of synthesized products and materials (nanoparticle, spectroscopy, X-ray diffraction, reaction mixture), which reflects the growing anchoring of microwaves in the synthesis of nanomaterials and the detailed understanding of reaction mechanisms. These clusters reveal a coherent structuring of the field between sustainability objectives, procedural efficiency and advanced product characterization.

The overlay map analysis reveals that the most recent topics are concentrated around terms such as heterocycle, medicinal chemistry, aromatic aldehyde, and keywords associated with the nanoparticle cluster. This indicates a refocusing of research toward pharmaceutical and biomedical applications, particularly the synthesis of heterocyclic bioactive compounds, while consolidating interest in functional nanomaterials. These trends suggest a shift in the field toward high-value approaches, targeting both therapeutic innovation and advanced materials.

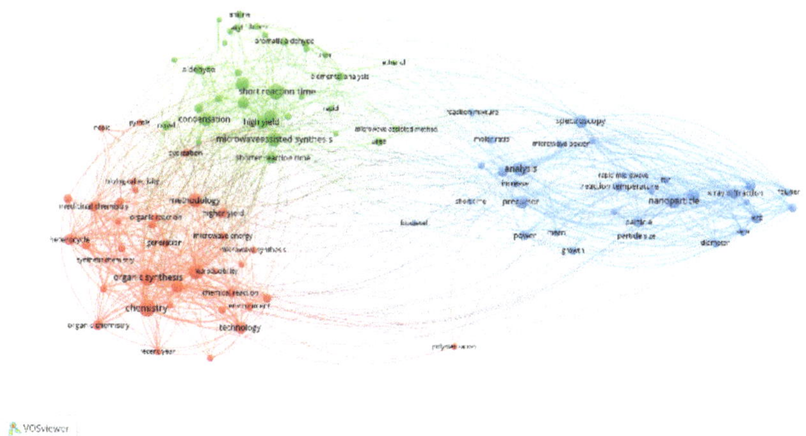

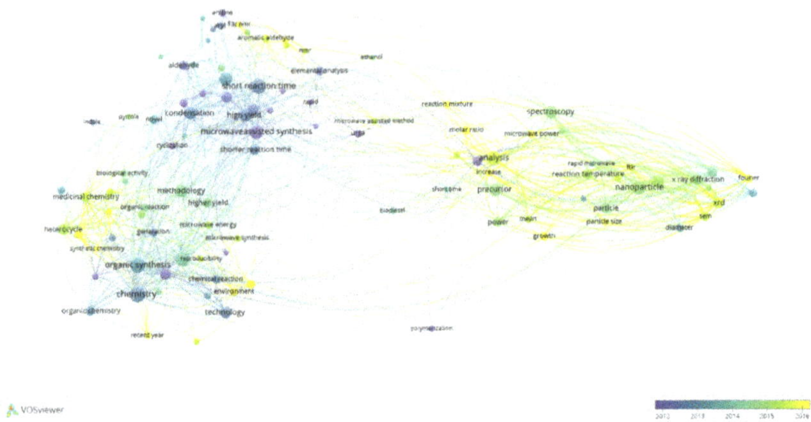

Figure 11. VOSviewer Co-Occurrence Text mining.

5. Performance comparison: microwave-assisted vs conventional synthesis

In conventional chemical synthesis, one of the major challenges is the optimization of reaction conditions to achieve high yields and purity of the desired products. This process often involves lengthy heating steps, making optimization tedious and time-consuming [128]. However, microwave-assisted heating has emerged as a revolutionary technology, significantly accelerating reaction times from hours or days to mere minutes or seconds. This rapid heating method enables efficient parallel or automated sequential synthesis, offering a powerful tool for modern chemists [129]. Time is a critical factor in scientific research, particularly in synthetic chemistry, where traditional trial-and-error approaches can be slow and labor-intensive. Microwave synthesis provides a faster alternative, allowing researchers to conduct more experiments in less time, thereby enhancing productivity and scientific innovation [130]. As Hayes aptly described, "MW synthesis is chemistry at the speed of light" a transformative advancement, that continues to reshape the field of chemistry [131]. Unlike conventional heating methods, which rely on slow heat transfer through conduction or convection, microwave heating delivers energy directly to the reactants, enabling rapid and uniform volumetric heating. This approach minimizes thermal gradients, improves reaction efficiency, and reduces energy consumption. Consequently, microwave-assisted synthesis presents a cost-effective, energy-efficient, and scalable solution for chemical production, overcoming many limitations of traditional heating methods [128].

Microwave heating has emerged as a highly efficient method in chemical processes due to its distinct advantages, including rapid heating kinetics, exceptional temperature homogeneity, and selective energy absorption. These characteristics not only enhance reaction efficiency but also reduce energy consumption and byproduct formation. Consequently, synthesis under MW-Ir is widely regarded as an encouraging approach for advancing sustainable and green chemistry practices, aligning with the principles of environmental responsibility and resource efficiency [2].

MW technology is appearing as a novel heat source providing fast volumetric heating, resulting in accelerated reaction conditions well as increased reaction rates, selectivity, and yields compared to conventional heating methods. Consequently, microwave heating facilitates fast chemical reactions and rapid material synthesis within minutes, rather than the hours or days typically required by traditional heating techniques, leading to lower costs, energy savings, and enhanced efficiency in material production[128]. Also, the rapid and selective heating provided by microwave irradiation often shortens reaction times from hours to minutes or even seconds, thereby reducing the overall reaction cost compared to conventional external heating methods[133]. Table 5, a comprehensive view in conventional and Microwave-Assisted synthesis comparison.

5.1 Rate and time acceleration via microwave heating

Numerous studies have demonstrated that MW can substantially accelerate chemical reaction rates by several orders of magnitude. Reactions that typically require hours or even days to complete under conventional heating methods can be achieved within minutes when subjected to MW heating [137]. Moreover, certain reactions that fail to proceed under traditional heating conditions have been successfully conducted using MW-heating. The data obtained from the experiments cannot be fully accounted for by the rapid heating effect alone, leading to the proposition of specific MW effects [138].

This remarkable enhancement in reaction efficiency is closely related to the fundamental differences in heating mechanisms between microwave and conventional methods. Microwave heating techniques provide an alternative thermal energy source that facilitates rapid chemical transformations and material synthesis within minutes, thereby replacing the extended durations commonly associated with conventional heating methods. In conventional heating (CH), reaction mixtures are generally heated via electric heaters or oil baths, where heat is transferred indirectly through the vessel walls to the reaction solvent [139]. Subsequently, heat distribution to the reactants occurs mainly through conduction and convection. This indirect mode of heating often generates thermal gradients, resulting in difficulties for the reaction mixture's core to attain the target temperature. Such gradients cause heterogeneous reaction progress and present substantial challenges for scaling up production [140]. Conversely, microwave (MW) heating delivers uniform volumetric heating throughout the entire reaction vessel, enabling rapid and consistent heat transfer [141]. In this section, several studies from different fields have been cited to illustrate the effects of microwave irradiation on reaction time and heating.

Using a continuous MW-assisted reactor, the transesterification of waste vegetable oil reached 96% conversion, whereas conventional heating gave 77.5%. In addition to achieving higher conversions, MW-heating improved the reaction rate [142]. Sung Hwa Jhung et al. [138] Silicalite-1 and VSB-5 were synthesized using both conventional heating (CE) and MW-Ir for comparison. Where, MW heating significantly enhanced the rates of nucleation and crystal growth, especially nucleation, reducing synthesis times by up to 30-fold for silicalite-1 and 12-fold for VSB-5. Microwave-nucleated precursors contained more numerous, smaller nuclei, resulting in faster growth and smaller crystals despite similar crystallinity. XRD analysis showed crystallization completed in 3 hours with MW versus 90 hours with CE, confirming the efficiency of microwave-assisted synthesis. Patrícia Silva et al. [143] studied a rapid and efficient microwave-assisted synthesis (MWAS) was developed for a series of isotypical 3D Ln^{3+}-based metal-organic frameworks (MOFs) derived from 2,5-pyridinedicarboxylic acid, including $[Ce_2(pydc)_3(H_2O)_2]$ and its lanthanide-doped analogs. The MWAS method reduced the reaction time drastically from the conventional 3 days under hydrothermal conditions to just 1 minute, while lowering the reaction temperature from 170 °C to 120 °C (or 90 °C with longer reaction time of 10 minutes). This approach provides an optimal, cost-effective route for the rapid isolation of these functional materials. Additionally, it enables straightforward preparation of optically active MOFs exhibiting strong photoluminescence in the visible spectrum.

Julie Smijova et al. [144] Examined the fast MW-assisted in situ synthesis of nano-ZnS/kaolinite nanocomposites and compared them with conventionally synthesized nano-ZnS/kaolinite nanocomposites. The study shows that, MW-assisted synthesis method can produce highly active nanocomposites in a much shorter time than the classical hydrothermal technique, a time reduction from 3 hours by conventional methods to 30 min by MW. Arelis Quirino-García et al. [145] ZnO-NE was successfully synthesized by direct precipitation, professional MW and domestic MW methods. Synthesis by Professional MW produced the largest crystal size (60 nm) and showed a better c/a ratio, resulting in a more defined hexagonal prism morphology for a reaction duration of 5 min.

Priyanka Srivastava et al. [146] they describe the MW-assisted synthesis and antimicrobial evaluation of novel heterocyclic sulfonamide derivatives. Compared to the common classical method, the MW-promoted synthetic approach requires short time (several hours from 5 to 8 for conventional method. On the other hand, with microwave-promoted synthesis the time is reduced a few minutes 7 à 15min) and produced better yield (about 65-76%) of the product.

Fatemeh Karimi et al. [147] compared the synthesis of an iridium-based electrocatalyst supported on antimony tin oxide (ATO) using conventional heating versus microwave irradiation heating. The polyol method was used for synthesis, which involves the use of an alcohol with several hydroxyl groups as both solvent and reducing agent. The main findings are that the microwave heating method can complete the primary synthesis step in 1 minute compared with over 10 hours for conventional heating, while producing catalysts with comparable physical and electrochemical properties.

5.2 Synthesis intensification by microwave irradiations

Microwave-assisted chemistry has emerged as a valuable technique for various applications in organic synthesis and transformations. While initially focused on reactions in organic solvents, current efforts emphasize more environmentally friendly approaches, including the use of greener solvents and supported reagents. Notable examples include microwave-assisted syntheses conducted in benign media, under solvent-free conditions, and employing solid-supported, reusable catalysts [148]. Microwave-assisted chemistry has become a valuable tool for diverse applications in organic synthesis and transformations. Initially centered on reactions in organic solvents, recent focus has shifted towards more environmentally friendly methods, including the use of greener solvents and supported reagents, often under solvent-free or aqueous conditions. Numerous recent studies highlight the rapid and high-quality formation of uniformly small nanomaterials under microwave irradiation, attributed to volumetric heating that minimizes thermal gradients. The development of continuous microwave systems is expected to play a crucial role in scaling up the production of valuable chemical materials [149]. Microwave-assisted dry-media reactions align with green chemistry principles by avoiding solvents, enhancing safety, simplifying work-up, reducing costs, and improving reactivity and selectivity. These advantages have driven increased interest in environmentally benign reagents and solvent-free microwave-assisted syntheses [150].

5.3 Microwave-Assisted Organic Synthesis

Microwave-Assisted Synthesis is widely used in organic chemistry to synthesize pharmaceutical compounds, polymers and chemical intermediates. For example: MW-synthesis is effective for diverse chemical reactions, providing both faster reaction times and increased product yields. Also, it can open new possibilities for conducting chemical reactions. Knowing that, MW can directly deliver energy to the reactive species. They enable transformations that cannot be achieved with conventional heating, opening a new advanced field in synthetic organic chemistry [140].

5.4 Synthesis of heterocyclic compounds

N-containing bicyclic heterocycles Fig.12, are organic compounds that consist of two fused or bridged rings, at least one of which is a heterocycle, and in which at least one nitrogen atom is present within the ring system. Compounds bearing nitrogen heterocycles occupy a central position in biochemistry, pharmaceutical and dye industries, and photographic applications, with their relevance progressively extending to fields such as polymers, adhesives, and molecular design [151].

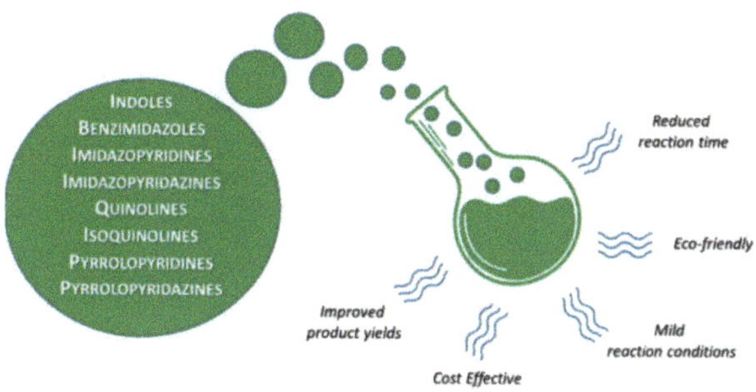

Figure 12. The positive aspects of MW-Ir techniques to synthesize bicyclic structures[151].

Synthesis of indole

Indole/ C_8H_7N serves as a widely favored and flexible scaffold for the construction of biologically active compounds exhibiting a wide range of pharmacological activities, including anti-inflammatory, antineoplastic, antibacterial, antitubercular, anti-hepatitis properties [152]. In a recent publication of Bellavita et al. [153] an efficient synthesis of a collection of functionalized 2-methyl-1H-indole-3-carboxylate compounds was reported, using commercially available anilines as starting materials, which were functionalized with a range of electron-deficient and electron-rich substituents, via a palladium (Pd)-catalyzed process. Specifically, N-aryl substituted

Materials Research Foundations 189 (2026) 1-44 https://doi.org/10.21741/9781644904039-1

enamine carboxylates bearing various substituents electrophilic substituent groups such as ($-NO_2$, -Cl, -Br), as well as electron-donating groups like ($-CH_3$, -OPh), Fig. 13 were synthesized and subsequently transformed into their corresponding indoles through Pd-catalyzed oxidation under both MW-assisted and conventional heating conditions.

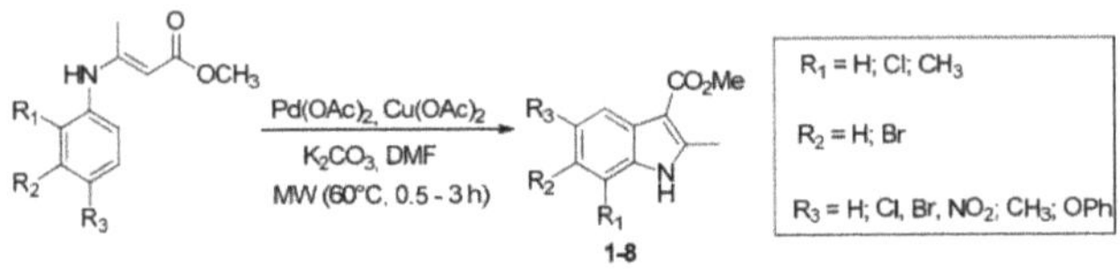

Figure 13. Microwave-assisted Pd-catalyzed cyclization of N-aryl enamine for the preparation of indoles [153].

Synthesis of Heterocyclic Compounds

Heterocyclic compounds are fundamental carbon-containing materials that play a crucial part in the synthesis of various organic pharmaceutical substances. These compounds are chemically complex due to the presence of heteroatoms, which confer a wide diversity of properties. Besides their significant importance in human health, heterocyclic compounds have broad applications in medicine, agriculture, polymer production, and various industrial fields. Several heterocyclic molecules are employed for medicinal use, such as antitumor agents, antiseptics, etc [154]. Benzimidazole and its derivatives have been synthesized using green methodologies as alternatives to conventional industrial or chemical processes. MW-Ir was employed to produce certain benzimidazole derivatives via the condensation of R–CHO with 1,2-diaminobenzene in the presence of a controlled proportion of a carboxylic acid. This approach offers several advantages, including shorter reaction times, the use of green solvents, and high reaction efficiency [155]. Several benzimidazole substituted compounds were synthesized via condensation of aldehydes with 1,2-diaminobenzene, catalyzed by nano-structured transition-metal catalysts. High yields (88–97%) of new benzimidazole compounds were obtained by Nasr-Esfahani et al. [156] using nanoparticle catalysts.

5.5 Microwave-Assisted Nanomaterial Synthesis

The use of microwave ovens has become an indispensable tool in modern organic synthesis. Nonetheless, it is widely recognized that microwave technology holds significant potential to advance all areas of synthetic chemistry. In particular, the synthesis of nanoparticles and nanostructures whose growth is highly sensitive to reaction conditions can greatly benefit from the efficient and controlled heating provided by microwave irradiation. Indeed, the number of publications focusing on microwave-assisted synthesis of inorganic nanomaterials is rapidly increasing Fig.14 encompassing nearly all classes of functional materials, including metals, oxides, sulfides, phosphates, and halides [157]. Beyond drastically reducing reaction times by several orders of magnitude, microwave heating also suppresses side reactions, thereby enhancing both the yield and reproducibility of specific synthetic protocols. These advantages represent critical factors to consider in the development of greener synthesis methodologies. Consequently, microwave technology is expected to play a fundamental role in the future of environmentally friendly synthetic nanotechnology [158,159].

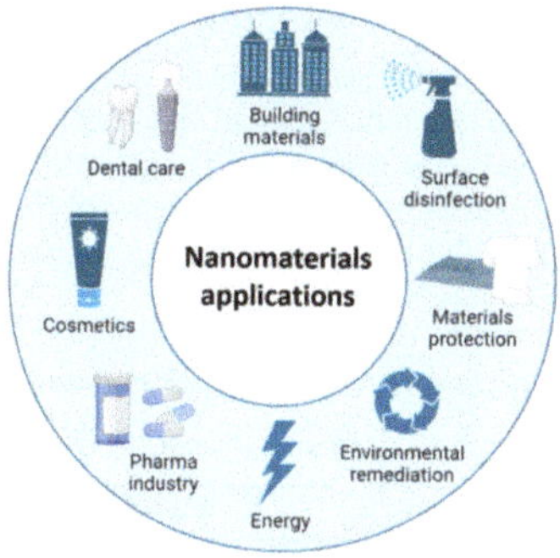

Figure 14. Potential antimicrobial applications of nanomaterials in different fields [171].

Nanoparticles can be synthesized either as isolated entities or organized into various structures by combining them with other compounds, metals, or materials. Currently, two primary approaches to nanoparticle synthesis are recognized: top-down and bottom-up. The top-down approach typically involves physical methods of synthesis, whereas the bottom-up approach generally encompasses chemical and biological techniques, the latter often referred to as "green" methods [159]. Minal Harde et al. [160] have reported an efficient MW-assisted method for producing fluorescent graphene quantum dots (FGQDs), using *Ferula Asafoetida* as a green precursor. However, the one-pot MW process enabled rapid synthesis process, without the need for any catalysts. Highlighting the advantages of volumetric heating and the efficiency of energy using in this process, related to MW methods. The resulting FGQDs, demonstrated excellent structural stability, uniform particle sizes between 4 and 7 nm, and strong blue luminescence with a yield of 18.92%. As a result, MW not only accelerated the synthesis but also enhanced the incorporation of oxygen containing functional groups, contributing to the high fluorescence and sensitivity of FGQDs for applications such as Fe^{3+} detection, drug delivery, and intracellular sensing.

6. Green chemistry and sustainability aspects

Reduced solvent volumes and the ability to use green solvents like water or ionic liquids align well with the principles of green chemistry. Microwave-Assisted Synthesis supports solvent-free or alternative solvent-based approaches, contributing to more environmentally sustainable processes.

7. Example of application of microwave-assisted synthesis of biodiesel

This study is part of a doctoral dissertation carried out in Algeria and defended in February 2024. One of the objectives of this work was to produce biodiesel from cooking oils collected from university residences by transforming them by basic transesterification, into methyl ester (biodiesel). The conventional reaction was optimized and the resulting product characterized, then the reaction was intensified by microwave and its operating conditions optimized and compared to conventional transesterification.

Materials Research Foundations 189 (2026) 1-44 https://doi.org/10.21741/9781644904039-1

7.1 Optimization of microwave-intensified trans-esterification reaction

The operational parameters studied in this section are reaction time, molar ratio of methanol to oil, and KOH catalyst concentration, with biodiesel yield and certain characteristics including density, saponification value, and acid value as responses. The process parameters and their respective study ranges are presented in Table 5. Microwave power was fixed at 800 W. The experimental and predicted response values for the three variables are shown in Table 6, based on a central composite (CC) design comprising 20 experiments.

Table 5. Independent variables and levels used for response surface methodology, CC design for microwave transesterification.

Optimal conditions		**Validation tests**				
% KOH	1		**Y %**	**Density**	**SI**	**AI**
Ratio methanol/oil	6	**Predicted**	94.94	0.8907	183.4491	1.3503
		Experimental	88.9	0.8932	182.384	1.6628
t(s)	40	**Error %**	6.79	0.2279	0.58398	18.79

Table 6. Experimental and Predicted Responses from the CC Design (Microwave).

Sample	R reel		Density		SI		AI	
	Exp	Predicted	exp	Predicted	exp	Predicted	exp	Predicted
1	0.970	0.947	0.899	0.900	190.037	188.250	1.581	1.519
2	0.961	0.980	0.907	0.909	188.033	187.217	1.673	1.610
3	0.957	0.955	0.901	0.902	193.758	194.741	1.398	1.321
4	0.979	0.962	0.917	0.917	172.259	172.306	1.476	1.406
5	0.936	0.952	0.893	0.896	176.906	177.888	1.395	1.344
6	0.927	0.928	0.892	0.893	195.225	195.271	1.395	1.351
7	0.912	0.893	0.884	0.884	195.290	197.134	1.117	1.059
8	0.822	0.844	0.885	0.887	190.301	193.116	1.119	1.059
9	0.938	0.953	0.888	0.886	196.100	195.395	1.314	1.403
10	0.955	0.940	0.899	0.896	191.892	191.146	1.397	1.480
11	0.936	0.928	0.911	0.908	183.769	185.202	1.389	1.462
12	0.854	0.863	0.905	0.904	191.732	188.848	0.952	1.050
13	0.972	0.985	0.915	0.913	180.317	181.749	1.429	1.532
14	0.902	0.890	0.887	0.884	193.419	190.535	1.025	1.094
15	0.865	0.825	0.901	0.902	198.204	198.246	1.398	1.393
16	0.817	0.825	0.901	0.902	198.204	198.246	1.398	1.393
17	0.817	0.825	0.901	0.902	198.204	198.246	1.398	1.393
18	0.817	0.825	0.901	0.902	198.204	198.246	1.398	1.393
19	0.817	0.825	0.901	0.902	198.204	198.246	1.398	1.393
20	0.817	0.825	0.908	0.902	198.204	198.246	1.398	1.393

7.2 Models generated from the experimental design

The results of the central composite (CC) experimental design were subjected to multiple regression analysis represented by the quadratic regression model of Eq.5 that provided the best results. Table 7 presents all the coefficients of the response models studied, as well as their determination coefficients R^2.

$$Y = a_0 + \sum_{i=1}^{n} a_0 X_i + \sum_{i=1}^{n} a_{ii} X_i^2 + \sum_{i=1}^{n-1} \sum_{j=2}^{n} a_{ij} X_i X_j + \varepsilon \quad (5)$$

The coefficient of determination (R^2) is 93.23%, indicating that 93.23% of the variability in biodiesel yield is explained by this model. This shows that the predicted yield values are close to those obtained from the central composite (CC) experimental design, confirming the model's reliability. The R^2 values are highly satisfactory for yield, density, and saponification index. However, for the acid value, it is below 90%, with an acceptable value of 86.88%.

Table 7. The coefficients of the models for the studied responses.

Responses	R real	Density (g/cm^3)	SI	AI
Constant	0.825378	0.902695	198.246	1.39325
t	-0.004153	0.003087	-1.263	0.02289
KOH	-0.019235	-0.001057	1.084	-0.12228
Ratio	-0.028240	-0.008537	2.612	-0.13034
t×t	0.043005	-0.003855	-1.759	0.01733
KOH×KOH	0.024890	0.001272	-3.967	-0.04834
Ratio×Ratio	0.039866	-0.001291	-4.279	-0.02825
t×koh	-0.006378	0.001437	-5.350	-0.00150
t×Ratio	-0.014137	-0.003063	4.604	-0.02116
KOH×Ratio	-0.016610	-0.003313	3.189	-0.02203
R^2	**93.23%**	**94.73%**	**96.50%**	**86.88%**

7.3 Analysis of variance

Table 8 shows the ANOVA results and the significance of the obtained models, main factors, and second-order interactions.

The results of the ANOVA analysis for the studied responses show that all quadratic models are significant (p-value < 0.05), and that each response involves significant main factors and interactions. Time is a non-significant factor for biodiesel yield, catalyst concentration is non-significant for density and acid value, and molar ratio is the most significant factor for all three responses.

Table 8. Analyse de variance du plan CC par microonde.

Parameter	**R reel**		**density**		**AI**		**SI**	
Source	**Adj MS**	**P-Value**	**Adj MS**	**P-Value**	**Adj MS**	**P-Value**	**Adj MS**	**P-Value**
Model	0.0077	0.000	0.0001	0.000	0.0556	0.002	119.179	0.000
Linear	0.0053	0.002	0.0003	0.000	0.1477	0.000	43.666	0.002
t	0.0002	0.511	0.0001	0.003	0.0071	0.354	21.787	0.040
KOH	0.0050	0.010	0.0000	0.219	0.2041	0.000	16.039	0.070
Ratio	0.0165	0.001	0.0009	0.000	0.2320	0.000	93.171	0.001
Square	0.0266	0.000	0.0000	0.002	0.0167	0.149	153.886	0.000
t×t	0.0089	0.000	0.0002	0.001	0.0043	0.467	44.600	0.007
KOH×KOH	0.0229	0.002	0.0000	0.136	0.0336	0.061	226.829	0.000
Ratio×Ratio	0.0013	0.000	0.0000	0.130	0.0115	0.246	263.919	0.000
2-Way Interaction	0.0003	0.101	0.0000	0.009	0.0024	0.804	159.985	0.000
t×KOH	0.0015	0.442	0.0000	0.202	0.0000	0.962	229.013	0.000
t×Ratio	0.0022	0.106	0.0000	0.016	0.0035	0.507	169.583	0.000
KOH×Ratio	0.0005	0.064	0.0000	0.010	0.0038	0.490	81.358	0.001
Lack-of-Fit	0.0003	0.290	0.0000	0.330			3.892	

☐ Significant model ☐ Significant main factor ☐ Significant second-order interaction.

7.4 Main factor effects on biodiesel yield, density, acid value, and saponification value

Fig.15 illustrates the main effects of reaction time, % KOH, and methanol-to-oil molar ratio on the main responses namely, biodiesel yield, density, acid index, and saponification index.

It can be observed that several factors exhibit optimal values, and not all factors influence all responses in the same way. For instance, reaction time shows a minimum effect on yield and acid value, and a maximum effect on density and saponification index. Interestingly, the highest yields are achieved at low reaction times, low molar ratios, and low catalyst concentrations, making the process more economical and environmentally friendly. In contrast, for the acid value, intermediate reaction times along with higher molar ratios and catalyst concentrations are required, highlighting the need for multi-objective optimization. As with conventional methods, excessive alcohol is undesirable due to increased solubility of methyl esters in the glycerol phase [161], and excess catalyst should also be avoided due to the increased risk of saponification, which complicates achieving a homogeneous product mixture [162].

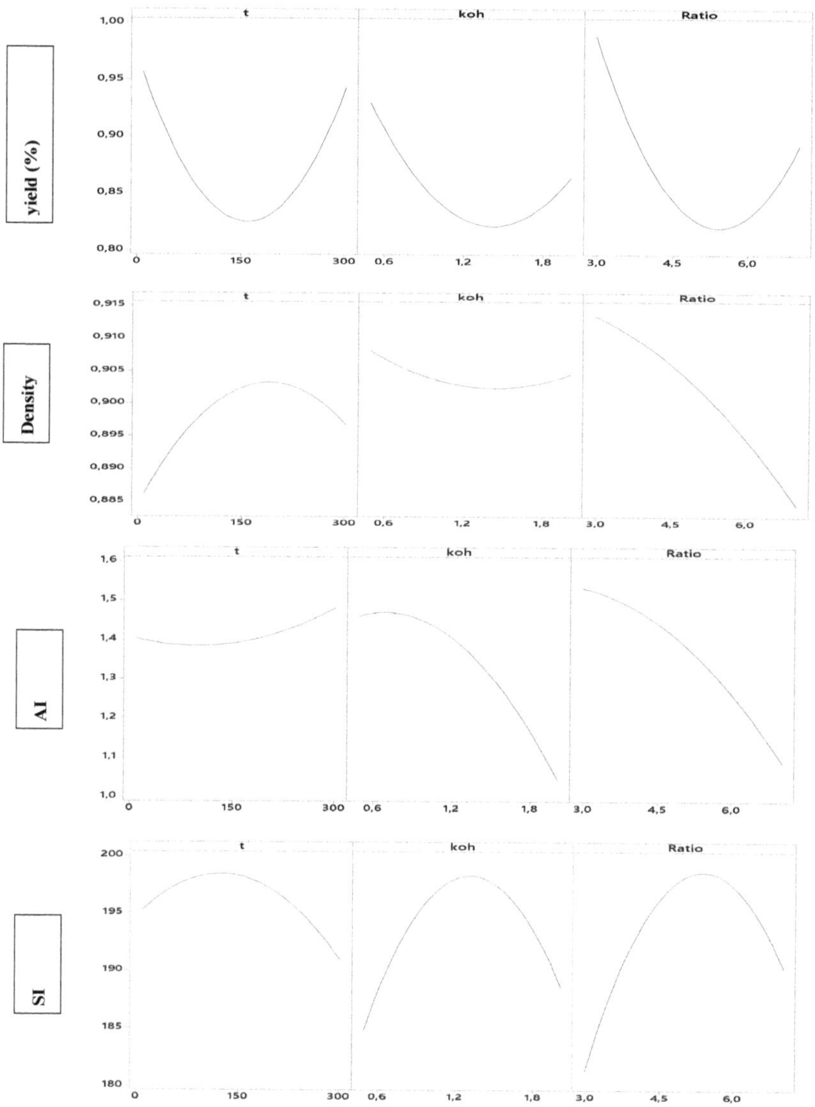

Figure 15. Effect of main factors on biodiesel yield.

Materials Research Foundations 189 (2026) 1-44 https://doi.org/10.21741/9781644904039-1

7.5 Contour plots of yield, density, acid index, and saponification index of biodiesel (Microwave-Assisted)

The contour plots corresponding to the microwave intensification process were generated using the same principle described earlier and are displayed in Fig.16.

The regions of low biodiesel yield are clearly visible, located at the center of the experimental domain. Biodiesel yield is minimal when the molar ratio is set to its minimum value, with a reaction time around 150 s and a KOH concentration around 1.25%. Similarly, when the KOH concentration is at 0.5%, the yield reaches its minimum for medium values of both reaction time and molar ratio.

Regarding density, it exceeds the standard limit values when both KOH percentage and molar ratio are set to their minimum levels, regardless of reaction time. Density decreases when time and molar ratio are at their maximum values, with KOH kept at the lowest level.

For acid value, minimal values were obtained with high KOH concentrations across the entire studied reaction time range when the molar ratio was at its minimum. Additionally, minimal acid values were also observed for maximum molar ratios across all reaction times when KOH concentration was at its maximum.

The saponification index is low when KOH percentage and reaction time are at their maximum values with the molar ratio at its minimum, as well as when reaction time is at its minimum and molar ratio at its maximum, with KOH percentage at its minimum.

Yield %

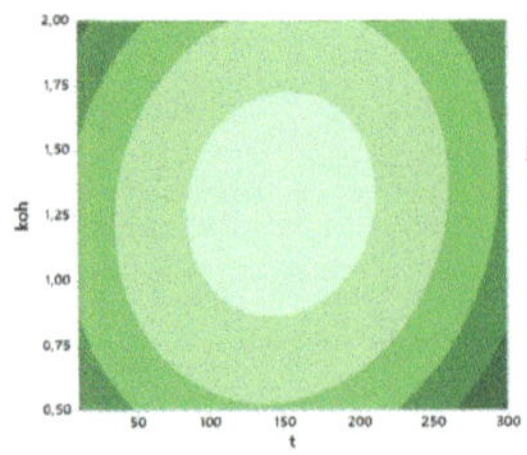

(a)

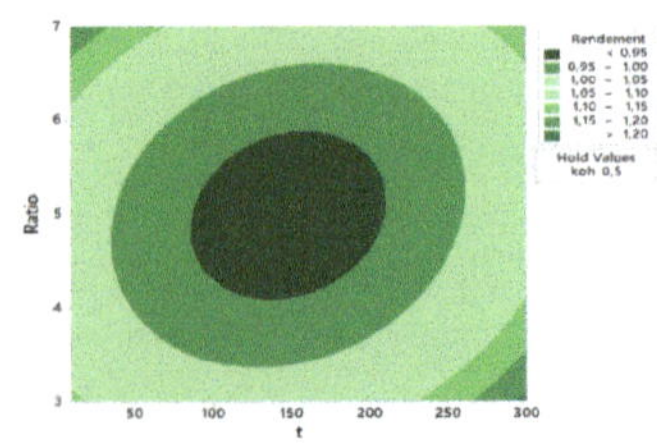

(b)

Density

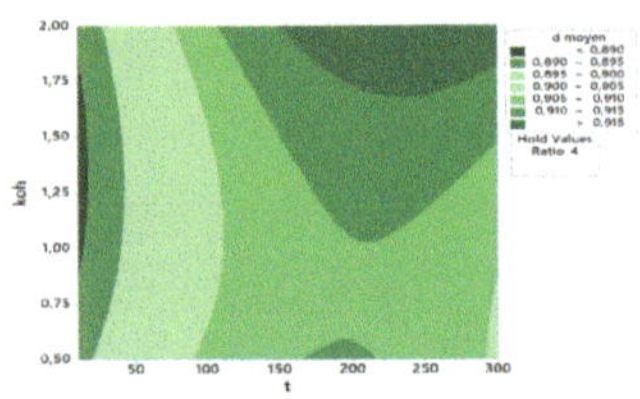

(c)

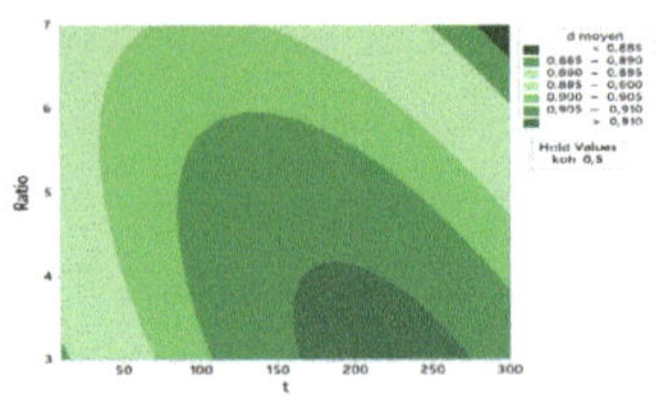

(d)

Materials Research Foundations 189 (2026) 1-44 https://doi.org/10.21741/9781644904039-1

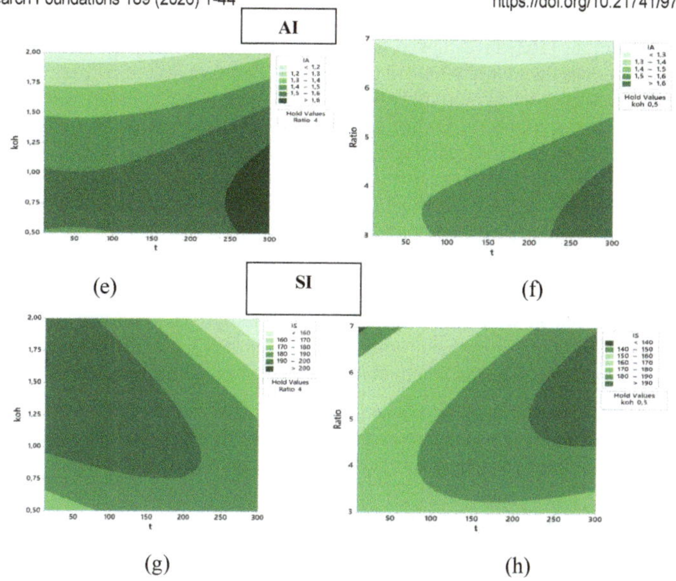

Figure 16. CCD Response Contour Plots (Microwave-Assisted).

7.6 Microwave-Assisted Optimization of the Central Composite Design

A multi-response optimization was carried out to predict all the studied responses within desirable ranges: maximizing biodiesel yield while minimizing density, saponification index, and acid index. The results from the Minitab optimizer are presented and compared with experimental validation in Table 10.

To validate the predicted model results, experiments were conducted under the optimized conditions. However, the experimental results showed a lower yield and a higher acid value than those predicted.

This intensification of the production of biodiesel production assisted by MW showed the first advantage, which is the reduction of the reaction time from 30min to 40 s, a reduction of 97% [26]. Another advantage lies in the improvement of the quality of the biodiesel obtained by the transesterification assisted by MW compared to the conventional method, such as the viscosity, as presented in the comparison Table 5, the first experimental observation was the appearance of the reaction medium, which presented a clear phase shift facilitating the separation step by decantation, which is quite long after a conventional reaction.

Several researchers have used microwave irradiation to optimize the operating conditions for biodiesel synthesis, often considering yield as the sole response. Kian Rokni et al. [161] applied a Box–Behnken design, and the optimal conditions obtained were a molar ratio of 6.91, a KOH concentration of 1.26%, and a reaction time of 5.85 minutes, resulting in a biodiesel yield of 95.31%.

Jassinnee Milano et al. [163] investigated the optimization of biodiesel production from waste cooking oil and Calophyllum Inophyllum oil using a Box–Behnken design. The optimal conditions achieved were a methanol-to-oil ratio of 59.60% (v/v), a catalyst concentration of 0.774% (w/w), a stirring speed of 600 rpm, and a reaction time of 7.15 minutes, leading to a biodiesel yield of 97.40%. N. Prajapati et al. [164] employed a central composite (CC) design for microwave-assisted optimization using waste cottonseed oil and a microwave power of 900 W. The optimal values were a molar ratio of 9.61:1, a KOH catalyst concentration of 1.34%, and a reaction time of 9.89 minutes, achieving a predicted maximum yield of 90.71%.

R. Selvaraj et al. [165] studied the optimization of biodiesel production using a Box–Behnken design combined with an artificial neural network (ANN) model to predict the yield. Waste cooking oil was pretreated with activated carbon, and the optimal conditions obtained were a catalyst concentration of 1%, a methanol-to-oil molar ratio of 6:1, a reaction time of 60 seconds, and a temperature of 75 °C. The maximum biodiesel yield was estimated at 93%.

The results obtained depend on the nature and characterization of the oil, the type of microwave equipment used, and its power setting. Nevertheless, these findings remain within the range of previously reported literature values, while achieving significantly shorter reaction times.

8. Challenges and Limitations of Microwave-Assisted Synthesis

Microwave-Assisted Synthesis (MAS) offers advantages such as rapid heating, improved selectivity, and energy efficiency; however, its widespread industrial application is hindered by several challenges. Scaling up MAS is complicated by uneven heating and limited microwave penetration in larger volumes, resulting in hot spots and poor temperature control. Industrial microwave reactors are costly, require specialized designs, and pose maintenance and safety concerns, particularly under high-pressure or high-temperature conditions. Additionally, MAS is less effective for non-polar solvents and gas-phase reactions, limiting its applicability. The lack of standardized protocols leads to variability and reproducibility issues across different systems, while integration with continuous flow reactors remains an engineering challenge. Together, these factors currently restrict MAS's broader adoption despite its promising potential [166,167].

8.1 Scale-Up Challenges

MW-assisted synthesis speeds up reactions and enhances yields compared to classical methods. Although scaling up to industrial levels remains challenging due to limited MW penetration, non-uniform heating, and uncontrolled reaction parameters, high equipment costs and strict safety requirements, also pose significant limitations. Even though domestic MW ovens are inexpensive and highly adaptable when the reaction vessel fits the cavity [167], their applicability becomes limited above the millimolar scale, when scaling to larger quantities, the constraints of this approach become increasingly evident [166].

8.1.1 Non-uniform heating at larger scales

One of the most significant challenges in scaling up microwave-assisted synthesis is achieving uniform heating across larger volumes. In laboratory settings, small-scale reactions benefit from rapid and uniform heating due to the efficient interaction of microwaves with the reaction mixture. However, as the volume increases, the penetration depth of microwaves becomes limited, leading to hot spots and cold spots within the reaction vessel.

The temperature is uncontrolled and rises continuously during irradiation. Controlling temperature by intermittent switching of the microwave field is possible but not fully effective [166].

8.1.2 Equipment limitations

Industrial-scale microwave reactors are expensive and require specialized designs to ensure uniform heating. Unlike conventional heating methods, which can be easily scaled up using larger reactors, microwave reactors face technical constraints related to microwave distribution and thermal management. To date, MW devices have maintained stringent safety standards for EM radiation, but have limited applicability in chemical reactions. Adjustment of reaction conditions is limited to the energy supplied and the duration of irradiation. Measuring pressure and temperature is highly challenging. As reaction control depends exclusively on energy input, temperature cannot be limited by powering off [166].

Future outlook

In chemistry, MW-assisted heat is approaching a transformative phase, supported by innovative selective energy distribution, catalyst design, and the integration of modern processing technologies. However, one of the major areas of development is materials synthesis. Where, MWs facilitate the control of over nanoparticle development. Which, resulting in uniformity in size distributions that enhance catalytic and electronic properties. Especially, valuable in energy storage and catalytic systems [168]. Moreover, coupling MW heating with 3D-printed complex geometries reactors offers a new way to optimize electromagnetic field distribution, by improving reaction efficiency. In bio-science and pharmaceutical applications, MW-assisted techniques show promise in accelerating bio-molecular transformations, like peptide synthesis, while preserving its molecular integrity. The emergence of continuous-flow microwave reactors enables large-scale, on-demand pharmaceutical production. Significantly, reducing time and purification requirement processing. From a sustainability perspective, MW instrument supported by renewable electricity are being explored for carbon neutral chemical transformation, providing lower energy consumption and reduced carbon dioxide emissions (CO_2). In addition, MW-assisted pyrolysis reactions and similar valorization techniques are being used to convert bio-mass material and waste material into valuable chemical feedstocks. Promoting circular economy routs. Concerning, industrial scale development is provided with flow chemistry platforms controlled with artificial intelligence systems. In the objective of promise faster reaction rates, smaller equipment sizes and real time process optimization. At the interface of science, efforts are currently underway to identify the correct non-thermal mechanism effect of MW heating , including activation energies and selective interactions, using advanced tools such as multiscale modeling [169,170]. Future catalysts could exploit these effects to overcome traditional thermodynamic limits.

Accessibility and education will also play an essential role: compact, automated MW reactors are set to become routine laboratory equipment, and university curricula will increasingly incorporate MW assisted synthesis to train a new generation in chemistry. Overall, is becoming a core technology for sustainable industrial chemistry, offering precise energy control, new reaction pathways and tight integration with green energy and intelligent manufacturing platforms.

Conclusion

The ability of microwave assisted synthesis to rapidly and uniformly heat the reaction media has made it a powerful alternative, to replace conventional synthetic techniques. This approach not only reduces reaction time, but also enhances product yields and selectivity, contributing to more sustainable energy in chemical processes. The results of the bibliometric analysis confirm that microwave-assisted synthesis is emerging as a high-potential scaling-up technique, particularly in the fields of organic and medicinal chemistry and nanomaterials. The most recent themes converge towards high value added applications, particularly in the synthesis of bioactive compounds and the design of functional materials. This scientific mapping provides a strategic basis for guiding future research, identifying opportunities for collaboration and better understanding the priorities of the scientific community around this technology.

References

[1] R. Gedye, F. Smith, K. Westaway, H. Ali, L. Baldisera, L. Laberge, J. Rousell, The use of microwave ovens for rapid organic synthesis, Tetrahedron Lett. 27 (1986) 279-282. https://doi.org/10.1016/S0040-4039(00)83996-9

[2] R.J. Giguere, T.L. Bray, S.M. Duncan, G. Majetich, Application of commercial microwave ovens to organic synthesis, Tetrahedron Lett. 27 (1986) 4945-4948. https://doi.org/10.1016/S0040-4039(00)85103-5

[3] R.N. Gedye, Organic synthesis using microwaves in homogeneous media, in: A. de la Hoz, A. Loupy (Eds.), Microwaves in Organic Synthesis, Wiley-VCH, Weinheim, 2002, pp. https://doi.org/10.1002/3527601775.ch4

[4] R.N. Gedye, W. Rank, K.C. Westaway, The rapid synthesis of organic compounds in microwave ovens. II, Can. J. Chem. 69 (1991) 706-711. https://doi.org/10.1139/v91-106

[5] G. Majetich, R. Hicks, Applications of microwave-accelerated organic synthesis, Radiat. Phys. Chem. 45 (1995) 567-579. https://doi.org/10.1016/0969-806X(94)00071-Q

[6] M. Gaba, N. Dhingra, Microwave chemistry: general features and applications, Ind. J. Pharm. Educ. Res. 45 (2011) 175-183.

[7] A. Laporterie, J. Marquie, J. Dubac, Microwave-assisted reactions on graphite, in: Microwaves in Organic Synthesis, Wiley-VCH, 2002, pp. 219-252. https://doi.org/10.1002/3527601775.ch7

[8] B.A. Roberts, C.R. Strauss, Toward rapid, "green", predictable microwave-assisted synthesis, Acc. Chem. Res. 38 (2005) 653-661. https://doi.org/10.1021/ar040278m

[9] Y.-J. Zhu, F. Chen, Microwave-assisted preparation of inorganic nanostructures in liquid phase, Chem. Rev. 114 (2014) 6462-6555. https://doi.org/10.1021/cr400366s

[10] M.Tsuji, Hashimoto, M., Nishizawa, Y., Kubokawa, Microwave-assisted synthesis of metallic nanostructures in solution, Chem. Eur. J. 11 (2005) 440-452. https://doi.org/10.1002/chem.200400417

[11] M. Baghbanzadeh, Carbone, L., Cozzoli, P. D, Microwave-assisted synthesis of colloidal inorganic nanocrystals, Angew. Chem. Int. Ed. 50 (2011) 11312-11359. https://doi.org/10.1002/anie.201101274

[12] C.A. Zito, M.O. Orlandi, D.P. Volanti, Accelerated microwave-assisted hydrothermal/solvothermal processing: fundamentals, morphologies, and applications, J. Electroceram. 40 (2018) 271-292. https://doi.org/10.1007/s10832-018-0128-z

[13] D. Dallinger, C.O. Kappe, Microwave-assisted synthesis in water as solvent, Chem. Rev. 107 (2007) 2563-2591. https://doi.org/10.1021/cr0509410

[14] H. Wang, J. Chen, Z. Pei, Enhancing lignocellulosic biorefinery sustainability: mechanisms and optimization of microwave-responsive deep eutectic solvents for rapid delignification, Biofuel Res. J. 12 (2025) 2306-2318. https://doi.org/10.18331/BRJ2025.12.1.3

[15] A.S. Lozano Pérez, J.J. Lozada Castro, C.A. Guerrero Fajardo, Application of microwave energy to biomass: a comprehensive review of microwave-assisted technologies, optimization parameters, and the strengths and weaknesses, J. Manuf. Mater. Process. 8 (2024) 121. https://doi.org/10.3390/jmmp8030121

[16] K.C. Panda, B.R. Kumar, J. Sruti, Microwave-assisted synthesis, antihypertensive activity, docking and SAR studies of some dihydropyrimidines, J. Appl. Chem. (2024).

[17] S. Nain, R. Singh, S. Ravichandran, Importance of microwave heating in organic synthesis, Adv. J. Chem. Sect. A 2 (2019) 94-104. https://doi.org/10.29088/SAMI/AJCA.2019.2.94104

[18] H.J. Kitchen, S.R. Vallance, J.L. Kennedy, Modern microwave methods in solid-state inorganic materials chemistry: from fundamentals to manufacturing, Chem. Rev. 114 (2014) 1170-1206. https://doi.org/10.1021/cr4002353

[19] V.L. Budarin, P.S. Shuttleworth, T.J. Farmer, The potential of microwave technology for the recovery, synthesis and manufacturing of chemicals from bio-wastes, Catal. Today 239 (2015) 80-89. https://doi.org/10.1016/j.cattod.2013.11.058

[20] A. Kumar, Y. Kuang, Z. Liang, Microwave chemistry, recent advancements, and eco-friendly microwave-assisted synthesis of nanoarchitectures and their applications: a review, Mater. Today Nano 11 (2020) 100076. https://doi.org/10.1016/j.mtnano.2020.100076

[21] M. Kohansal, Advances in green chemistry: sustainable approaches in organic synthesis, Int. J. New Chem. 12 (2025) 726-737.

[22] L.S. Fletcher, Recent developments in contact conductance heat transfer, J. Heat Transf. (1988) 1059-1070. https://doi.org/10.1115/1.3250610

[23] Y.L. He, W.Q. Tao, Convective heat transfer enhancement: mechanisms, techniques, and performance evaluation, Adv. Heat Transf. 46 (2014) 87-186. https://doi.org/10.1016/bs.aiht.2014.09.001

[24] S. Horikoshi, R.F. Schiffmann, J. Fukushima, Microwave chemical and materials processing, Microwave Chem. Mater. Process. 1 (2018) 33-45. https://doi.org/10.1007/978-981-10-6466-1_3

[25] Á. Martín, A.Navarrete, Microwave-assisted process intensification techniques, Curr. Opin. Green Sustain. Chem. 11 (2018) 70-75. https://doi.org/10.1016/j.cogsc.2018.04.019

[26] H. Kerras, N. Outili, A.H. Meniai, Waste cooking oil pretreatment using microwave and ultrasound methods, C. R. Chim. 26 (S1) (2023) 63-76. https://doi.org/10.5802/crchim.229

[27] S. Caddick, R. Fitzmaurice, Microwave enhanced synthesis, Tetrahedron 65 (2009) 3325-3355. https://doi.org/10.1016/j.tet.2009.01.105

[28] P. Priecel, J.A. Lopez-Sanchez, Advantages and limitations of microwave reactors: from chemical synthesis to the catalytic valorization of biobased chemicals, ACS Sustainable Chem. Eng. 7 (2018) 3-21. https://doi.org/10.1021/acssuschemeng.8b03286

[29] E.O. Aluko, A.S. Adekunle, J.A. Oyekunle, A review of nanomaterials and microwave synthesized metal oxides nanoparticles in schistosomiasis diagnosis, J. Fluoresc. (2025) 1-27. https://doi.org/10.1007/s10895-024-04113-w

[30] F. Narin, K.S. Hamilton, Bibliometric performance measures, Scientometrics 36 (1996) 293-310. https://doi.org/10.1007/BF02129596

[31] S. McReynolds, Research Guides: Lens Guide: How to Use Lens.org . Consulté le: 24 mai 2025. [En ligne]. Disponible sur: https://researchguides.library.syr.edu/lensguide/use

[32] A. Tarazi, Comparative analysis of the bibliographic data sources using PubMed, Scopus, Web of Science, and Lens: comparative analysis of bibliographic data sources, High Yield Med. Rev. 2 (2024). https://doi.org/10.59707/hymrUNHW4628

[33] J.T. McAllister, L. Lennertz, Z. Atencio Mojica, Mapping a discipline: a guide to using VOSviewer for bibliometric and visual analysis, Sci. Technol. Libr. 41 (2022) 319-348. https://doi.org/10.1080/0194262X.2021.1991547

[34] P. Veronesi, An introduction to dielectric heating, in: Microwave Chemistry, De Gruyter, 2017, pp. 18-30. https://doi.org/10.1515/9783110479935-002

[35] X. Jiao, H. Diao, T. Liu, Single-mode microwave heating for food science research: understanding specific microwave effects and reliability concerns, Food Phys. 2 (2025) 100048. https://doi.org/10.1016/j.foodp.2025.100048

[36] J. Jacob, L.H.L. Chia, F.Y.C. Boey, Thermal and non-thermal interaction of microwave radiation with materials, J. Mater. Sci. 30 (1995) 5321-5327. https://doi.org/10.1007/BF00351541

[37] N. Yoshikawa, Mechanism of microwave heating of matter, in: S. Horikoshi, N. Serpone (Eds.), RF Power Semiconductor Generator Application in Heating and Energy Utilization, Springer, Singapore, 2020. https://doi.org/10.1007/978-981-15-3548-2_3

[38] J. Sun, W. Wang, Q. Yue, Review on microwave-matter interaction fundamentals and efficient microwave-associated heating strategies, Mater. 9 (2016) 231. https://doi.org/10.3390/ma9040231

[39] S. Fan, Physical properties and basic theory of dielectric, IOP Conf. Ser.: Earth Environ. Sci. 692 (2021) 022122. https://doi.org/10.1088/1755-1315/692/2/022122

[40] M. Qin, L. Zhang, H. Wu, Dielectric loss mechanism in electromagnetic wave absorbing materials, Adv. Sci. 9 (2022) 2105553. https://doi.org/10.1002/advs.202105553

[41] T.W. Dalkin, Conduction and polarization mechanisms and trends in dielectrics, IEEE Electr. Insul. Mag. 22 (2006) 11-28. https://doi.org/10.1109/MEI.2006.1705854

[42] J. Anwar, U. Shafique, R. Rehman, Microwave chemistry: effect of ions on dielectric heating in microwave ovens, Arab. J. Chem. 8 (2015) 100-104. https://doi.org/10.1016/j.arabjc.2011.01.014

[43] S. Horikoshi, T. Sumi, N. Serpone, Unusual effect of the magnetic field component of the microwave radiation on aqueous electrolyte solutions, J. Microw. Power Electromagn. Energy 46 (2012) 215-228. https://doi.org/10.1080/08327823.2012.11689838

[44] J.A. Menéndez, A. Arenillas, B. Fidalgo, Microwave heating processes involving carbon materials, Fuel Process. Technol. 91 (2010) 1-8. https://doi.org/10.1016/j.fuproc.2009.08.021

[45] N. Wang, W. Zou, X. Li, Study and application status of the nonthermal effects of microwaves in chemistry and materials science-a brief review, RSC Adv. 12 (2022) 17158-17181. https://doi.org/10.1039/D2RA00381C

[46] Z. Li, K. Peng, N. Ji, Advanced mechanisms and applications of microwave-assisted synthesis of carbon-based materials: a brief review, Nanoscale Adv. 7 (2025) 419. https://doi.org/10.1039/D4NA00701H

[47] I. Morales, D. Archilla, P. de la Presa, Colossal heating efficiency via eddy currents in amorphous microwires with nearly zero magnetostriction, Sci. Rep. 10 (2020) 602. https://doi.org/10.1038/s41598-020-57434-8

[48] C. Feng, X. Liu, Y. Sun, Enhanced microwave absorption of flower-like FeNi@C nanocomposites by dual dielectric relaxation and multiple magnetic resonance, RSC Adv. 4 (2014) 22710-22715. https://doi.org/10.1039/C4RA01437E

[49] Z. Ma, Q. Liu, J. Yuan, Analyses on multiple resonance behaviors and microwave reflection loss in magnetic Co microflowers, Phys. Status Solidi B 249 (2012) 575-580. https://doi.org/10.1002/pssb.201147382

[50] Y. Sani, R.S. Azis, I. Ismail, Recent developments in synthesis techniques and microwave absorption performance of materials based on molybdenum disulfide (MoS_2) with metal oxides: a review, Recent Prog. Mater. 7 (2025) 1-54. https://doi.org/10.21926/rpm.2501003

[51] N.O. Sokal, Class-E RF power amplifiers, QEX 204 (2001) 9-20.

[52] S. Ahmed, B.L. Rajak, M. Gogoi, Magnetic nanoparticles mediated cancer hyperthermia, in: Smart Healthcare for Disease Diagnosis and Prevention, Academic Press, 2020, pp. 153-173. https://doi.org/10.1016/B978-0-12-817913-0.00016-X

[53] M. Gupta, E. W. W. Leong, Microwaves and Metals, 1st ed., John Wiley & Sons, Singapore, 2008. https://doi.org/10.1002/9780470822746

[54] J. Cheng, R. Roy, D. Agrawal, Radically different effects on materials by separated microwave electric and magnetic fields, Mater. Res. Innov. 5 (2002) 170-177. https://doi.org/10.1007/s10019-002-8642-6

[55] S. Horikoshi, N. Serpone (Eds.), Microwaves in Nanoparticle Synthesis: Fundamentals and Applications, John Wiley & Sons, Weinheim, 2013. https://doi.org/10.1002/9783527648122

[56] S. Moratal, R. Benavente, M.D. Salvador, Microwave sintering study of strontium-doped lanthanum manganite in a single-mode microwave with electric and magnetic field at 2.45 GHz, J. Eur. Ceram. Soc. 42 (2022) 5624-5630. https://doi.org/10.1016/j.jeurceramsoc.2022.05.060

[57] J. Xu, T. Gao, S. Liu, Enhanced microwave absorption and thermal conductivity of biomass-derived BCNO materials, J. Mater. Sci.: Mater. Electron. 36 (2025) 417. https://doi.org/10.1007/s10854-025-14486-8

[58] G.P. Kannan, K.V. Nagaraja, Madhukesh, An effective higher order finite element computation method for analyzing the eddy current losses in induction motors using subparametric transformations, Partial Differ. Equ. Appl. Math. 9 (2024) 100652. https://doi.org/10.1016/j.padiff.2024.100652

[59] R. Pei, Y. Chen, H. Rao, Anion doping as a "trigger" to modulate defect-tailored dielectric coupling for ultrathin microwave absorber, Small 21 (2025) 2408538. https://doi.org/10.1002/smll.202408538

[60] W. Li, J. Zhang, M. Li, Analysis of loss mechanisms due to eddy currents and hysteresis in two-phase amorphous hybrid soft magnetic composites, IEEE Trans. Magn. (2025). https://doi.org/10.1109/TMAG.2025.3545294

[61] S.J.C. Masuku, J.Z. Msomi, T.S. Mahule, Observations of the novel low-field microwave absorption in $Zn_xCo_{1-x}Fe_{2-x}Al_xO_4$ ferrites, J. Supercond. Novel Magn. 38 (2025) 128. https://doi.org/10.1007/s10948-025-06954-3

[62] F. Li, R. Zhao, Y. Liu, Magnetic shielding properties of an iron-based nanocrystalline alloy for induction heating systems, Heliyon.10 (2024) e37119. https://doi.org/10.1016/j.heliyon.2024.e37119

[63] S.V. Jadhav, B.M. Kim, H.Y. Lee, Induction heating and in vitro cytotoxicity studies of $MnZnFe_2O_4$ nanoparticles for self-controlled magnetic particle hyperthermia, J. Alloys Compd. 745 (2018) 282-291. https://doi.org/10.1016/j.jallcom.2018.02.174

[64] Y. Huo, Z. Wang, Y. Zhang, High-entropy ferrite with tunable magnetic properties for excellent microwave absorption, Int. J. Miner. Metall. Mater. 32 (2025) 668-677. https://doi.org/10.1007/s12613-024-2883-y

[65] J. Sun, W. Wang, Q. Yue, Review on microwave-matter interaction fundamentals and efficient microwave-associated heating strategies, Mater. 9 (2016). https://doi.org/10.3390/ma9040231

[66] T. Basak, Influence of various shapes of annular metallic support on microwave heating of 2D cylinders, Chem. Eng. Sci. 61 (2006) 2023-2034. https://doi.org/10.1016/j.ces.2005.10.050

[67] Z. Peng, J.Y. Hwang, J. Mouris, Microwave penetration depth in materials with non-zero magnetic susceptibility, ISIJ Int. 50 (2010) 1590-1596. https://doi.org/10.2355/isijinternational.50.1590

[68] Z. Peng, J.Y. Hwang, C.-L. Park, Numerical analysis of heat transfer characteristics in microwave heating of magnetic dielectrics, Metall. Mater. Trans. A 43 (2012) 1070-1078. https://doi.org/10.1007/s11661-011-1014-3

[69] A. Schönhals, F. Kremer, Theory of dielectric relaxation, in: F. Kremer, A. Schönhals (Eds.), Broadband Dielectric Spectroscopy, Springer, Berlin, Heidelberg, 2003. https://doi.org/10.1007/978-3-642-56120-7_1

[70] S. Chandrasekaran, S. Ramanathan, T. Basak, Microwave material processing-a review, AIChE J. 58 (2012) 330-363. https://doi.org/10.1002/aic.12766

[71] R. Rosa, P. Veronesi, C. Leonelli, A review on combustion synthesis intensification by means of microwave energy, Chem. Eng. Process.: Process Intensif. 71 (2013) 2-18. https://doi.org/10.1016/j.cep.2013.02.007

[72] R. Rosa, P. Veronesi, C. Leonelli, A review on combustion synthesis intensification by means of microwave energy, Chem. Eng. Process.: Process Intensif. 71 (2013) 2-18. https://doi.org/10.1016/j.cep.2013.02.007

[73] Z. Peng, J.Y. Hwang, C.-L. Park, Numerical analysis of heat transfer characteristics in microwave heating of magnetic dielectrics, Metall. Mater. Trans. A 43 (2012) 1070-1078. https://doi.org/10.1007/s11661-011-1014-3

[74] Z. Peng, J.Y. Hwang, J. Mouris, 41, ISIJ Int. 50 (2010) 1590-1596. https://doi.org/10.2355/isijinternational.50.1590

[75] T. Basak, Influence of various shapes of annular metallic support on microwave heating of 2D cylinders, Chem. Eng. Sci. 61 (2006) 2023-2034. https://doi.org/10.1016/j.ces.2005.10.050

[76] F.J. Keil, Process intensification, Rev. Chem. Eng. 34 (2018) 135-200. https://doi.org/10.1515/revce-2017-0085

[77] T. Basak, Influence of various shapes of annular metallic support on microwave heating of 2D cylinders, Chem. Eng. Sci. 61 (2006) 2023-2034. https://doi.org/10.1016/j.ces.2005.10.050

[78] V. Kumar, K.D.P. Nigam, Process intensification in green synthesis, Green Process. Synth. 1 (2012). https://doi.org/10.1515/greenps-2011-0003

[79] E. Santacesaria, et al., Biodiesel process intensification by using static mixers tubular reactors, Ind. Eng. Chem. Res. 51 (2012) 8777-8787. https://doi.org/10.1021/ie201640w

[80] A.A. Kiss, Novel catalytic reactive distillation processes for a sustainable chemical industry, Top. Catal. 62 (2019) 1132-1148. https://doi.org/10.1007/s11244-018-1052-9

[81] H. Goyal, T.-Y. Chen, W. Chen, A review of microwave-assisted process intensified multiphase reactors, Chem. Eng. J. 430 (2022) 133183. https://doi.org/10.1016/j.cej.2021.133183

[82] L. Estel, M. Poux, N. Benamara, Continuous flow-microwave reactor: Where are we?, Chem. Eng. Process.: Process Intensif. 113 (2017) 56-64. https://doi.org/10.1016/j.cep.2016.09.022

[83] R.S. Abiev, Process intensification in chemical engineering: General trends and Russian contribution, Rev. Chem. Eng. 37 (2021) 69-97. https://doi.org/10.1515/revce-2018-0040

[84] M. Baldea, T.F. Edgar, Dynamic process intensification, Curr. Opin. Chem. Eng. 22 (2018) 48-53. https://doi.org/10.1016/j.coche.2018.08.003

[85] J.K. Adewole, A.M. Yeneneh, M.Y. Khan, Membrane engineering and process intensification, Front. Chem. Eng. 5 (2024) 1360708. https://doi.org/10.3389/fceng.2023.1360708

[86] T.K. Fahim, I.S.M. Zaidul, M.R.A. Bakar, Particle formation and micronization using non-conventional techniques-review, Chem. Eng. Process.: Process Intensif. 86 (2014) 47-52. https://doi.org/10.1016/j.cep.2014.10.009

[87] F. Gallucci, J. Antonio Medrano, L. Roses, Process intensification via membrane reactors, the DEMCAMER project, Processes 4 (2016) 16. https://doi.org/10.3390/pr4020016

[88] T. Kopac, Emerging applications of process intensification for enhanced separation and energy efficiency, environmentally friendly sustainable adsorptive separations: A review, Int. J. Energy Res. 45 (2021) 15839-15856. https://doi.org/10.1002/er.6944

[89] V. Palma, D. Barba, M. Cortese, Microwaves and heterogeneous catalysis: A review on selected catalytic processes, Catalysts 10 (2020) 246. https://doi.org/10.3390/catal10020246

[90] E.T. Thostenson, T.-W. Chou, Microwave processing: fundamentals and applications, Compos. Part A Appl. Sci. Manuf. 30 (1999) 1055-1071. https://doi.org/10.1016/S1359-835X(99)00020-2

[91] M. A. Rostagno, J. M. P. Prado (Eds.), Natural Product Extraction: Principles and Applications, RSC Green Chemistry Series, The Royal Society of Chemistry, Cambridge, 2013. https://doi.org/10.1039/9781849737579

[92] H. Ahmad, M.K. Hossain, Supported nanocatalysts: recent developments in microwave synthesis for application in heterogeneous catalysis, Mater. Adv. 3 (2022) 859-887. https://doi.org/10.1039/D1MA00840D

[93] A. De la Hoz, A. Diaz-Ortiz, A. Moreno, Microwaves in organic synthesis. Thermal and non-thermal microwave effects, Chem. Soc. Rev. 34 (2005) 164-178. https://doi.org/10.1039/B411438H

[94] F. Marken, U.K. Sur, B.A. Coles, Focused microwaves in electrochemical processes, Electrochim. Acta 51 (2006) 2195-2203. https://doi.org/10.1016/j.electacta.2005.04.081

[95] A. Ramirez, J.L. Hueso, R. Mallada, In situ temperature measurements in microwave-heated gas-solid catalytic systems. Detection of hot spots and solid-fluid temperature gradients in the ethylene epoxidation reaction, Chem. Eng. J. 316 (2017) 50-60. https://doi.org/10.1016/j.cej.2017.01.077

[96] I. Ahmed, M. Duchesne, Y. Tan, Electrically heated fluidized beds—a review, Ind. Eng. Chem. Res. 63 (2024) 4205-4235. https://doi.org/10.1021/acs.iecr.3c04232

[97] I. Ahmed, M. Duchesne, Y. Tan, Electrically heated fluidized beds—a review, Ind. Eng. Chem. Res. 63 (2024) 4205-4235. https://doi.org/10.1021/acs.iecr.3c04232

[98] T. Sickert, I. Kalinke, J. Christoph, Microwave-assisted freeze-drying with frequency-based control concepts via solid-state generators: A simulative and experimental study, Processes 11 (2023) 327. https://doi.org/10.3390/pr11020327

[99] H. Mokhtari, S. Barkhan, D. Haseli, A bibliometric analysis and visualization of the Journal of Documentation: 1945-2018, J. Doc. 77 (2020) 69-92. https://doi.org/10.1108/JD-08-2019-0165

[100] J.E. Roberts, H.F. Cook, Microwaves in medical and biological research, Br. J. Appl. Phys. 3 (1952) 33. https://doi.org/10.1088/0508-3443/3/2/301

[101] R.B. Nelson, Tubes for dielectric heating at 915 megacycles, Trans. Am. Inst. Electr. Eng., Part I: Commun. Electron. 71 (1952) 72-80. https://doi.org/10.1109/TCE.1952.6371890

[102] R.J. Giguere, T.L. Bray, S.M. Duncan , Application of commercial microwave ovens to organic synthesis, Tetrahedron Lett. 27 (1986) 4945-4948. https://doi.org/10.1016/S0040-4039(00)85103-5

[103] L.K. Jie, M.S.F., C.Y.K., The use of a microwave oven in the chemical transformation of long chain fatty acid esters, Lipids 23 (1988) 367-369. https://doi.org/10.1007/BF02537351

[104] D.R. Baghurst, D.M.P. Mingos, Application of microwave heating techniques for the synthesis of solid state inorganic compounds, J. Chem. Soc., Chem. Commun. 12 (1988) 829-830. https://doi.org/10.1039/c39880000829

[105] S.-T. Chen, S.-H. Chiou, Y.-H. Chu, Rapid hydrolysis of proteins and peptides by means of microwave technology and its application to amino acid analysis, Int. J. Pept. Protein Res. 30 (1987) 572-576. https://doi.org/10.1111/j.1399-3011.1987.tb03367.x

[106] M. Ali, S.P. Bond, S.A. Mbogo, Use of a domestic microwave oven in organometallic chemistry, J. Organomet. Chem. 371 (1989) 11-13. https://doi.org/10.1016/0022-328X(89)85201-5

[107] University of Cape Town, Design of microwave heating equipment for laboratory applications, https://open.uct.ac.za/items/f7b3df86-08d2-444e-8088-303a0ecb7efa, 2025 (accessed 23 June 2025).

[108] Rate enhancement of organic reactions by microwaves at atmospheric pressure, ProQuest, https://www.proquest.com/openview/d6bc01756742c7101b7cac30577dde99/1?pq-origsite=gscholar&cbl=47615, 2025 (accessed 23 June 2025).

[109] L. Dubois, J.-P. Sozanski, V. Tessier, Temperature control and thermal dosimetry by microwave radiometry in hyperthermia, IEEE Trans. Microwave Theory Techn. 44 (1996) 1755-1761. https://doi.org/10.1109/22.539932

[110] M. Pagnotta, A. Nolan, L. Kim, A simple modification of a domestic microwave oven for improved temperature control, J. Chem. Educ. 69 (1992) 599. https://doi.org/10.1021/ed069p599

[111] C.H. Tong, R.R. Lentz, D.B. Lund, A microwave oven with variable continuous power and a feedback temperature controller, Biotechnol. Prog. 9 (1993) 488-496. https://doi.org/10.1021/bp00023a600

[112] E. Belotserkovsky, O. Shenfeld, A. Katzir, Infrared fiberoptic temperature control of the heating process in a microwave oven, IEEE Trans. Microwave Theory Techn. 42 (1994) 901-903. https://doi.org/10.1109/22.293547

[113] C.O. Kappe, Controlled microwave heating in modern organic synthesis, Angew. Chem. Int. Ed. 43 (2004) 6250-6284. https://doi.org/10.1002/anie.200400655

[114] S. Maheshwari, A. Verma, A. Singh, Carbon Quantum Dots: Green Catalysts for Organic Synthesis, Curr. Microwave Chem. (2025). https://doi.org/10.2174/0122133356368085250304074014

[115] F. Ashamary, M. Elancheziyan, R. Atchudan, Incorporation of Fe_2O_3 Spacer Molecules in Microwave-Exfoliated Graphene Oxide as Efficient Electrodes for Simultaneous Detection of Cd^{2+}, Pb^{2+}, and Hg^{2+} in Water, Adv. Eng. Mater. 27.15 (2025) 2500503. https://doi.org/10.1002/adem.202500503

[116] Microwave Assisted Synthesis and Surface Characterization of GO-Sb_2O_3 NanocompositesforAAElectrochemicalSensor, ProQuest,https://www.proquest.com/openview/8efcb3eff755cbae2add093c1075f9c3/1?pq-origsite=gscholar&cbl=18750&diss=y, 2025 (accessed 23 June 2025). [117] C.R. Sagandira, C. Len, Microwave-assisted continuous flow synthesis of propofol, Sustain.Chem.Pharm.42 (2024) 101793.

[118] C.-N. Chen, C. Pang, F. Xiao, Clinical features and risk factors for pelvic metastasis after microwave ablation of hepatocellular carcinoma: a nested case-control study, Int. J. Hyperthermia 42.1 (2025) 2519350.[119] P. Liang, Y. Wang, Microwave ablation of hepatocellular carcinoma, Oncology 72.Suppl. 1 (2007) 124-131. https://doi.org/10.1080/02656736.2025.2519350

[120] P. Liang, B. Dong, X. Yu, Prognostic factors for survival in patients with hepatocellular carcinoma after percutaneous microwave ablation, Radiology 235.1 (2005) 299-307. [121] W. Li, L.-B. Zhang, J.-H. Peng, Preparation of high surface area activated carbons from tobacco stems with K_2CO_3 activation using microwave radiation, Ind. Crops Prod. 27.3 (2008) 341-347. https://doi.org/10.1016/j.indcrop.2007.11.011

[122] W. Li, J. Peng, L. Zhang, Preparation of activated carbon from coconut shell chars in pilot-scale microwave heating equipment at 60 kW, Waste Manage. 29.2 (2009) 756-760. https://doi.org/10.1016/j.wasman.2008.03.004

[123] Y.R. Wu, H. Wen, B.C. Zhang, Current developments of sparse microwave imaging, 雷达学报 3.4 (2014) 383-395. https://doi.org/10.3724/SP.J.1300.2014.14105

[124] Synthetic Aperture Radar Three-dimensional Imaging-From TomoSAR and Array-InSAR-to-Microwave-Vision:(in English), https://radars.ac.cn/en/article/doi/10.12000/JR19090, 2025 (accessed 23 June 2025).

[125] A.L. Vikharev, A.M. Gorbachev, M.A. Lobaev, Novel microwave plasma-assisted CVD reactor for diamond delta doping, phys. status solidi (RRL)-Rapid Res. Lett. 10.4 (2016) 324-327. https://doi.org/10.1002/pssr.201510453

[126] A.I. Klimov, O.B. Kovalchuk, V.V. Rostov, Measurement of parameters of X-band high-power microwave superradiative pulses, IEEE Trans. Plasma Sci. 36.3 (2008) 661-664. https://doi.org/10.1109/TPS.2008.917300

[127] ChemInform (Wiley) 627246 Publications | 825737 C Related journals, SciSpace - Journal, https://scispace.com/journals/cheminform-1rkhv9xt, 2025 (accessed 23 June 2025).

[128] Y.-J. Zhu, F. Chen, Microwave-assisted preparation of inorganic nanostructures in liquid phase, Chem. Rev. 114.12 (2014) 6462-6555. https://doi.org/10.1021/cr400366s

[129] F. Wiesbrock, R. Hoogenboom, U. S. Schubert, Microwave-assisted polymer synthesis: state-of-the-art and future perspectives, Macromol. Rapid Commun. 25.20 (2004) 1739-1764. https://doi.org/10.1002/marc.200400313

[130] H. Wang, J. Chen, Z. Pei, Enhancing lignocellulosic biorefinery sustainability: mechanisms and optimization of microwave-responsive deep eutectic solvents for rapid delignification, Biofuel Res. J. 12.1 (2025) 2306-2318. https://doi.org/10.18331/BRJ2025.12.1.3

[131] R.S. Varma, Microwave Technology - Applications in Chemical Synthesis, in: Kirk-Othmer Encyclopedia of Chemical Technology, John Wiley & Sons, New York, 2005.

[132] M.B. Gawande, S.N. Shelke, R. Zboril, Microwave-assisted chemistry: synthetic applications for rapid assembly of nanomaterials and organics, Acc. Chem. Res. 47.4 (2014) 1338-1348. https://doi.org/10.1021/ar400309b

[133] T. Yamada, W. Teranishi, N. Sakurada, Microwave-assisted C-C bond formation of diarylacetylenes and aromatic hydrocarbons on carbon beads under continuous-flow conditions, Commun. Chem. 6.1 (2023) 78. https://doi.org/10.1038/s42004-023-00880-y

[134] M. P. Á. Mingos, Tilden Lecture. Applications of microwave dielectric heating effects to synthetic problems in chemistry, Chem. Soc. Rev. 20.1 (1991) 1-47. https://doi.org/10.1039/cs9912000001

[135] C. O. Kappe, Controlled microwave heating in modern organic synthesis, Angew. Chem. Int. Ed. 43.46 (2004) 6250-6284. https://doi.org/10.1002/anie.200400655

[136] M. Ravi, S. Song, K. Gu, Electrical properties of biodegradable poly(ε-caprolactone): lithium thiocyanate complexed polymer electrolyte films, Mater. Sci. Eng. B 195 (2015) 74-83. https://doi.org/10.1016/j.mseb.2015.02.003

[137] B. L. Hayes, Recent advances in microwave-assisted synthesis, Aldrichimica Acta 37.2 (2004) 66-77.

[138] S.H. Jhung, T. Jin, Y.K. Hwang, Microwave effect in the fast synthesis of microporous materials: which stage between nucleation and crystal growth is accelerated by microwave irradiation?, Chem.-Eur. J. 13.16 (2007) 4410-4417. https://doi.org/10.1002/chem.200700098

[139] V. Pandey, V. Chawla, S. K. Saraf, Comparative study of conventional and microwave-assisted synthesis of some Schiff bases and their potential as antimicrobial agents, Med. Chem. Res. 21.6 (2012) 844-852. https://doi.org/10.1007/s00044-011-9592-6

[140] D. Patel, B. Patel, Microwave assisted organic synthesis: an overview, J. Pharm. Res. 4.7 (2011) 2090-2092.

[141] A. Kumar, Y. Kuang, Z. Liang, Microwave chemistry, recent advancements, and eco-friendly microwave-assisted synthesis of nanoarchitectures and their applications: a review, Mater. Today Nano 11 (2020) 100076. https://doi.org/10.1016/j.mtnano.2020.100076

142] M.A.M. Ali, J. Gimbun, K.L. Lau, Biodiesel synthesized from waste cooking oil in a continuous microwave assisted reactor reduced PM and NOx emissions, Environ. Res. 185 (2020) 109452. https://doi.org/10.1016/j.envres.2020.109452

[143] P. Silva, D. Ananias, S.M. Bruno, Photoluminescent metal-organic frameworks-rapid preparation, catalytic activity, and framework relationships, Eur. J. Inorg. Chem. 2013.32 (2013) 5576-5591. https://doi.org/10.1002/ejic.201300637

[144] J. Smijova, J. Tokarský, K.M. Kutláková, Rapid microwave-assisted hydrothermal in situ synthesis of nano-ZnS/kaolinite nanocomposite: A non-toxic photocatalyst active under UV and sunlight, Appl. Clay Sci. 265 (2025) 107669. https://doi.org/10.1016/j.clay.2024.107669

[145] A. Quirino-García, C. Martinez-Alonso, J.E.S. López, Synthesis of nano-zinc oxide by conventional chemical precipitation vs microwaves and its effect on the germination of cucumber (Cucumis sativus) seeds, Results Chem. 10 (2024) 101720. https://doi.org/10.1016/j.rechem.2024.101720

[146] P. Srivastava, K. P. Srivastava, Microwave-assisted expeditious synthesis and antimicrobial evaluation of novel heterocyclic derivatives of sulfonamides, IOSR J. Appl. Chem. 17 (6, Ser. I) (2024) 26-31.

[147] F. Karimi, B.A. Peppley, Comparison of conventional versus microwave heating for polyol synthesis of supported iridium based electrocatalyst for polymer electrolyte membrane water electrolysis, Int. J. Hydrogen Energy 42.8 (2017) 5083-5094. https://doi.org/10.1016/j.ijhydene.2017.01.090

[148] V. Polshettiwar, R.S. Varma, Microwave-assisted organic synthesis and transformations using benign reaction media, Acc. Chem. Res. 41.5 (2008) 629-639. https://doi.org/10.1021/ar700238s

[149] V. Polshettiwar, M.N. Nadagouda, R.S. Varma, Microwave-assisted chemistry: a rapid and sustainable route to synthesis of organics and nanomaterials, Aust. J. Chem. 62.1 (2009) 16-26. https://doi.org/10.1071/CH08404

[150] F.A. Bassyouni, S.M. Abu-Bakr, M.A. Rehim, Evolution of microwave irradiation and its application in green chemistry and biosciences, Res. Chem. Intermed. 38.2 (2012) 283-322. https://doi.org/10.1007/s11164-011-0348-1

[151] F. Frecentese, F. Sodano, A. Corvino, The application of microwaves, ultrasounds, and their combination in the synthesis of nitrogen-containing bicyclic heterocycles, Int. J. Mol. Sci. 24.13 (2023) 10722. https://doi.org/10.3390/ijms241310722

[152] A. Kumari, R.K. Singh, Medicinal chemistry of indole derivatives: Current to future therapeutic prospectives, Bioorg. Chem. 89 (2019) 103021. https://doi.org/10.1016/j.bioorg.2019.103021

[153] R. Bellavita, M. Casertano, N. Grasso, Microwave-Assisted Synthesis of 2-Methyl-1 H-indole-3-carboxylate Derivatives via Pd-Catalyzed Heterocyclization, Symmetry 14.3 (2022) 435. https://doi.org/10.3390/sym14030435

[154] L.A. Mohammed, M.A. Farhan, S.A. Dadoosh, A review on benzimidazole heterocyclic compounds: synthesis and their medicinal activity applications, SynOpen 7.04 (2023) 652-673. https://doi.org/10.1055/a-2155-9125

[155] D. Azarifar, M. Pirhayati, B. Maleki, Acetic acid-promoted condensation of o-phenylenediamine with aldehydes into 2-aryl-1-(arylmethyl)-1H-benzimidazoles under microwave irradiation, J. Serbian Chem. Soc. 75.9 (2010) 1181-1189. https://doi.org/10.2298/JSC090901096A

[156] M. Nasr-Esfahani, I. Mohammadpoor-Baltork, A.R. Khosropour, Synthesis and characterization of Cu(II) containing nanosilica triazine dendrimer: A recyclable nanocomposite material for the synthesis of benzimidazoles, benzothiazoles, bis-benzimidazoles and bis-benzothiazoles, J. Mol. Catal. A: Chem. 379 (2013) 243-254. https://doi.org/10.1016/j.molcata.2013.08.009

[157] A. de la Hoz, A. Loupy (Eds.), Microwaves in Organic Synthesis, 2nd ed., Wiley-VCH, Weinheim, 2012. https://doi.org/10.1002/9783527651313

[158] A. O. Adeola, M. P. Duarte, R. Naccache, Microwave-assisted synthesis of carbon-based nanomaterials from biobased resources for water treatment applications: emerging trends and prospects, Front. Carbon 2 (2023) 1220021. https://doi.org/10.3389/frcrb.2023.1220021

[159] S. Gulati, A. Baul, A. Amar, Eco-friendly and sustainable pathways to photoluminescent carbon quantum dots (CQDs), Nanomaterials 13 (2023) 554. https://doi.org/10.3390/nano13030554

[160] M. Harde, S. Lakade, A. Patil, Microwave-assisted synthesis of fluorescent graphene quantum dots from ferula asafoetida: a green approach for efficient nanomaterial production and selective detection of serum Fe^{3+} ions, Adv. Nat. Sci.: Nanoscience Nanotechnol. 16 (2025) 025001. https://doi.org/10.1088/2043-6262/adb566

[161] K. Rokni, M. Mostafaei, M.D. Soufi, Microwave-assisted intensification of transesterification reaction for biodiesel production from camelina oil: Optimization by Box-Behnken Design, Bioresour. Technol. Rep. 17 (2022) 100928. https://doi.org/10.1016/j.biteb.2021.100928

[162] M. C. Hsiao, et al., Enhancement of biodiesel production from high-acid-value waste cooking oil via a microwave reactor using a homogeneous alkaline catalyst, Energies 14 (2021) 437. https://doi.org/10.3390/en14020437

[163] J. Milano, H.C. Ong, H.H. Masjuki, Optimization of biodiesel production by microwave irradiation-assisted transesterification for waste cooking oil-Calophyllum inophyllum oil via response surface methodology, Energy Convers. Manag. 158 (2018) 400-415. https://doi.org/10.1016/j.enconman.2017.12.027

[164] N. Prajapati, P. Kodgire, S.S. Kachhwaha, Comparison of RSM based FFD and CCD Methods for biodiesel production using microwave technique, Mater. Today: Proc. 62 (2022) 6985-6991. https://doi.org/10.1016/j.matpr.2021.12.379

[165] R. Selvaraj, I.G. Moorthy, R.V. Kumar, Microwave mediated production of FAME from waste cooking oil: Modelling and optimization of process parameters by RSM and ANN approach, Fuel 237 (2019) 40-49. https://doi.org/10.1016/j.fuel.2018.09.147

[166] M. Nüchter, B. Ondruschka, W. Bonrath, Microwave assisted synthesis-a critical technology overview, Green Chem. 6 (2004) 128-141. https://doi.org/10.1039/B310502D

[167] T. Mooney, M. Ilamanova, B. Török, Microwave-assisted flow chemistry for green synthesis and other applications, Curr. Microw. Chem. 9 (2022) 65-69. https://doi.org/10.2174/2213335610666221208163107

[168] C. Sun, M. Ge, S. Tan, Research Progress on Microwave Synthesis of 3d Transition Metal (Mn, Fe, Co, and Ni) Oxide Nanomaterials for Supercapacitors, Molecules 30 (2025) 1843. https://doi.org/10.3390/molecules30081843

[169] L. Han, M. Zhang, H. Li, Rapid Fabrication of ZSM-5/AlPO4-5 Composites via Microwave-Ionothermal Strategy for Enhanced Methanol-to-Olefins Catalysis, Catalysts 15 (2025) 605. https://doi.org/10.3390/catal15060605

[170] J. Liu, Z. Gong, D. Wu, MOFs as platforms for multiscale structural regulation: new frontiers in photocatalyst design, Microstructures 5 (2025) N-A. https://doi.org/10.20517/microstructures.2024.171

[171] J. van der Geer, J.A.J. Hanraads, R.A. Lupton, The art of writing a scientific article, J. Sci. Commun. 163 (2010) 51-59

Materials Research Foundations 189 (2026) 45-66 https://doi.org/10.21741/9781644904039-2

Chapter 2

Microwave-Assisted Bioactive Heterocycles: Applications in Medicinal Chemistry

Shraddha Gautam[1], Kalpana Swain[2], Satyanarayan Pattnaik[2]*

[1]Faculty of Pharmacy, Sanskriti University, Mathura, India

[2]Calcutta Institute of Pharmaceutical Technology and Allied Health Sciences, Howrah, India

Abstract

This chapter explores the use of microwave-assisted organic synthesis (MAOS) for the rapid and eco-friendly synthesis of bioactive heterocycles, including pyrimidines, quinazolines, indoles, thiazoles, oxazoles, and imidazoles. MAOS enhances reaction rates, yields, and selectivity while reducing energy consumption and solvent use, aligning with green chemistry principles. Each class of heterocycle discussed plays a significant role in medicinal chemistry, with applications spanning anticancer, antiviral, antimicrobial, and CNS therapies. Comparative advantages over conventional synthesis are highlighted, along with key methodologies and reaction mechanisms, underscoring MAOS as a transformative tool in the design of novel therapeutic agents.

Keywords

Microwave-Assisted Synthesis, Bioactive Heterocycles, Medicinal Chemistry, Green Chemistry, Pharmacological Applications

Contents

Materials Research Foundations 189 (2026) 45-66 https://doi.org/10.21741/9781644904039-2

1. Introduction

Heterocyclic compounds represent an important category of organic molecules that serve as the foundational structure for a wide range of biologically active substances, such as antibiotics, antiviral drugs, cancer-fighting agents, and stimulants for the central nervous system [1]. Their presence in natural and synthetic drugs highlights their indispensable role in medicinal chemistry. The synthesis of these heterocyclic compounds has traditionally relied on conventional thermal methods, which often involve prolonged reaction times, harsh reaction conditions, and relatively low yields. Such methods not only limit efficiency but also contribute to high energy consumption and environmental concerns due to excessive solvent use and waste generation [2,3].

Microwave-assisted organic synthesis (MAOS) has emerged as a powerful tool to overcome these limitations, providing a sustainable, efficient, and rapid approach to heterocyclic compound synthesis [4,5]. This technique leverages microwave radiation (300-300,000 MHz) to accelerate chemical reactions, leading to enhanced reaction rates, improved product yields, and reduced environmental impact [6-8]. Microwave radiation exploitation in heterocyclic chemistry has gained significant attention due to its alignment with green chemistry principles, which emphasize reducing hazardous reagents, minimizing solvent consumption, and optimizing reaction efficiency as given in Figure 1.

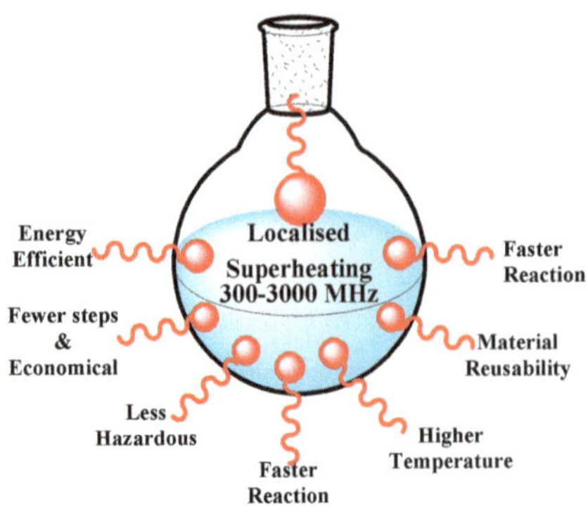

Figure 1: Significance of Microwave assisted reactions.

Materials Research Foundations 189 (2026) 45-66 https://doi.org/10.21741/9781644904039-2

Recent advancements in MAOS have demonstrated its transformative potential in medicinal chemistry by enabling the efficient synthesis of heterocycles with diverse pharmacological activities. Notable examples include pyrimidines, quinazolines, indoles, thiazoles, oxazoles, and imidazoles, all of which serve as core structures in many drug substances. The rapid and selective formation of these bioactive heterocycles under microwave irradiation has facilitated the feasible production of novel drug candidates [10, 11].

This chapter explores the application of microwave-assisted production of bioactive heterocyclic compounds, highlights key reaction methodologies, and discusses their applications in medicinal chemistry as given in Figure 2. Furthermore, a comparative analysis with conventional methods is provided to emphasize the advantages of this approach.

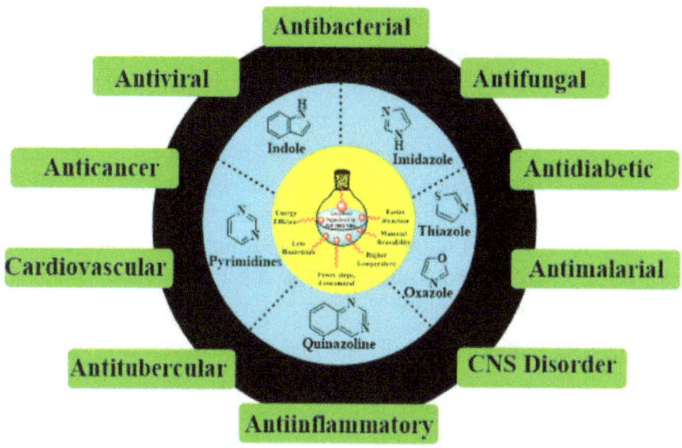

Figure 2: Application of Microwave Assisted Bioactive heterocyclic compounds.

2. Microwave-Assisted Synthesis of Bioactive Heterocycles

Several classes of heterocyclic compounds exhibit significant pharmacological activities. The synthesis of these compounds via microwave-assisted methods has revolutionized their accessibility and structural diversity.

2.1 Pyrimidines

Pyrimidines represent an essential group of heterocyclic compounds with extensive uses in the fields of pharmaceuticals, agrochemicals, and materials science [12]. Conventionally, pyrimidines are synthesized via condensation reactions involving β-dicarbonyl compounds (e.g., ethyl acetoacetate) and various nitrogen nucleophiles such as urea, thiourea, or guanidine derivatives [12-14]. However, these traditional methods often require prolonged heating, resulting in lower yields and by-product formation. The advent of microwave-assisted organic synthesis (MAOS) offers an efficient alternative, providing rapid reaction rates, improved yields, solvent-free syntheses, and cleaner reactions. The microwave-assisted synthesis of pyrimidines typically involves the condensation of a β-dicarbonyl compound (such as ethyl acetoacetate) with a

Materials Research Foundations 189 (2026) 45-66 https://doi.org/10.21741/9781644904039-2

nitrogen-containing nucleophile under controlled microwave irradiation at 100-160°C for 10 minutes (scheme 1) [15].

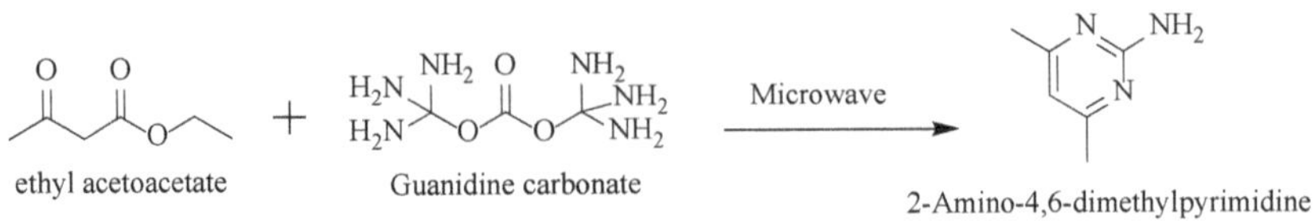

Scheme 1: Microwave Assisted synthesis of Pyrimidine.

The efficient internal heating through dipolar polarization and ionic conduction accelerates the reaction kinetics compared to conventional heating [16]. Pyrimidines produced through this approach are employed in creating antimalarial, anticancer, and antiviral drugs. This technique is especially advantageous for swiftly generating compound libraries in the field of medicinal chemistry.

Pyrimidines form the basic framework of several critical drugs, especially those targeting nucleic acid metabolism. Their most well-known role is as the backbone of nucleic acid bases cytosine, thymine, and uracil highlighting their indispensable nature in genetics and molecular biology [17]. This biological significance has been harnessed in drug design, leading to the development of a wide array of pyrimidine-based pharmaceuticals. The most prominent areas of application in medicinal chemistry are discussed as follows.

Pyrimidine analogs interfere with DNA synthesis, making them effective against tumors. Drugs like 5-fluorouracil (5-FU) and capecitabine inhibit thymidylate synthase, leading to DNA damage and cancer cell death. Other agents, such as gemcitabine, are incorporated into DNA during replication and cause chain termination [18,19]. Pyrimidines form the backbone of several antiviral drugs. Zidovudine (AZT), a thymidine analog, is used in the treatment of HIV by inhibiting reverse transcriptase. Other examples include lamivudine and emtricitabine, which are cytidine analogs used in combination therapies for HIV and hepatitis B [20,21].

Some pyrimidine derivatives act as antibiotics or antifolates. Trimethoprim, a diaminopyrimidine, inhibits bacterial dihydrofolate reductase (DHFR), blocking folate synthesis essential for DNA replication in bacteria [22]. Pyrimidines have been incorporated into the design of modern antimalarial agents. For example, pyrimethamine, a 2,4-diaminopyrimidine, selectively inhibits *Plasmodium* DHFR and is used in combination therapies to combat malaria [23].

Several pyrimidine-containing compounds modulate ion channels and neurotransmitter receptors. Pyrimidine derivatives are found in calcium channel blockers, anticonvulsants, and antidepressants. For instance, diazepam, though structurally a benzodiazepine, contains a fused pyrimidine ring and is widely used for anxiety and epilepsy [24].

2.2 Quinazolines

Microwave-assisted synthesis of quinazolines using one-pot condensation reactions has provided high yields and reduced reaction times compared to conventional approaches [25]. Microwave-assisted condensation of anthranilic acid and isocyanates efficiently yields quinazolinone derivatives in under 10 minutes with >90% yield [26,27].

Materials Research Foundations 189 (2026) 45-66 https://doi.org/10.21741/9781644904039-2

The synthesis of quinazoline derivatives via microwave-assisted methods has emerged as a highly efficient and environmentally benign approach, offering numerous advantages over conventional heating techniques. It involves the cyclo-condensation of 2-aminobenzophenone or 2-aminobenzonitrile derivatives with aldehydes or ortho-esters in the presence of an ammonia source such as ammonium acetate or urea. These reactions may be carried out in a suitable solvent like ethanol or acetic acid, or under solvent-free conditions, further enhancing their green chemistry appeal. The reactants are combined in a microwave-sealed vessel and subjected to microwave irradiation at temperatures ranging from 130°C to 160°C, with power settings between 300 to 540 watts for a duration of approximately 5 to 20 minutes as shown in Scheme 2 [25,26]. The rapid heating provided by microwave energy significantly accelerates the reaction rate, often leading to higher yields and cleaner product profiles compared to traditional methods.

After the reaction is complete, the mixture is allowed to cool to room temperature before being poured into water or ice to precipitate the product. This product is then collected through filtration and purified by recrystallization or chromatographic methods [28]. This microwave-assisted method serves as a flexible, scalable, and high-yielding approach for synthesizing biologically significant quinazoline frameworks, making it especially appealing for use in medicinal chemistry and drug discovery [29].

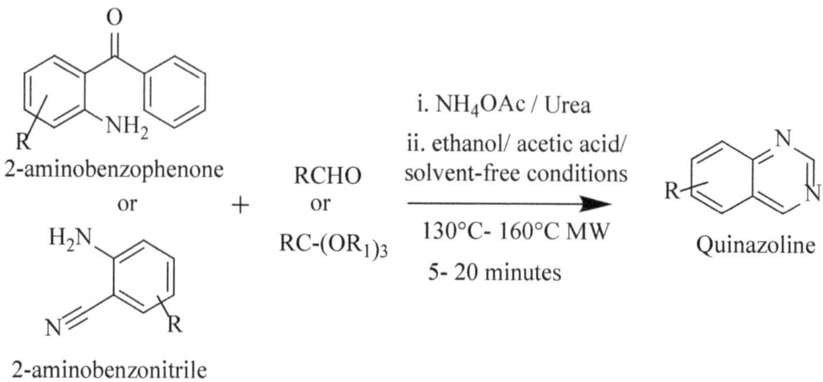

Scheme 2: *Microwave Assisted synthesis of Quinazoline.*

Quinazoline-derived compounds are pivotal in current cancer treatments, largely because they target and inhibit receptor tyrosine kinases. These receptors are often overactive in certain cancers, promoting uncontrolled cell division. Clinically approved drugs like gefitinib and erlotinib exemplify this mechanism, offering therapeutic benefit in conditions such as non-small cell lung cancer. The structural versatility of quinazolines makes them a promising scaffold for designing next-generation anticancer agents [30,31].

Quinazolines also exhibit potent α1-adrenergic receptor blocking activity, which makes them valuable in managing hypertension and benign prostatic hyperplasia (BPH). Several drugs in clinical use prazosin (Minipress®), terazosin (Hytrin®), and doxazosin (Cardura®) are quinazoline-based α1-antagonists [32]. These agents function by relaxing vascular smooth muscle,

Materials Research Foundations 189 (2026) 45-66 https://doi.org/10.21741/9781644904039-2

thereby lowering blood pressure, and are also effective in improving urinary flow by reducing smooth muscle tone [33,34].

Quinazoline derivatives have shown promising results against Mycobacterium tuberculosis, including drug-resistant strains. Scientists are investigating experimental compounds that incorporate quinazoline structures to evaluate their effectiveness as inhibitors of DprE1 [35]. Additionally, certain quinazoline analogues exhibit broad-spectrum antibacterial activity, with derivatives showing efficacy against Staphylococcus aureus and Escherichia coli. Although none are yet clinically approved for antibacterial indications, ongoing studies highlight their potential for future development [36,37].

Research has identified several quinazoline-based compounds with significant activity against Plasmodium falciparum [38]. One well-known experimental compound, WR99210, a dihydrotriazine derivative structurally related to quinazolines, targets dihydrofolate reductase (DHFR) in the malaria parasite. Such compounds disrupt folate metabolism, which is crucial for DNA synthesis and cell survival in the parasite [39].

Some quinazoline derivatives show activity in the central nervous system (CNS) and are being explored for potential use in epilepsy, neurodegenerative diseases, and psychotic disorders [40]. For example, riluzole, although not a direct quinazoline, shares similar structural features and is used for treating amyotrophic lateral sclerosis (ALS). Experimental quinazoline compounds have shown acetylcholinesterase inhibitory activity, which may benefit conditions like Alzheimer's disease [41]. Additionally, preclinical studies suggest potential for anticonvulsant and anxiolytic effects [40].

Quinazoline derivatives possess notable anti-inflammatory properties. Experimental molecules have demonstrated the ability to reduce levels of pro-inflammatory cytokines and nitric oxide, indicating their potential use in conditions like rheumatoid arthritis, osteoarthritis, and other chronic pain syndromes. Although no quinazoline-based NSAIDs are currently in clinical use, this remains a promising area of research [42,43].

Several quinazoline compounds have been identified as potential antiviral agents. For example, certain analogues have shown inhibitory activity against HIV-1 reverse transcriptase, HCV NS5B polymerase, and SARS-CoV-2 proteases. These mechanisms involve the interruption of viral replication or interference with viral protein synthesis. While these agents are largely in preclinical or early clinical stages, their development is being accelerated by the ongoing demand for broad-spectrum antiviral drugs [44,45].

Quinazoline derivatives are also being investigated for their role in managing type 2 diabetes mellitus. Some experimental compounds act by promoting incretin hormone activity and glucose-dependent insulin secretion [46,47]. While these applications remain experimental, they represent a novel approach in addressing both diabetes and its metabolic complications.

The wide-ranging therapeutic potential of quinazoline derivatives, underscored by several clinically approved drugs and many more under investigation, reflects their importance in modern medicinal chemistry. Their chemical tractability, tunable pharmacophores, and compatibility with various targets make them indispensable scaffolds in the discovery and development of next-generation pharmaceuticals.

Materials Research Foundations 189 (2026) 45-66 https://doi.org/10.21741/9781644904039-2

2.3 Indoles

The microwave assisted synthesis shortened the synthetic pathway and enhances the reaction rate, yield, and selectivity [48,49]. The Fischer Indole Synthesis is a classical method for preparing indoles from phenylhydrazines and aldehydes or ketones, typically under acidic conditions. Microwave-assisted synthesis has modernized this reaction by offering significant advantages over conventional methods [50].

Under microwave irradiation, the reaction proceeds rapidly often within 5 to 20 minutes at 600 W with improved yields, selectivity, and purity by Scheme 3. Microwave energy enables uniform and efficient heating, which reduces reaction times and eliminates the need for harsh conditions. The reaction is commonly carried out in acetic acid, ethanol, or even under solvent-free conditions, using acids like HCl, acetic acid, or polyphosphoric acid as catalysts [51].

Scheme 3: *Microwave Assisted synthesis of Indole by Fischer Indole Synthesis.*

This technique aligns with green chemistry principles by minimizing solvent use, energy consumption, and waste, making it highly valuable for the eco-friendly synthesis of bioactive indole derivatives in medicinal chemistry.

The synthesis of Indole from o-alkynylanilines under microwave irradiation is a powerful and efficient method. This transformation typically involves 5-endo-dig cyclization of ortho-alkynyl substituted anilines, catalyzed by acidic or metal catalysts such as Au(I), Pt(II), Iodine I_2 (10 mol%), or Bronsted acids [52].

Microwave irradiation greatly improves this reaction by delivering quick and even heating, which leads to reduced reaction times (5–15 minutes), increased yields, and purer product profiles compared to traditional thermal methods, as illustrated in Scheme 4.The reaction can proceed under solvent-free or green solvent conditions, aligning with green chemistry principles [53].

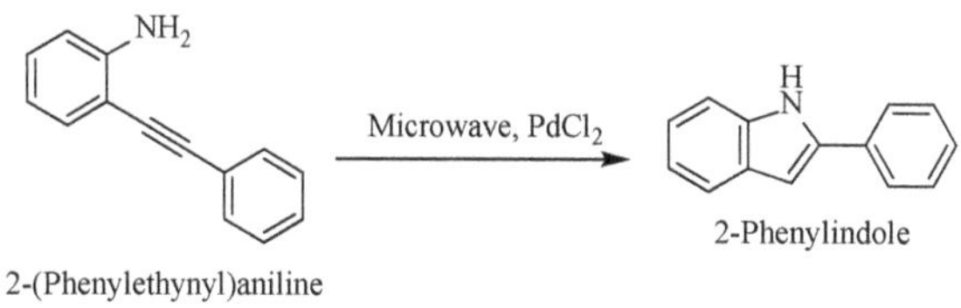

Scheme 4: *Microwave Assisted synthesis of Indole by cyclisation.*

Palladium-catalyzed cross-coupling reactions is used in C–H activation or C–N bond formation leading to indole frameworks by utilising **Scheme 5.**

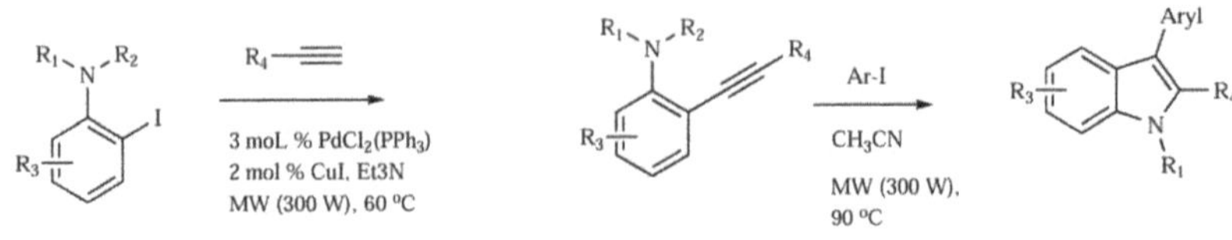

Scheme 5: *Microwave Assisted synthesis of Indole by Palladium-Catalysed Cross-Coupling Reactions.*

Indoles are a privileged scaffold in medicinal chemistry and their applications span across anticancer, anti-inflammatory, antimicrobial, antiviral, CNS-active, and hormonal therapies.

Indole derivatives are crucial in cancer treatment because they can engage with a variety of cellular targets. Many indole-based compounds exhibit antitumor activity by inhibiting tubulin polymerization, inducing apoptosis, or modulating kinase signaling pathways[54]. Notable examples include *vincristine* and *vinblastine*, which are indole alkaloids used widely in chemotherapy for leukemia and lymphomas [55]. Another prominent compound, *indole-3-carbinol*, is known to modulate estrogen metabolism and suppress tumor growth [56]. Additionally, synthetic indole derivatives like *sunitinib* function as tyrosine kinase inhibitors and are used in the treatment of renal cell carcinoma and gastrointestinal stromal tumors [57].

Indoles are also crucial in the development of anti-inflammatory drugs. Their capacity to block crucial enzymes and mediators associated with inflammation, including cyclooxygenase-2 (COX-2), inducible nitric oxide synthase (iNOS), and tumor necrosis factor-alpha (TNF-α), renders them potent therapeutic agents, such as Indomethacin [58-60]. Compounds that contain indole have shown considerable effectiveness against various microbes. They often work by disrupting cell wall synthesis, inhibiting key enzymes, or interfering with DNA replication. Delavirdine, for instance, is an indole-based non-nucleoside reverse transcriptase inhibitor (NNRTI) used in the treatment of HIV-1 infection[61]. Other synthetic indole derivatives have shown promise in combating multidrug-resistant bacteria like MRSA (Methicillin-resistant *Staphylococcus aureus*) and even *Mycobacterium tuberculosis*, the causative agent of tuberculosis [62].

The indole nucleus is structurally similar to several key neurotransmitters, making it highly relevant in CNS drug development. *Serotonin* (5-hydroxytryptamine), a crucial neurotransmitter regulating mood, appetite, and sleep, is itself an indole compound [63]. Many synthetic drugs mimicking serotonin's structure target serotonin receptors to treat migraines, depression, and anxiety. For example, *sumatriptan* is a selective 5-HT1 receptor agonist used to relieve migraine headaches [64]. Indole derivatives also form the basis for drugs influencing melatonin pathways, such as in the treatment of sleep disorders [65].

In hormonal therapy, indole derivatives are employed as modulators of endocrine pathways. Compounds mimicking or interfering with estrogen and androgen receptors have been designed using indole scaffolds. Some indole-based agents act as selective estrogen receptor modulators (SERMs), which are useful in managing conditions like breast cancer and osteoporosis [66]. Additionally, synthetic analogs of *indole-3-acetic acid,* a plant hormone are being explored for their potential effects on endocrine-related processes in humans, including thyroid regulation.

Beyond their major therapeutic uses, indoles also serve various other roles in medicinal chemistry. Indole derivatives act as antioxidants by neutralizing reactive oxygen species (ROS) and shielding cells from oxidative harm, a trait advantageous in treating neurodegenerative and cardiovascular conditions [67]. Some indole-based molecules have shown antidiabetic activity by acting on targets like peroxisome proliferator-activated receptor gamma (PPAR-γ) or inhibiting carbohydrate-digesting enzymes such as α-glucosidase [68]. Additionally, indole derivatives are being developed as antihypertensive agents through their inhibitory action on the angiotensin-converting enzyme (ACE), a key player in blood pressure regulation [69].

2.4 Thiazoles and Oxazoles

Oxazoles and thiazoles are aromatic heterocyclic compounds with five-membered rings, where oxazoles contain oxygen and nitrogen, and thiazoles include sulfur and nitrogen. The conventional methods for synthesizing these heterocycles typically require severe conditions, extended reaction durations, and multiple steps.

2.4.1 Microwave-Assisted Synthesis of Oxazoles and Thiazoles

Oxazoles are typically synthesized via cyclodehydration or condensation reactions involving α-hydroxyketones, acylamino ketones, or isocyanides [70]. Under microwave irradiation, these reactions proceed much faster due to uniform energy distribution and efficient internal heating. A common approach involves the reaction of α-hydroxyketones with formamide or ammonium acetate under solvent-free conditions, leading to 2,5-disubstituted oxazoles with high efficiency as given by Scheme 6.

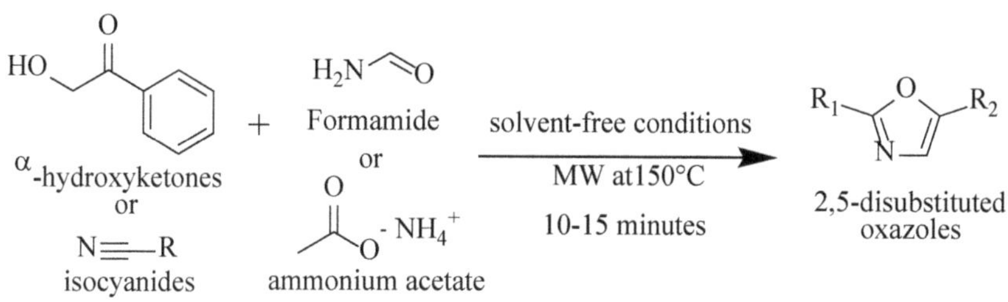

Scheme 6: *Microwave Assisted synthesis of Oxazole.*

Multicomponent reactions (MCRs), such as the Ugi reaction followed by cyclization, also benefit significantly from microwave heating. These strategies reduce the number of synthetic steps while increasing atom economy and minimizing waste. Microwave irradiation not only accelerates bond formation but also improves regioselectivity and overall product purity, making it highly suitable for medicinal chemistry applications.

In one method, α-bromoacetophenone reacts with formamide under microwave irradiation at 150 °C for 10–15 minutes, yielding 2-phenyl-5-methyloxazole with excellent purity and yield [71].

The synthesis of thiazoles is often accomplished using the Hantzsch method, which involves condensation between α-haloketones and thioamides or thioureas [72]. Under conventional conditions, this reaction requires prolonged heating; however, microwave-assisted conditions drastically reduce reaction times to a few minutes and often proceed under solvent-free or minimal-solvent conditions [73].

Materials Research Foundations 189 (2026) 45-66 https://doi.org/10.21741/9781644904039-2

Microwave irradiation enhances reaction kinetics by increasing molecular collisions and accelerating cyclization [74]. Furthermore, one-pot methods have been developed where aldehydes, thiourea, and other components are combined in a single reaction vessel under microwave irradiation, simplifying synthesis and improving throughput especially important for combinatorial chemistry and drug screening. As the microwave-assisted synthesis of 2-amino-4-phenylthiazole from benzaldehyde, thiourea, and α-bromoketone is completed within 10 minutes with yields exceeding 85%.

Hantzsch thiazole synthesis:

The reaction between bis-nucleophile (N-C=S) as thioamide and an α-halocarbonyl compound as α-chlorocarbonyl in presence of acetonitrile at 70°C for 3 minutes cyclises to yield thiazole by condensation reaction by **Scheme** 7 [75].

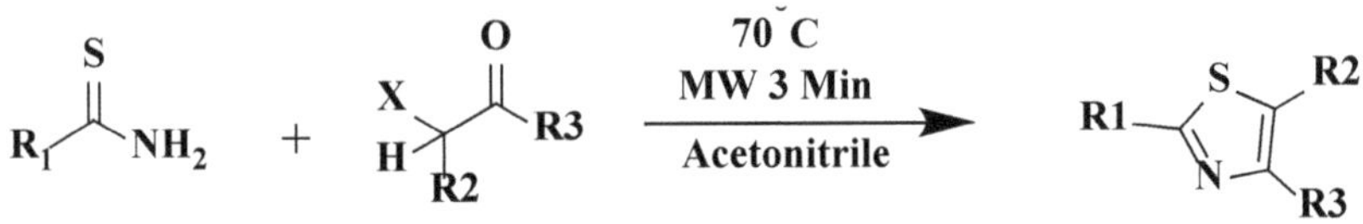

Scheme 7: *Microwave Assisted synthesis of thiazole by Hantzsch method.*

Robinson-Gabriel Thiazole Synthesis

Chiral thiazoles with N-Fmoc protection were produced through the cyclization of N-protected thioamide ester under microwave irradiation, as depicted in Scheme 8 [75].

Oxazoles and thiazoles, as heterocyclic frameworks, are pivotal in medicinal chemistry because they are commonly found in numerous pharmacologically active substances. Their structural adaptability, aromatic characteristics, and advantageous electronic attributes make them essential in crafting and developing medications aimed at various diseases. These heterocycles enhance drug effectiveness by boosting bioavailability, metabolic stability, and the ability to bind effectively to biological targets.

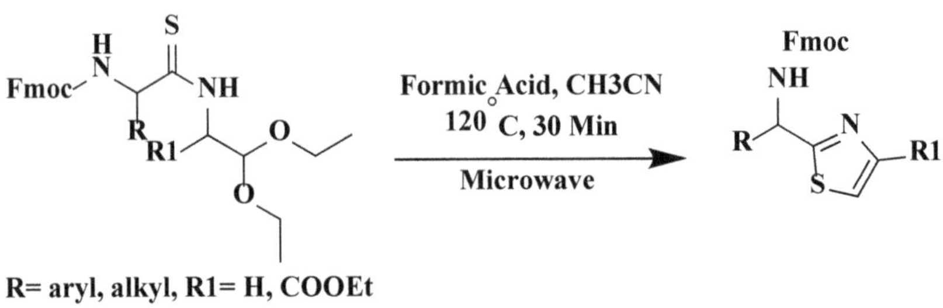

Scheme 8: *Microwave Assisted synthesis of thiazole by Robinson-Gabriel method.*

Materials Research Foundations 189 (2026) 45-66 https://doi.org/10.21741/9781644904039-2

2.4.2 Antimicrobial Activity

Oxazole derivatives have demonstrated potent antibacterial and antifungal activities. Their mode of action often involves inhibition of bacterial protein synthesis or interference with microbial DNA replication [76]. Certain oxazole compounds act by targeting bacterial enzymes such as DNA gyrase or topoisomerase, thereby inhibiting cell proliferation. For example, synthetic oxazole analogs have shown strong effect against *S. aureus*, *E. coli*, and *C. albicans*, making them valuable in treating infections resistant to traditional antibiotics [77].

Thiazole-based compounds exhibit broad-spectrum antimicrobial activity. Many thiazole derivatives act by disrupting bacterial cell wall synthesis or inhibiting critical enzymes like dihydrofolate reductase [78]. Notably, thiazole-containing agents are found in sulfa drugs and other antibiotics. Compounds like sulfathiazole represent early examples, but newer synthetic derivatives continue to be explored for activity against multidrug-resistant strains [77].

2.4.3 Antiviral Applications

Certain oxazoles have been investigated for their antiviral potential, particularly against RNA viruses. These compounds may inhibit viral replication enzymes or interfere with viral entry mechanisms. Oxazole derivatives have been studied for their effects against hepatitis C virus (HCV), influenza virus, and other pathogens by modulating viral proteases or polymerases [79].

Thiazole rings are present in multiple antiviral agents, most notably ritonavir, a potent HIV protease inhibitor. Ritonavir works by preventing viral polyprotein cleavage, which is essential for viral maturation. Other thiazole analogs are being developed to treat hepatitis B and herpes viruses. Their ability to mimic nucleoside structures or enzyme substrates makes them ideal scaffolds for targeting viral replication machinery [80,81].

2.4.4 Anticancer Properties

Oxazole derivatives show significant anticancer activity through various mechanisms. Some act as kinase inhibitors, blocking signaling pathways essential for tumor growth (e.g., EGFR, VEGFR, and PI3K pathways) [82]. Others induce apoptosis or disrupt microtubule dynamics. Their ability to intercalate DNA or inhibit topoisomerases also makes them strong candidates for chemotherapeutic agents [82,83].

Several thiazole derivatives are active against cancer cells by targeting tubulin polymerization or inhibiting enzymes such as histone deacetylases (HDACs) and tyrosine kinases [84]. The drug dasatinib, used in chronic myeloid leukemia (CML), contains a thiazole moiety and acts as a multi-targeted kinase inhibitor [85]. Additionally, novel thiazole-based compounds are under development to overcome resistance to conventional chemotherapy and to target tumor-specific pathways [86].

2.4.5 Antitubercular and Antimalarial Activity

Thiazole-based molecules are prominent in the development of antitubercular agents, particularly against drug-resistant *Mycobacterium tuberculosis*. These compounds often inhibit enzymes such as enoyl-ACP reductase, essential for mycolic acid synthesis in the bacterial cell wall. In addition, thiazole derivatives like pyrimethamine exhibit antimalarial activity by targeting *Plasmodium* dihydrofolate reductase, disrupting DNA synthesis and cell division [87- 90].

Materials Research Foundations 189 (2026) 45-66 https://doi.org/10.21741/9781644904039-2

Although less common, oxazole derivatives have shown promising results against *Plasmodium falciparum*, the parasite responsible for malaria. Their structural flexibility allows for functional modifications that enhance selectivity and potency against the parasite's metabolic enzymes [91].

2.4.6 Central Nervous System (CNS) Activity

Some oxazole-based compounds exhibit neuroprotective, anticonvulsant, or anxiolytic properties. They are believed to modulate GABAergic or serotonergic pathways in the CNS. Research is ongoing to develop oxazole derivatives that cross the blood-brain barrier and show potential in treating epilepsy, depression, and neurodegenerative diseases such as Alzheimer's [92].

Thiazole-containing compounds also demonstrate CNS activity and are being explored for their potential as antipsychotic, antidepressant, and anticonvulsant agents. Their interaction with neurotransmitter receptors and ion channels plays a role in modulating brain chemistry and electrical signaling [93].

2.4.7 Enzyme Inhibition and Receptor Binding

Both oxazoles and thiazoles are frequently incorporated into enzyme inhibitors and receptor ligands due to their rigid, planar structures that facilitate effective binding. These rings enhance pharmacodynamic properties by fitting into enzyme active sites or binding pockets on receptors, thereby modulating biological responses. Their incorporation also improves the lipophilicity, binding affinity, and selectivity of the drug molecules [94].

2.4.8 Bioisosteres in Drug Design

Medicinal chemists often employ oxazole and thiazole rings as bioisosteres for phenyl rings, amides, or esters. Substituting such groups with heterocycles improves metabolic stability, reduces off-target effects, and enhances binding interactions with the biological target. This strategy has proven effective in optimizing lead compounds and developing safer, more effective medications [95,96].

2.5 Imidazole

Imidazoles serve as a fundamental component in numerous biologically active substances and pharmaceutical drugs [97]. The use of microwave irradiation significantly speeds up many traditional reactions employed in the synthesis of imidazoles [98]. Some notable methods include:

2.5.1 Debus–Radziszewski Imidazole Synthesis

This is one of the most classical methods for synthesizing imidazole derivatives. It involves a three-component condensation of 1,2-dicarbonyl compound (e.g., glyoxal), an aldehyde, and ammonia or a primary amine [99]. Under conventional conditions, this reaction typically requires refluxing for several hours. However, microwave irradiation allows this reaction to proceed in just a few minutes, often with improved yields and fewer by-products as shown in Scheme 9.

Materials Research Foundations 189 (2026) 45-66 https://doi.org/10.21741/9781644904039-2

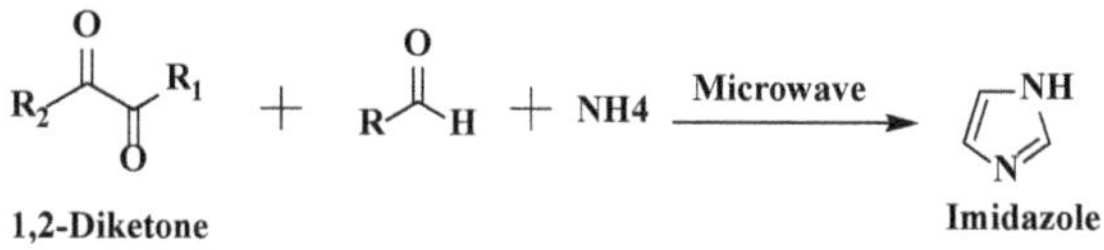

Scheme 9: *Microwave Assisted synthesis of Imidazole by Debus–Radziszewski Synthesis.*

2.5.2 Multicomponent Reactions (MCRs)

MCRs are widely used in modern drug discovery for their simplicity and efficiency. Microwave-assisted MCRs for imidazole synthesis can be performed in solvent-free conditions or in environmentally benign solvents, and are often catalyst-free or utilize recyclable solid-supported catalysts.

Example components include A β-diketone (such as benzil), An aldehyde, Ammonium acetate. Microwave irradiation not only enhances the reaction rate but also promotes a one-pot synthesis, which is highly desirable for pharmaceutical applications.

2.5.3 Transition-Metal Catalyzed Methods

Transition-metal catalysts such as copper, iron, or palladium can be used under microwave irradiation to facilitate C–N or C–C bond formation in imidazole synthesis. Microwave energy enhances the efficiency of these catalysts, often enabling lower catalyst loading and shorter reaction times.

2.5.4 Solvent-Free and Green Protocols

Microwave-assisted synthesis aligns well with the principles of green chemistry. Many protocols have been developed that use no solvents, biodegradable ionic liquids, or recyclable solid catalysts (e.g., silica, montmorillonite K10). These methods reduce the environmental footprint of chemical synthesis [100,101].

2.5.6 Pharmaceutical Applications of Imidazoles

Imidazole derivatives are involved in a wide array of therapeutic applications. The presence of two nitrogen atoms in the ring system allows imidazoles to interact with diverse biological targets, including enzymes, receptors, and DNA.

Imidazole-based drugs like clotrimazole, miconazole, and ketoconazole are among the most widely used antifungal agents. They inhibit lanosterol 14-α-demethylase, a key enzyme in fungal ergosterol biosynthesis, leading to disrupted cell membrane integrity [102].

Several imidazole derivatives exhibit potent antibacterial activity against both Gram-positive and Gram-negative bacteria. These compounds can act by disrupting bacterial enzymes, inhibiting DNA synthesis, or interfering with metabolic pathways.

Imidazole derivatives can act as cytotoxic agents by binding to DNA, interfering with mitosis, or inhibiting topoisomerases. For example, some imidazole-containing heterocycles show promising results against breast, prostate, and colon cancers [103,104].

Imidazoles have been incorporated into molecules that inhibit cyclooxygenase enzymes (COX-1 and COX-2), which are crucial for prostaglandin synthesis and inflammatory responses [105].

Some imidazole derivatives have shown activity against HIV, hepatitis, and herpes simplex viruses by inhibiting viral replication enzymes or interfering with nucleic acid synthesis [106].

Imidazoles are also studied for their potential in treating CNS disorders. Compounds with imidazole cores have shown activities are Anxiolytics, antidepressants, and anticonvulsants. They may act synthesis on GABA receptors, serotonin receptors, or ion channels [107,108].

Click or tap here to enter text.Conclusion

Microwave-assisted organic synthesis (MAOS) has revolutionized the field of heterocyclic chemistry by offering an efficient, sustainable, and versatile alternative to conventional thermal methods. The integration of microwave energy in organic transformations ensures rapid, uniform heating, enabling significant reductions in reaction times, enhancement in yields, and improvement in product selectivity. This technology aligns perfectly with the principles of green chemistry by minimizing solvent use, reducing hazardous by-products, and conserving energy, thereby promoting environmentally responsible synthesis.

The chapter demonstrates that microwave irradiation has been successfully applied in the synthesis of a wide variety of bioactive heterocycles, including pyrimidines, quinazolines, indoles, thiazoles, oxazoles, and imidazoles—each possessing substantial pharmacological potential. These heterocyclic frameworks form the backbone of many therapeutic agents exhibiting anticancer, antimicrobial, antiviral, antimalarial, anti-inflammatory, and central nervous system activities. The adoption of MAOS has also facilitated the rapid generation of diverse compound libraries, accelerating the discovery of lead molecules in medicinal chemistry and drug design.

Furthermore, the capacity of microwave-assisted methods to support solvent-free and multicomponent reactions underscores their practical and industrial relevance. The precision, reproducibility, and scalability of these techniques make them highly suitable for modern pharmaceutical synthesis. Future advancements in reactor design, catalyst development, and microwave-driven combinatorial chemistry are expected to further broaden the applicability of this technique.

In conclusion, microwave-assisted synthesis stands as a transformative approach in contemporary medicinal chemistry. Its ability to combine speed, efficiency, and environmental stewardship marks a paradigm shift toward more sustainable drug discovery and development. By enabling the rapid and green synthesis of pharmacologically valuable heterocycles, MAOS continues to redefine the boundaries of innovation in synthetic and medicinal chemistry.

References

[1] R.J. Obaid, E.U. Mughal, N. Naeem, M.M. Al-Rooqi, A. Sadiq, R.S. Jassas, Z. Moussa, S.A. Ahmed, Pharmacological significance of nitrogen-containing five and six-membered heterocyclic scaffolds as potent cholinesterase inhibitors for drug discovery, Process Biochem. 120 (2022) 250-259. https://doi.org/10.1016/j.procbio.2022.06.009

[2] M.A.P. Martins, C.P. Frizzo, D.N. Moreira, L. Buriol, P. Machado, Solvent-free heterocyclic synthesis, Chem. Rev. 109 (2009) 4140-4182. https://doi.org/10.1021/cr9001098

[3] A. Daştan, A. Kulkarni, B. Török, Environmentally benign synthesis of heterocyclic compounds by combined microwave-assisted heterogeneous catalytic approaches, Green Chem. 14 (2012) 17-37. https://doi.org/10.1039/C1GC15837F

[4] L. Xochicale-Santana, C.C. Vidyasagar, B. M. Muñoz-Flores, V. M. J. Pérez, Microwave-assisted organic syntheses (MAOS): The green synthetic method, in: B. Kharisov, O. Kharissova (Eds.), Handbook of Greener Synthesis of Nanomaterials and Compounds, Elsevier, New York, 2021, pp. 491-542. https://doi.org/10.1016/B978-0-12-821938-6.00015-3

[5] N. Sharma, U.K. Sharma, E.V. Van der Eycken, Microwave-assisted organic synthesis: Overview of recent applications, in: W. Zhang, B.W. Cue (Eds.), Green Techniques for Organic Synthesis and Medicinal Chemistry, 2nd ed., John Wiley & Sons, NJ, 2018, pp. 441-468. https://doi.org/10.1002/9781119288152.ch17

[6] A. Jha, Microwave assisted synthesis of organic compounds and nanomaterials, in: Nanofibers-Synthesis, Properties and Applications, IntechOpen, 2021. https://doi.org/10.5772/intechopen.98224

[7] S. Maramai, M. Taddei, Microwave Irradiation, in: Stefano Protti, Andrea Palmieri (Eds.), Sustainable Organic Synthesis: Tools and Strategies, The Royal Society of Chemistry, Cambridge, 2021, pp. 488-521. https://doi.org/10.1039/9781839164842-00488

[8] A.K. Mahato,B. M. Sahoo, B. K. Banik, B. C. Mohanta, Microwave-assisted synthesis: Paradigm of green chemistry, J. Indian Chem. Soc. 95 (2018) 1327-1339.

[9] F.A. Bassyouni, S.M. Abu-Bakr, M.A. Rehim, Evolution of microwave irradiation and its application in green chemistry and biosciences, Res. Chem. Intermed. 38 (2012) 283-322. https://doi.org/10.1007/s11164-011-0348-1

[10] O. Ebenezer, M. A. Jordaan, G. Carena, T. Bono, Mi. Shapi, J. A. Tuszynski, An overview of the biological evaluation of selected nitrogen-containing heterocycle medicinal chemistry compounds, Int. J. Mol. Sci. 23 (2022) 8117. https://doi.org/10.3390/ijms23158117

[11] P. Prabitha, K. Kolachi, A.B. HM, S. Dubey, Fundamental chemistry of N-heterocycles: Pyrrole and pyrazole chemistry, imidazole chemistry, triazole chemistry advances in heterocyclic ring systems, in: Prashant Tiwari, Popat Mohite (Eds.), Paving the Path to Discoveries and Unlocking the Secrets of N-Heterocycles, Cambridge Scholars Publishing, United Kingdom, 2024, pp. 64-102.

[12] P.K. Maji, Recent progress in the synthesis of pyrimidine heterocycles: A review, Curr. Org. Chem. 24 (2020) 1055-1096. https://doi.org/10.2174/1385272824999200507123843

[13] M. Grazia Martina, L. Giannessi, M. Radi, Multicomponent synthesis of purines and pyrimidines: From the origin of life to new sustainable approaches for drug-discovery applications, Eur. J. Org. Chem. 26 (2023) e202201288. https://doi.org/10.1002/ejoc.202201288

[14] S.N. Anjirwala, P.S. Parmar, S.K. Patel, Synthetic protocols for non-fused pyrimidines, Synth. Commun. 52 (2022) 2079-2121. https://doi.org/10.1080/00397911.2022.2137682

[15] G. Tiwari, A. Khanna, V. K. Mishra and R. Sagar, Recent developments on microwave-assisted organic synthesis of nitrogen- and oxygen-containing preferred heterocyclic scaffolds, RSC Adv. 13 (2023) 32858-32892. https://doi.org/10.1039/D3RA05986C

[16] H. Ruan, J. Jiang, B. Sun, W. Zhang, W. Gao, L. Y. Wang, Z. Ma, A rapid low-temperature internal heating strategy with optimal frequency based on constant polarization voltage for lithium-ion batteries, Appl. Energy 177 (2016) 771-782. https://doi.org/10.1016/j.apenergy.2016.05.151

[17] R.K. Kumarachari, S. Peta, A.S. Surur, Y.T. Mekonnen, Synthesis, characterization and in silico biological activity of some 2-(N,N-dimethyl guanidinyl)-4,6-diaryl pyrimidines, J. Pharm. Bioallied Sci. 8(2016) 181-187. https://doi.org/10.4103/0975-7406.171678

[18] E. Zarenezhad, M. Farjam, A. Iraji, Synthesis and biological activity of pyrimidines-containing hybrids: Focusing on pharmacological application, J. Mol. Struct. 1230 (2021) 129833. https://doi.org/10.1016/j.molstruc.2020.129833

[19] G. Pizzorno, R.B. Diasio, Y.-C. Cheng, Pyrimidine analogs, in: D.W. Kufe, R.E. Pollock, R.R. Weichselbaum, R.C. Bast Jr., T.S. Gansler, J.F. Holland, E. Frei III (Eds.), Holland-Frei Cancer Medicine, 6th ed., BC Decker, Hamilton, 2003, pp. 1327-1339.

[20] C.M. Perry, D. Faulds, Lamivudine: A review of its antiviral activity, pharmacokinetic properties and therapeutic efficacy in the management of HIV infection, Drugs 53 (1997) 657-680. https://doi.org/10.2165/00003495-199753040-00008

[21] S. EG, B. Lakshmanan, A review on therapeutic potential of heterocyclic pyrimidine derivatives as potent antiviral agents, Asian J. Pharm. Clin. Res. (2020) 30-34. https://doi.org/10.22159/ajpcr.2020.v13i7.37430

[22] A. Kumar, A. Mishra, Role of heterocyclic compounds in pharmaceuticals and medicines, J. Exp. Zool. India 26 (2023) 1-?. https://doi.org/10.51470/jez.2023.26.1.1

[23] K. Somandi, Synthesis and biological evaluation of pyrimidine and isoquinoline inhibitors as potential antimalarial antifolates and transmission-blocking agents, University of the Witwatersrand, Johannesburg, South Africa, 2023.

[24] S. Pant, S. Nain, Recent advances in the development of pyrimidine-based CNS agents, Curr. Drug Discov. Technol. 20 (2023) 14-28. https://doi.org/10.2174/1570163819666221003094402

[25] T. Besson, E. Chosson, Microwave-assisted synthesis of bioactive quinazolines and quinazolinones, Comb. Chem. High Throughput Screen. 10 (2007) 903-917. https://doi.org/10.2174/138620707783220356

[26] L. Mohammadkhani, M.M. Heravi, Microwave-assisted synthesis of quinazolines and quinazolinones: An overview, Front. Chem. 8 (2020) 580086. https://doi.org/10.3389/fchem.2020.580086

[27] S.K. Wahan, S. Sharma, P.A. Chawla, Synthesis of quinazolinone and quinazoline derivatives using green chemistry approach, Phys. Sci. Rev. 8 (2023) 3079-3094. https://doi.org/10.1515/psr-2021-0223

[28] L. Mohammadkhani, M.M. Heravi, Microwave-assisted synthesis of quinazolines and quinazolinones: An overview, Front. Chem. 8 (2020) 580086. https://doi.org/10.3389/fchem.2020.580086

[29] S. Kumar, A. Maurya, A. Agarwal, Microwave-assisted synthesis of heterocyclic scaffolds, SynOpen 8 (2024) 138-152. https://doi.org/10.1055/s-0043-1775379

[30] H.T. Abdel-Mohsen, H. Mohamed, S. Ali, M.A. Ahmed, Recent advances in structural optimization of quinazoline-based protein kinase inhibitors for cancer therapy (2021-present), Molecules 29 (2024) 875. https://doi.org/10.3390/molecules29040875

[31] A. Sonousi, R.A. Hassan, E.O. Osman, A.M. Abdou, S.H. Emam, Design and synthesis of novel quinazolinone-based derivatives as EGFR inhibitors with antitumor activity, J. Enzyme Inhib. Med. Chem. 37 (2022) 2644-2659. https://doi.org/10.1080/14756366.2022.2118735

[32] C.M. Benning, N. Kyprianou, Quinazoline-derived α1-adrenoceptor antagonists induce prostate cancer cell apoptosis via an α1-adrenoceptor-independent action, Cancer Res. 62 (2002) 597-602.

[33] I. Anglin, D. Glassman, N. Kyprianou, Induction of prostate apoptosis by α1-adrenoceptor antagonists: Mechanistic significance of the quinazoline component, Prostate Cancer Prostatic Dis. 5 (2002) 88-95. https://doi.org/10.1038/sj.pcan.4500561

[34] B. Akduman, E.D. Crawford, Terazosin, doxazosin, and prazosin: Current clinical experience, Urology 58 (2001) 49-54. https://doi.org/10.1016/S0090-4295(01)01302-4

[35] J. Gawad, C. Bonde, Design, synthesis and biological evaluation of novel 6-(trifluoromethyl)-N-(4-oxothiazolidin-3-yl) quinazoline-2-carboxamide derivatives as potential DprE1 inhibitors, J. Mol. Struct. 1217 (2020) 128394. https://doi.org/10.1016/j.molstruc.2020.128394

[36] J. Dai, L. Li, H. Zhang, X. Liu, Discovery of quinazolone pyridiniums as potential broad-spectrum antibacterial agents, Molecules 30 (2025) 243. https://doi.org/10.3390/molecules30020243

[37] Z. Li, X. Wang, Y. Chen, The antibacterial activity of quinazoline and quinazolinone hybrids, Curr. Top. Med. Chem. 22 (2022) 1035-1044. https://doi.org/10.2174/1568026622666220307144015

[38] C. Mendoza-Martínez, P. López, R. Hernández, Design, synthesis and biological evaluation of quinazoline derivatives as anti-trypanosomatid and anti-plasmodial agents, Eur. J. Med. Chem. 96 (2015) 296-307. https://doi.org/10.1016/j.ejmech.2015.04.028

[39] A. Nzila, B. Okumu, C. Mwangi, Preclinical evaluation of the antifolate QN254, 5-chloro-N′6′-(2,5-dimethoxy-benzyl)-quinazoline-2,4,6-triamine, as an antimalarial drug candidate, Antimicrob. Agents Chemother. 54 (2010) 2603-2610. https://doi.org/10.1128/AAC.01526-09

[40] R.S. Cheke, P.A. Chanda, M.F. Olaleye, Quinazoline: An update on current status against convulsions, J. Mol. Struct. 1248 (2022) 131384. https://doi.org/10.1016/j.molstruc.2021.131384

[41] Z. Haghighijoo, M. Khodadadi, R. Mohammadi, Therapeutic potential of quinazoline derivatives for Alzheimer's disease: A comprehensive review, Eur. J. Med. Chem. 227 (2022) 113949. https://doi.org/10.1016/j.ejmech.2021.113949

[42] M.F. Zayed, Medicinal chemistry of quinazolines as analgesic and anti-inflammatory agents, ChemEngineering 6 (2022) 94. https://doi.org/10.3390/chemengineering6060094

[43] V. Alagarsamy, S. Thangavel, R. Subramanian, An overview of quinazolines: Pharmacological significance and recent developments, Eur. J. Med. Chem. 151 (2018) 628-685. https://doi.org/10.1016/j.ejmech.2018.03.076

[44] G. Ahmad, S. Khan, M.A. Farooq, N.H. Siddiqui, N-Heterocycles as promising antiviral agents: A comprehensive overview, Molecules 29 (2024) 2232. https://doi.org/10.3390/molecules29102232

[45] T.A. Farghaly, M.S. El-Sayed, R.M. Ali, A literature review focusing on the antiviral activity of [1,2,4] and [1,2,3]-triazoles, Mini Rev. Med. Chem. 24 (2024) 1602-1629. https://doi.org/10.2174/0113895575277122231108095511

[46] M.A. El-Zahabi, R. Hassan, A. Mostafa, Design, synthesis, molecular modeling and anti-hyperglycemic evaluation of quinazoline-sulfonylurea hybrids as peroxisome proliferator-activated receptor gamma (PPARγ) and sulfonylurea receptor (SUR) agonists, Int. J. Mol. Sci. 23 (2022) 9605. https://doi.org/10.3390/ijms23179605

[47] S. A. Kumar, S. S. Patel, S. Patel, P. Parikh, Future treatment of diabetes - tyrosine kinase inhibitors, J. Diabetes Metab. Disord. 22 (2022) 61-71. https://doi.org/10.1007/s40200-022-01164-3

[48] S.A. Patil, R. Patil, D.D. Miller, Microwave-assisted synthesis of medicinally relevant indoles, Curr. Med. Chem. 18 (2011) 615-637. https://doi.org/10.2174/092986711794480195

[49] A.K. Rathi, P. Sharma, R.K. Gupta, Microwave-assisted synthesis-catalytic applications in aqueous media, Coord. Chem. Rev. 291 (2015) 68-94. https://doi.org/10.1016/j.ccr.2015.01.011

[50] E.C. Creencia, M. Tsukamoto, T. Horaguchi, One-pot-one-step, microwave-assisted Fischer indole synthesis, J. Heterocycl. Chem. 48 (2011) 1095-1102. https://doi.org/10.1002/jhet.689

[51] P. Acosta-Guzmán, Microwave as a greener alternative in the synthesis of organic compounds, in: A. Nag (Ed.), Greener Synthesis of Organic Compounds, CRC Press, Boca Raton, 2022, pp. 57-73. https://doi.org/10.1201/9781003089162-4

[52] O.S. Kamble, M. Khatravath, R. Dandela, Applications of ethynylanilines as substrates for construction of indoles and indole-substituted derivatives, ChemistrySelect 6 (2021) 7408-7427. https://doi.org/10.1002/slct.202101437

[53] M. Bilal, S. Ahmed, K. Hussain, Synthesis of indoles via intermolecular and intramolecular cyclization by using palladium-based catalysts, Catalysts 11 (2021) 1018. https://doi.org/10.3390/catal11091018

[54] Y. Hong, H. Zhao, X. Li, Indole derivatives as tubulin polymerization inhibitors for the development of promising anticancer agents, Bioorg. Med. Chem. 55 (2022) 116597. https://doi.org/10.1016/j.bmc.2021.116597

[55] P. Dhyani, C. Quispe, E. Sharma, Anticancer potential of alkaloids: A key emphasis to colchicine, vinblastine, vincristine, vindesine, vinorelbine and vincamine, Cancer Cell Int. 22 (2022) 206. https://doi.org/10.1186/s12935-022-02624-9

[56] B.B. Aggarwal, H. Ichikawa, Molecular targets and anticancer potential of indole-3-carbinol and its derivatives, Cell Cycle 4 (2005) 1201-1215. https://doi.org/10.4161/cc.4.9.1993

[57] M.R. Mohamed, S. Patel, H. Rahman, Anti-tumor activity of indole: A review, Lett. Drug Des. Discov. 21 (2024) 3332-3348. https://doi.org/10.2174/0115701808288928240226104445

[58] S. Singh, N. Sharma, R. Chandra, The indole nucleus as a selective COX-2 inhibitor and anti-inflammatory agent (2011-2022), Org. Chem. Front. 9 (2022) 3624-3639. https://doi.org/10.1039/D2QO00534D

[59] Í.T. Jacob, R. Fernandes, M. Ahamed, Anti-inflammatory activity of novel thiosemicarbazone compounds indole-based as COX inhibitors, Pharmacol. Rep. 73 (2021) 907-925. https://doi.org/10.1007/s43440-021-00221-7

[60] B. Portevin, J. Dupont, F. Martin, 1,3-Diaryl-4,5,6,7-tetrahydro-2H-isoindole derivatives: A new series of potent and selective COX-2 inhibitors in which a sulfonyl group is not a structural requisite, J. Med. Chem. 43 (2000) 4582-4593. https://doi.org/10.1021/jm990965x

[61] Q. Chen, Y. Li, X. Wang, Therapeutic potential of indole derivatives as anti-HIV agents: A mini-review, Curr. Top. Med. Chem. 22 (2022) 993-1008. https://doi.org/10.2174/1568026621666211012111901

[62] H.-L. Qin, Z. Chen, J. Liu, Indole-based derivatives as potential antibacterial activity against methicillin-resistant Staphylococcus aureus (MRSA), Eur. J. Med. Chem. 194 (2020) 112245. https://doi.org/10.1016/j.ejmech.2020.112245

[63] A.R. Green, D.G. Grahame-Smith, 5-Hydroxytryptamine and other indoles in the central nervous system, in: L.L. Iversen, S.D. Iversen, S.H. Snyder (Eds.), Biochemistry of Biogenic Amines, Plenum Press, New York, 1975, pp. 169-245. https://doi.org/10.1007/978-1-4684-3171-1_4

[64] L. Pellesi, T.P. Do, A. Hougaard, Pharmacological management of migraine: Current strategies and future directions, Expert Opin. Pharmacother. 25 (2024) 673-683. https://doi.org/10.1080/14656566.2024.2349791

[65] S.-Y. Wang, X.-C. Shi, P. Laborda, Indole-based melatonin analogues: Synthetic approaches and biological activity, Eur. J. Med. Chem. 185 (2020) 111847. https://doi.org/10.1016/j.ejmech.2019.111847

[66] R. Singla, P. Sharma, A. Gupta, Identification of novel indole-based heterocycles as selective estrogen receptor modulators, Bioorg. Chem. 79 (2018) 72-88. https://doi.org/10.1016/j.bioorg.2018.04.002

[67] E. Barresi, L. Giannessi, M. Radi, Indole-based compounds in the development of anti-neurodegenerative agents, Molecules 29 (2024) 2127. https://doi.org/10.3390/molecules29092127

[68] Y. Zhu, J. Li, X. Zhang, Research progress of indole compounds with potential antidiabetic activity, Eur. J. Med. Chem. 223 (2021) 113665. https://doi.org/10.1016/j.ejmech.2021.113665

[69] C. Tan, H. Li, P. Wang, Antihypertensive activity of indole and indazole analogues: A review, Arab. J. Chem. 15 (2022) 103756. https://doi.org/10.1016/j.arabjc.2022.103756

[70] N. Kaur, R. Sharma, S. Gupta, Synthesis of five-membered O, N-heterocycles using metal and nonmetal, Synth. Commun. 49 (2019) 1345-1384. https://doi.org/10.1080/00397911.2019.1594308

[71] S. Lokesh Kumar, C. Binish, A comprehensive review of one-pot multicomponent strategies for oxazole synthesis, ChemistrySelect 10 (2025) e01172. https://doi.org/10.1002/slct.202501172

[72] S. Kamila, K. Mendoza, E.R. Biehl, Microwave-assisted Hantzsch thiazole synthesis of N-phenyl-4-(6-phenylimidazo[2,1-b]thiazol-5-yl)thiazol-2-amines from the reaction of 2-chloro-1-(6-phenylimidazo[2,1-b]thiazol-5-yl)ethanones and thioureas, Tetrahedron Lett. 53 (2012) 4921-4924. https://doi.org/10.1016/j.tetlet.2012.06.116

[73] E. Deau, F. Langa, M. Rossi, Microwave-assisted synthesis of novel N-(4-phenylthiazol-2-yl)-benzo[d]thiazole-, thiazolo[4,5-b]pyridine-, thiazolo[5,4-b]pyridine- and benzo[d]oxazole-2-carboximidamides inspired by marine topsentines and nortopsentines, Tetrahedron 70 (2014) 5532-5540. https://doi.org/10.1016/j.tet.2014.06.102

[74] F. Langa, E. Deau, M. Rossi, Microwave irradiation: More than just a method for accelerating reactions, Contemp. Org. Synth. 4 (1997) 373-386. https://doi.org/10.1039/CO9970400373

[75] A.M. Borcea, I. Ionuț, O. Crișan, O. Oniga, An overview of the synthesis and antimicrobial, antiprotozoal, and antitumor activity of thiazole and bisthiazole derivatives, Molecules 26 (2021) 624. https://doi.org/10.3390/molecules26030624

[76] H.-Z. Zhang, Z.-L. Zhao, C.-H. Zhou, Recent advances in oxazole-based medicinal chemistry, Eur. J. Med. Chem. 144 (2018) 444-492. https://doi.org/10.1016/j.ejmech.2017.12.044

[77] A. Ansari, R. Sharma, P. Kumar, Facile one-pot multicomponent synthesis and molecular docking studies of steroidal oxazole/thiazole derivatives with effective antimicrobial, antibiofilm and hemolytic properties, Steroids 134 (2018) 22-36. https://doi.org/10.1016/j.steroids.2018.04.003

[78] J. Xia, H. Li, S. Chen, Discovery of quaternized pyridine-thiazole-pleuromutilin derivatives with broad-spectrum antibacterial and potent anti-MRSA activity, J. Med. Chem. 66 (2023) 5061-5078. https://doi.org/10.1021/acs.jmedchem.2c02135

[79] I.M. Lagoja, E. De Clercq, Anti-influenza virus agents: Synthesis and mode of action, Med. Res. Rev. 28 (2008) 1-38. https://doi.org/10.1002/med.20096

[80] I.P. Singh, S. Gupta, S. Kumar, Thiazole compounds as antiviral agents: An update, Med. Chem. 16 (2020) 4-23. https://doi.org/10.2174/1573406415666190614101253

[81] J. Khan, A. Rani, M. Aslam, R.S. Maharia, G. Pandey, B. Nand, Exploring triazole-based drugs: Synthesis, application, FDA approvals, and clinical trial updates - A comprehensive review, Tetrahedron 162 (2024) 134122. https://doi.org/10.1016/j.tet.2024.134122

[82] S. Kulkarni, K. Kaur, V. Jaitak, Recent developments in oxazole derivatives as anticancer agents: Review on synthetic strategies, mechanism of action and SAR studies, Anti-Cancer Agents Med. Chem. 22 (2022) 1859-1882. https://doi.org/10.2174/1871520621666210915095421

[83] M.A. Chiacchio, L. Bianchi, F. Rossi, Oxazole-based compounds as anticancer agents, Curr. Med. Chem. 26 (2019) 7337-7371. https://doi.org/10.2174/0929867326666181203130402

[84] K.E. Arnst, S. Banerjee, H. Chen, S. Deng, D.J. Hwang, W. Li, D.D. Miller, Current advances of tubulin inhibitors as dual acting small molecules for cancer therapy, Med. Res. Rev. 39(4) (2019) 1398-1426. https://doi.org/10.1002/med.21568

[85] Y. Zhang, X. Li, H. Wang, The progress of small molecule targeting Bcr-Abl in the treatment of chronic myeloid leukemia, Mini Rev. Med. Chem. 24 (2024) 642-663. https://doi.org/10.2174/0113895575218335230926070130

[86] M.I. Shosha, F.Z. El-Ablack, E.A. Saad, New thiazole derivative as a potential anticancer and topoisomerase II inhibitor, Sci. Rep. 15 (2025) 710. https://doi.org/10.1038/s41598-024-81294-1

[87] F.C. Biaggio, G. Rinaldi, L. Moretti, Synthesis and biological activity of prostaglandin analogs containing heteroatoms in the cyclopentane ring, Curr. Org. Chem. 9 (2005) 419-457. https://doi.org/10.2174/1385272053174912

[88] K.H. Narasimhamurthy, T.R. Swaroop, K.S. Rangappa, A review on progress of thiazole derivatives as potential anti-inflammatory agents, Eur. J. Med. Chem. Rep. 100225 (2024). https://doi.org/10.1016/j.ejmcr.2024.100225

[89] M.K. Kumawat, Thiazole-containing heterocycles with antimalarial activity, Curr. Drug Discov. Technol. 15 (2018) 196-200. https://doi.org/10.2174/1570163814666170725114159

[90] T. Kalita, P. Sharma, R. Kaur, A review on synthetic thiazole derivatives as an antimalarial agent, Curr. Drug Discov. Technol. 21 (2024) 10-42. https://doi.org/10.2174/0115701638276379231223101625

[91] R.K. Singh, A. Patel, S. Gupta, Antimicrobial and antimalarial evaluation of some novel 1,3-oxazole derivatives, Chem. Biol. Interface 7 (2017) 116-123.

[92] D. Patel, R. Sharma, P. Singh, Review on therapeutic diversity of oxazole scaffold: An update, ChemistrySelect 9 (2024) e202403179. https://doi.org/10.1002/slct.202403179

[93] S.A. Saganuwan, Functional chemical groups that may likely become a source for the synthesis of novel central nervous system (CNS) acting drugs, CNS Agents Med. Chem. 17 (2017) 178-186. https://doi.org/10.2174/1871524917666170502153752

[94] A. Dorababu, Synthesis, pharmacological evaluation and structure-activity relationship of recently discovered enzyme antagonist azoles, Heliyon 6 (2020) e03656. https://doi.org/10.1016/j.heliyon.2020.e03656

[95] M.A. Subbaiah, N.A. Meanwell, Bioisosteres of the phenyl ring: Recent strategic applications in lead optimization and drug design, J. Med. Chem. 64 (2021) 14046-14128. https://doi.org/10.1021/acs.jmedchem.1c01215

[96] M. Kaspady, R. Sharma, A. Singh, Synthesis, antibacterial activity of 2,4-disubstituted oxazoles and thiazoles as bioisosteres, Lett. Drug Des. Discov. 6 (2009) 21-28. https://doi.org/10.2174/157018009787158481

[97] A. Siwach, P.K. Verma, Synthesis and therapeutic potential of imidazole containing compounds, BMC Chem. 15 (2021) 1-69. https://doi.org/10.1186/s13065-020-00730-1

[98] X.-L. Yu, Y. Zhang, H. Li, Synthesis of imidazole-based molecules under ultrasonic irradiation approaches, Molecules 28 (2023) 4845. https://doi.org/10.3390/molecules28124845

[99] T.T. Hieu, L. Nguyen, P. Tran, Recent achievement in the synthesis of imidazoles, Curr. Org. Chem. 27 (2023) 1398-1446. https://doi.org/10.2174/0113852728259414231010050749

[100] I. Abdellah, S. Ahmed, M. Hassan, Advancements in green chemistry: Microwave-assisted synthesis of poly-heterocyclic compounds in aqueous media, Int. J. Cur. Res. Sci. Eng. Tech. 7 (2024) 16-26. https://doi.org/10.30967/IJCRSET/Mohamed-R-Eletmany/127

[101] A. Kokel, C. Schäfer, B. Török, Application of microwave-assisted heterogeneous catalysis in sustainable synthesis design, Green Chem. 19 (2017) 3729-3751. https://doi.org/10.1039/C7GC01393K

[102] A. Vitiello, F. Ferrara, M. Boccellino, A. Ponzo, C. Cimmino, E. Comberiati, A. Zovi, S. Clemente, M. Sabbatucci, Antifungal drug resistance: An emergent health threat, Biomedicines 11(4) (2023) 1063. https://doi.org/10.3390/biomedicines11041063

[103] N. Kumar, N. Goel, Recent development of imidazole derivatives as potential anticancer agents, Phys. Sci. Rev. 8 (2023) 2903-2941. https://doi.org/10.1515/psr-2021-0041

[104] I. Ali, M.N. Lone, H.Y. Aboul-Enein, Imidazoles as potential anticancer agents, Med. Chem. Comm. 8 (2017) 1742-1773. https://doi.org/10.1039/C7MD00067G

[105] A.-M. Alaa, K.E. ElTahir, Y.A. Asiri, Synthesis, anti-inflammatory activity and COX-1/COX-2 inhibition of novel substituted cyclic imides. Part 1: Molecular docking study, Eur. J. Med. Chem. 46 (2011) 1648-1655. https://doi.org/10.1016/j.ejmech.2011.02.013

[106] R. Giugliano, G. Della Sala, C. Buonocore, C. Zannella, P. Tedesco, F. Palma Esposito, C. Ragozzino, A. Chianese, M.V. Morone, V. Mazzella, L. Núñez-Pons, V. Folliero, G. Franci, A. De Filippis, M. Galdiero, D. de Pascale, New imidazolium alkaloids with broad spectrum of action from the marine bacterium Shewanella aquimarina, Pharmaceutics 15(8) (2023) 2139. https://doi.org/10.3390/pharmaceutics15082139

[107] S.A. Saganuwan, Conversion of benzimidazoles, imidazothiazoles and imidazoles into more potent central nervous system acting drugs, CNS Agents Med. Chem. 20 (2020) 3-12. https://doi.org/10.2174/1871524919666190621160323

[108] K.K. Goel, P. Sharma, R. Kumar, Imidazoles as serotonin receptor modulators for treatment of depression: Structural insights and structure-activity relationship studies, Pharmaceutics 15 (2023) 2208. https://doi.org/10.3390/pharmaceutics15092208

Materials Research Foundations 189 (2026) 67-102 https://doi.org/10.21741/9781644904039-3

Chapter 3

Microwave-Assisted Synthesis of Heterocycles: Applications in Medicinal Chemistry

N. Dinesh Kumar[1], R. Velmurugan[2] and M. Swaminathan[3]*

[1]Department of Chemistry, Erode Sengunthar Engineering College, Erode-638057, Tamilnadu, India

[2]Department of Chemistry, Ramakrishna Mission Vivekananda College, Chennai – 600028, India

[3]Nanomaterials laboratory, Department of Chemistry, Kalasalingam Academy of Research and Education, Krishnankoil, Tamil Nadu-626126, India

m.swaminathan@klu.ac.in

Abstract

Microwave-assisted organic synthesis (MAOS), a potent and eco-friendly method for synthesizing a variety of bioactive heterocycles, greatly improves reaction efficiency, selectivity, and yield. This chapter focuses on the strategic use of microwave irradiation in the synthesis of nitrogen heterocyclic frameworks with five, six, and seven membered rings, which form essential scaffolds in medicinal chemistry. Microwave-assisted reactions provide a solvent-efficient way to create structurally complicated compounds. This method has demonstrated significant potential in the development of bioactive substances with strong anti-inflammatory, antiviral, antifungal, anticancer, and antibacterial properties. The developments in microwave-assisted heterocycle synthesis and their significant uses in the search for and creation of medicinal drugs are compiled in this chapter.

Keywords

Microwave Irradiation, Eco-Friendly Synthesis, N-Heterocycles, Bioactivity, Medicinal Characteristics

Contents

Materials Research Foundations 189 (2026) 67-102 https://doi.org/10.21741/9781644904039-3

1. Introduction

Continuous improvements in medicinal chemistry are fueled by the ongoing global research priority to improve human health and sanitation. The design and synthesis of bioactive compounds to treat a wide range of illnesses and stop the transmission of infectious agents are the main areas of intensive emphasis for researchers in this subject. [1,2] The discovery of affordable medicinal compounds with great efficacy and few side effects is a primary goal. In order to guarantee desirable pharmacological profiles and low toxicity, a great deal of research has been focused on the synthesis, optimization, and biological screening of new compounds. Modern drug development has been significantly altered by this changing paradigm, with heterocyclic scaffolds and now it is serving as the fundamental structural building blocks of most pharmacologically active organic molecules [3].

Since heterocyclic ring structures are frequently found in a variety of therapeutically active compounds, which make up as much as 80% of all manufactured medications, they have garnered a lot of interest in medicinal chemistry. Among these, nitrogen-containing heterocycles stand out as some of the most prevalent and pharmacologically significant scaffolds present in bioactive compounds [4]. These N-heterocyclic compounds are adaptable building blocks for substances with a broad range of biological activity in various therapeutic domains, which made these N-heterocycles essential to drug discovery. Their wide range of uses, from the creation of vitamins and herbicides to the creation of strong antifungal, antibacterial, and anticancer medicines, demonstrates their versatility [5]. These moieties are essential structural components of synthetic bioactive compounds as well as natural products, especially in vitamins, phenylpropanoids, alkaloids, and polyketides. Because of their importance, organic chemistry has placed a lot of emphasis on heterocycle synthesis, which has resulted in the creation of numerous named reactions [6]. Both natural and artificial bioactive chemicals contain mostly heterocyclic structures. These heterocyclic frameworks' derivatives have shown a range of pharmacological effects, highlighting the crucial role of heterocycles in the development of therapeutically useful compounds [7-9]. Synthetic chemists have discovered a number of ways to create heterocyclic molecules utilizing costly catalysts like palladium, gold, or other similarly costly and environmentally dangerous materials [10].

The increasing need for sustainable and effective heterocyclic synthesis has led to the development of numerous contemporary methods. These include, among other things, ball milling, solvothermal techniques, photochemical and solvent-free reactions, metal-free procedures, electrochemical synthesis, and ultrasonication [11,12]. Developments in reaction engineering have greatly shortened reaction times and increased overall efficiency, whereas traditional thermal reactions frequently call for lengthy periods of time from several hours to days [13]. Over the past few decades, microwave oven using microwave energy has become a potent instrument in many scientific domains, such as bioactive organic and heterocyclic synthesis, materials science, biochemistry, polymer chemistry, and nanotechnology. It has outperformed traditional techniques due to its profoundly shorter reaction times, higher yields, and improved product purity and selectivity. In order to achieve quick and consistent heating, this development, known as microwave-assisted organic synthesis (MAOS), uses dielectric heating, in which solvents or reagents directly absorb microwave radiation. MAOS resulted in a high-performance, effective, and sustainable substitute for contemporary chemical synthesis [14].

Microwave-assisted synthesis has become one of the most potent tools among the many cutting-edge techniques developed to improve synthetic sustainability and efficiency. Microwave radiation, which had been reported first as a heating method in 1946, had been used practically in chemical synthesis in 1980s. Since then, there has been a notable increase in the use of MAOS and microwave-induced organic reaction enhancements (MOREs). Numerous classic name reactions were performed successfully using microwave irradiation. In addition to increasing reaction speeds, microwave conditions improve chemo-, regio-, and stereoselectivity. Microwave irradiation, in contrast to traditional heating, enables quick and consistent energy absorption, frequently elevating the reaction mixture's temperature quickly over the solvent's boiling point. In many situations, solvent-free conditions are also feasible, allowing the reactants to absorb energy directly, thus promoting environmentally friendly and sustainable synthesis methods [15]. The key concepts of green chemistry are most effectively demonstrated by microwave-assisted synthesis, which ensures remarkable energy efficiency while facilitating decreased waste and by-product generation, higher yields, enhanced atom economy, and the usage of fewer hazardous substances and little to no toxic solvents. Several instances demonstrate its ecologically benign character by demonstrating its capacity to support catalyst-free, solvent-free, and highly selective (regio-, enantio-, and chemoselective) reactions. These characteristics demonstrate the revolutionary potential of microwave heating in accomplishing quick, sustainable, and clean chemical reactions with a significantly lower energy input than traditional heating techniques [16].

Microwave-assisted synthesis is a very efficient way to synthesize the bioactive heterocycles, which can fulfill the growing need for quick, effective, and sustainable medication development. This technology greatly outperforms traditional approaches by combining high selectivity, faster reaction rates, and environmental friendliness. A wide range of N-, O-, and S-containing heterocyclic frameworks, such as spiro, fused, and simple ring systems, can be quickly synthesized from easily accessible feedstocks by employing a transition metal catalyst under microwave conditions. The one-pot, multicomponent nature of MAOS reduces solvent usage, waste, and energy consumption, and all these are in perfect harmony with the principles of green chemistry. Furthermore, in this method, the Active Pharmaceutical Ingredient (API) characteristics like stability and solubility can be adjusted without changing their inherent bioactivity, facilitating the quick development of innovative treatments [17].

When it comes to creating bioactive heterocycles, MAOS is a potent and environmentally friendly technology that offers one-pot reactions that are quicker, more selective, and give higher yields with less of an impact on the environment than traditional techniques. This chapter emphasizes significant progress in MAOS, highlighting its revolutionary relevance in contemporary drug development by some selected N-heterocyclic synthesis.

2. Synthesis of 5-Membered Bioactive N-heterocycles

Nitrogen containing heterocycles are considered most important in drug synthesis due to their bioactive characteristics.

2.1 Five-membered rings with one nitrogen

A regioselective microwave-driven protocol enabled the preparation of polyfunctionalized 3-ferrocenyl-1*H*-pyrroles through the reaction between ferrocenyl vinyl azide and1,3-dicarbonyl compounds such as pentane-2,4-dione (Scheme 1). Under optimized conditions (0.5 equivalents of vinyl azide and 0.55 equivalents of diketone in DCE at 100°C for 30 minutes), the method produced moderate yields (32–75%). Notably, the approach required no metal catalyst or strong acid/base, offering a rapid, mild, and environmentally friendly alternative for accessing tetrasubstituted pyrroles. This method underscores the utility of azides in cycloaddition chemistry under microwave irradiation and demonstrates significant potential for constructing redox-active organometallic pyrrole systems [18].

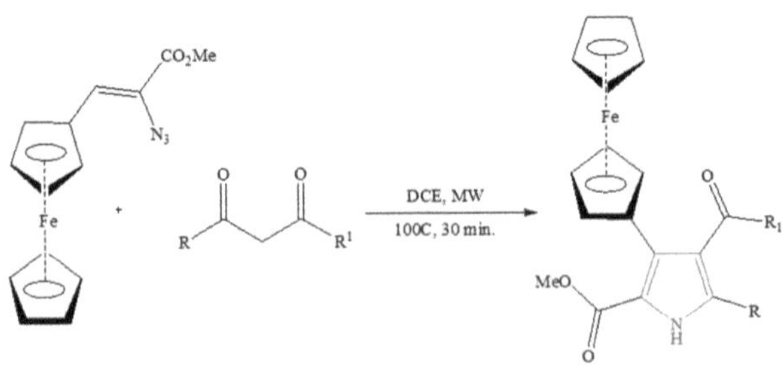

***Scheme 1**: MW aided regioselective synthesis of 3-ferrocenylpyrroles*

Materials Research Foundations 189 (2026) 67-102 https://doi.org/10.21741/9781644904039-3

A microwave mediated protocol is employed for producing 2-amino-4,5-diphenylpyrrole-3-carbonitriles from benzoin and substituted anilines (Scheme 2). Good yields of the desired pyrrole derivatives were obtained under the optimum reaction conditions and these compounds have outstanding *in vitro*anti-inflammatory properties, underscoring their potential as medicinal agents [19].

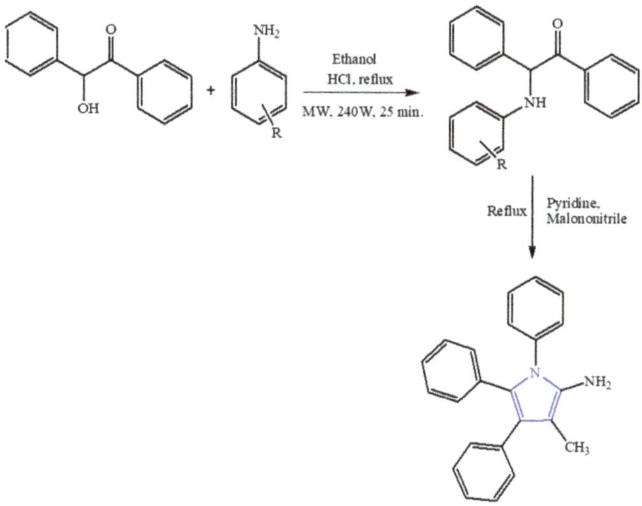

Scheme 2: *MW synthesis of diphenylpyrrole-3-carbonitriles*

By using primary amines, aromatic aldehydes, and sodium diethyl oxaloacetate to synthesize substituted pyrrole derivatives in multi component microwave assisted method is shown in Scheme 3.A pharmacological analysis revealed that a number of the resultant pyrrole derivatives have strong antiviral and cytostatic properties, highlighting their potential for therapeutic development [20].

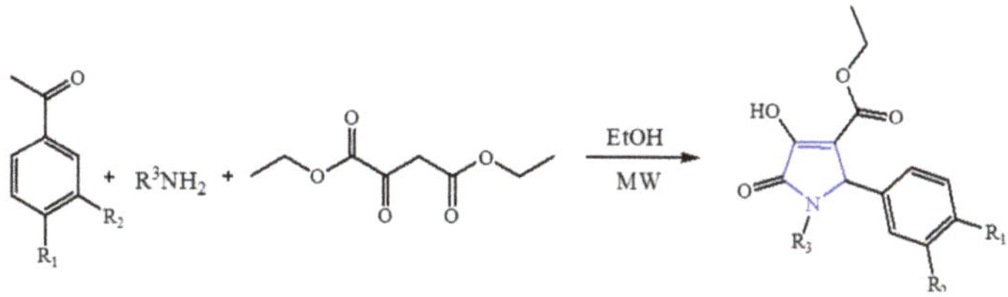

Scheme 3:*MW-aided 3-component synthesis of pyrroles*

An innovative $FeCl_3$-catalyzed one-pot strategy was employed in a microwave-assisted approach for the efficient synthesis of 2*H*-chromene–fused pyrroles, achieving excellent yields (83–95%) (Scheme 4) [21].

Materials Research Foundations 189 (2026) 67-102 https://doi.org/10.21741/9781644904039-3

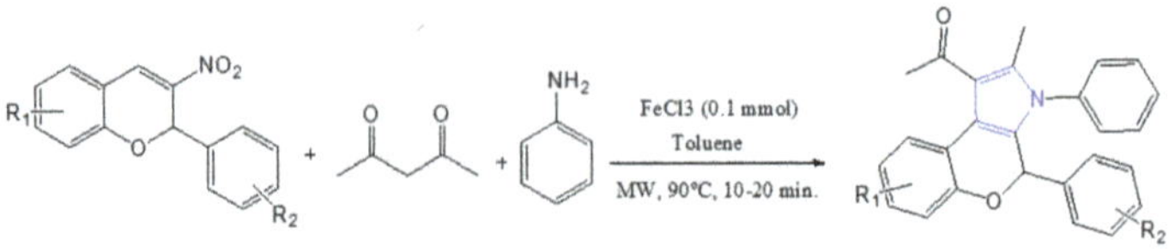

***Scheme 4**: MW synthesis of 2H-chromene–fused pyrroles*

The potential of pyrroles in drug design and therapeutic applications continues to pique scientific curiosity, as they are physiologically favored scaffolds [22]. A green, efficient Van-Leusen-type [3+2] cycloaddition between chalcones and p-toluenesulfonylmethyl isocyanide (TosMIC) was developed using the microwave conditions (Scheme 5). Utilizing a task-specific ionic liquid system—[PAIM][Cl] as base and [BMIM][PF_6] as solvent, the reaction delivered up to 90% yield across 26 different substrates.

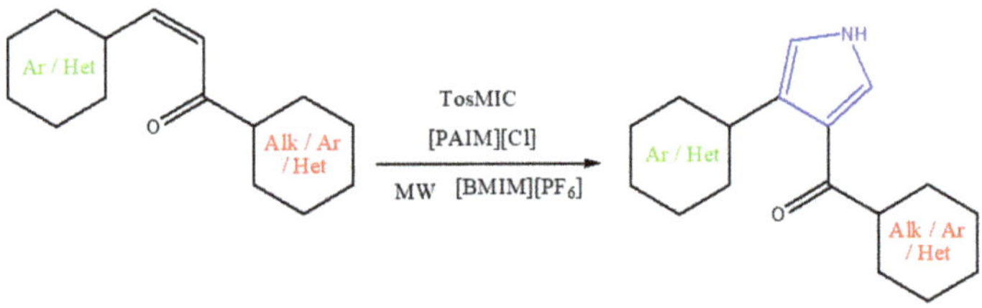

***Scheme 5:** Pyrrole synthesis by MW [3+2] cycloaddition reactions*

The microwave-assisted method extended the intramolecular cyclocondensation method for the effective formation of 2,3,4,5-tetrasubstituted pyrroles (86% for benzyl-substituted derivatives) from Schiff bases of the enamine type that are obtained from 2-amino acids (Scheme 6) [23]. The study highlighted the importance of substituent effects and enamine precursors on cyclization efficiency and offered tactical advice for creating physiologically relevant pyrrole frameworks for applications in materials science and medicine.

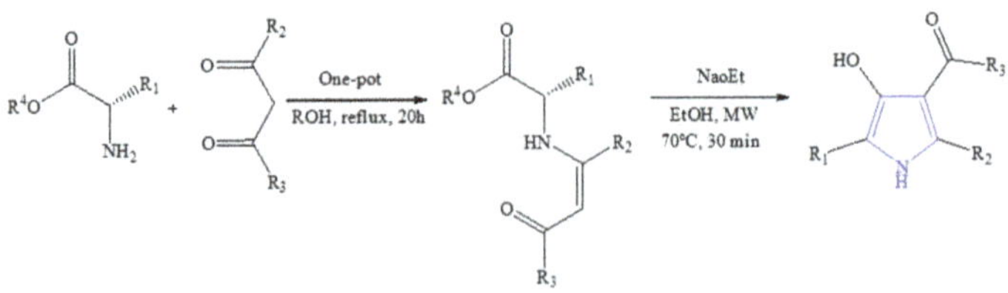

***Scheme 6**: Tetra-substituted pyrrole synthesis by MW intermolecular cyclocondensation*

2.2 Five-membered rings with two nitrogen atoms

2.2.1 Pyrazole

Ferrocene–pyrazole hybrids were effectively produced in 58–75% yield by microwave-assisted cyclocondensation of phenylhydrazine with ferrocene chalcones in an acetic acid–water combination at 100°Cand 150W for 5–30 minutes (Scheme 7) [24]. By employing the MTT assay, the prepared compounds were tested for anticancer activity. Of these, hybrid (R = NH_2) showed the strongest activity, with IC_{50} values of 60.44 μM (SNB19), 6.81 μM (HL60), 124.40 μM (PC-3), and 3.12 μM (HCT-116).

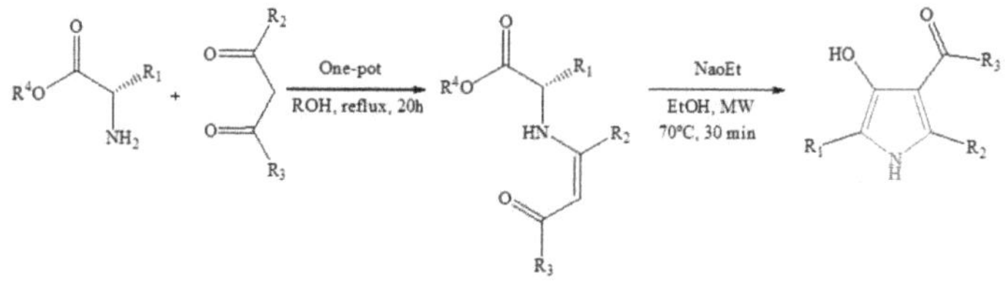

***Scheme 7:** MW synthesis of ferrocene pyrazoles*

Using microwave irradiation, Anwer et al. [25] cyclized arylhydrazines with ethyl 2-cyano-2-{[4-(phenyldiazenyl)phenyl]diazenyl}acetate in ethanol to create pyrazolone compounds in better yields in a concise amount of time of only 0.5 to 3 minutes (Scheme 8). This strategy was effectively applied to similar circumstances in reactions with carbohydrazide derivatives and hydrazine hydrate. With Sorafenib and Erlotinib serving as standard medications, the resultant compounds were tested for anticancer activity using the MTT assay against liver, colon, breast, and lung cancer cell lines.

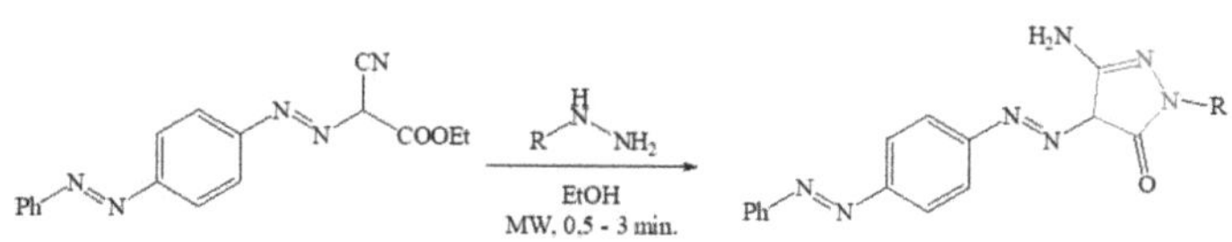

***Scheme 8**: MW-aided synthesis of pyrazolone derivatives*

In 2025, Peitzika et al.[26] synthesized dihydro-pyrazole derivatives via microwave-assisted reactions using substituted dibenzalacetones and hydrazine derivatives (Scheme 9). Key intermediates were prepared through the Claisen–Schmidt reaction. Several compounds exhibited strong antioxidant activity and potent inhibition of soybean lipoxygenase. Notably, these compounds showed excellent anti-lipid peroxidation (98% and 97%), and docking studies confirmed favorable enzyme–ligand interactions, supporting their potential as bioactive agents.

Materials Research Foundations 189 (2026) 67-102 https://doi.org/10.21741/9781644904039-3

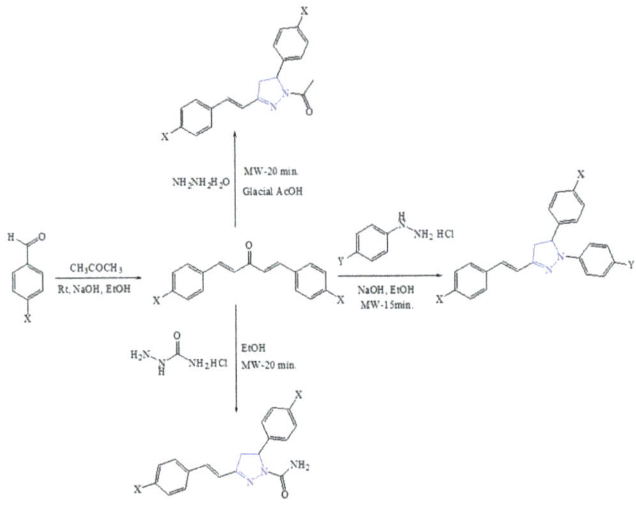

Scheme 9: *MW synthesis of dihydro-pyrazole hybrids*

A 5-membered ring with two neighboring nitrogen atoms, pyrazole is a flexible scaffold renowned for its widespread application in materials research, medicine, and agriculture [27].Using the Vilsmeier–Haack reaction, Selvam et al. reported synthesizing a sequence of 1-(4-substitutedphenyl)-3-phenyl-1*H*-pyrazole-4-carbaldehydes in 2014 [28](Scheme 10).Superior efficacy was shown by the *p*-substituted compounds, suggesting that they could be attractive and leads for the development of safer analgesic and anti-inflammatory drugs.

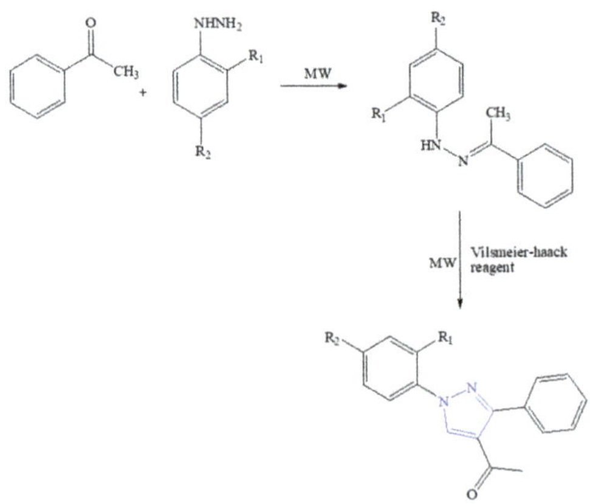

Scheme 10: *MW-driven synthesis of pyrazole derivatives*

Materials Research Foundations 189 (2026) 67-102 https://doi.org/10.21741/9781644904039-3

In the same year (2018), Pattanashetty et al. reported a rapid microwave-heated synthesis of quinolinebased pyrazoline compounds synthesized through 10 minutes MW irradiation of substituted chalconated 4-hydroxyquinolin-2(1*H*)-one and thiosemicarbazide in presence of sulphuric acid, offering significantly better yields in less reaction times compared to classical methods (Scheme 11) [29]. The products were tested for anti-tubercular and antimicrobial activities. Notably, compounds (*p*-OMe-Ph, *p*-$N(CH_3)_2$-Ph, and m-NO_2-Ph substituents) showed potent anti-tubercular activity with MIC values as low as 0.05 μg/mL. Compounds also exhibited strong antibacterial and antifungal properties, with *p*-OMe-Ph substituted compound outperforming standard tetracycline against B. subtilis. Computational and ADME studies confirmed favorable receptor binding and drug-like properties, highlighting these derivatives as potential lead candidates for new anti-tubercular and antimicrobial agents.

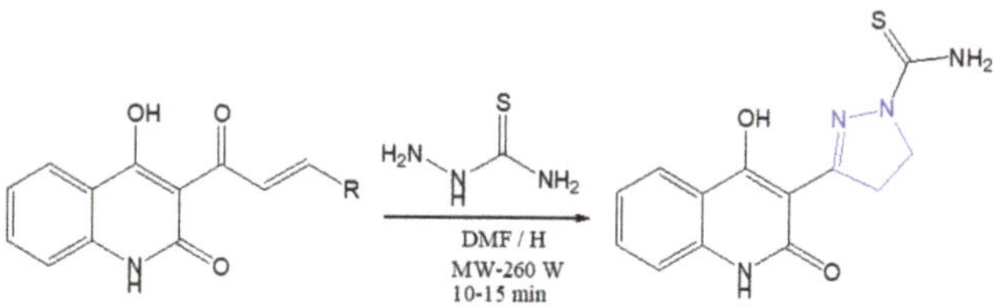

***Scheme 11**: MW synthesis of quinoline-based pyrazoline derivatives*

2.2.2 Imidazole

Imidazole compounds have attracted a lot of attention because of their powerful and varied medicinal qualities. Xia et al. (2007) [30] reported three-component, microwave-heated synthesis of fourteen distinct 2,4,5-trisubstituted imidazoles with impressive yields ranging from 74to 93% (Scheme 12). Microwave irradiation significantlyshortened the reaction time, which used to be many hours under conventional heating but was now only a few minutes.

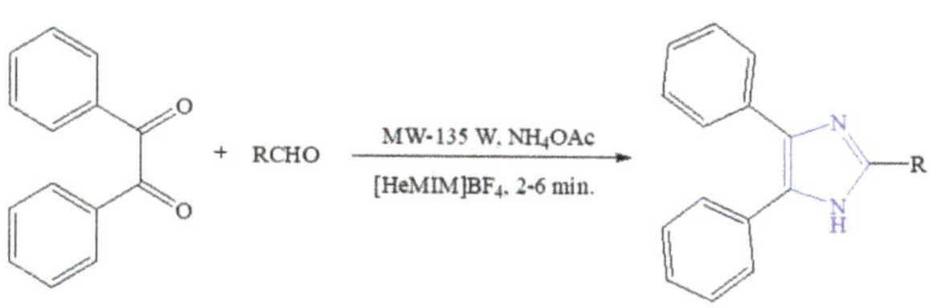

***Scheme 12**: MWsynthesis of 2-3-5 trisubstituted imidazoles*

Employing a one-pot condensation of benzil, aromatic aldehydes, a primary amine, and ammonium acetate using microwave irradiation and solvent-less conditions, Mohsen Esmaeilpour et al. (2015) [31] synthesized physiologically active tri- and tetrasubstituted imidazolesusing $Fe_3O_4@SiO_2$-Imid-PMA^n, a recyclable nanocatalyst (Scheme 13).

Materials Research Foundations 189 (2026) 67-102 https://doi.org/10.21741/9781644904039-3

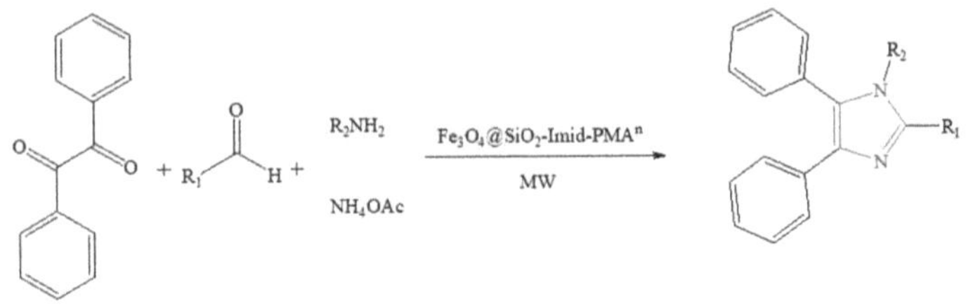

***Scheme 13:** $Fe_3O_4@SiO_2$-Imid-PMA^{n}promotes the MW one-pot synthesis of tetrasubstituted imidazoles*

In 2018, a quick and effective microwave-assisted one-pot approach was employed to synthesize a series of triaryl imidazole-substituted pyrrolyl derivatives (Scheme 14) [32]. Sustainable chemistry is further supported by the use of bismuth nitrate as a green catalyst and this approach provides a productive path to imidazole-pyrrole hybrids that are structurally varied and possibly bioactive.

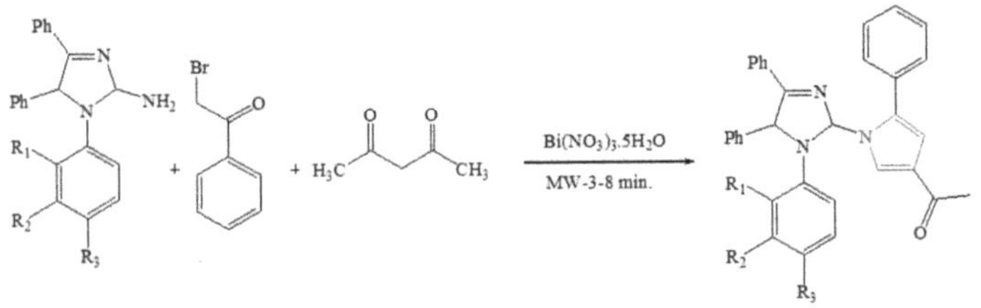

***Scheme 14:** MWsynthesis of imidazole substituted pyrroles*

The green, microwave-heatedformation of a novel fluorescent imidazolylseries designed to combat microbial resistance (Scheme 15) wasreported [33]. These results demonstrate imidazolyl dye's potential as dual-purpose compounds with antibacterial and optoelectronic uses.

Materials Research Foundations 189 (2026) 67-102 https://doi.org/10.21741/9781644904039-3

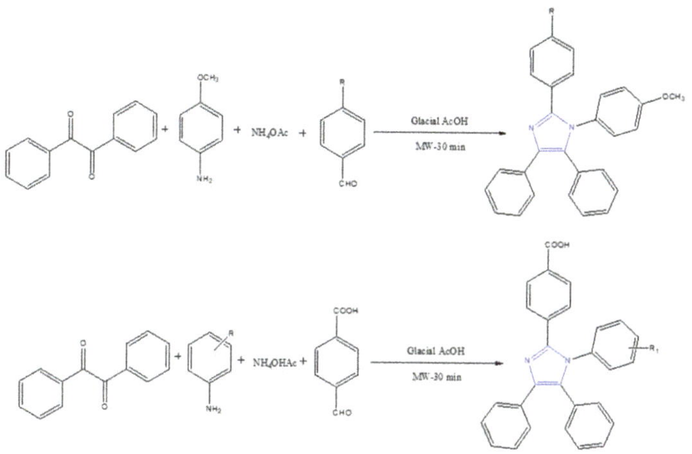

Scheme 15: MW synthesis of Imidazolyl

2.3 Five-membered ringwiththree nitrogen atoms

2.3.1 Triazole

Triazole is a nitrogen-rich five-membered heterocycle that has become an essential scaffold in medicinal chemistry. Isomers 1,2,3-triazole and 1,2,4-triazole are its two forms. Its wide-ranging uses in drug discovery have encouraged the development of a number of synthetic techniques, of which microwave-heated synthesis has garnered significant interest because of its efficacy [34].

In 2020, a novel, metal-free method for creating *N*-unsubstituted 1,2,3-triazoles by a multicomponent reaction combining aldehydes, sodium azide, and nitroalkanes under microwave irradiation (Scheme 16) was proposed by Garg et al. [35]. This is the first work to achieve the direct synthesis of 4-aryl-N*H*-1,2,3-triazoles under such eco-friendly conditions using anthranilic acid as a green organocatalyst. Notably, gram-scale synthesis of physiologically significant chemicals verified its scalability. This creative method opens up effective and sustainable avenues for green and medicinal chemistry.

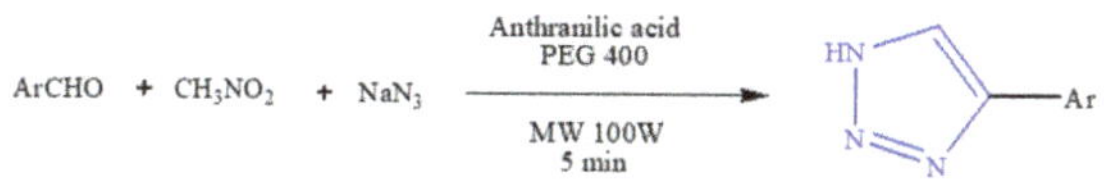

Scheme 16: MWsynthesis of 4-aryl-NH-1,2,3-triazoles

An efficient MW synthesis of benzotriazole–1,2,3-triazole hybrids developed using both conventional and microwave-heated conditions with Cu(I)-catalyzed click reactions (Scheme 17) [36]. Microwave synthesis drastically reduced reaction times (4–8 min vs. 6–12 h) and improved yields (89–95%), highlighting its eco-friendly and economical advantages. The hybrids showed

Materials Research Foundations 189 (2026) 67-102 https://doi.org/10.21741/9781644904039-3

strong anticancer activity against A549 and H1299 lung cancer cells (up to 90%) and exhibited high DNA-binding affinity.

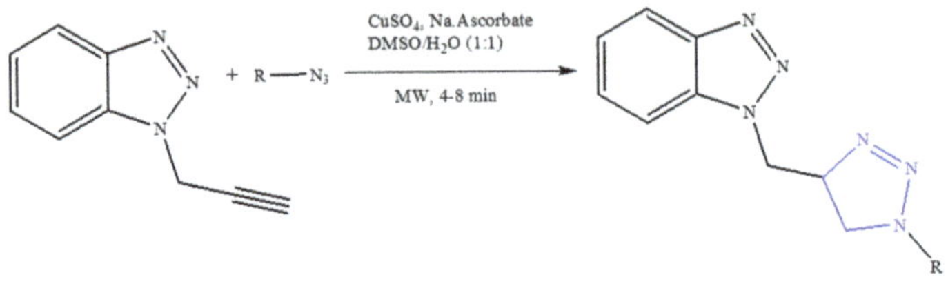

Scheme 17: *MWclick synthesis of benzotriazole-1,2,3-triazole hybrids*

Click chemistry was used as a strategic strategy to create a regioselective and efficient synthesis of quinolin-3-yl-methyl-1,2,3-triazolyl-1,2,4-triazol-3(4*H*)-ones (Scheme 18). The reaction employed a $CuSO_4 \cdot 5H_2O$-catalyzed [3+2] azide–alkyne cycloaddition in a THF/water medium, offering a green and atom-economical route [37]. This method exemplifies the utility of click reactions in constructing complex heterocyclic frameworks under environmentally benign conditions.The last cycloaddition with terminal alkynes produced single regioisomers of the target molecules. These triazole-appended quinolines were further emphasized as promising scaffolds for the creation of novel therapeutic drugs by *in vitro* and *in silico* assessments. This work demonstrates the effectiveness of click chemistry with microwave assistance in producing high-yielding, quick, and sustainable heterocycle synthesis.

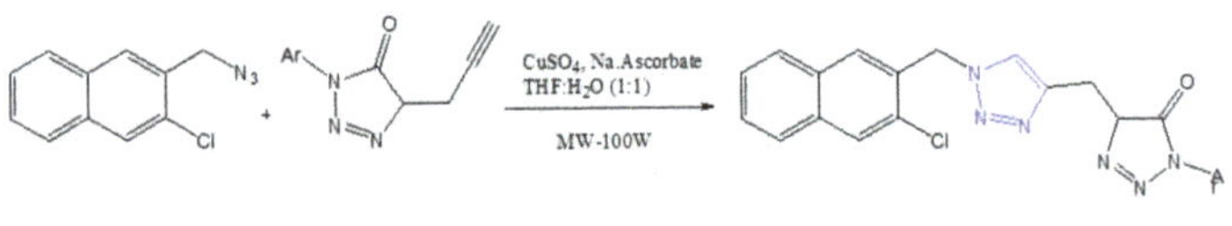

Scheme 18: *MW-assisted synthesis of N-substituted 1,2,3-triazolyl-1,2,4-triazol-3(4H)-ones*

A new and effective one-pot synthesis of benzo[1,3]thiazine-fused 1,2,3-triazole compounds was described using Cu-catalyst promoted intramolecular C–H arylation of *in situ* produced 5-iodotriazoles in an ionic liquid medium under microwave irradiation (Scheme 19) [38].Both substances also showed notable biofilm inhibition, which makes them promising options for treating bacterial infections. Their promise as drug leads was further validated by preliminary ADME profiling and pharmacokinetic analysis, opening the door for additional time-kill research and therapeutic development.

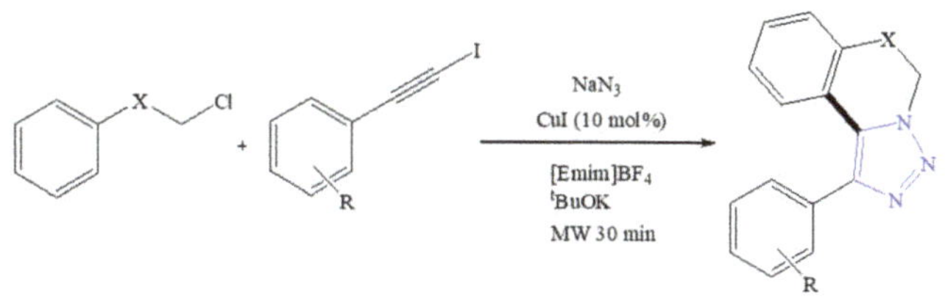

Scheme 19: *MWsynthesis of N-substituted 1,2,3-triazolyl-1,2,4-triazol-3(4H)-ones*

The both traditional and microwave-assisted methods to combine pharmacologically active indole and triazole moieties into a single molecular framework, resulting in a unique series of *N*-substituted 1,2,3-triazolylmethyl indole compounds (Scheme 20) [39]. These compounds have demonstrated promise as lead candidates for antibacterial, antioxidant, and anticancer medicines; the molecular docking results and their observed biological effects are in good agreement.

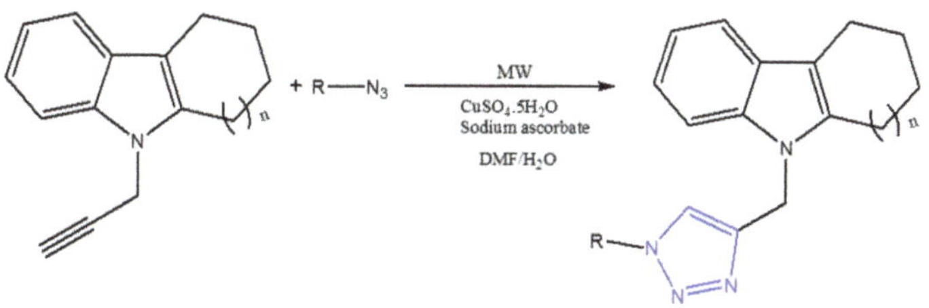

Scheme 20: *MWsynthesis of N-substituted 1,2,3-triazolylmethyl indole derivatives*

Rapid synthesis of 3-aryl-4-arylmethylideneamino-4,5-dihydro-1*H*-1,2,4-triazol-5-one derivatives in a much shorter reaction time was made possible by a microwave-heated green synthesis approach (Scheme 21) [40]. These compounds demonstrated their potential in antifungal therapy by demonstrating considerable mycostatic activity against a variety of *Candida* species.

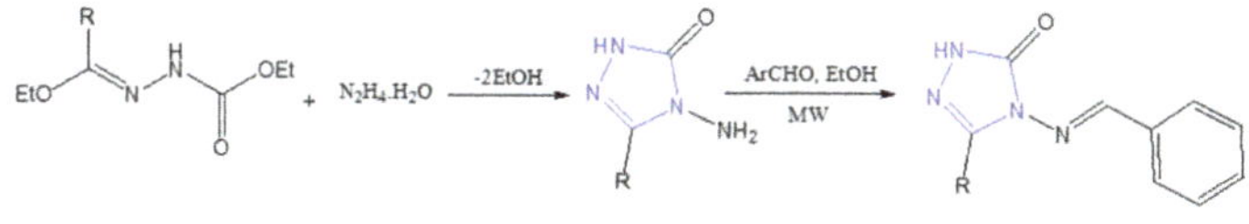

Scheme 21: *MWsynthesis of 3-aryl-4-arylmethylideneamino-4,5-dihydro-1H-1,2,4-triazol-5-ones*

According to Jha and colleagues (2010), 3,5-dibenzyl-4-amino-1,2,4-triazole was successfully synthesized using a microwave in just 8–9 minutes at 250 W by combining 2,4-pentanedione with a diazotized intermediate (Scheme 22) [41]. This study also explored the synthesis of metal complexes, which were tested for anticancer activity. This innovative method not only improves synthetic efficiency but also supports the development of promising anticancer agents.

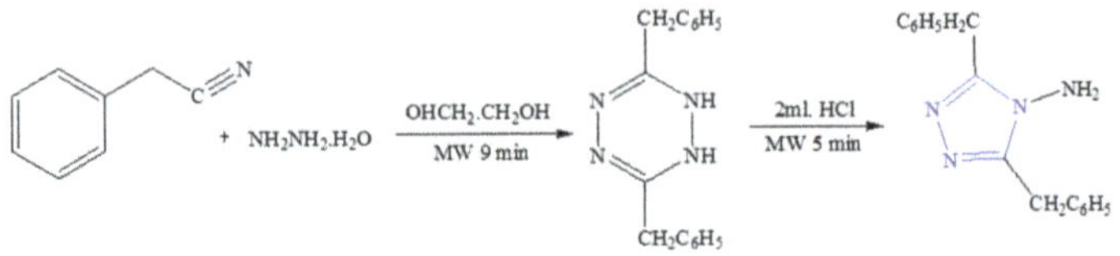

Scheme 22: *MWsynthesis of 3,5-dibenzyl-4-amino-1,2,4-triazole*

2.4 Five-membered rings with four nitrogen atoms

2.4.1 Tetrazole

Tetrazoles are a well-known class of nitrogen-rich heterocycles with a wide range of structural variations and important therapeutic applications. Their bio isosteric similarity to carboxylic acids [42] and distinct electrical properties make them effective structures for modern drug design. The variety of biological activities of tetrazole derivatives, such as their antiallergic, antibacterial, antihypertensive, anti-inflammatory, and anticancer qualities, highlight their potential as adaptable therapeutic agents [43].

Trifluoromethylated 5-substituted-1*H*-tetrazoles were synthesized utilizing *N*-aryl-2,2,2-trifluoroacetimidoyl chlorides and sodium azide in acetonitrile via a catalyst-free, microwave-heated 1,3-dipolar cycloaddition (Scheme23) [44]. These tetrazole compounds are excellent candidates for therapeutic development because the trifluoromethyl group improves metabolic stability, membrane permeability, and bioavailability.

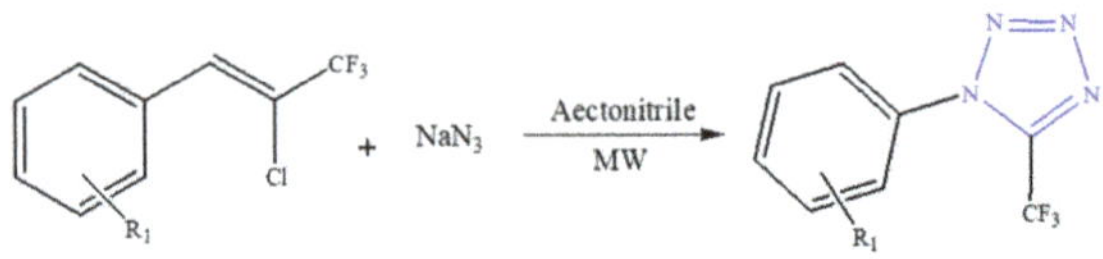

Scheme 23*: Synthesis of trifluoromethylated tetrazoles*

In 2018, a quick and effective solvent-free technique was established for creating physiologically significant derivatives of 5-substituted-1*H*-tetrazoles via a multicomponent domino Knoevenagel condensation/1,3-dipolar cycloaddition of sodium azide, aromatic aldehydes,and malononitrile (Scheme 24) [45]. The method provides excellent yields, quick reaction times, and environmentally favorable settings by heating the reactants with Fe_3O_4 magnetic nanoparticles in a microwave or conventional manner.

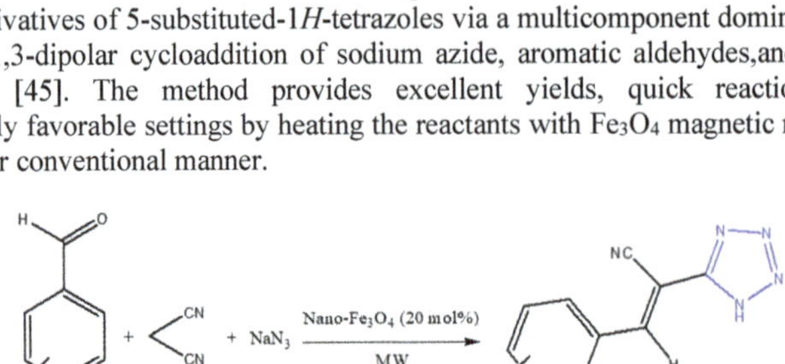

Scheme 24: *MW-assisted synthesis of various tetrazoles*

Materials Research Foundations 189 (2026) 67-102 https://doi.org/10.21741/9781644904039-3

An implemented sodium azide (NaN_3) and $POCl_3$ as reagent and solvent to create an efficacious and adaptable one-pot direct synthesis of 1,5-disubstituted tetrazoles and diarylureas from secondary amides (Scheme 25) under both conventional and microwave-assisted settings. In addition to simplifying the synthesis process, this methodical approach provides a practical platform for the production of medicinally significant tetrazole and urea molecules [46].

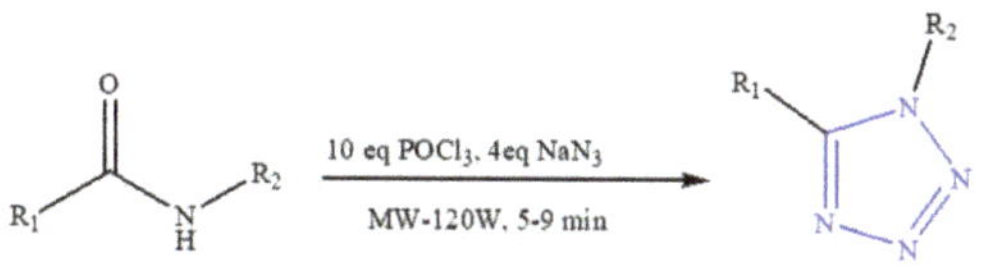

Scheme 25: *MW-assisted synthesis of various-tetrazoles from amides*

3. Microwave-Assisted synthesis of 6-Membered Bioactive heterocycles

3.1 Six-membered rings with one nitrogen

3.1.1 Pyridine

Scottish scientist Thomas Anderson discovered pyridine, a crucial six-membered heterocyclic ring, in 1846 by pyrolyzing bone, which resulted in the isolation of picoline, the first known pyridine derivative. Pyridines are basic structural motifs that are frequently found in biologically active substances. They have a number of promising medicinal applications, which includes as antibacterial and anticancer, as well as therapeutic advantages for conditions like spinal cord injuries, myasthenia gravis, multiple sclerosis, and botulism. Numerous synthetic methods, such as ring-forming [4+2] or [3+3] cycloaddition reactions, aminolysis of 2-alkoxypyridines, condensation of α,β-unsaturated ketones with malononitrile in the presence of ammonium acetate, and nucleophilic substitution of 2-halopyridines with amines, have been developed to prepare 2-aminopyridines. Even with this abundance of approaches, there is still a need for synthetic ways that are more sustainable, green, and cost-effective. This pyridine synthesis was carried out under a microwave oven, employing bentonite as a catalyst (Scheme 26) [47]. This study swiftly produced a series of symmetrical and unsymmetrical pyridines by utilizing the reaction's nonselectivity.

Materials Research Foundations 189 (2026) 67-102 https://doi.org/10.21741/9781644904039-3

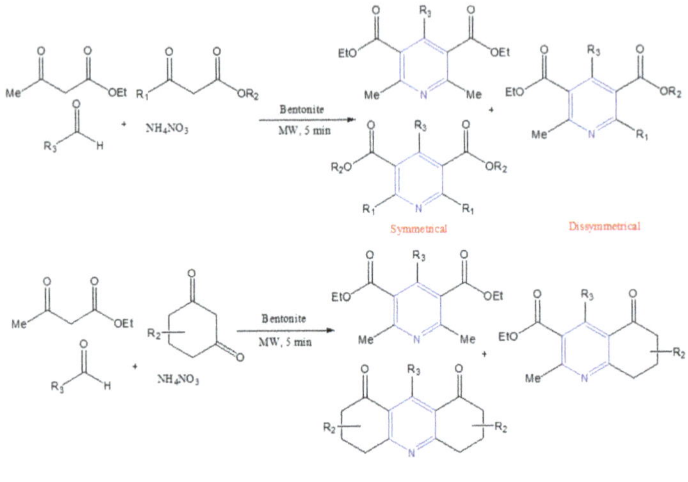

***Scheme 26**: MWsynthesis of symmetrical and dissymmetrical pyridines*

To yield symmetrical pyridines in good yields with significantly shorter reaction durations with ammonium acetate, ethyl acetoacetate, and a variety of aliphatic or aromatic aldehydes, a modified Hantzsch reaction was carried out using palladium on charcoal as an oxidizing agent and K-10 montmorillonite as a catalyst under solvent-free microwave-assisted conditions (Scheme 27) [48].

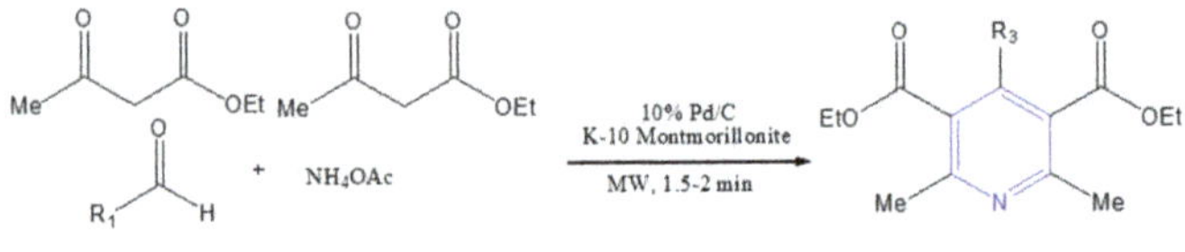

***Scheme 27**: Hantzsch synthesis of pyridines by MW assistance*

Microwave-assisted three-component processes to create thiazolo[3,2-a]pyridine compounds in a chemoselective and environmentally benign manner using water as a solvent was reported by Shi et al. (2009) [49](Scheme 28). Under mild conditions, this quick method that used 2-mercaptoacetic acid, aromatic aldehydes, and malononitrile produced effective results. According to biological examination, the majority of compounds demonstrated potent antioxidant qualities and cytotoxic effects, albeit without selectivity, against mouse lymphocytes and HCT-116 cancer cells.

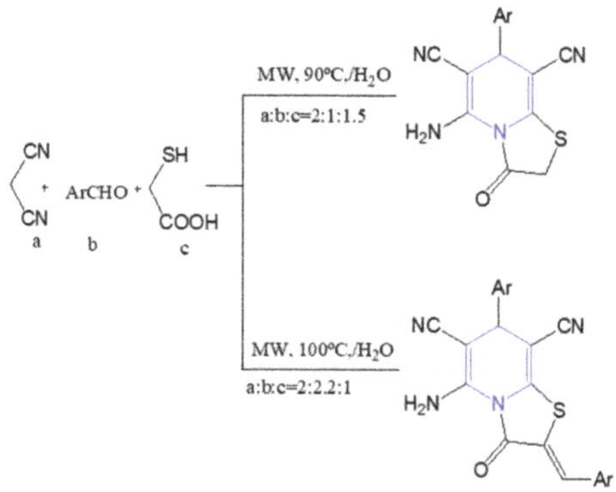

***Scheme 28**: MW synthesis of thiazolo[3,2-a]pyridines*

Employing 4-formylcoumarins, various ketones, malononitrile, and ammonium acetate, a catalyst-free, microwave-heated one-pot four-component synthesis of new bioactive 2-amino-4-(2-oxo-2*H*-chromen-4-yl)-6-arylpyridine-3-carbonitriles without the need for solvents (Scheme 29) was reported by Chougala et al. (2017) [50].

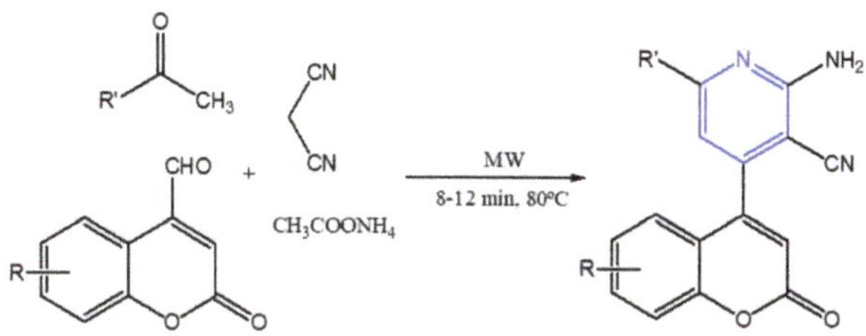

***Scheme 29**: MW synthesis of substituted 2-amino-arylpyridines*

An effective recyclable heterogeneous catalyst was prepared for the environmentally friendly, one-pot synthesis of eleven new dihydropyrazolo[3,4-b]pyridinecompounds (Scheme 30) [51].Because to its favorable circumstances, quick reaction time, high efficiency, easy workup, environmentally acceptable solvent use, and removal of chromatography purification, this method is a great choice for sustainable heterocyclic synthesis. This work paves the road for additional drug design research by highlighting a sustainable synthetic pathway that combines biological relevance with green chemistry.

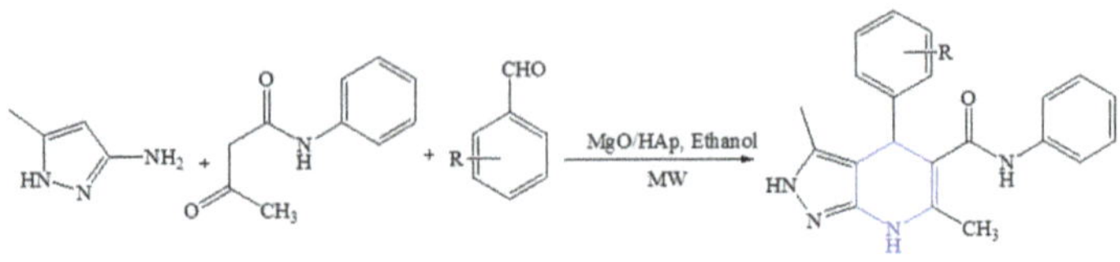

***Scheme 30**: MWsynthesis of dihydropyrazolo[3,4-b]pyridines*

A novel, environmentally friendly, and extremely effective one-pot synthesis approach for 1,4-dihydropyridine compounds by the reaction of diethyl acetylenedicarboxylate, malononitrile, morpholin-4-amine, and many substituted benzaldehydes using potassium carbonate as a safe catalyst under microwave radiation (Scheme 31) was described by Kumar et al. (2023) [52].

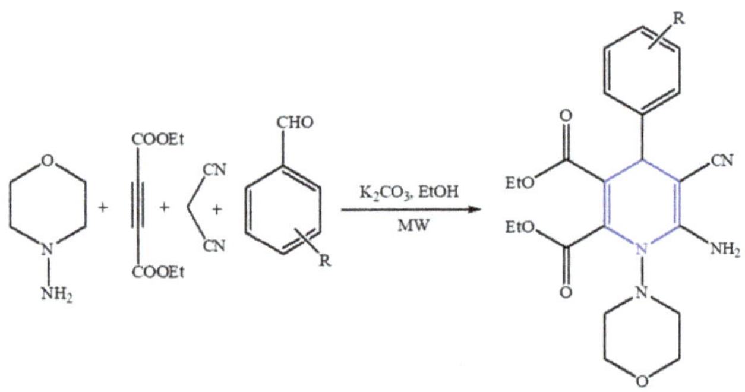

***Scheme 31**: MW-assisted synthesis of 3-cyanopyridines*

3.1.2 Quinoline

Quinoline is an aromatic benzene ring fused pyridine and is known to have a significant influence on the dye, agrochemical, and pharmaceutical sectors. Quinoline and its derivatives, which were initially isolated from coal tar in 1834, have shown remarkable pharmacological plasticity and are essential scaffolds for the creation of bioactive compounds. Quinoline's reactivity and therapeutic potential can be aligned with sustainable MW methods to promote greener, more responsible chemistry and speed up drug development.

Microwave-assisted synthesis of 3-unsubstituted 4-hydroxyquinolin-2(1*H*)-one derivatives bearing fluoro and trifluoromethyl substituents, key scaffolds in antibacterial fluoroquinolones and selective NMDA receptor modulators (Scheme 32) was reported by Arya et al. (2007) [53]. The synthesized derivatives were tested for phototoxicity, cytotoxicity, and antimicrobial activities. Notably, the hydroxy substituted compound demonstrated strong potential for photochemotherapy and antimicrobial applications, while CF_3-substituted derivatives exhibited the highest activity against Mycobacterium tuberculosis H37Rv.

Materials Research Foundations 189 (2026) 67-102 https://doi.org/10.21741/9781644904039-3

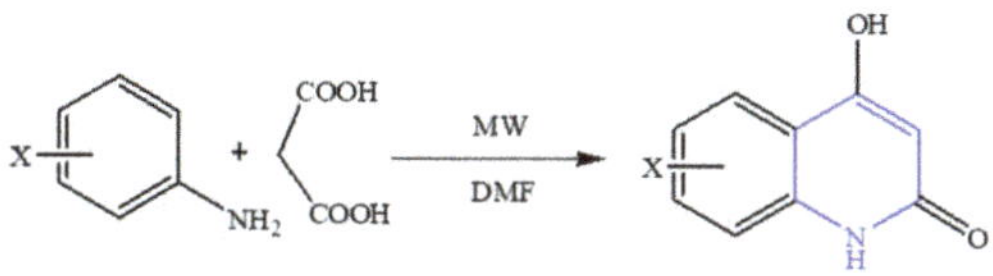

***Scheme 32**: MW-assisted synthesis of fluorinated4-hydroxyquinolinones*

Microwave irradiation was utilized to create quinaldine derivatives from aniline derivatives and acetaldehyde in an effective, solvent-free process (Scheme 33) [54]. With its quick reaction times, high product yields, easy work-up, and use of molecular oxygen as a green oxidant, this procedure highlights the benefits of microwave-heated synthesis in contemporary organic chemistry in a practical and environmentally responsible manner.

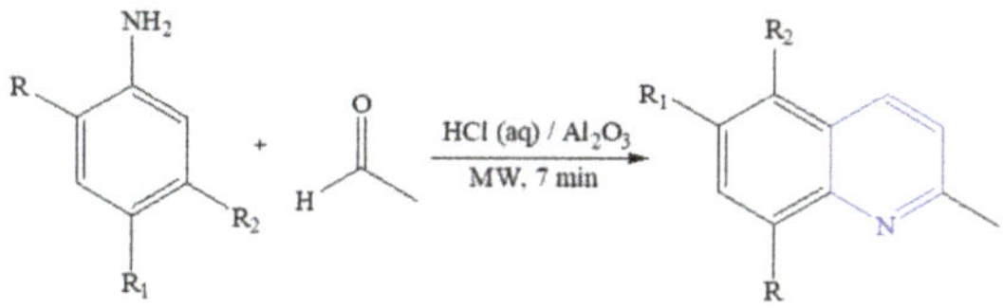

***Scheme 33**: MW-assisted synthesis of quinaldine derivatives*

Quinoline derivatives using p-sulfonic acid calix[4]arene ($CX4SO_3H$) as a catalyst were synthesized under microwave conditions [55] (Scheme 34). The resulting compounds were evaluated for their antineoplastic potential against NCI–H226 and Hep-2C cancer cell lines. Among them, the 4-chloro and 4-fluoro substituted compound exhibited potent anticancer activity against the NCI–H226 and Hep-2C cell line.

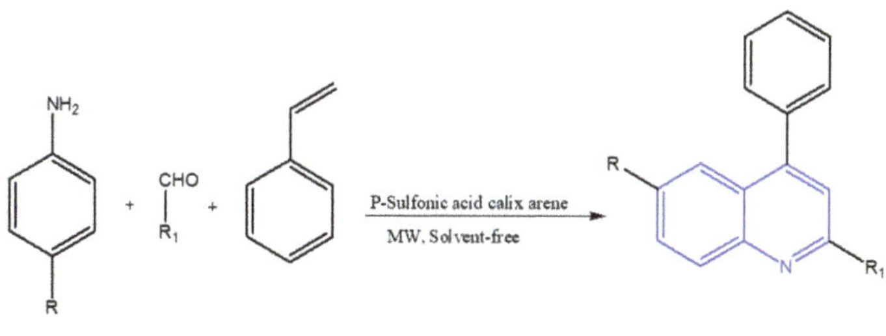

***Scheme 34**: 2,4,6-Trisubstituted quinolines synthesis by MW-assistance*

3.2 Six-membered rings with two nitrogens

3.2.1 Pyrimidine

Pyrimidine, a six-membered unsaturated ring with two nitrogen atoms, is an aromatic heterocyclic molecule that shares structural similarities with pyridine and is categorized as one of the three diazines. Because they are found in many natural and artificial bioactive compounds, compounds with pyrimidine rings are of great interest. Numerous of these substances have a wide variety of pharmacological characteristics and are crucial in clinical settings [56].

Microwave-heated synthesis of benzimidazo[1,2-a]pyrimidinecompounds provided a quicker, safer, and more effective substitute for traditional techniques (Scheme 35) [57].2,4-substitution on the pyrimidine ring increases antifungal potency more efficiently than 2,3,4-substitution, according to structure–activity relationship (SAR) study. With continued research targeted at increasing efficacy through calculated structural alterations, this work presents a promising new class of antibacterial drugs.

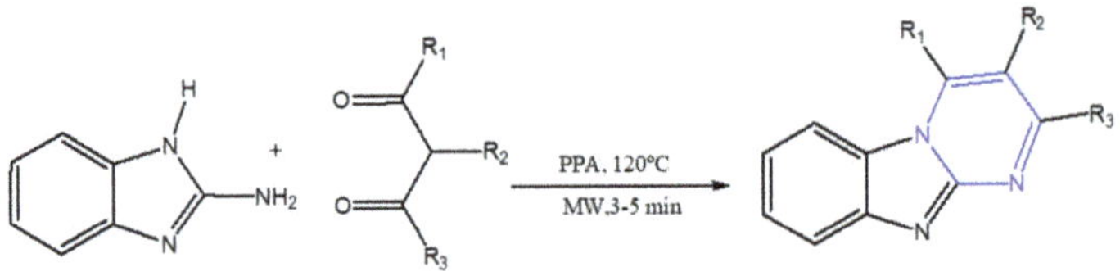

***Scheme 35**: MW-assisted synthesis of diand trisubstituted benzimidazopyrimidines*

A green, microwave-driven one-step technique for the formation of 2,6-diaryl-4(3*H*)-pyrimidinones and 2,6-diaryl-4-aminopyrimidines was developed by L. Xavier et al. (2013) [58] in an effort to generate bioactive pyrimidines in a sustainable manner. This novel approach, which uses water as a solvent and potassium carbonate as a greener base, completes the reaction in approximately 40 minutes instead of the two days and two steps that are typically required for ethanol and triethylamine (Scheme 36). A total of twenty pyrimidine derivatives with significant antinociceptive action were produced using this environmentally friendly method. With its anti-inflammatory, antibacterial, antiviral, and analgesic qualities, the study highlights the pharmacological significance of pyrimidines and shows how green chemistry can efficiently aid in drug discovery.

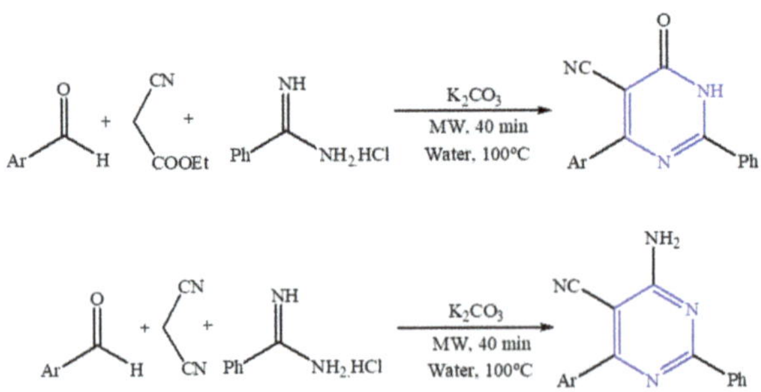

***Scheme 36**: MW-heated three-component pyrimidine synthesis*

Double Mannich cyclization of β-nitroenamines with formaldehyde and aniline derivatives, by microwave-assisted method without metal catalysts produced polysubstituted nitrotetrahydro pyrimidines(Scheme 37) [59], smoothly in 30 to 40 minutes, indicating environmental compatibility and ease of operation.

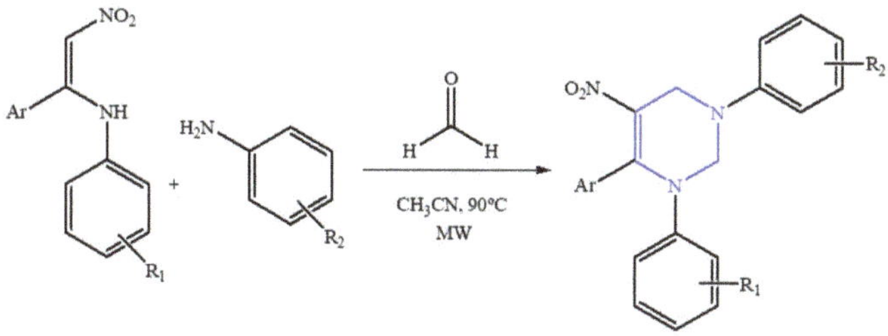

***Scheme 37**: MW synthesis of tetrahydropyrimidines*

A feasible and efficient microwave-assisted three-component synthesize of N^3,N^4-disubstituted 3,4-diaminopyrazolo[3,4-*d*]pyrimidines(Scheme 38) was studied by Jia Hui Ng et al. (2023)[60].The most powerful molecule demonstrated a GI_{50} of 0.5 μM against Jurkat T cells and a high selectivity index of 65, indicating excellent therapeutic potential. The biological evaluation also revealed notable anti-leukemic action.

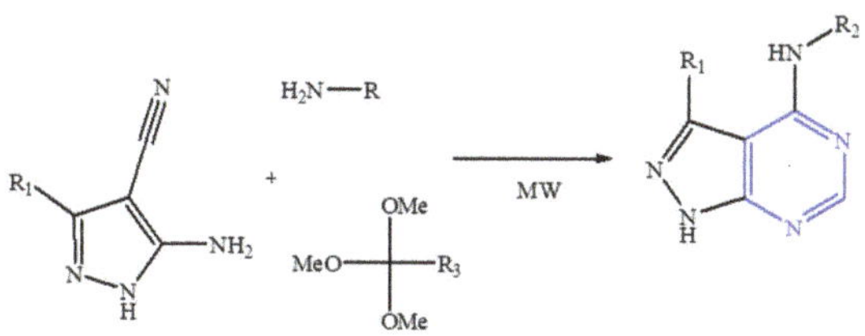

***Scheme 38**: Pyrazolo[3,4-d]pyrimidinessynthesisby MW-assistance*

Through the implementation of aldol condensation and ring closure with guanidine hydrochloride, Phan et al. (2025) discussed a straightforward and gentle two-step microwave-heated synthesis of a novel series of 2-amino-4,6-diarylpyrimidine derivatives (Scheme 39) [61]. Significant cytotoxicity was demonstrated by these compounds.

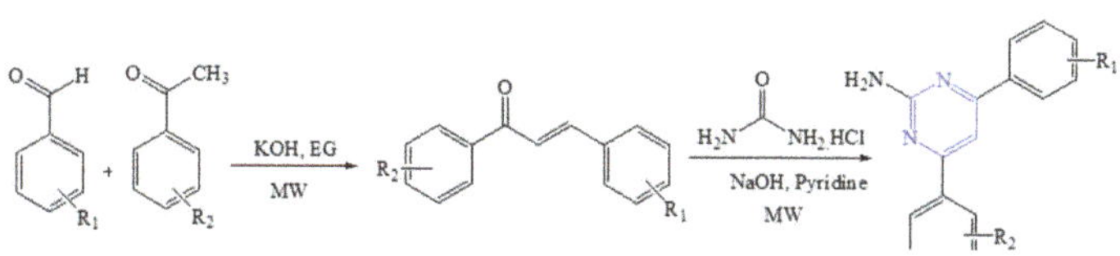

***Scheme 39**: MW-assisted synthesis of aminodiarylpyrimidine*

Materials Research Foundations 189 (2026) 67-102 https://doi.org/10.21741/9781644904039-3

3.2.2 Phenazine

A green and efficient microwave-driven method for synthesizing quinoxaline and dipyridophenazine derivatives from 1,2-phenylenediamines using TiO_2–P25–SO_4^{2-} as solid acid catalyst was reported (Scheme 40) [62].

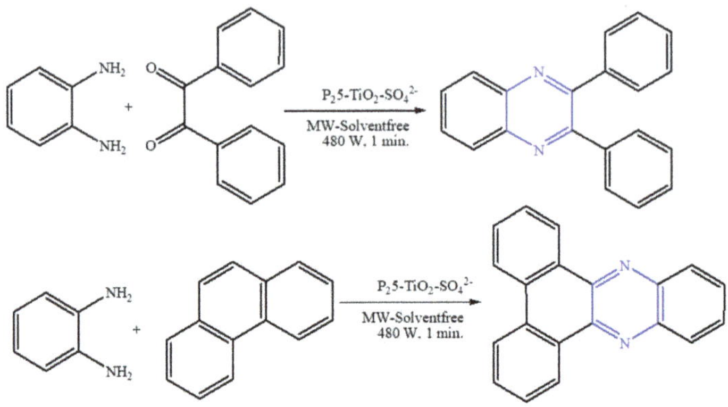

***Scheme 40**: MW-assisted synthesis of quinoxaline and dihydrophenazine*

3.2.3 Pyridazine

Microwave-assisted method for synthesizing novel 1-thiazolylpyridazine derivatives, achieving high yields with reduced reaction times was investigated by Melha et al. (2021) (Scheme 41) [63]. The antimicrobial potential of the compounds was evaluated against *S. aureus*, *P. aeruginosa*, and *C. albicans* using the XTT assay.

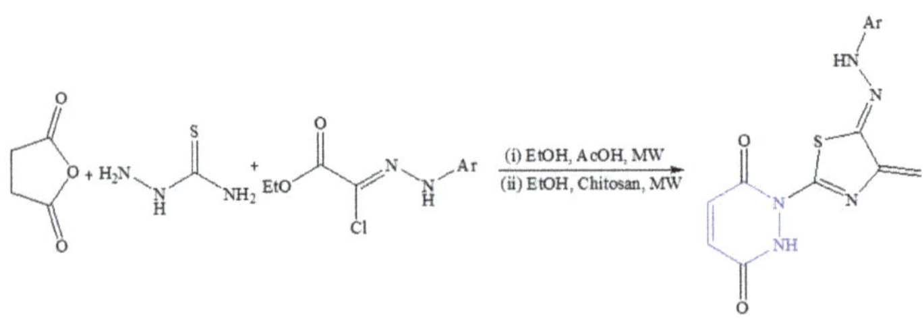

***Scheme 41**: MW-assisted synthesisof thiazolylpyridazines*

3.2.4 Quinoxalines

An important class of nitrogen-containing heterocyclic compounds is quinoxalines, also known as benzo[a]pyrazines, which are present in a variety of natural goods, biologically active synthetic medicine possibilities, and optoelectronic materials.

A green and efficient four-component cascade synthesis of spiro indeno[1,2-*b*]quinoxaline-11,3'-pyrrolizines utilizing phenylenediamine, ninhydrin, proline, and nitrostyrene compounds under traditional and microwave conditions was developed by Akondi et al. (2013) (Scheme 42) [64]. The 1,3-dipolar cycloaddition proceeded with high regio- and stereoselectivity. Biological evaluation revealed notable AChE inhibitory activity, demonstrating their potential as neuroactive agents.

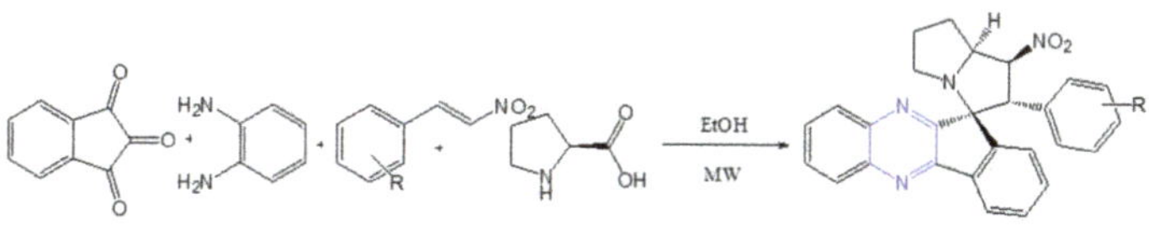

***Scheme 42**: MW-assisted synthesis of spiroquinoxalinepyrrolizines*

Azzam Alanazi et al. (2022) reported a green and effective synthesis method, which employed benzene-1,2-diamine and bromomalononitrile under the microwave heating to yield a novel 3-amino-1,4-dihydroquinoxaline-2-carbonitrile scaffold (Scheme 43). Schiff bases were synthesized using this crucial intermediate, and they were subsequently cyclized with thioglycolic acid to produce thiazolidinone derivatives in high yields (89–96%), which possess insecticidal potential [65].

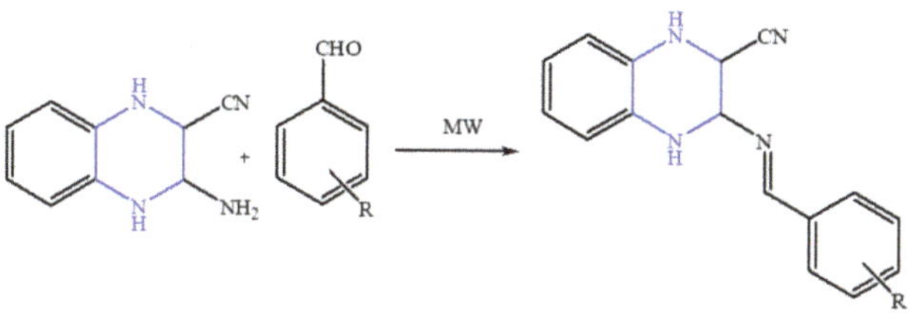

***Scheme 43**: MW-assisted synthesis of quinoxaline with insecticidal potential*

A number of indolo[2,3-*b*]quinoxaline derivatives that are strong acetylcholinesterase (AChE) inhibitors (Scheme 44) to treat Alzheimer's disease were synthesized under microwave conditions [66]. The work successfully constructed the tetracyclic scaffold by molecularly hybridizing the indole and quinoxaline moieties using microwave-assisted synthesis.

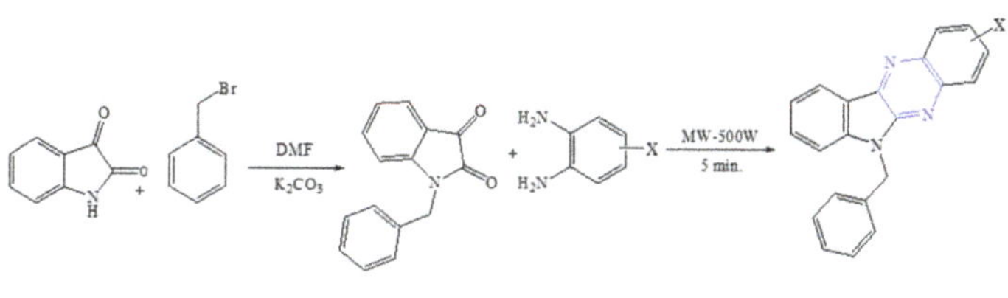

***Scheme 44:** MW-assisted synthesis of indolo quinoxaline*

3.2.5 Quinazoline

Quinazolinone scaffold synthesis is essential for building and enhancing novel pharmaceutical compounds. A green, quick, and effective microwave-assisted technique for creating quinazolinone derivatives through iron-catalyzed C–N cyclization of substituted 2-halobenzoic acids with amidines was presented by Zhang et al. (2009) (Scheme 45) [67]. The technique yielded moderate to high yields and showed remarkable adaptability, tolerating even less reactive substrates like guanidines. Pharmaceuticals can benefit from the development of this technique for quinazolinone compound synthesis.

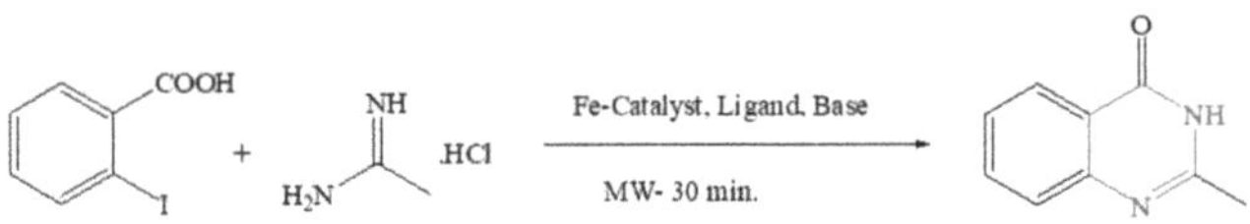

Scheme 45: *MW-assisted synthesis of quinazolinones*

Microwave-heated synthesis of quinazolinones utilizing water as a green solvent (Scheme 46) was proposed by Kanti Bera et al. (2024) [68]. A key innovation in this work is the use of aldehyde bisulfite adducts (Bertagnini's salts) as stable, solid surrogates for aldehydes, offering advantages in reactivity and purification. The process eliminates the need for column chromatography, simplifying product isolation and reducing waste. This approach is a valuable contribution to sustainable pharmaceutical synthesis.

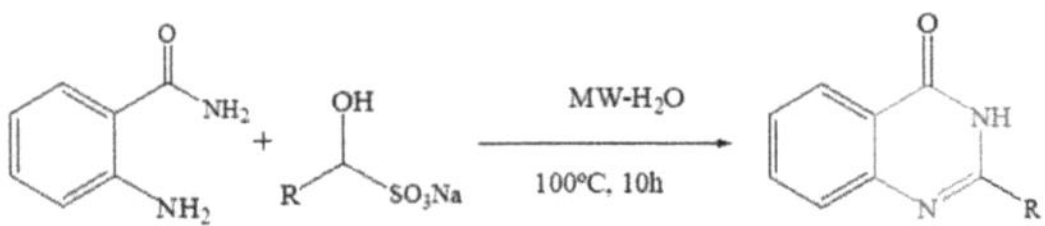

Scheme 46: *MW-assisted synthesis of quinazolinones in water as solvent*

3.2.6 Cinnoline

Because cinnoline and its derivatives have a distinctive fused bicyclic structure, a pyridazine ring fused with a benzene ring and the stiff, flat framework provided by a unique scaffold, perfect for biological interactions.

Synthesis of several 2-arylcinnolin-6(2*H*)-one compounds via a one-pot tandem annulation of 3-oxo-2-arylhydrazonopropanals with acetoacetanilide, using triethylamine in ethanol was studied by M. Al-Matar et al. (2018)(Scheme 47) [69].Three different heating techniques were used to conduct the reactions: conventional, ultrasonic, and microwave irradiation. The most effective of these was microwave irradiation, which produced better reaction rates and yields.

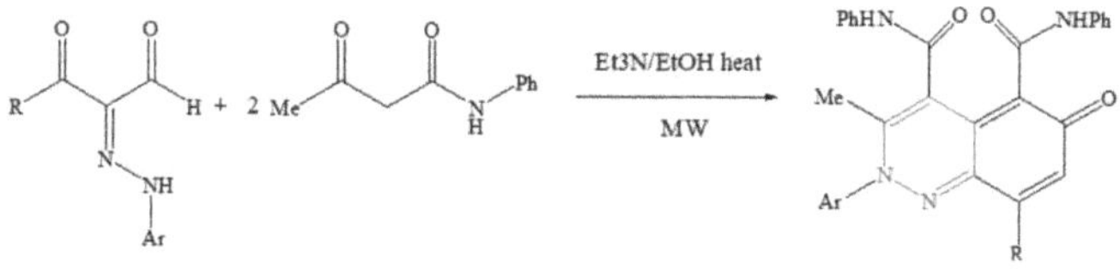

Scheme 47: MW-assisted two step synthesis of cinnolines

4. Seven membered N-heterocycles

Developing seven-membered and larger heterocyclic rings holds considerable importance in medicinal chemistry, as these frameworks are commonly embedded in biologically active compounds, including central nervous system agents, antibiotics, and anticancer drugs. Despite their relevance, synthesizing such medium-to-large-sized rings using conventional thermal methods remains challenging, often plagued by low product yields, extended reaction durations, and labor-intensive purification steps. To overcome these synthetic hurdles, MAOS has emerged as a promising and environmentally conscious approach. By enabling rapid and uniform heating throughout the reaction mixture, MAOS significantly enhances reaction kinetics, minimizes solvent usage, and frequently improves efficiency and selectivity in forming complex heterocyclic systems.

Litina et al. employed 2,3-diaminophenol and other ketones with acid catalysts to synthesize amino-1,5-benzoxazepines and hydroxy-1,5-benzodiazepines in a single pot without the need for solvents. This efficient microwave-assisted method notably was used to prepare previously inaccessible 2,2-dimethyl-4-aryl-1,5-benzodiazepines under analogous conditions [70]. The synthesized compounds demonstrated promising in vitro antioxidant activity, lipid peroxidation inhibition, and anti-inflammatory properties, indicating their potential therapeutic applications. The fact that product selectivity is highly influenced by the type of substituents on the ketone substrates is a drawback to this approach and can make reaction condition tuning challenging. Furthermore, using an acid catalyst may restrict the substrate scope, as certain functional groups sensitive to acidic conditions could undergo degradation or side reactions. Despite these challenges, the protocol remains a versatile and environmentally friendly approach for synthesizing biologically significant benzoxazepine and benzodiazepine derivatives (Scheme 48).

Materials Research Foundations 189 (2026) 67-102 https://doi.org/10.21741/9781644904039-3

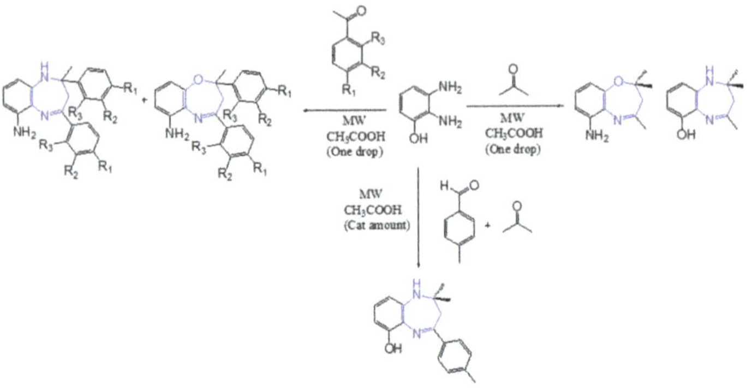

Scheme 48: *MWsynthesis of benzoxazepines and benzodiazepines*

Shah et al. explained a quick synthesis of new 1,5-benzodiazepine derivatives with strong antibacterial properties, aided by microwaves. Using a catalytic amount of trifluoroacetic acid under microwave irradiation, 3-[(1*E*)-*N*-(2-aminophenyl)ethanimidoyl]-4-hydroxy-2*H*-chromen-2-one derivatives were condensed with different pyrazole aldehydes resulting in a new series of highly functionalized 1,5-benzodiazepines [71].Biological evaluation revealed that several synthesized benzodiazepines exhibited significant antimicrobial activity against bacterial strains such as Staphylococcus typhi, Staphylococcus pyogenes, and Vibrio cholerae, as well as antifungal activity against Candida albicans(Scheme 49).

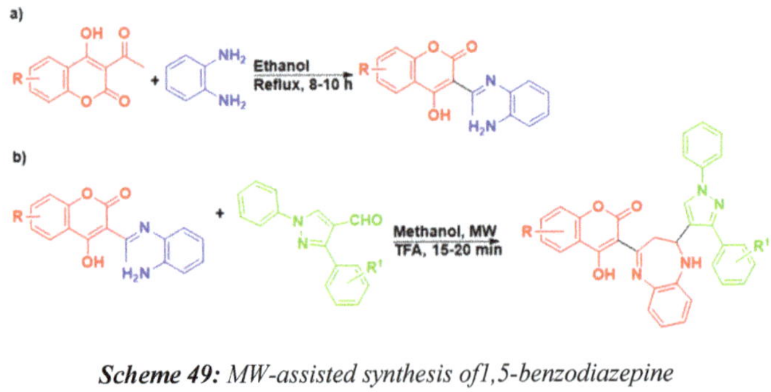

Scheme 49: *MW-assisted synthesis of1,5-benzodiazepine*

Majeed et al. explored the structural modification of ibuprofen via microwave-assisted organic synthesis to generate novel five- and seven-membered heterocycles with enhanced biological potential (Scheme 50) [72]. Recognizing that the carboxylic acid moiety of ibuprofen is responsible for several undesirable gastrointestinal effects, the authors converted this functional group into an acid hydrazide through reaction with hydrazine hydrate using solvent-less microwave irradiation. This transformation afforded the hydrazide intermediate, which subsequently underwent condensation with aromatic aldehydes to yield Schiff bases. These

Schiff bases were further reacted with selected anhydrides (phthalic, malic, succinic) and N-hydroxy succinimide to form seven-membered heterocycles

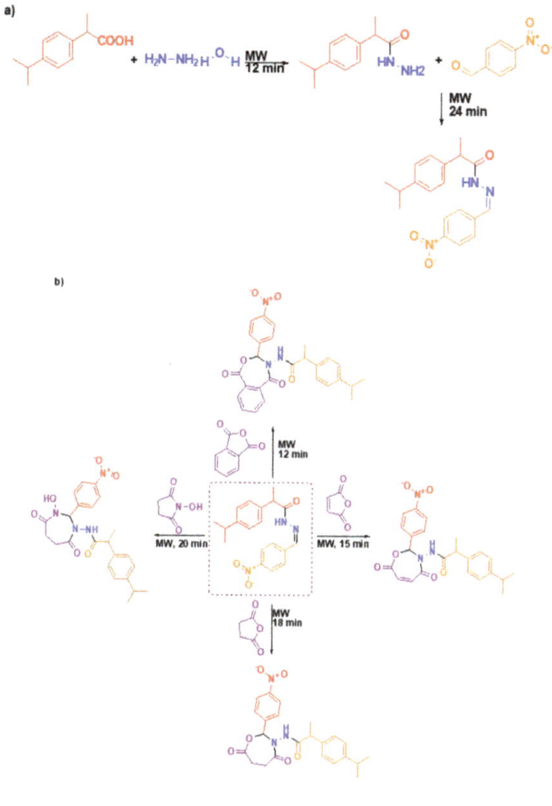

Scheme 50: *MW-assisted synthesis of seven-membered heterocycles*

The synthesized heterocycles exhibited promising antibacterial activity, whereas unmodified ibuprofen showed no effect.

Ramkumar et al. reported the synthesis of oxazepinediones via azine intermediates for the first time, employing a two-step protocol. Substituted aldehydes were reacted with hydrazine hydrate under microwave irradiation for 3 to 5 minutes in a solvent- and catalyst-free environment to yield symmetrical azine derivatives (Scheme 51) [73]. Subsequently, these azines were treated with phthalic anhydride in tetrahydrofuran (THF) to afford the target oxazepinedione compounds. Compounds with 2,4-dichloro and 4-bromo substituents shown higher antioxidant activity *in vitro* when compared to vitamin C, the standard reference.

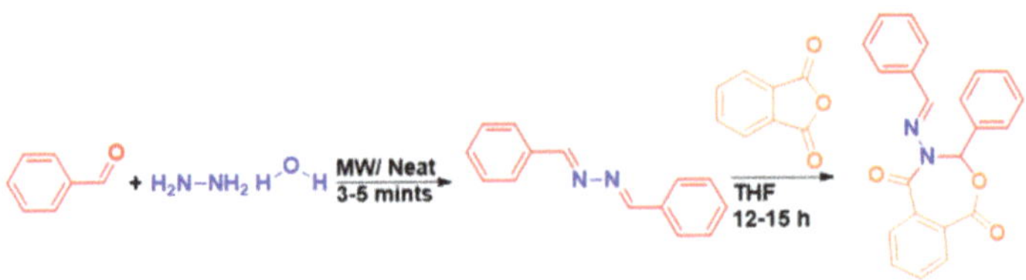

Scheme 51: *MW synthesis of oxazepinediones*

Application of MW-assisted Heterocycles in Medicinal Chemistry

Heterocyclic scaffolds with direct applications in medicinal chemistry have been discovered and optimized more quickly through microwave-assisted organic synthesis (MAOS). Microwave methods' quick heating, higher yields, and more environmentally friendly reaction profiles speed up structure–activity relationship (SAR) research and make it easier to create targeted libraries. Exemplary MW techniques and compounds, representative of microwave-enabled heterocyclic classes, and their main medical uses are compiled in the table below.

Table 1: Representative Applications of Microwave-Assisted Heterocycles in Medicinal Chemistry

Heterocyclic Class	Synthesis Method	Medicinal Application	Reference
Pyrrole Derivatives	MW-Assisted	Antioxidant, anti-tubercular, and anticancer	[18-19]
Pyrazole/ Pyrazoline hybrids	MW-Assisted	Antioxidant, anti-tubercular, and anticancer	[24-29]
Imidazole	MW-Assisted	Antimicrobial, fluorescence-based sensors	[30-33]
Triazole	MW-Assisted	Anticancer, antibacterial, anti-biofilm	[36-39]
Tetrazole	MW-Assisted	Antihypertensive, anti-inflammatory	[44-46]
Pyridine	MW-Assisted	Antioxidant, anticancer, CNS-active agents	[47-50]
Quinoline	MW-Assisted	Antibacterial, anti-tubercular, antineoplastic	[53-55]
Pyrimidine	MW-Assisted	Anti-inflammatory, antinociceptive, antiviral	[56-60]
Quinoxaline & Derivatives	MW-Assisted	AChE inhibition, anti-Alzheimer activity	[64-66]
Benzodiazepine/ Oxazepine	MW-Assisted	Antioxidant, antibacterial, antifungal	[70-73]

Materials Research Foundations 189 (2026) 67-102 https://doi.org/10.21741/9781644904039-3

Discussion

Microwave-assisted synthesis (MAOS) plays a crucial role in modern medicinal chemistry, particularly in the design of drug-like heterocyclic frameworks with improved pharmacokinetic and pharmacodynamic profiles. The method's rapid reaction rates enable the exploration of extensive compound libraries essential for structure–activity relationship (SAR) studies. For instance, pyrrole and pyrazole derivatives synthesized under microwave irradiation show enhanced anticancer and antimicrobial activities due to higher purity and regioselectivity. Similarly, MW-assisted synthesis of triazoles and quinolines has yielded lead molecules with strong binding affinity toward kinase and enzyme receptors. Moreover, the ability to conduct solvent-free or water-mediated reactions aligns MAOS with the principles of green chemistry, minimizing toxic waste generation. These attributes collectively make MAOS a sustainable and high-throughput strategy for the synthesis of bioactive heterocycles with clinical potential.

Table 2 provides a detailed explanation of the benefits of the microwave method over the traditional approach.

Table 2: Comparison of MW-heating and conventional heating methods

Feature	MW-Heating Synthesis	Conventional Heating Synthesis
Time Efficiency	Reactions complete in minutes	Reactions may require several hours
Energy Input	Lower, due to rapid and focused heating	Higher, due to prolonged and gradual heating
Solvent Usage	Often solvent-free or employs green solvents	Typically requires solvents, often toxic
Reaction Control	Instantaneous and uniform heating	Slower heat transfer; less precise control
Yield and Purity	Frequently higher due to reduced side reactions	Variable, often compromised by degradation
Reproducibility	High, once optimized	Moderate to low; dependent on thermal gradient
Safety Profile	Closed systems with fewer byproducts	Open systems, risk of decomposition, or hazards

Conclusions

Microwave-assisted organic synthesis (MAOS) has rapidly evolved into a powerful strategy for constructing complex heterocyclic frameworks. This technique presents a compelling alternative to conventional thermal methods by substantially improving efficiency, sustainability, and product quality. MAOS's remarkable time efficiency is one of its greatest advantages. Reaction times are frequently shortened from hours to minutes or even seconds because of this process, which speeds up reaction kinetics. This significant reduction in processing time directly translates into decreased energy consumption, making microwave-assisted methods cost-effective and environmentally favorable. Moreover, many microwave protocols operate under solvent-free or minimal solvent conditions, frequently using green solvents like water, ethanol, or polyethylene glycol (PEG-400). These features contribute to the reduced environmental footprint of the method by minimizing waste and avoiding hazardous reagents.

Regarding product outcomes, microwave irradiation enhances molecular selectivity and suppresses side reactions, leading to higher yields and improved product purity. Additionally, MAOS often supports one-pot syntheses that maximize atom economy and reduce the need for intermediate steps such as protection and deprotection, further improving synthetic efficiency.

Nevertheless, MAOS is not devoid of limitations. The scalability of reactions remains a significant bottleneck, as most commercially available microwave reactors are limited in reaction volume. Although continuous-flow microwave reactors promise to overcome this constraint, their widespread industrial application is still in development. Additionally, specialized microwave equipment's initial cost and availability may limit access for laboratories with constrained resources. Another challenge is the rapid nature of microwave-induced transformations, which can hinder real-time monitoring and necessitate rigorous reaction optimization to ensure reproducibility and consistency.
Despite these challenges, the collective advantages of MAOS, ranging from shorter reaction times and improved yields to environmental compatibility, underscore its value in modern synthetic strategies, especially in constructing complex heterocyclic scaffolds.

Reference

[1] M. Zhu, L. Ma, J. Wen, B. Dong, Y. Wang, Z. Wang, J. Zhou, G. Zhang, J. Wang, Y. Guo, C. Liang, Rational design and Structure Activity relationship of coumarin derivatives effective on HIV-1 protease and partially on HIV-1 reverse transcriptase, Eur. J. Med. Chem. 186 (2020) 11900. https://doi.org/10.1016/j.ejmech.2019.111900

[2] A.A. Altaf, A. Shahzad, Z. Gul, N. Rasool, A. Badshah, B. Lal, E. A. Khan, A Review on the Medicinal Importance of Pyridine Derivatives, J. Drug Des. Med. Chem. 1 (2015) 1-11. https://doi.org/10.1155/2015/913435

[3] R. Javahershenas, S. Nikzat, Recent advances in the multicomponent synthesis of heterocycles using tetronic acid, RSC Adv. 13 (2023)16619-16629. https://doi.org/10.1039/D3RA02505E

[4] P. Taylor, R.P. Robinson, Y.M. Fobian, D.C. Blakemore, L.H. Jones, O. Fadeyi, Modern advances in heterocyclic chemistry in drug discovery, Org. Biomol. Chem. 14 (2016) 6611-6637. https://doi.org/10.1039/C6OB00936K

[5] M. Henary, C. Kananda, L. Rotolo, B. Savino, E.A. Owens, G. Cravotto, Benefits and applications of microwave-assisted synthesis of nitrogen-containing heterocycles in medicinal chemistry, RSC Adv. 10 (2020) 14170-14197. https://doi.org/10.1039/D0RA01378A

[6] A. Dastan, A. Kulkarni, B. Török, Environmentally benign synthesis of heterocyclic compounds by combined microwave-assisted heterogeneous catalytic approaches, Green Chem. 14 (2012) 17-37. https://doi.org/10.1039/C1GC15837F

[7] A. Harvey, R. Edrada-Ebel, R.J. Quinn, The re-emergence of natural products for drug discovery in the genomics era, Nat. Rev. Drug Discovery, 14 (2015) 111-129. https://doi.org/10.1038/nrd4510

[8] T. Rodrigues, D. Reker, P. Schneider, G. Schneider, Counting on natural products for drug design, Nat. Chem. 8 (2016) 531-541. https://doi.org/10.1038/nchem.2479

[9] B. Shen, A new golden age of natural products drug discovery, Cell 163 (2015) 1297-1300. https://doi.org/10.1016/j.cell.2015.11.031

[10] T.D. Montgomery, V.H. Rawal, Palladium-Catalyzed Modular Synthesis of Substituted Piperazines and Related Nitrogen Heterocycles, Org. Lett. 18 (2016) 740-743. https://doi.org/10.1021/acs.orglett.5b03708

[11] D. Dolles, M. Decker, 5 - Dual-Acting Compounds Acting as Receptor Ligands and Enzyme Inhibitors, M. Decker (Ed), Design of Hybrid Molecules for Drug Development, Elsevier, Oxford 2017, pp137-165. https://doi.org/10.1016/B978-0-08-101011-2.00005-2

[12] B. Nişancı, Electrochemical activation in organic synthesis applications, B. Török, C. Schäfer (Eds.), In Advances in Green and Sustainable Chemistry, Nontraditional Activation Methods in Green and Sustainable Applications, Elsevier, 2021, pp 329-347. https://doi.org/10.1016/B978-0-12-819009-8.00009-8

[13] J. Yoon, J. S. Ryu, A rapid synthesis of lavendustin-mimetic small molecules by click fragment assembly, Bioorg. Med. Chem. Lett. 20 (2010) 3930-3935. https://doi.org/10.1016/j.bmcl.2010.05.014

[14] G. Tiwari, A. Khanna, V.K. Mishra, R. Sagar, Recent developments on microwave-assisted organic synthesis of nitrogen- and oxygen-containing preferred heterocyclic scaffolds, RSC Adv. 13 (2023) 32858-32892. https://doi.org/10.1039/D3RA05986C

[15] S.S. Panda, D.W. Hansen, Microwave-assisted synthesis of five-membered azaheterocyclic systems using nanocatalysts, Nanoparticles in Green Organic Synthesis, Elsevier, (2023) 255-278. https://doi.org/10.1016/B978-0-323-95921-6.00007-X

[16] J.P. Soni, K.S. Chemitikanti, S.V. Joshi, N. Shankaraiah, The microwave-assisted syntheses and applications of non-fused single-nitrogen-containing heterocycles, Org. Bio. Chem. 18 (2020) 9737-9761. https://doi.org/10.1039/D0OB01779E

[17] A. Adhikari, S. Bhakta, T. Ghosh, Microwave-assisted synthesis of bioactive heterocycles: An overview, Tetrahedron 126 (2022) 133085. https://doi.org/10.1016/j.tet.2022.133085

[18] H.G. Bonacorso, F.M. Libero, G.M. Dal Forno, E.P. Pittaluga, D.F. Back, M. Hörner, M.A. Martins, N. Zanatta, New regioselective synthesis of polyfunctionalized 3-ferrocenyl-1H-pyrroles under microwave irradiation, Tetrahedron Lett. 57 (2016) 4568-4573. https://doi.org/10.1016/j.tetlet.2016.08.088

[19] D. Vyankatesh, D., T. Swapnali, A. Rahul, M. Shrinivas, and M. Chandrakant, Microwave Assisted Synthesis of 2 Amino-4, 5-diphenyl-1-Substituted-1 H-Pyrrole-3-Carbonitrile for Anti-Inflammatory and Antifungal Activity, Der Pharmacia Lettre 9 (2017) 51-58.

[20] S. Manta, N. Tzioumaki, N. Kollatos, P. Andrea, M. Margaritouli, A. Panagiotopoulou, I. Papanastasiou, C. Mitsos, A. Tsotinis, D. Schols, D. Komiotis, Polyfunctionalized pyrrole derivatives: Easy three-component microwave-assisted synthesis, cytostatic and antiviral evaluation, Curr. Microw. Chem. 5 (2018) 23-31. https://doi.org/10.2174/2213335605666180221155915

[21] N. Baral, D.R. Mishra, N.P. Mishra, S. Mohapatra, B.P. Raiguru, P. Panda, S. Nayak, M. Nayak, P.S. Kumar, Microwave-assisted rapid and efficient synthesis of chromene-fused pyrrole derivatives through multicomponent reaction and evaluation of antibacterial activity

with molecular docking investigation, J. Het. Chem. 57 (2020) 575-589. https://doi.org/10.1002/jhet.3773

[22] V. Jadhav, A. Anchi, S. Kurahatti, M.R. Rathod, S.S. Malunavar, P. Prabhala, R.G. Kalkhambkar, S.D Joshi, M. Arshad, H.I. Mohamed, A Facile and Green synthesis of Van Leusen 3, 4-disubstituted Pyrroles-A Microwave-assisted IL mediated [3+2] Cycloaddition of TosMIC and Chalcones: Molecular docking, Biological and DFT studies, J. Mol. Struct. 1339 (2025) 142108. https://doi.org/10.1016/j.molstruc.2025.142108

[23] D. Quiroga, S. Torres-Cortés, E. Coy-Barrera, Microwave-Assisted Synthesis of 1-(5-Substituted-4-hydroxy-2-methyl-1H-pyrrol-3-yl)ethan-1-ones from 2-Amino Acid-Derived Enamine-Type Schiff Bases, Molbank, 1 (2025) M1975. https://doi.org/10.3390/M1975

[24] E.V. Filho, J.W. Pina, M.K. Antoniazi, L.B. Loureiro, M.A. Ribeiro, C.B. Pinheiro, C.J. Guimarães, F.C. de Oliveira, C. Pessoa, A.G. Taranto, S.J. Greco, Synthesis, docking, machine learning and antiproliferative activity of the 6-ferrocene/heterocycle-2-aminopyrimidine and 5-ferrocene-1H-Pyrazole derivatives obtained by microwave-assisted Atwal reaction as potential anticancer agents, Bioorg. Med. Chem. Lett. 48 (2021) 128240. https://doi.org/10.1016/j.bmcl.2021.128240

[25] K.E. Anwer, S.S. El-Hddad, N.E. Abd El-Sattar, A. El-Morsy, F. Khedr, S. Mohamady, D.E. Keshek, S.A. Salama, K. El-Adl, N.S. Hanafy, Five and six membered heterocyclic rings endowed with azobenzene as dual EGFRT790M and VEGFR-2 inhibitors: design, synthesis, in silico ADMET profile, molecular docking, dynamic simulation and anticancer evaluations, RSC Adv. 13 (2023) 35321-35338. https://doi.org/10.1039/D3RA06614B

[26] S.C. Peitzika, E. Tsiampakari, E. Pontiki, Microwave Assisted Synthesis of Antioxidant Dihydro-Pyrazole Hybrids as Possible Lipoxygenase Inhibitors, Molecules 30 (2025) 2224. https://doi.org/10.3390/molecules30102224

[27] T.P. Selvam, P.V. Kumar, G. Saravanan, C.R. Prakash, Microwave-assisted synthesis, characterization and biological activity of novel pyrazole derivatives, J. Saudi Chem. Soc. 18 (2014) 1015-1021. https://doi.org/10.1016/j.jscs.2011.12.006

[28] D. Becerra, J.C. Castillo, Recent advances in the synthesis of anticancer pyrazole derivatives using microwave, ultrasound, and mechanochemical techniques, RSC Adv. 15 (2025) 7018-7038. https://doi.org/10.1039/D4RA08866B

[29] S.H. Pattanashetty, K.M. Hosamani, D.A. Barretto, Microwave assisted synthesis, computational study and biological evaluation of novel quinolin-2 (1H)-one based pyrazoline hybrids, Chem. Data Collect. 15 (2018) 184-196. https://doi.org/10.1016/j.cdc.2018.06.003

[30] M. Xia, Y.D. Lu, A novel neutral ionic liquid-catalyzed solvent-free synthesis of 2,4,5-trisubstituted imidazoles under microwave irradiation, J. Mol. Catal. A Chem. 265 (2007) 205-208. https://doi.org/10.1016/j.molcata.2006.10.004

[31] M. Esmaeilpour, J. Javidi, M. Zandi, One-pot synthesis of multi substituted imidazoles under solvent-free conditions and microwave irradiation using Fe3O4@ SiO2-imid-PMA n magnetic porous nanospheres as a recyclable catalyst, New J. Chem. 39 (2015) 3388-3398. https://doi.org/10.1039/C5NJ00050E

[32] A. Chawla, V.K. Kapoor, Microwave Assisted One Pot Synthesis and Antimicrobial Activity of 2-(3′-acetyl-2′-methyl-5′-phenyl)-pyrrol-1-yl-1, 4,5-triphenyl-1H-imidazole Derivatives, Der Pharma Chem. 10 (2018) 27-31.

[33] C. Bathula, M.K. Ravindra, K.A. Kumar, H. Yadav, S. Ramesh, S. Shinde, N.K. Shrestha, K.M. Mahadevan, V. Reddy, A. Mohammed, Microwave assisted synthesis of imidazolyl fluorescent dyes as antimicrobial agents, J. Mater. Res. Technol. 9 (2020) 6900-6908. https://doi.org/10.1016/j.jmrt.2020.01.011

[34] S. Swami, N. Sharma, G. Sharma, R. Shrivastava, Recent advances in microwave-assisted synthesis of triazoles and their derivatives: a green approach toward sustainable development methods, RSC Adv. 15 (2025) 2361-2415. https://doi.org/10.1039/D4RA06886F

[35] A. Garg, D. Sarma, A.A. Ali, Microwave assisted metal-free approach to access 1,2,3-triazoles through multicomponent synthesis, Curr. Res. Green Sust. Chem. 3 (2020) 100013. https://doi.org/10.1016/j.crgsc.2020.100013

[36] S.Y. Alraqa, K. Alharbi, A. Aljuhani, N. Rezki, M.R. Aouad, I. Ali, Design, click conventional and microwave syntheses, DNA binding, docking and anticancer studies of benzotriazole-1,2,3-triazole molecular hybrids with different pharmacophores, J. Mol. Struct. 1225 (2021) 129192. https://doi.org/10.1016/j.molstruc.2020.129192

[37] A.R. Nesaragi, R.R. Kamble, P.K. Bayannavar, S.K.J. Shaikh, S.R. Hoolageri, B. Kodasi, S.D. Joshi, V.M. Kumbar, Microwave assisted regioselective synthesis of quinoline appended triazoles as potent anti-tubercular and antifungal agents via copper (I) catalyzed cycloaddition, Bioorg. Medi. Chem. Lett. 41 (2021) 127984. https://doi.org/10.1016/j.bmcl.2021.127984

[38] E.R. Sucharitha, T.M. Krishna, R. Manchal, G. Ramesh, S. Narsimha, Fused benzo [1,3] thiazine-1,2,3-triazole hybrids: Microwave-assisted one-pot synthesis, in vitro antibacterial, antibiofilm, and in silico ADME studies, Bioorg. Medi. Chem. Lett. 47 (2021) 128201. https://doi.org/10.1016/j.bmcl.2021.128201

[39] D. Ashok, G. Thara, B.K. Kumar, G. Srinivas, D. Ravinder, T. Vishnu, M. Sarasija, B. Sushmitha, Microwave-assisted synthesis, molecular docking studies of 1,2,3-triazole-based carbazole derivatives as antimicrobial, antioxidant and anticancer agents, RSC Adv. 13 (2023) 25-40. https://doi.org/10.1039/D2RA05960F

[40] B. Kahveci, M. Özil, E. Menteşe, O. Bekircan, K. Buruk, Microwave-assisted synthesis and antifungal activity of some new 1H-1, 2, 4-triazole derivatives, Russ. J. Chem. 44 (2008) 1816-1820. https://doi.org/10.1134/S1070428008120178

[41] A. Jha, Y.L.N. Murthy, G. Durga, T.T. Sundari, Microwave-Assisted Synthesis of 3,5-Dibenzyl-4-amino-1,2,4 -triazole and its Diazo Ligand, Metal Complexes Along with Anticancer Activity, J. Chem. 7 (2010) 1571-1577. https://doi.org/10.1155/2010/569605

[42] F.H. Allen, C.R. Groom, J.W. Liebeschuetz, D.A. Bardwell, T.S. Olsson, P.A. Wood, The hydrogen bond environments of 1H-tetrazole and tetrazolate rings: The structural basis for tetrazole-carboxylic acid bioisosterism, J. Chem. Inf. Model. 52 (2012) 857-866. https://doi.org/10.1021/ci200521k

[43] M.K. Pasha, S. Raoufi, M. Ghobadi, M. Kazemi, Biologically active tetrazole scaffolds: Catalysis in magnetic nanocomposites, Synth. Commun. 50 (2020) 3685-3716. https://doi.org/10.1080/00397911.2020.1811872

[44] F. Rahmani, A. Darehkordi, B. Notash, Microwave-assisted synthesis of 1-aryl-5-(trifluoromethyl)-1H-tetrazoles, J. Fluor. Chem. 166 (2014) 84-87. https://doi.org/10.1016/j.jfluchem.2014.07.025

[45] P. Akbarzadeh, N. Koukabi, E. Kolvari, Three-component solvent-free synthesis of 5-substituted-1 H-tetrazoles catalyzed by unmodified nanomagnetite with microwave irradiation or conventional heating, Res. Chem. Intermed. 45 (2019) 1009-1024. https://doi.org/10.1007/s11164-018-3657-9

[46] R. Sribalan, A. Lavanya, M. Kirubavathi, V. Padmini, Selective synthesis of ureas and tetrazoles from amides controlled by experimental conditions using conventional and microwave irradiation, J. Saudi Chem. Soc. 22 (2018) 198-207. https://doi.org/10.1016/j.jscs.2016.03.004

[47] I.C. Cotterill, A.Y. Usyatinsky, J.M. Arnold, D.S. Clark, J.S. Dordick, P.C. Michels, Y.L. Khmelnitsky, Microwave assisted combinatorial chemistry synthesis of substituted pyridines, Tetrahedron Lett. 39 (1998) 1117-1120. https://doi.org/10.1016/S0040-4039(97)10796-1

[48] O. De Paolis, J. Baffoe, S.M. Landge, B. Torok, Multicomponent domino cyclization-oxidative aromatization on a bifunctional Pd/C/K-10 catalyst: an environmentally benign approach toward the synthesis of pyridines, Synthesis 21 (2008) 3423-3428. https://doi.org/10.1055/s-0028-1083177

[49] F. Shi, C. Li, M. Xia, K. Miao, Y. Zhao, S. Tu, W. Zheng, G. Zhang, N. Ma, Green chemoselective synthesis of thiazolo [3,2-a] pyridine derivatives and evaluation of their antioxidant and cytotoxic activities, Bioorg. Med. Chem. Lett. 19 (2009) 5565-5568. https://doi.org/10.1016/j.bmcl.2009.08.046

[50] B.M. Chougala, S. Samundeeswari, M. Holiyachi, N.S. Naik, L.A. Shastri, S. Dodamani, S. Jalalpure, S.R. Dixit, S.D. Joshi, V.A. Sunagar, Microwave synthesis of coumarinyl substituted pyridine derivatives as potent anticancer agents and molecular docking studies, Chemistry Select 2 (2017) 5234-5242. https://doi.org/10.1002/slct.201700358

[51] N. Kerru, L. Gummidi, S. Maddila, S.B. Jonnalagadda, Efficient synthesis of novel functionalized dihydro-pyrazolo [3,4-d] pyridines via the three-component reaction using MgO/HAp as a sustainable catalyst, Inorg. Chem. Communi. 123 (2017) 108321. https://doi.org/10.1016/j.inoche.2020.108321

[52] K.S. Kumar, A.R. Robert, N. Kerru, S. Maddila, A novel, swift, and effective green synthesis of morpholino-pyridine analogues in microwave irradiation conditions, Results Chem. 5 (2023) 100692. https://doi.org/10.1016/j.rechem.2022.100692

[53] K. Arya, M. Agarwal, Microwave prompted multigram synthesis, structural determination, and photo-antiproliferative activity of fluorinated 4-hydroxyquinolinones, Bioorg. Med. Chem. Lett. 17 (2008) 86-93. https://doi.org/10.1016/j.bmcl.2006.09.082

[54] J. Safari, S.H. Banitaba, S.S. Samiei, One-pot synthesis of quinaldine derivatives by using microwave irradiation without any solvent-A green chemistry approach, J. Chem. Sci. 121 (2009) 481-484. https://doi.org/10.1007/s12039-009-0057-0

[55] N.A. Liberto, J.B. Simões, S. de Paiva Silva, C.J. da Silva, L.V. Modolo, A. de Fátima, L.M. Silva, M. Derita, S. Zacchino, O.M.P. Zuñiga, G.P. Romanelli, Quinolines: Microwave-assisted synthesis and their antifungal, anticancer and radical scavenger properties, Bioorg. Med. Chem. 25 (2017) 1153-1162. https://doi.org/10.1016/j.bmc.2016.12.023

[56] H. ur Rashid, M.A.U. Martines, A.P. Duarte, J. Jorge, S. Rasool, R. Muhammad, N. Ahmad, M.N. Umar, Research developments in the syntheses, anti-inflammatory activities and structure-activity relationships of pyrimidines, RSC Adv. 11(2021) 6060-6098. https://doi.org/10.1039/D0RA10657G

[57] O. Algul, A. Meric, S. Polat, N.D. Yuksek, M. Serin, Comparative studies on conventional and microwave-assisted synthesis of a series of 2,4-di and 2,3,4-trisubstituted benzimidazo [1, 2-a] pyrimidines and their antimicrobial activities, Cent. Eur. J. Chem. 7 (2009) 337-342. https://doi.org/10.2478/s11532-009-0023-1

[58] A.L. Xavier, A.M. Simas, E.P.D.S. Falcão, J.V. dos Anjos, Antinociceptive pyrimidine derivatives: aqueous multicomponent microwave assisted synthesis, Tetrahedron Lett. 54 (2013) 3462-3465. https://doi.org/10.1016/j.tetlet.2013.04.099

[59] C. Gülbenek, M. Yıldırım, A. Yıldırım, Microwave-mediated approach to highly substituted nitropyrimidines via double Mannich reactions and their biological properties, Tetrahedron 120 (2022) 132904. https://doi.org/10.1016/j.tet.2022.132904

[60] J. Hui Ng, A.V. Dolzhenko, One-pot synthesis of novel pyrazolo [4,3-e][1,2,4] triazolo [1,5-c] pyrimidines under microwave irradiation in a green solvent, Tetrahedron 141 (2023) 133521. https://doi.org/10.1016/j.tet.2023.133521

[61] T.T. Mai, H.D. Nguyen, T.T. Duong, L.P. Tran, T.T. Duong, T.K.C. Huynh, E.V. Koroleva, Z.V. Ignatovich, A.L. Ermolinskaya, H.P. Nguyen, T.H.A. Nguyen, A. Ton, T-H. Do, T.D. Hoang, 2-Amino-4, 6-diarylpyrimidines as potential chronic myeloid leukemia cell inhibitors targeting anti-ABL1 kinase: microwave-assisted synthesis, biological evaluation, molecular docking, and dynamics studies, RSC Adv. 15 (2025) 4458-4471. https://doi.org/10.1039/D4RA08330J

[62] B. Krishnakumar, M. Swaminathan, Solvent free synthesis of quinoxalines, dipyridophenazines and chalcones under microwave irradiation with sulfated Degussa titania as a novel solid acid catalyst, J. Mol. Cat. A Chem. 350 (2011) 16-25. https://doi.org/10.1016/j.molcata.2011.08.026

[63] S. Abu-Melha, S., S.M. Gomha, A.S. Abouzied, M.M. Edrees, A.S. Abo Dena, Z.A. Muhammad, Microwave-assisted one pot three-component synthesis of novel bioactive thiazolyl-pyridazinediones as potential antimicrobial agents against antibiotic-resistant bacteria, Molecules 26 (2021) 4260. https://doi.org/10.3390/molecules26144260

[64] A.M. Akondi, S. Mekala, M.L. Kantam, R. Trivedi, L.R. Chowhan, A. Das, An expedient microwave assisted regio-and stereoselective synthesis of spiroquinoxalinepyrrolizine derivatives and their AChE inhibitory activity, New J. Chem. 41 (2017) 873-878. https://doi.org/10.1039/C6NJ02869A

[65] M.A. Alanazi, W.A. Arafa, I.O. Althobaiti, H.A. Altaleb, R.B. Bakr, N.A. Elkanzi, Green design, synthesis, and molecular docking study of novel Quinoxaline derivatives with insecticidal potential against Aphis craccivora, ACS Omega 7 (2022) 27674-27689. https://doi.org/10.1021/acsomega.2c03332

[66] M. Elhag, H.E. Abdelwahab, M.M. El Sadek, E.A. Hamed, P. Dambruoso, A. Casapullo, K. Mansour, A.Z. Omar, R.O. El-Zawawy, Microwave-assisted synthesis of indoloquinoxaline derivatives as promising anti-alzheimer agents: DFT and molecular docking study, J. Mol. Struct. 1321 (2025) 140126. https://doi.org/10.1016/j.molstruc.2024.140126

[67] X. Zhang, D. Ye, H. Sun, D. Guo, J. Wang, H. Huang, X. Zhang, H. Jiang, H. Liu, Microwave-assisted synthesis of quinazolinone derivatives by efficient and rapid iron-catalyzed cyclization in water, Green Chem. 11 (2009) 1881-1888. https://doi.org/10.1039/b916124b

[68] S.K. Bera, S. Behera, L.D. Luca, F. Basoccu, R. Mocci, A. Porcheddu, Unveiling the Untapped Potential of Bertagnini's Salts in Microwave-Assisted Synthesis of Quinazolinones, Molecules 29 (2024) 1986. https://doi.org/10.3390/molecules29091986

[69] H.M. Al-Matar, K.M. Dawood, W.M. Tohamy, Tandem one-pot synthesis of 2-arylcinnolin-6-one derivatives from arylhydrazonopropanals and acetoacetanilides using sustainable ultrasound and microwave platforms, RSC Adv. 8 (2018) 34459-34467. https://doi.org/10.1039/C8RA06494F

[70] C.G. Neochoritis, C.A. Tsoleridis, J. S. Stephanatou, C.A. Kontogiorgis, D.J.H. Litina, 1,5-Benzoxazepines vs 1,5-benzodiazepines. One-pot microwave-assisted synthesis and evaluation for antioxidant activity and lipid peroxidation inhibition, J. Med. Chem. 23 (2010) 8409-8420. https://doi.org/10.1021/jm100739n

[71] S. Thakrar, A. Bavishi, A. Radadiya, S. Parekh, D. Bhavsar, H. Vala, N. Pandya, A. Shah, Microwave-assisted rapid synthesis of novel 1, 5-benzodiazepine derivatives as potent antimicrobial agent, J. Het. Chem. 50 (2013) E73-E79. https://doi.org/10.1002/jhet.1065

[72] A.A. Ali, N.S. Majeed, Microwave-Assisted Synthesis of Five and Seven-Membered Heterocycles from Ibuprofen and Study Their Biological Activity, J. K. Chem. Sci. 3 (2024) 223-239. https://doi.org/10.36329/jkcm/2024/v3.i3.15885

[73] S. Ramkumar, R. Ramarajan, Design, Synthesis, Spectral Characterization, Antioxidant Activity, Molecular Docking and in silico ADMET Studies of 1, 3 Oxazepines, Chemistry Select 8 (2023) 202204818. https://doi.org/10.1002/slct.202204818

Materials Research Foundations 189 (2026) 103-133 https://doi.org/10.21741/9781644904039-4

Chapter 4

Microwave-Assisted Synthesis of Polypyrrole and Potential Applications

D. Mahanta[1]*, K.K. Mudoi[1], S. Saikia[1]

[1]Department of Chemistry, Gauhati University, Guwahati, Assam, 781014, India

debam@gauhati.ac.in, debam2@gmail.com

Abstract

Polypyrrole is a widely studied conducting polymer known for its easy synthetic routes, high environmental stability, high conductivity, a wide range of applications, etc. Owing to such versatility, Polypyrrole has attracted the attention of various researchers for a long time. The synthesis procedures of Polypyrrole have also improved over time. With the recent advancements in microwave-assisted synthesis, researchers are now exploring its application in the field of Polypyrrole synthesis. This book chapter focuses on the basics of microwave-assisted synthesis, the application of microwave in the synthesis of Polypyrrole, its advantages and limitations over other conventional methods, and potential opportunities in various fields.

Keywords

Conducting Polymer, Microwave Irradiation, Polymerization Reaction, Energy Storage Application, Energy, Sensor Application; Biomedicine

Contents

Materials Research Foundations 189 (2026) 103-133 https://doi.org/10.21741/9781644904039-4

1. Introduction

Over the last century and a half, a remarkable class of materials called polymers has been developed and produced, reshaping the world of modern technology. These materials haven't just challenged traditional classical ones; they have become one of the bases of various important industrial goods. Besides social factors, the demand to replace classical materials has become the key factor in the high production of polymers [1].

Polyaniline (PANI), Polypyrrole (PPy), Polythiophene (PTh), and their derivative polymers are extensively studied in the field of research [2]. PPy stands out among various conducting polymers due to its high electrical conductivity (100-1000 S cm^{-1}), environmental stability, simple and low-cost synthesis, and excellent biocompatibility compared to polymers like PANI, PTh, and Poly(3,4-ethylenedioxythiophene) (PEDOT) [3]. PPy is synthesized through various synthesis methods such as oxidative-polymerization [4], electrodeposition [5], interfacial-polymerization [6], ultra-fast microwave (MW)-assisted process [7], etc. Additionally, PPy exists in various morphological forms, such as nanoparticles, nanotubes, nanofibers, and films [8]. PPy has found extensive applications in fields such as sensors [9], energy storage devices [10], drug delivery systems [11], and antimicrobial coatings [12] due to its tunable electrical and surface properties. Its versatility is further enhanced by combining with other materials to form nanocomposites, improving its mechanical, electrical, and functional characteristics.

Among the various synthesis procedures of polymer materials, one of the recently explored nonconventional processes is MW-assisted synthesis. A report on MW-assisted synthesis was

Materials Research Foundations 189 (2026) 103-133 https://doi.org/10.21741/9781644904039-4

published by Gedye et al. and Giguere et al. in 1986. MW-assisted synthetic strategy has shown various advantages, such as high reaction rate, selectivity, high yield production, etc., compared to conventional methods in many organic reactions [13,14]. Notable that the energy of MW quanta is not very high. Hence, it can't break the typical chemical bonds to initiate a chemical reaction directly [15]. These advantages have sparked increasing interest in broadening and improving the use of this method. This also leads to the recent implementation of this technique in the field of conducting polymer synthesis.

2. Fundamentals of Polypyrrole

2.1 Chemical Structure and Properties

The pyrrole (Py) monomer is the primary building unit of the PPy conducting polymer. Py is a five-membered heterocyclic ring with the formula $C_4H_4N_9$, containing four carbon atoms and one nitrogen atom in a heterocyclic aromatic ring. The monomer units in PPy chains are primarily bonded at α-α positions (Fig. 1a). However, a small fraction of Py is linked through α-β and β-β positions (Fig. 1b,c) [16]. PPy is formed when Py monomers link together to create a conjugated polymer backbone, typically at the C-2 and C-5 positions of the ring (Fig. 1d,e). This π-conjugated system enables the electrical conductivity in the polymeric backbone with alternating sp^2 and sp^3 carbon atoms [2].

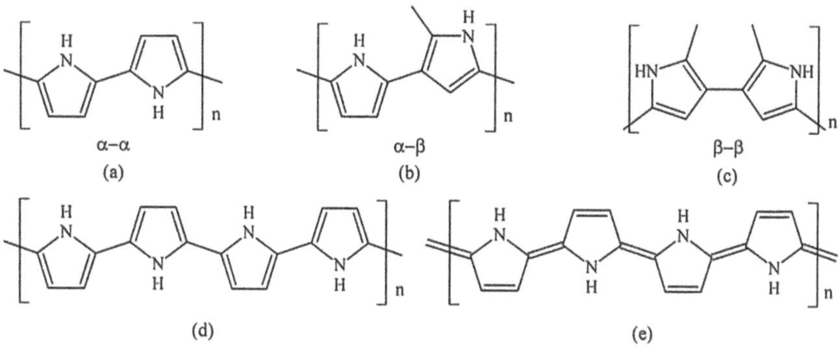

Figure 1. Secondary structural monomer units of (a–c) PPy and its (d) aromatic and (e) quinoid forms. Reproduced from ref. [16], copyright (2020), with permission from Elsevier.

The undoped PPy possesses low electrical conductivity in the range of 10^{-5} to 10^{-10} S cm^{-1} because the undoped PPy is neutral, where the conjugated backbone lacks significant charge carriers to enhance the electrical conduction. To improve the conductivity, PPy is doped employing chemical or electrochemical oxidation (*p*-doping) or reduction (*n*-doping). *p*-Doping requires oxidation of the PPy backbone using oxidizing agents like $FeCl_3$, iodine, etc., creating positively charged carriers along the polymeric chain. Meanwhile, *n*-doping involves the reduction of the polymer by introducing negatively charged carriers into the polymeric matrix. *n*-Doping is typically achieved using strong reducing agents like alkali, lithium, or sodium metals. Usually, *p*-doping is more common and widely used due to its higher conductivity (1 to 10^5 S cm^{-1}) and stability in the ambient environment than that of *n*-doping, which lies in the range of 10^{-3} to 1 S cm^{-1} (Fig. 2) [17].

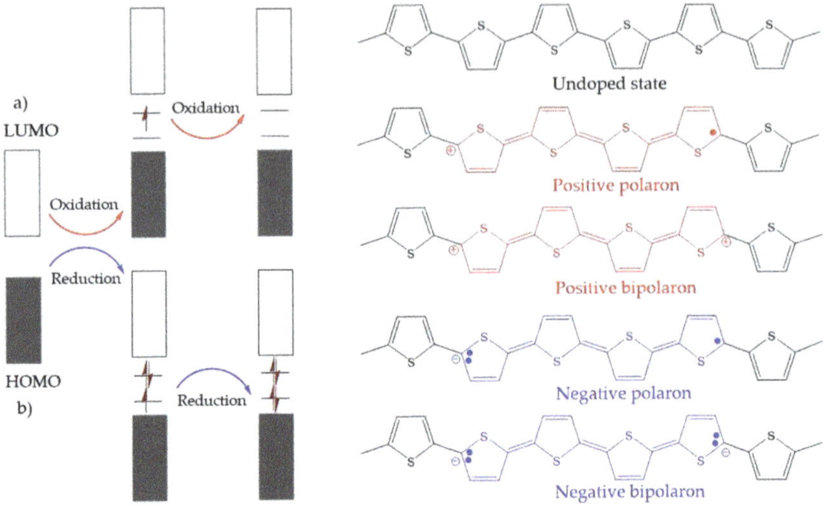

*Figure 2. The electronic band and chemical structures of PTh with **a)** p-type doping, and **b)** n-type doping [18]. This figure is reproduced under the terms of Creative Commons Attribution License (CC BY)*

2.2 Traditional Synthesis Methods

2.2.1 Chemical Oxidative Polymerization

Chemical oxidative polymerization is a rapid, economical, efficient approach, and the most common method for synthesizing PPy. This approach allows for the easy production of fine PPy powders [9,19]. Typically, the process involves the oxidation of the Py monomer using a chemical oxidant in either water-based or non-water-based solutions. The usual method consists of preparing two separate mixtures, one with the Py monomer and the other with the oxidizing agent, which are then combined to initiate polymerization [20].

In a typical chemical oxidative polymerization process, Py monomers generate a reactive radical cation *via* oxidation. Subsequently, these charged intermediates then couple together to form soluble bipyrrole structures by releasing protons. These bipyrroles are further oxidized and produce oligomers with radical cations [21]. Oligomers then grow in size, eventually forming high-molecular-weight PPy through continuous coupling and chain extension. The polymerization continues until the growing PPy chains lose solubility and precipitate out from the aqueous medium [22]. The schematic diagram and flowchart of chemical oxidative polymerization are shown in Fig. 3a,b.

PPy doped with different anionic surfactants can be synthesized. The results confirm that both surfactant type and molar ratios strongly influence morphology and conductivity. Maximum conductivity can be achieved at an $FeCl_3$:Py ratio of 1.1 and a surfactant:monomer ratio of 0.3. Notably, PPy-β-NSA displays the highest conductivity, 33.3 S cm^{-1}. Hence, doping level,

Materials Research Foundations 189 (2026) 103-133 https://doi.org/10.21741/9781644904039-4

counterion organization, and surfactant structure are decisive factors for optimizing PPy's electrical performance [23].

PPy nanotubes were synthesized *via* chemical oxidative polymerization of Py using $FeCl_3$ in an aqueous medium containing methyl red sodium salt. Reaction temperature and methyl red concentration significantly influence the conductivity of the resulting nanotubes. Lowering the temperature to -50 °C enhances conductivity, reaching up to 1.04×10^2 S cm^{-1}. Methyl red sodium salt also influences the conductivity by increasing it from 5 S cm^{-1} to 84 S cm^{-1} at 20 °C [24]. Moreover, dopant concentration plays an important role in determining the PPy's conductivity. Camphor sulfonic acid (CSA) doped PPy films exhibited structural, morphological, and optoelectronic modifications with increasing CSA content. Electrical conductivity improved significantly, rising from 429×10^{-7} S cm^{-1} to 156×10^{-6} S cm^{-1} at 50% CSA [25].

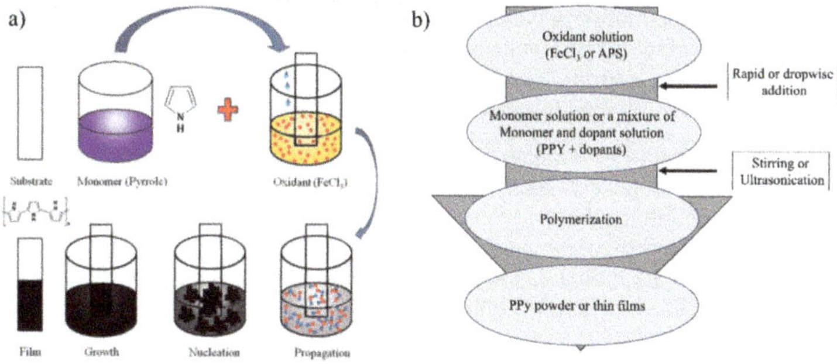

Figure 3. a) Schematic illustration of the chemical oxidative polymerization of PPy, Reproduced from ref. [21], copyright (2015), with permission from Elsevier. b) flow diagram outlining the standard synthesis route of PPy using chemical oxidative polymerization

2.2.2. Interfacial Polymerization

Interfacial polymerization has emerged as a powerful strategy for preparing nanostructured conducting polymers, particularly PPy. Interfacial polymerization confines reactions at phase boundaries and thereby generates unique morphologies. Interfacial routes exploit the liquid-liquid boundary as a confined reaction zone, unlike bulk oxidative polymerization, which often produces irregular granular powders with limited processability [26]. At the oil-water interface, Py monomers dissolved in the organic phase encounter oxidants such as ferric chloride ($FeCl_3$) or ammonium persulfate (APS) in the aqueous phase. Polymerization is thus confined to the narrow interfacial region, and the resulting PPy chains adopt extended conformations due to interfacial protonation and orientation effect. The interfacial polymerization can take place overnight at room temperature to yield PPy nanofibers that exhibit adequate polymerization with nanofibrous morphologies (Fig. 4) [27]. The pH of the aqueous phase plays a decisive role in the interfacial polymerization of PPy, as it directly governs monomer protonation, chain propagation, and the doping level of the resulting polymer. Under acidic conditions, Py monomers are readily oxidized at the oil-water interface, leading to efficient radical cation formation and subsequent polymer growth. Protonation not only facilitates oxidative coupling but also enhances the incorporation of counter anions, thereby improving electrical conductivity through stabilized polaron and bipolaron states [28]. Furthermore, while interfacial polymerization yields well-dispersed nanostructures with diameters of 10 to 80 nm, integrating these nanostructures into macroscopic devices without losing their intrinsic properties remains nontrivial [29,30].

Surfactants play a pivotal role in tailoring the morphology and conductivity of PPy synthesized *via* interfacial polymerization. Surfactants mainly act as soft templates and form micelles or interfacial aggregates that guide the anisotropic growth of PPy chains, promoting nanofibrous or nanotubular architectures instead of irregular granular morphologies [27]. For instance, anionic surfactants such as sodium dodecyl sulfate (SDS) or p-toluenesulfonic acid (p-TSA) have been shown to improve chain ordering and electrical conductivity compared to conventional mineral acids [31]. The choice and concentration of surfactant significantly influence fiber diameter, porosity, and dispersion stability of the resulting PPy nanostructures. Excess surfactant, however, can hinder polymerization kinetics or lead to structural collapse due to excessive micelle crowding [27]. Overall, surfactants not only direct nanostructure formation but also act synergistically as dopants, making them indispensable for optimizing morphology and electronic performance of PPy produced at liquid-liquid interfaces.

Interfacial polymerization can overcome the limitations of acid-induced routes by yielding highly conducting and morphologically controlled PPy nanostructures. Yet, challenges remain in balancing reaction conditions, dopant selection, and reproducibility. The key advantages of this method lie in its scalability, mild processing conditions, and ability to generate functional nanostructures, whereas its limitations include sensitivity to pH, potential side reactions under strong acidity, and the need for careful control of interfacial dynamics [32,33].

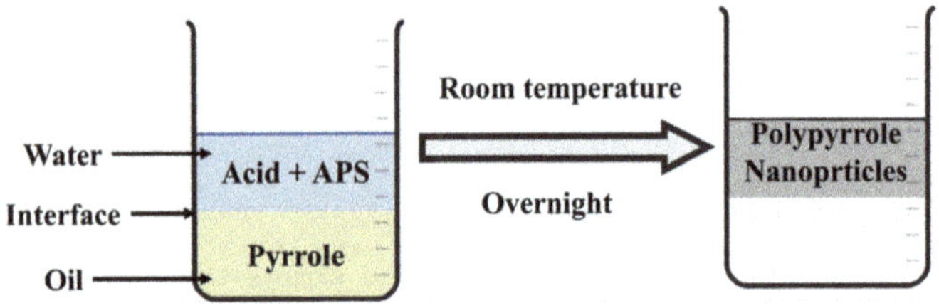

Figure 4. Schematic of the synthesis of PPy nanoparticles by interfacial polymerization.

2.2.3 Electrochemical Polymerization

Electrochemical polymerization is a productive method to synthesize PPy with controllable thickness, surface area, and conductivity. Electrochemical polymerization can be performed in a three-electrode cell with the working electrode, reference electrode, and counter electrode, and the electrolytic solution is prepared by dissolving Py monomer in aqueous or organic solvents [34]. By tuning the input parameters (potential, current, and deposition period) passing through the electrochemical cell, electrodeposited PPy can be prepared with desired properties. Electrochemical polymerization of PPy has attracted significant attention for energy storage applications, particularly in supercapacitors and batteries [2]. The influence of surfactants on the structure of PPy has also been investigated. Results showed that in the presence of surfactant, cone-

Materials Research Foundations 189 (2026) 103-133 https://doi.org/10.21741/9781644904039-4

shaped PPy was formed. While a flat film of PPy was formed in the absence of surfactant. The conductivity of PPy synthesized in the presence of surfactant is also higher [35,36].

Electrochemical synthesis of PPy involves the oxidation of Py monomers at the electrode to form radical cations. The growth of the PPy backbone is accompanied by radical coupling and undergoes deprotonation to gradually form oligomers like bipyrrole and tripyrrole. Despite of various advantages, the electrochemical polymerization process is complex and not fully understood [37]. The reactivity of radical cations significantly influences the polymerization yield. Excessive reactivity can lead to side reactions with electrolyte species, while excessive stability can cause diffusion away from the electrode, preventing film formation. Hence, using high-purity chemicals and well-prepared electrode surfaces is crucial for obtaining quality electroconductive polymer films [2]. Electrochemical polymerization eliminates binders and extra steps by forming polymer films directly on the electrode. It offers better control over thickness, morphology, and doping with higher purity and conductivity. The process is simpler, cleaner, and produces electrode-ready materials for energy devices [38]. Although electrochemical polymerization offers significant advantages over other synthesis methods, it is challenging to scale up the process for industrial-level production. A schematic representation and a flow chart for the synthesis of PPy are shown in Fig. 5.

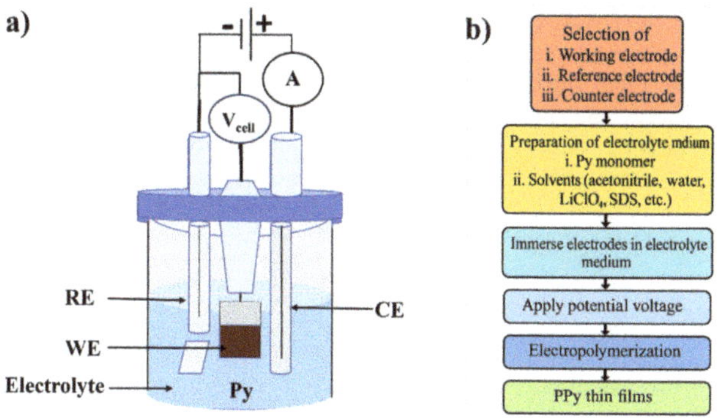

Figure 5. a) Schematic diagram illustrating the electrochemical polymerization of PPy, b) Flow diagram describing the synthesis pathway of PPy through electrochemical polymerization using a conventional three-electrode setup.

2.2.4 Electrospinning

Electrospinning is a versatile technique for producing ultrafine polymer fibers, with diameters ranging from several micrometers to nanometers, by employing high electrostatic forces. Electrospinning is a fiber fabrication technique where a high voltage is applied to a polymer solution or melt delivered through a fine needle. The electric field causes the droplet at the needle tip to elongate into a Taylor cone, and a charged jet is released once the electrostatic forces exceed the surface tension. As the jet travels toward a grounded collector, it undergoes stretching and solvent evaporation (or cooling in the case of a melt), ultimately forming continuous ultrafine

Materials Research Foundations 189 (2026) 103-133 https://doi.org/10.21741/9781644904039-4

fibers (Fig. 6) [27,39]. The electrospinning parameters, such as viscosity, applied voltage, surface tension, temperature, relative humidity, needle size, and flow rate, should be optimized carefully to obtain nanofibers with appropriate structures, morphologies, thickness, surface area, and pore size distribution. Electrospinning is a flexible technique used for mass production of PPy nanofibers, and these electrospun nanofibers exhibit ultrafine fiber morphology, high surface area, and tunable porous structures with controlled composition [16,40].

However, it is quite a sophisticated technique, and pure PPy cannot be obtained directly *via* the electrospinning technique due to its low solubility and dispersibility. Fiber formation can be facilitated by incorporating additives such as surfactants, co-solvents, or polymer carriers into the spinning solution. However, these additives may influence the conductivity of the resulting PPy nanofibers [41,42]. Multiple studies have described the production of PPy nanofibers *via* electrospinning. Previously, electrospinning was employed to create conductive PPy nanofibers. Both pure PPy fibers and those blended with polyethylene oxide (PEO) as a carrier were produced using this method in the presence of sodium di(2-ethylhexyl) sulfosuccinate (NaDEHS) as a dopant. The electrospinning process was carried out at ambient conditions with an applied voltage of 30 kV. Their results indicated that the pure PPy nanofibers exhibited an electrical conductivity of approximately 27×10^{-3} S cm^{-1}. In contrast, the PPy@PEO composite nanofibers demonstrated significantly lower conductivity, ranging from 49×10^{-9} S cm^{-1} to 12×10^{-6} S cm^{-1} [43].

Although electrospinning is a versatile technique for producing continuous PPy nanofibers at a large scale, it also possesses several drawbacks. The process demands careful adjustment of multiple parameters, often through trial and error, and typically involves high voltages and solvents, raising energy and environmental concerns. To aid fiber formation, additives are frequently used, which can significantly decrease the conductivity. Additionally, the technique is mainly limited to polymers that are both soluble and thermoplastic, restricting material options [27].

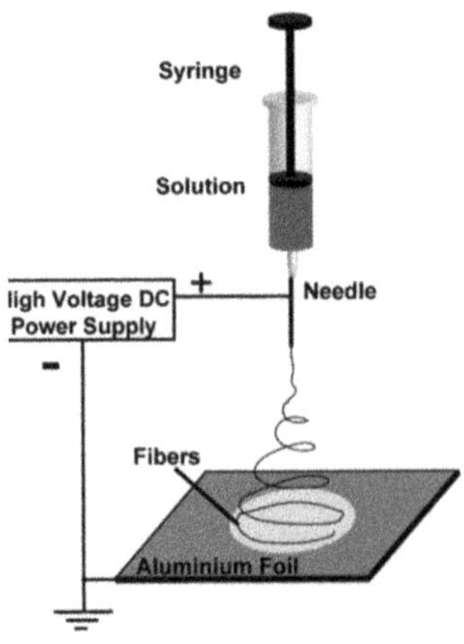

Figure 6. Schematic diagram of the electrospinning method. Reproduced from ref. [27], copyright (2011), with permission from Elsevier.

Materials Research Foundations 189 (2026) 103-133 https://doi.org/10.21741/9781644904039-4

Table 1. Comparison of the pros and cons of conventional synthesis methods

Synthesis Procedure	Pros	Cons
Chemical Oxidative Polymerization	• Rapid and straightforward preparation • Effective for large-scale production as powder, coating, and dispersion	• Hard to achieve uniform PPy structure • Produces thick films with limited thickness control • Not ideal for rigid, insoluble polymers
Electrochemical Polymerization	• High accuracy and purity • Thin film production • Morphology and electrochemical properties are easily tuned • Good mechanical and electrical properties	• Not scalable • Some monomers can't be electropolymerized • High oxidation may overoxidize PPy • Expensive and time-intensive
Interfacial Polymerization	• Facile and fast reaction time • Requires mild conditions • Allows formation of thin films, capsules, or membranes with controlled morphology • Scalable for industrial applications, especially in membrane production	• Limited to monomers that are soluble in immiscible solvents • Difficult to achieve uniform polymer thickness and structure • Use of organic solvents may raise environmental and safety concerns • Post-processing and purification steps can be complex and time-consuming
Electrospinning	• Produces nanofibers with uniform diameter • High surface area and porosity • Controllable composition	• Limited to soluble and thermoplastic polymers

3. Microwave Heating

MW heating is an alternate heating process that is non-contacting. MW radiation has a frequency range of 300 GHz to 300 MHz and a wavelength of 1 mm to 1 m. Although it propagates at the speed of light, its energy is lower than the threshold required for bond dissociation. This describes the ineffective nature of MW heating on the chemical framework of the reactant molecules [44]. In commercial systems, the most commonly used MW irradiation is at a frequency of 2.45 GHz. This particular frequency is selected to avoid disruption to other communication devices. The

Materials Research Foundations 189 (2026) 103-133 https://doi.org/10.21741/9781644904039-4

corresponding electric field oscillation frequency of this irradiation is around 4.9×10^9 times per second, leading to the continuous shifts in the orientation of dipolar species, ionic particles, holes, and electrons. This results in the noncontact, approximately uniform, fast heating of the system [45].

4. Principle of Microwave Heating

MW heating is a dielectric heating-based instantaneous localized superheating technique that responds to the two basic fundamental mechanisms of energy transfer from MW to substance: dipole rotation and ionic conduction.

4.1 Dipolar Polarization Mechanism

This mechanism is based on the polar part of the solvents and reactants and is fundamentally controlled by the electric field of the MW radiation [7,46]. In normal conditions, polar molecules of the reactants and solvents are in random, and chaotic motion. When the electric field is applied, they start to rotate so that they can align themselves with the applied electric field. However, when the applied electric field is strong enough, polar molecules do not have enough time to react to the applied electric field. Resultantly, they try to match the magnetic field. This leads to collisions, rotation, friction, dielectric loss, and finally the release of thermal energy [47,48]. If the essential condition of having sufficient time to realign with the applied electric field (as in the case of high MW frequencies) is not satisfied by reaction dipoles or to reorient rapidly (as in the case of low MW frequencies), MW heating cannot be observed. Thus, depending on the MW frequency, when the dipoles lack adequate time for realignment or rapid adjustment with the electric field, microwave heating may not occur [44].

4.2 Ionic Conduction Mechanism

The Ionic conduction mechanism is a more effective process than the Dipolar polarization mechanism in terms of heat generation. In this phenomenon, dissolved ions oscillate in the presence of MW, leading to collisions with each other and heat generation [7]. If the reaction mixture consists of free ions, due to their ionic motion, they try to align themselves with the direction of the applied field, leading to the instantaneous super heat generation [44]. Both Dipolar polarization and Ionic conduction mechanisms are demonstrated in Fig. 7.

Briefly, MW penetrates the bulk of the material where molecules with a permanent dipole moment or ions attempt to align with the applied electromagnetic field. This leads to the collision, rotation, and friction of molecules, resulting in output heat. Hence, it can be summarized that MW irradiation's efficiency and heating rate depend on the MW absorption capability of the solvent, dielectric properties, and relaxation time of the reaction mixture [50–52].

Materials Research Foundations 189 (2026) 103-133 https://doi.org/10.21741/9781644904039-4

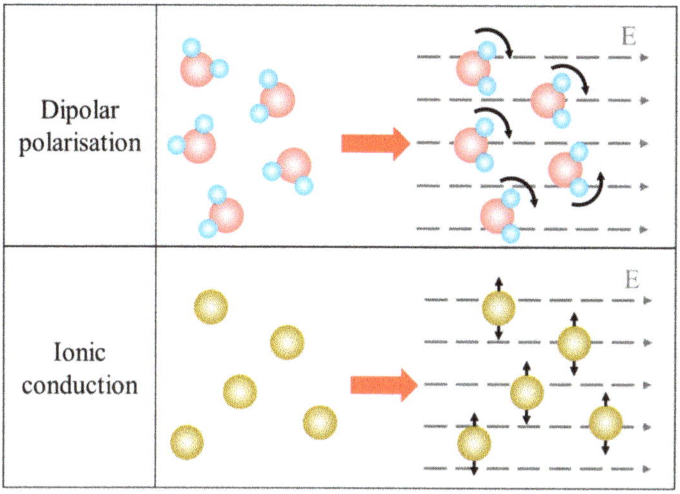

Figure 7. Schematic diagrams of dipolar polarization and ionic conduction. Reproduced from ref. [49], copyright (2022), with permission from Elsevier.

5. Advantages and Disadvantages of Microwave System

5.1 Increase of Reaction Rate by Microwave

In a chemical reaction between two molecules, the process starts when they collide in a specific way and absorb a certain amount of energy, i.e., activation energy ($\boldsymbol{E_a}$) to reach a higher energy state before forming stable products. MWs can accelerate reactions by providing energy efficiently and also without changing the required $\boldsymbol{E_a}$. At the time of each electromagnetic wave cycle, MW energy is rapidly transferred, leading to high localized temperatures and faster reaction rates with better product yields [53]. MW heats materials extremely quickly, transferring energy even faster than molecules can relax. It creates a non-equilibrium condition and high instantaneous temperature. This rapid heating increases reaction rates and product yields. Even though MWs don't alter $\boldsymbol{E_a}$ or molecular orientation, they pronouncedly raise temperature, which increases the number of effective molecular collisions. Therefore, MW heating can speed up reactions many times compared to conventional methods. This fast temperature rise is key to MW applications in nanomaterial synthesis and organic reactions [54].

5.2 High Yield and High Purity Product Formation

The main reason for high-purity product formation with high yield percentage is the shorter reaction time of MW-assisted synthesis. Shorter reaction time significantly decreases the chances of an unwanted side product, leading to the maximization of the formation of high-purity main product [44,55].

Materials Research Foundations 189 (2026) 103-133 https://doi.org/10.21741/9781644904039-4

5.3 Greater Control Over the Reaction

MW strategy offers higher selectivity, scalability, higher control over the heating time and catalyst, etc., compared to batch processes. Modern MW-assisted synthesis offers the singular ability to cool the reaction vessel while simultaneously delivering more MW energy to the reaction. This allows the significant enhancement of reaction rates through higher MW power input, all while maintaining the desired bulk temperature, an especially essential attribute for reactions with high activation energies [56].

On the other hand, MW heating has some limitations compared to conventional heating system. The first one is the MW absorption capacity of the materials. Lower absorption capability leads to a lower heating effect by the MW, which is not a limitation in the case of traditional heating process. Again, additional investment and expertise needed for the MW heating, as its instrumentation is complex and expensive than that of a traditional heating system [7].

6. Components of Microwave Heating System

MW heating systems consist primarily of four main components: (1) the power source (MW source: magnetron), (2) the reaction vessel, (3) the waveguide feed, and (4) the broiler hole. The constituents of the power source are the magnetron tube, relay, transformer, choke, and controls. Magnetron has a high-power output (roughly 85%) and a long lifespan of around 5000 hours. A waveguide directs the MW energy to the applicator. It features smart partitions that anticipate and prevent MW leakage. An insulating conveyor belt is used to transport materials through the oven for heating in a continuous-type MW heating system. The oven cavity serves as the reaction platform, where a vessel containing the chemical reaction mixture absorbs the supplied electric energy. This vessel is usually built from microwave-transparent materials such as polystyrene or Teflon [57,58]. A schematic diagram of MW heating system is illustrated in Fig. 8.

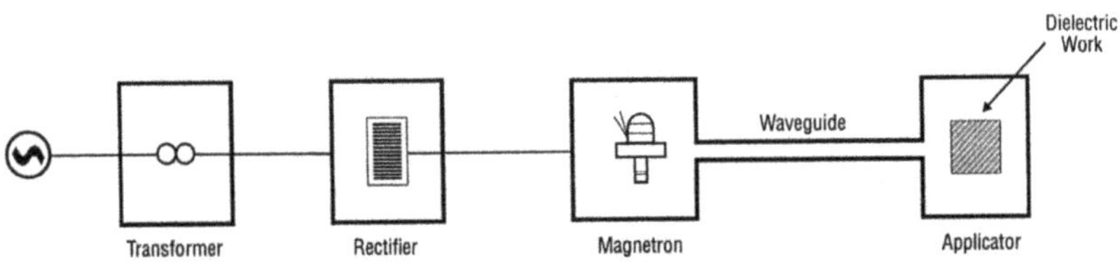

Figure 8. Schematic of the MW heating system. Reproduced from ref. [57], copyright (1999), with permission from Elsevier.

MW Reaction system can also be classified as open reaction system (OR) and close reaction system (CR). The difference between these systems is that OR system uses high boiling point solvents, whereas CR system uses water, pure organic, or mixed organic solvent systems. OR system offers safe, basic, and cost-effective benefits over CR system [44].

7. Solvents of Microwave System

Solvent selection is an important part of the green chemistry approach for more sustainable, safer chemistry. The role of solvent selection is also crucial in MW-assisted synthesis, particularly for nanostructured materials. It influences not only the efficiency of the process but also the

Materials Research Foundations 189 (2026) 103-133 https://doi.org/10.21741/9781644904039-4

environmental, economic, and societal effects in the context of green chemistry. The solvent's loss tangent is related to the efficiency of the process. It is the ability of solvent molecules to convert electromagnetic energy into thermal energy at a specific MW frequency and temperature. In other words, it provides a direct measure of heat generation based on a specific level of polarization. The loss tangent value can be designated as tan δ. According to the tan δ values, solvents can be grouped into three categories as shown in Table 2 [59].

Table 2. Classes of the solvents used in the MW systems based on their tan δ value

Serial No.	Class	tan δ value	MW absorption ability
1	Very low	< 0	Very poor
2	High	> 0.5	Good
3	Meddle	≈ 0.1 – 0.5	Medium

Another type of solvent system is a mixed solvent system. The interaction between different solvents alters their MW absorption behaviour, providing an effective strategy for tuning nanomaterial properties. Similarly, other trial variables such as MW frequency, temperature, reaction time, reactant nature, etc., can greatly affect the pathway of the reaction, as well as the shape, size, and uniformity of the resulting product. One more effective strategy for the enhancement of MW heating involves the addition of a small amount of a strongly MW-absorbing agent, such as an ionic liquid, to a non-polar solvent. This approach enables the use of poor MW absorber non-polar media as viable reaction environments for MW-assisted processes [44,60].

8. Effect of Microwave in Various Types of Polymerization Processes

There are various types of polymerizations, such as Step-Growth polymerization, Ring Opening polymerization (ROP), and Radical polymerization. MW-assisted polymerization influences the product quality, reaction rate, and in some cases, reaction pathways noticeably.

8.1 Step-Growth Polymerization

In Step-growth polymerization, multifunctional monomers, commonly bifunctional, couple or condense repeatedly, leading to the formation of polymers. Due to the production of water as a byproduct, this type of polymerization is also known as polycondensation. Common examples of polymers synthesized through this process include polyamides, polyimides, polyethers, polyesters, etc. MW irradiation significantly reduces reaction times, from hours to minutes. An increase of the polymerization rate was found to be ten times during the MW-assisted synthesis of poly(aspartic acid) [50]. Different works claimed higher yield and efficiency, faster polymerization, and higher molecular weight product [61,62].

8.2 Ring-Opening Polymerization

ROP is an effective technique for producing polymers with high molecular weights. It proceeds *via* the opening of the ring structure of the cyclic monomers to form long-chain polymers, which are of controlled and well-defined structures. For instance, a domestic MW oven was used in the polymerization of ε-caprolactone. By comparing the traditional heating and MW heating, it was demonstrated that MW promotes chain growth in the initial stage, whereas traditional heating favours growing centre formation in the initial stage. This leads to higher molecular weight

Materials Research Foundations 189 (2026) 103-133 https://doi.org/10.21741/9781644904039-4

polymers in the case of MW-assisted polymerization [63]. Also, in MW-assisted ring-opening polymerization to ε-caprolactone, higher grafting can be obtained compared to that from conventional heating [64]. From various studies, information about some other advantages of MW-assisted ring-opening polymerization can be found, such as higher control over polymerization, MW acceleration of the reaction, etc. [65,66].

8.3 Radical Polymerization

Radical polymerization, which involves the generation of reactive radical species, is widely used for making polymer materials like polystyrene, poly(methyl methacrylate), and various copolymers. It can be of various types, such as free radical polymerization, emulsion/dispersion polymerization, controlled radical polymerization, etc. In case of free radical polymerization, MW-synthesis helps to accelerate some polymerization and in some special cases, like grafting, allows polymerization without added initiators [67,68]. In the aqueous system of emulsion radical polymerization, MW irradiation increases initiator decomposition and reduces the activation energy of some polymer synthesis, leading to a faster polymerization rate [69,70]. MW synthesis provides better control. Hence, well-defined polymers can be synthesized at a higher rate in controlled radical polymerization using MW irradiation [71,72].

9. Microwave assisted Polypyrrole Synthesis

Among the various synthesis procedures of PPy, MW-assisted synthesis is a relatively recent approach. It is gaining attention due to its advantages, like faster reaction times, better control over the reaction, and better energy efficiency. PPy nanoparticles can be synthesized *via* an easy and fast MW-assisted process in a MW synthesizer. Here, $FeCl_3 \cdot 6H_2O$ can be used as an oxidising agent with PVA as a stabilizer. A traditional polymerization process of PPy at room temperature takes around 24 hours. MW irradiation with a power density of 10 W can reduce the polymerization time to only 2 min. PPy produced by such a method exhibits an average uniform size of around 50 nm. This size is a little bit smaller than the PPy particle size produced *via* the traditional way. Moreover, compared to traditional PPy, MW-assisted PPy has more near-infrared absorption capability and better photothermal effect [73]. There is another method called MW-assisted chemical bath deposition (MW-CBD) for depositing PPy as thin films on various surfaces, like glass. The whole procedure can be operated using a MW oven operating at 2.45 GHz. For the preparation of the chemical bath, 0.1 M Py monomer and 0.025 M ammonium persulfate are dissolved separately in 2 vol% H_2SO_4 in 100 ml beakers. The first beaker with the glass substrate is placed inside a second beaker filled with water to absorb excess heat generated during MW irradiation. Then this entire setup is placed on the rotating disk inside the MW oven cavity. An interconnected web-like structure of PPy on the glass surface can be observed [74].

Another such example of polymerization on the glass surface is metal oxide doped PPy-polyvinyl alcohol blend thin film by in situ chemical oxidative polymerization in a MW oven [75].

Coating of PPy on nanoparticles can be done using MW. An example of such PPy-coated magnetic nano-composite is Fe_3O_4@PPy. Fe_3O_4@PPy nano-composite can be prepared using the ultrasonic-MW cooperative synthesis method. This method utilizes both MW heating and ultrasonication. Such well-shaped beads of nanoparticles have a uniform and smooth surface morphology, shown in field emission scanning electron microscopy (SEM) in Fig. 9. The diameter distribution of the beads is from 5 to 10 nm [76].

Materials Research Foundations 189 (2026) 103-133 https://doi.org/10.21741/9781644904039-4

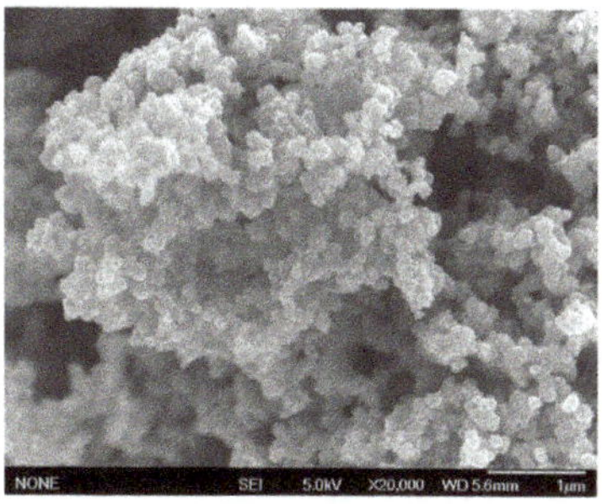

Figure 9. FESEM image of Fe_3O_4@PPy nano-composite. Reproduced from ref. [76], copyright (2016), with permission from Elsevier.

Similarly, PPy/graphene composite can also be synthesized using direct MW treatment. PPy distributed over the graphene sheet homogeneously forming a three-dimensional cross-linked structure of the composite [77]. PPy composite with carbon nanotubes (CNT) is capable of being synthesized using MW irradiation. It shows a high specific capacitance of 304 F g^{-1} at a current density of 1 mA cm^{-2} [78]. MW irradiation can produce PPy/Ag composite using $AgNO_3$ and even without any external oxidising or reducing agent [79]. Chitosan grafted PPy (PPy-g-CS) and copolymer poly (pyrrole-N-(1-naphthyl) ethylenediamine (COP) have been synthesized by a MW-assisted procedure. In these reactions, carbon dots can be used as initiators. Even carbon dot initiator can lead to the polymerization of Py in the presence of MW irradiation within two minutes. Fig. 10 shows the FESEM images of PPy, PPy-g-CS, and COP. From the images, it can be seen that PPy-g-CS contains irregular micro-nanoscale PPy spherical particles. COP also exhibits uneven spherical formations along with a few plate-like particles [80].

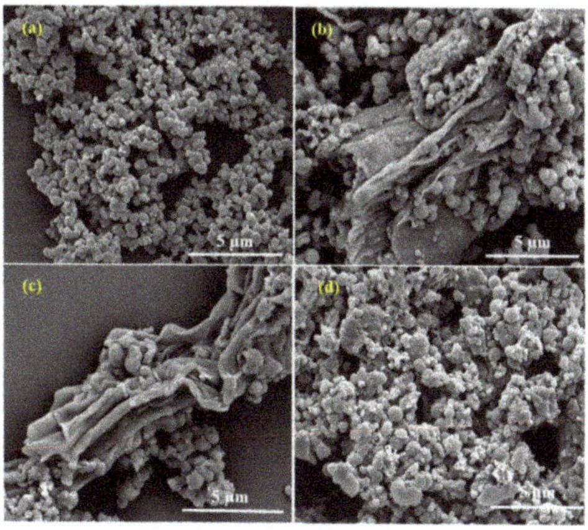

Figure 10. FESEM images of (a) PPy; (b) and (c) PPy-g-CS; and (d) COP. Reproduced from ref. [80], copyright (2020), with permission from Elsevier.

In the case of PPy derivative, poly[(pyrrole-2,5-diyl) co-(benzylidene)] (PPCB) is synthesized using MW irradiation polymerization within 35 s which is very fast as compared to traditional methods [81]. Some PPy composite synthesis procedures use MW irradiation during the final product formation. Here components of the composites are formed using traditional methods, but final composite formation is done using the MW irradiation. Some such examples of composites are PPy/Pt nanocomposite and PPyNF/NiO_x nanocomposites [82,83].

10. Potential Applications of Microwave-Synthesized Polypyrrole

Among the conducting polymers, PPy is of particular interest due to its high conductivity, environmental stability, good redox properties, easy and simple synthesis routes from commercially available monomers [84]. PPy has found use in a variety of applications, including energy storage and conversion applications, biosensors, mechanical actuators, drug delivery systems, corrosion resistance, and light-emitting diodes, as a result of these intriguing inherent qualities [85–88]. PPy synthesized *via* the MW-assisted method has drawn significant attention in various research fields, emerging as a potential synthesis method for PPy as well as other conducting polymers.

10.1 Sensor

PPy is a versatile, cost-effective, and easily tunable material known for its strong response to various analytes through electrical, chemical, and mechanical changes. It is well-suited for a various type of sensing applications. The MW-assisted synthesis of PPy offers additional benefits, such as precise control over its morphology, particle size, and conductivity [27]. These structural advantages enhance the material's performance in various sensor platforms. MW-mediated PPy-based sensors, including electrochemical, gas, and biosensors, show great promise for commercial use due to PPy's excellent environmental stability, responsiveness to external stimuli, and compatibility with solution-based processing techniques [30]. In recent work, the MW-assisted chemical bath technique was employed for NH_3 gas sensing. The porous thin film PPy shows high sensitivity to NH_3 gas of 85% with response and recovery time periods of 70 and 110 s for NH_3. The porous morphological feature is a valuable criterion for the adsorption of gases on the surface of the thin film [74]. In another work, metal-oxide (TiO_2) mixed porous, granular, and globular PPy-PVA thin films were synthesized through the MW-assisted method, showing good response and recovery time at 5-ppm lower-level ammonia and trimethylamine gas vapours at room temperature. These works are applicable for environmental pollution monitoring [75].

A novel rGO/PPy-Pt bimetallic electrochemical biosensor was developed *via* MW-assisted reduction for the simultaneous and selective detection of ascorbic acid (AA), dopamine (DA), and uric acid (UA). The sensor exhibited distinct electrocatalytic responses to each analyte without cross-interference. Under optimized conditions, linear detection ranges were established as 0.8-2.1 mM (AA), 0.03-1.4 mM (DA), and 0.1-0.35 mM (UA), with detection limits of 0.12 μM, 0.071 μM, and 0.16 μM, respectively. Enhanced sensor performance is attributed to synergistic effects within the nanocomposite, yielding improved selectivity, reduced interfacial resistance, and higher response currents. These findings support the sensor's potential for practical use in routine analysis [89]. In a recent work, the MW-assisted technique was used to prepare the PPy/Pt nanocomposite, employed in the fabrication of a non-enzymatic electrochemical H_2O_2 sensor. The nanocomposite sensor exhibited good selectivity, high stability, good repeatability, good reproducibility, and significantly high sensitivity [83]. In another research aspect, a PPy-induced graphene oxide hydrogel with 3D PPy/graphene network was synthesized *via* rapid MW-mediated method. The resulting composite reveals outstanding electrocatalytic activity and selectivity for detecting dihydroxybenzenes, specifically hydroquinone and catechol, within the detection limit of 0.88 μM and 0.69 μM, respectively. High sensitivity, stability, and efficient fabrication of PPy-induced graphene oxide hydrogel make it a promising candidate for electrochemical sensing applications [77]. Various types of PPy-based sensors synthesized *via* the MW-assisted method have listed on Table 3.

Thus MW-assisted PPy synthesis method plays an important role in the field gas sensors, electrochemical sensors, and biosensors. Gaining a deeper understanding of the underlying synthesis mechanisms will further expand the application potential of such sensors across a broader range of fields.

Table 3. A review of the performance of MW-assisted PPy or PPy nanocomposites as sensor applications

Material	Sensor type	Concentration for sensing (Sensitivity)	Response time and Recovery time	Reference
PPy/Pt	Electrochemical (H_2O_2)	25-500 μM (305.45 μA mM^{-1} cm^2) 500–6300 μM (164.94 μA mM^{-1} cm^2)	–	[83]
rGO/PPy-Pt	Electrochemical (AA, DA, UA)	AA-0.12 μM DA-0.07 μM UA-0.16 μM	Fast	[89]
PPy/Graphene Hydrogel	Electrochemical (HQ, CT)	HQ-0.88 μM CT-0.69 μM	–	[77]
PPy (MW-CBD)	NH_3 gas sensor	50-250 ppm	70 s and 110 s	[74]
Metal oxides doped PPy-PVA	NH_3 gas sensor	5-800 ppm	–	[75]

10.2. Supercapacitor

PPy is frequently employed in energy storage devices like supercapacitors and batteries because of its sufficiently large surface area, excellent conductivity, and enhanced electrochemical performance, which make it perfect for improving energy storage and conversion efficiency. The pure PPy-based supercapacitors exhibit specific capacitance in the range 100–400 F g^{-1}. However, with dopant level of 0.33, PPy has a theoretical specific capacitance up to 620 F g^{-1} [30,90]. A serious drawback of pure PPy is that it suffers from decomposition of the structure during charge/discharge cycles. PPy comprises carbon-based composites (graphene, graphene-oxide, CNT, activated carbon, etc.), have shown significantly enhanced the capacitance as well as cyclic

Materials Research Foundations 189 (2026) 103-133 https://doi.org/10.21741/9781644904039-4

stability as supercapacitor electrode materials. Uniform, non-agglomerated PPy nanospheres have been synthesized through a straightforward chemical oxidative polymerization of Py. This process uses 3-chloroperbenzoic acid as the oxidant and does not require any templates, additional oxidizing agents, or surfactants. The resulting PPy nanospheres demonstrate notable electrical conductivity (ranging from 10 to 100 S cm^{-1}) along with impressive electrochemical properties, achieving a specific capacitance of 176 F g^{-1} at a current density of 1 A g^{-1}. This efficient and uncomplicated method opens up a promising pathway for fabricating conductive polymers and nitrogen-enriched carbon nanospheres [91].

In the conventional oxidative polymerization method, CNT/PPy-based nanocomposites with a hierarchical mesoporous nanoclay structure have been synthesized. In this procedure, PPy is coated over CNTs, and the nanoclay structure significantly improves electron transport *via* hopping mechanisms. It revealed a high specific capacitance of 425 F g^{-1} and an energy density of 37.81 Wh kg^{-1} at 10 mV s^{-1} in 1 M KCl. Further validation through charge–discharge analysis at 5 A g^{-1} showed strong rate performance and durability, with the material retaining 94% of its capacitance after 2,000 cycles [92].

In another work, the ultrafast MW irradiation technique successfully synthesized the PPyNF/NiO_x nanocomposite. This composite exhibited an impressive specific capacitance of 657 F g^{-1} at a current density of 0.5 A g^{-1}. Additionally, the composite retained cyclic stability of 80% of the initial value after 1000 cycles, highlighting its strong potential for high-performance supercapacitor applications [82].

As shown in Fig. 11, the MW-assisted method, PPy/CNT composites were synthesized *via* a MW-assisted hydrothermal approach using an MO-$FeCl_3$ complex as both oxidant and morphology-directing template. The microwave treatment enhanced CNT functionalization and enabled uniform PPy deposition. The electrochemical tests showed a capacitance of 304 F g^{-1} in 1 M KCl electrolyte, underscoring the effectiveness of MW-assisted synthesis in producing high-performance electrode materials for supercapacitor applications [78].

MW-mediated Cu_xO@PPyNF nanocomposites showed excellent electrochemical properties and high specific capacitance (464.6 F g^{-1}). The electrodes exhibited excellent cycling stability (95.6% retention after 2,000 cycles) at a current density of 1 A g^{-1} [93]. A review of the electrochemical performances of PPy and its composites synthesized *via* various synthesis methods is listed in Table 4.

Despite the potential of MW-assisted methods for synthesizing PPy and its composites remains relatively underexplored in the energy storage domain. This technique allows for better control over the structure of nanocomposites, including their size, shape, and morphology. MW-assisted synthesis needs further investigation could significantly push the boundaries of energy storage performance. Future studies must give more attention to this simple yet powerful approach.

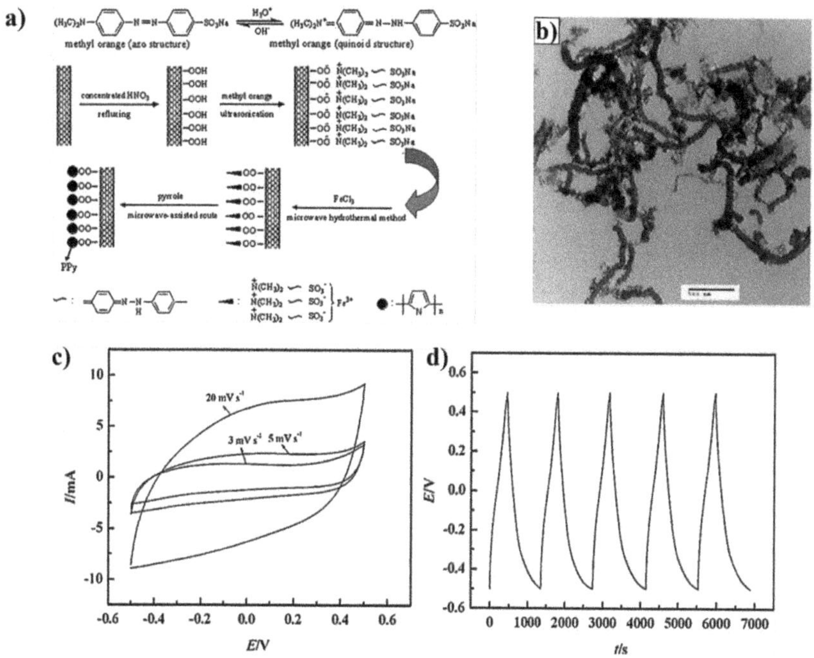

Figure 11.a) Schematic representation of the MO structure and the proposed mechanism for forming CNT/MO-FeCl$_3$ and PPy/CNT composites, b) TEM micrographs of CNT/MO-FeCl$_3$ composites synthesized via the microwave hydrothermal approach, and the electrochemical evaluation of PPy/CNT composites: c) cyclic voltammetry at scan rates of 3, 5, and 20 mV s^{-1}; d) galvanostatic charge-discharge at a current density of 1 mA cm^{-2}. Reproduced from ref. [78], copyright (2010), with permission from Elsevier.

Table 4. A review of the electrochemical performance of PPy and its composites synthesized via different synthesis methods

Synthesis method	Material	Morpholo-gy	Specific Capacitance	Capacitanc-e Retention	Refere-nce
Chemical oxidative polymerization	CNT/PPy	Mesoporo-us framewor-k	425 F g^{-1} at 1 A g^{-1}	94% after 2000 cycles	[92]
Electrochemical Polymerization	PPy/SWCNT	PPy-coated nanotubes	144 F g^{-1} at 200 mV s^{-1}	NA	[94]

Chemical oxidative polymerization	PPy	Nanospheres	176 F g^{-1} at 1 A g^{-1}	NA	[91]
Electrochemical Polymerization	PPy/SWCNT	PPy-coated nanotubes	265 F g^{-1}	NA	[95]
Electrochemical Polymerization	PPy/G	PPy-nanoparticles	237 F g^{-1} 10 mV s^{-1}	NA	[96]
Gravity-assisted Chemical Oxidative Polymerization	GA-hPPy	Films with hollow horn arrays	360 F g^{-1} 10 mV s^{-1}	88.2% after 10,000 cycles at 30 A g^{-1}	[97]
In-situ chemical polymerization	PPy/GO	Nanowire	728 F g^{-1} at 0.5 A g^{-1}	93% after 1000 cycles	[98]
MW assisted	PPyNF/NiO_x	Nanofiber	657 F g^{-1} at 0.5 A g^{-1}	80% after 1000 cycles	[82]
MW assisted	PPy/CNT	Thorns grown nanocomposite	304 F g^{-1} at 1 A g^{-1}	NA	[78]
MW assisted	Cu_xO@PPyNF	Nanofiber	464.6 F g^{-1} at 1 A g^{-1}	95.6% after 2000 cycles	[93]

10.3 Battery

There are no significant reports that have been published yet in the field of batteries regarding MW-assisted PPy synthesis. However, PPy and its composites synthesized *via* other conventional methods stand as potential materials for batteries with excellent energy density and improved charge/discharge cyclability.

PPy has been extensively explored as an electrode material for both primary active material and as a conductive matrix in composite systems, particularly for lithium-ion and sodium-ion batteries. One of the major advantages of PPy lies in its intrinsic conductivity and porous morphology, which enables rapid electron transport and ion diffusion during charge/discharge cycles. Its mechanical flexibility and structural tunability also make it an attractive component in flexible and wearable energy storage devices [99,100].

For instance, using a combination of soft-template polymerization and MW-assisted solvothermal synthesis, a novel p-SnO_2@PPy nanotube (NT) was developed as a rechargeable anode for lithium-ion batteries (LIBs). The resulting electrodes exhibit high specific discharge capacity of ~880 mA h g^{-1} at specific current of 500 mA g^{-1}, ~860 mA h g^{-1} at 1000 mA g^{-1}, and ~770 mA h g^{-1} at 2000 mA g^{-1} with remarkable cycling stability. These improvements are attributed to the hybrid nanocable design, which enhances electrical conductivity, provides a large surface area with abundant ion transport pathways, and accommodates volume fluctuations during cycling. This approach shows significant promise for advancing LIB anode materials toward higher energy and power performance in commercial applications [101].

Lots of works have been published regarding PPy or PPy nanocomposites as superlative cathode material in batteries. The synthesis method plays a vital role in the electrochemical performance of the material. MW-assisted synthesis offers precise control over key parameters such as particle morphology, crystallinity, size and nucleation and growth of the particles. When these aspects are carefully regulated, the resulting PPy nanoparticles can exhibit properties highly favourable for energy storage applications.

10.4 Biomedicine

PPy is widely used in biomedical applications such as drug delivery and photothermal therapy due to its excellent biocompatibility, electrical conductivity, chemical stability, thermal stability, and strong near-infrared (NIR) absorbance [102]. These properties of PPy allowed controlled stimulative drug release and efficient photothermal conversion for targeted cancer or bacterial cells. MW-assisted synthesized PPy is highly suitable for biomedical applications. The method provides rapid, uniform heating that yields ordered PPy nanostructure, which is important for cellular uptake, targeted delivery, and biomedical reproducibility [103]. Additionally, the purity of the product is crucial for in vivo applications, and the MW-assisted method reduces the need for toxic chemicals and improves the purity of the product. Recently, a work demonstrates a rapid MW-assisted route for synthesizing uniform PPy nanoparticles within just two minutes, achieving a synthesis time reduction of over 100-fold compared with conventional methods. The nanoparticles exhibited strong NIR absorption, excellent photothermal stability, low cytotoxicity, and high biocompatibility. Both in vitro and in vivo evaluations confirmed efficient tumor cell destruction under safe laser irradiation, with complete tumor ablation achieved in animal models. These findings highlight MW irradiation as an effective strategy for scalable production of high-performance nanomaterials with great promise for photothermal therapy and other biomedical applications [73]. This comprehensive study suggests that MW-assisted synthesis is poised to soon become a focal point in biomedical research due to its efficiency, simplicity, and potential for developing high-performance nanomaterials.

10.5 Biocidal effect

PPy has demonstrated biocidal (antimicrobial) properties. Its biocidal effects mainly arise because positively charged PPy could interact with negatively charged bacterial cell membranes, causing membrane disruption and cell lysis. It also possesses other effects, like oxidative stress and electron transfer interference with microbial metabolism. PPy is a versatile candidate for antimicrobial applications due to aforementioned properties, combined with its conductivity and biocompatibility, especially useful when enhanced through nanocomposite strategies or functionalization [104]. The MW-assisted technique is a very facile technique to obtain PPy with desired morphology, particle size, and conductivity with much less reaction time. Thus, PPy or its composites synthesized *via* the MW-assisted approach offer a promising route for developing effective biocidal materials [7]. In recent studies, a straightforward one-step microwave-assisted method was used to fabricate PPy-functionalized chitosan and poly(pyrrole-N-(1-naphthyl)ethylenediamine) (COP), with carbon dots serving as the only initiators. These polymers demonstrated notable antimicrobial activity against *E. coli* and *S. aureus*. The observed bactericidal effects are attributed to multiple synergistic mechanisms, including electrostatic disruption of microbial membranes, nutrient chelation, and physical interference with cellular metabolism. Specifically, the polymers hinder nutrient uptake and mass transfer, while chitosan interactions with teichoic acids in Gram-positive bacteria trigger additional bacteriolytic responses. The ease and cost-efficiency of this synthetic method, combined with its strong antimicrobial efficacy, highlight the potential of these materials as promising candidates for next-generation antibacterial agents [80].

10.6 Others

The ultra-fast MW-assisted method for synthesizing PPy and its nanocomposites has shown great potential for applications such as dye-sensitized solar cells (DSSCs) and solid-phase extraction. This technique offers rapid, energy-efficient, and uniform heating, producing PPy nanoparticles with controlled morphology, enhanced electrical conductivity, and improved dispersibility [105]. Platinum nanoparticles supported on PPy nanospheres were successfully synthesized *via* a MW-assisted polyol process and employed as counter electrodes in DSSCs. The hybrid PPy-Pt structure provided high surface area, mesoporosity, and uniform Pt dispersion, leading to improved catalytic activity for the I_3^-/I^- redox reaction. Electrochemical studies confirmed lower charge-transfer resistance and enhanced redox kinetics compared with conventional Pt electrodes. As a result, DSSCs using PPy-Pt counter electrodes achieved a higher PCE of 8.6% compared with 8.1% for standard Pt. These findings highlight the promise of polymer-metal nanostructures for efficient, cost-effective photovoltaic devices [106]. Solid phase extraction (SPE) has recently become the most extensively used sample pretreatment technology. In recent years, SPE technology in combination with fast-developing nanomaterials to achieve efficient and rapid sample pretreatment has become a prime research interest [107]. For instance, an ultrasonic-MW technique was employed to prepare Fe_3O_4@PPy nanocomposites, which is a time-efficient process in comparison to conventional string-heating. The prepared nanocomposite demonstrates excellent magnetic sensitivity and high extraction efficiency for vitamin D_2 and D_3 from milk samples. The detection limits for vitamin D ranged from 0.02 to 0.05 mg mL^{-1}, with both intra and interday variations below 8.4% and recovery rates between 71.9% and 90.3%. The extraction process was completed within 15 minutes, requiring minimal use of organic solvents. Hence, this approach offers a fast, straightforward, and reliable strategy for determining vitamin D in milk matrices [76].

Conclusion

In past years, PPy has emerged as a promising material with broad applicability across sectors such as energy storage, sensing technologies, biomedical applications, and beyond. It continues to garner widespread scientific interest long after its initial discovery, due to its unique combination of electrical, chemical, and mechanical properties. Persistent efforts are made to improve PPy synthesis techniques to gain greater control over its structure, morphology, size, and to enhance its various properties essential for different fields of study. One such innovation is the MW-assisted synthesis, a relatively recent advancement that offers significant benefits over conventional synthesis strategies. This method offers several opportunities for refinement, and it continues to achieve remarkable feats such as increased reaction rates, improved reaction control, the production of high-purity products, etc. Despite its potential, the application of MW-assisted synthesis in PPy research has been minimal so far. Hence, a vast field of research lies ahead regarding the implementation of MW-assisted synthesis in PPy production. Especially considering its advantages over other synthesis methods, this method may provide us with high-quality PPy for the preparation of energy storage devices like supercapacitors, batteries with higher capacitance, capacity, energy density, and power density; sensors for various gases, chemicals, biomolecules, ions, etc.; different types of medicinal uses; or various unexplored areas of application. This chapter provides an overview of the fundamental synthesis techniques for PPy, with a particular emphasis on microwave-assisted methods. A comparative analysis between MW-assisted and conventional synthesis routes is presented, followed by a discussion of current and prospective applications of PPy in various domains.

References

[1] D. Feldman, Polymer history, Des. Monomers Polym. 11 (2008) 1-15. https://doi.org/10.1163/156855508X292383

[2] T. V. Vernitskaya, O.N. Efimov, Polypyrrole: A conducting polymer (synthesis, properties, and applications), Russ. Chem. Rev. 66 (1997) 443-457. https://doi.org/10.1070/RC1997v066n05ABEH000261

[3] Y. Sood, K. Singh, H. Mudila, P.E. Lokhande, L. Singh, D. Kumar, A. Kumar, N.M. Mubarak, M.H. Dehghani, Insights into properties, synthesis and emerging applications of polypyrrole-based composites, and future prospective: A review, Heliyon 10 (2024) e33643. https://doi.org/10.1016/j.heliyon.2024.e33643

[4] E. Tavakkol, H. Tavanai, A. Abdolmaleki, M. Morshed, Production of conductive electrospun polypyrrole/poly(vinyl pyrrolidone) nanofibers, Synth. Met. 231 (2017) 95-106. https://doi.org/10.1016/j.synthmet.2017.06.017

[5] Z. Chen, X. Zhao, R. Lu, R. Hong, X. Yang, Electrodeposition of polypyrrole as binder-free and high mass-loading electrodes for flexible supercapacitors, Synth. Met. 296 (2023) 117378. https://doi.org/10.1016/j.synthmet.2023.117378

[6] H. Albaris, G. Karuppasamy, Inspection of room temperature hydrogen sensing property of nanostructured polypyrrole/polyaniline hetero-junctions synthesized by one-pot interfacial polymerization, Mater. Chem. Phys. 250 (2020) 123153. https://doi.org/10.1016/j.matchemphys.2020.123153

[7] A. Jabbar Khan, L. Gao, A. Numan, S. Khan, I. Hussain, M. Sajjad, S.S. Shah, A. Mateen, G. Zhao, Recent advancements in the tailoring of nanomaterials via microwave-assisted synthesis: A comprehensive review, Crit. Rev. Solid State Mater. Sci. 50 (2025) 1-24. https://doi.org/10.1080/10408436.2024.2446049

[8] J. Stejskal, M. Trchová, Conducting polypyrrole nanotubes: A review, Chem. Pap. 72 (2018) 1563-1595. https://doi.org/10.1007/s11696-018-0394-x

[9] A. Ramanavičius, A. Ramanavičiene, A. Malinauskas, Electrochemical sensors based on conducting polymer-polypyrrole, Electrochim. Acta 51 (2006) 6025-6037. https://doi.org/10.1016/j.electacta.2005.11.052

[10] L. Yuan, B. Yao, B. Hu, K. Huo, W. Chen, J. Zhou, Polypyrrole-coated paper for flexible solid-state energy storage, Energy Environ. Sci. 6 (2013) 470-476. https://doi.org/10.1039/c2ee23977a

[11] K. Kontturi, P. Pentti, G. Sundholm, Polypyrrole as a model membrane for drug delivery, J. Electroanal. Chem. 453 (1998) 231-238. https://doi.org/10.1016/S0022-0728(98)00246-0

[12] A. Nautiyal, M. Qiao, J.E. Cook, X. Zhang, T.S. Huang, High performance polypyrrole coating for corrosion protection and biocidal applications, Appl. Surf. Sci. 427 (2018) 922-930. https://doi.org/10.1016/j.apsusc.2017.08.093

[13] R.J. Giguere, T.L. Bray, S.M. Duncan, G. Majctich, Application of commercial microwave ovens to organic synthesis., Tetrahedron Lett. 27 (1986) 4945-4948. https://doi.org/10.1016/S0040-4039(00)85103-5

[14] M.M. Shams, B. Zamiri, H. Fatoorehchi, Microwave-assisted synthesis of fuel cell catalyst, in: Proc. 14th Int. Electron. Conf. Synth. Org. Chem., MDPI, 2010. https://doi.org/10.3390/ecsoc-14-00406

[15] M. Nüchter, B. Ondruschka, W. Bonrath, A. Gum, Microwave-assisted synthesis - A critical technology overview, Green Chem. 6 (2004) 128-141. https://doi.org/10.1039/B310502D

[16] A. Mahun, S. Abbrent, P. Bober, J. Brus, L. Kobera, Effect of structural features of polypyrrole (PPy) on electrical conductivity reflected on 13C ssNMR parameters, Synth. Met. 259 (2020) 116250. https://doi.org/10.1016/j.synthmet.2019.116250

[17] M. In, A. Ringopening, "Synthetic metals" : A novel role for organic polymers, Makromol. Chem., Macromol. Symp. 51 (1991) 11-28. https://doi.org/10.1002/masy.19910510104

[18] T.H. Le, Y. Kim, H. Yoon, Electrical and electrochemical properties of conducting polymers, Polymers (Basel). 9 (2017) 150. https://doi.org/10.3390/polym9040150

[19] M.E.A. Mohsin, M. Elias, A. Arsad, K.C. Yong, O. Alothman, Z.Y. Zakaria, Electrical conductivity, thermal, rheological and morphological characteristics of grafted blend of polypyrrole and polypropylene, Chem. Eng. Trans. 56 (2017) 1375-1380.

[20] M. Mahmoodian, B. Pourabbas, S. Mohajerzadeh, Effect of anionic dopants on thickness, morphology and electrical properties of polypyrrole ultra-thin films prepared by in situ chemical polymerization, Thin Solid Films 583 (2015) 255-263. https://doi.org/10.1016/j.tsf.2015.03.043

[21] A.A. Jatratkar, J.B. Yadav, S. V. Kamat, V.S. Patil, D.B. Mahadik, H.C. Barshilia, V. Puri, R.K. Puri, Consequence of oxidant to monomer ratio on optical and structural properties of polypyrrole thin film deposited by oxidation polymerization technique, J. Phys. Chem. Solids 80 (2015) 78-83. https://doi.org/10.1016/j.jpcs.2015.01.004

[22] Y. Tan, K. Ghandi, Kinetics and mechanism of pyrrole chemical polymerization, Synth. Met. 175 (2013) 183-191. https://doi.org/10.1016/j.synthmet.2013.05.014

[23] M. Li, W. Li, J. Liu, J. Yao, Preparation and characterization of PPy doped with different anionic surfactants, Polym. Eng. Sci. 53 (2013) 2465-2469. https://doi.org/10.1002/pen.23538

[24] I.M. Minisy, P. Bober, I. Šeděnková, J. Stejskal, Methyl red dye in the tuning of polypyrrole conductivity, Polymer (Guildf). 207 (2020) 122854. https://doi.org/10.1016/j.polymer.2020.122854

[25] S.T. Navale, A.T. Mane, A.A. Ghanwat, A.R. Mulik, V.B. Patil, Camphor sulfonic acid (CSA) doped polypyrrole (PPy) films : Measurement of microstructural and optoelectronic properties, Measurement 50 (2014) 363-369. https://doi.org/10.1016/j.measurement.2014.01.012

[26] Y. Song, J.B. Fan, S. Wang, Recent progress in interfacial polymerization, Mater. Chem. Front. 1 (2017) 1028-1040. https://doi.org/10.1039/C6QM00325G

[27] Y. Liu, F. Wu, Synthesis and application of polypyrrole nanofibers: A review, Nanoscale Adv. 5 (2023) 3606-3618. https://doi.org/10.1039/D3NA00138E

[28] Q. Pei, R. Qian, Protonation and deprotonation of polypyrrole chain in aqueous solutions, Synth. Met. 45 (1991) 35-48. https://doi.org/10.1016/0379-6779(91)91845-2

[29] D.K. Beaman, E.J. Robertson, G.L. Richmond, Ordered polyelectrolyte assembly at the oil-water interface, Proc. Natl. Acad. Sci. 109 (2012) 3226-3231. https://doi.org/10.1073/pnas.1200244109

[30] L. Hao, C. Dong, L. Zhang, K. Zhu, D. Yu, Polypyrrole nanomaterials: Structure, preparation and application, Polymers (Basel). 14 (2022) 5139. https://doi.org/10.3390/polym14235139

[31] J. Hazarika, A. Kumar, Scalable and low cost synthesis of highly conducting polypyrrole nanofibers using oil-water interfacial polymerization under constant stirring, J. Phys. Chem. B 121 (2017) 6926-6933. https://doi.org/10.1021/acs.jpcb.7b03179

[32] S.J. Hawkins, N.M. Ratcliffe, A study of the effects of acid on the polymerization of pyrrole, on the oxidative polymerization of pyrrole and on polypyrrole, J. Mater. Chem. 10 (2000) 2057-2062. https://doi.org/10.1039/b001912g

[33] A.L. Pang, A. Arsad, M. Ahmadipour, Synthesis and factor affecting on the conductivity of polypyrrole: A short review, Polym. Adv. Technol. 32 (2021) 1428-1454. https://doi.org/10.1002/pat.5201

[34] M.M. Gvozdenović, B.Z. Jugović, J.S. Stevanović, B.N. Grgur, Elektrohemijska sinteza elektroprovodnih polimera, Hem. Ind. 68 (2014) 673-684. https://doi.org/10.2298/HEMIND131122008G

[35] M. Wysocka-Żołopa, K. Winkler, Electrochemical synthesis and properties of conical polypyrrole structures, Electrochim. Acta 258 (2017) 1421-1434. https://doi.org/10.1016/j.electacta.2017.12.005

[36] J.A. Raj, R. Sasikumar, M. Sethupathy, C. Vedhi, P. Manisankar, Electrochemical synthesis of nanosize polypyrrole in presence of aqueous surfactant solutions, Adv. Mater. Res. 678 (2013) 244-247. https://doi.org/10.4028/www.scientific.net/AMR.678.244

[37] R.B. Choudhary, S. Ansari, B. Purty, Robust electrochemical performance of polypyrrole (PPy) and polyindole (PIn) based hybrid electrode materials for supercapacitor application: A review, J. Energy Storage 29 (2020) 101302. https://doi.org/10.1016/j.est.2020.101302

[38] Y. Huang, H. Li, Z. Wang, M. Zhu, Z. Pei, Q. Xue, Y. Huang, C. Zhi, Nanostructured polypyrrole as a flexible electrode material of supercapacitor, Nano Energy 22 (2016) 422-438. https://doi.org/10.1016/j.nanoen.2016.02.047

[39] A. Frenot, I.S. Chronakis, Polymer nanofibers assembled by electrospinning, Curr. Opin. Colloid Interface Sci. 8 (2003) 64-75. https://doi.org/10.1016/S1359-0294(03)00004-9

[40] T. Blachowicz, A. Ehrmann, Conductive electrospun nanofiber mats, Materials (Basel). 13 (2019) 152. https://doi.org/10.3390/ma13010152

[41] Y. Cong, S. Liu, H. Chen, Fabrication of conductive polypyrrole nanofibers by electrospinning, J. Nanomater. 2013 (2013) 148347. https://doi.org/10.1155/2013/148347

[42] I. Masmur, S. Perangin-angin, H. Sembiring, A.H. Tarigan, J. Ginting, D.A. Barus, M. Ginting, Effect of different compositions of polyethylene oxide-polypyrrole (PEO-PPy) nanofiber mats on antibacterial and biocompatibility properties, ChemistrySelect 7 (2022) e202201346. https://doi.org/10.1002/slct.202201346

[43] I.S. Chronakis, S. Grapenson, A. Jakob, Conductive polypyrrole nanofibers via electrospinning: Electrical and morphological properties, Polymer (Guildf). 47 (2006) 1597-1603. https://doi.org/10.1016/j.polymer.2006.01.032

[44] A. Kumar, Y. Kuang, Z. Liang, X. Sun, Microwave chemistry, recent advancements, and eco-friendly microwave-assisted synthesis of nanoarchitectures and their applications: A review, Mater. Today Nano 11 (2020) 100076. https://doi.org/10.1016/j.mtnano.2020.100076

[45] F. Wiesbrock, R. Hoogenboom, U.S. Schubert, Microwave-assisted polymer synthesis: State-of-the-art and future perspectives, Macromol. Rapid Commun. 25 (2004) 1739-1764. https://doi.org/10.1002/marc.200400313

[46] J. Arun, S. Nachiappan, G. Rangarajan, R.P. Alagappan, K.P. Gopinath, E. Lichtfouse, Synthesis and application of titanium dioxide photocatalysis for energy, decontamination and viral disinfection: A review, Environ. Chem. Lett. 21 (2023) 339-362. https://doi.org/10.1007/s10311-022-01503-z

[47] J. Sun, W. Wang, Q. Yue, Review on microwave-matter interaction fundamentals and efficient microwave-associated heating strategies, Materials. 9 (2016) 231. https://doi.org/10.3390/ma9040231

[48] C. Gabriel, S. Gabriel, E.H. Grant, B.S.J. Halstead, D. M. P. Mingos, Dielectric parameters relevant to microwave dielectric heating, Chem. Soc. Rev. 27 (1998) 213-223. https://doi.org/10.1039/a827213z

[49] O. Gerard, A. Numan, S. Krishnan, M. Khalid, R. Subramaniam, R. Kasi, A review on the recent advances in binder-free electrodes for electrochemical energy storage application, J. Energy Storage 50 (2022) 104283. https://doi.org/10.1016/j.est.2022.104283

[50] R. Hoogenboom, U.S. Schubert, Microwave-assisted polymer synthesis: Recent developments in a rapidly expanding field of research, Macromol. Rapid Commun. 28 (2007) 368-386. https://doi.org/10.1002/marc.200600749

[51] L. Zong, S. Zhou, N. Sgriccia, M.C. Hawley, L.C. Kempel, A review of microwave-assisted polymer chemistry (MAPC), J. Microw. Power Electromagn. Energy 38 (2003) 49-74. https://doi.org/10.1080/08327823.2003.11688487

[52] B.L. Hayes, Recent advances in microwave-assisted synthesis, Aldrichimica Acta 37 (2004) 66-76.

[53] P.G. Jessop, S. Trakhtenberg, J. Warner, The twelve principles of green chemistry, in: W. H. Flank, M.A. Abraham, M.A. Matthews (Eds.), Innovations in Industrial and Engineering Chemistry: A Century of Achievements and Prospects for the New Millennium, American Chemical Society, Washington, DC, 2009, pp. 401-436. https://doi.org/10.1021/bk-2009-1000.ch012

[54] M.R. Rosana, Y. Tao, A.E. Stiegman, G.B. Dudley, On the rational design of microwave-actuated organic reactions, Chem. Sci. 3 (2012) 1240-1244. https://doi.org/10.1039/c2sc01003h

[55] V.L. Budarin, P.S. Shuttleworth, M. De Bruyn, T.J. Farmer, M.J. Gronnow, L. Pfaltzgraff, D.J. Macquarrie, J.H. Clark, The potential of microwave technology for the recovery, synthesis and manufacturing of chemicals from bio-wastes, Catal. Today 239 (2015) 80-89. https://doi.org/10.1016/j.cattod.2013.11.058

[56] A.K. Rathi, M.B. Gawande, R. Zboril, R.S. Varma, Microwave-assisted synthesis - Catalytic applications in aqueous media, Coord. Chem. Rev. 291 (2015) 68-94. https://doi.org/10.1016/j.ccr.2015.01.011

[57] K.E. Haque, Microwave energy for mineral treatment processes - A brief review, Int. J. Miner. Process. 57 (1999) 1-24. https://doi.org/10.1016/S0301-7516(99)00009-5

[58] J.M. Kremsner, A. Stadler, C.O. Kappe, The scale-up of microwave-assisted organic synthesis, in: M. Larhed, K. Olofsson (Eds.), Microwave Methods in Organic Synthesis (Topics in Current Chemistry), Springer, Berlin Heidelberg, 2006, pp. 233-278. https://doi.org/10.1007/128_048

[59] J. Robinson, S. Kingman, D. Irvine, P. Licence, A. Smith, G. Dimitrakis, D. Obermayer, C.O. Kappe, Understanding microwave heating effects in single mode type cavities - Theory and experiment, Phys. Chem. Chem. Phys. 12 (2010) 4750-4758. https://doi.org/10.1039/b922797k

[60] V. Polshettiwar, R.S. Varma, Microwave-assisted organic synthesis and transformations using benign reaction media, Acc. Chem. Res. 41 (2008) 629-639. https://doi.org/10.1021/ar700238s

[61] A.A. Caouthar, A. Loupy, M. Bortolussi, J.C. Blais, L. Dubreucq, A. Meddour, Synthesis and characterization of new polyamides based on diphenylaminoisosorbide, J. Polym. Sci. Part A Polym. Chem. 43 (2005) 6480-6491. https://doi.org/10.1002/pola.21116

[62] K. Faghihi, M. Hagibeygi, New polyamides containing azobenzene unites and hydantoin derivatives in main chain: Synthesis and characterization, Eur. Polym. J. 39 (2003) 2307-2314. https://doi.org/10.1016/S0014-3057(03)00178-2

[63] Z.J. Yu, L.J. Liu, Effect of microwave energy on chain propagation of poly(ε- caprolactone) in benzoic acid-initiated ring opening polymerization of ε-caprolactone, Eur. Polym. J. 40 (2004) 2213-2220. https://doi.org/10.1016/j.eurpolymj.2004.05.007

[64] L. Liu, Y. Li, Y.E. Fang, L. Chen, Microwave-assisted graft copolymerization of ε-caprolactone onto chitosan via the phthaloyl protection method, Carbohydr. Polym. 60 (2005) 351-356. https://doi.org/10.1016/j.carbpol.2005.01.009

[65] S. Sinnwell, H. Ritter, Microwave accelerated polymerization of 2-phenyl-2-oxazoline, Macromol. Rapid Commun. 26 (2005) 160-163. https://doi.org/10.1002/marc.200400477

[66] F. Wiesbrock, R. Hoogenboom, M.A.M. Leenen, M.A.R. Meier, U.S. Schubert, Investigation of the living cationic ring-opening polymerization of 2-methyl-, 2-ethyl-, 2-nonyl-, and 2-phenyl-2-oxazoline in a single-mode microwave reactor, Macromolecules 38 (2005) 5025-5034. https://doi.org/10.1021/ma0474170

[67] V. Singh, A. Tiwari, D.N. Tripathi, R. Sanghi, Microwave assisted synthesis of guar-g-polyacrylamide, Carbohydr. Polym. 58 (2004) 1-6. https://doi.org/10.1016/j.carbpol.2004.04.010

[68] C.M. Fellows, Preliminary observations on the copolymerization of acceptor monomer: Donor monomer systems under microwave irradiation, Cent. Eur. J. Chem. 3 (2005) 40-52. https://doi.org/10.2478/BF02476236

[69] J. Bao, A. Zhang, Poly(methyl methacrylate) nanoparticles prepared through microwave emulsion polymerization, J. Appl. Polym. Sci. 93 (2004) 2815-2820. https://doi.org/10.1002/app.20758

[70] M.A. Aldana-García, J. Palacios, E. Vivaldo-Lima, Modeling of the microwave initiated emulsion polymerization of styrene, J. Macromol. Sci. - Pure Appl. Chem. 42 (2005) 1207-1225. https://doi.org/10.1080/10601320500189505

[71] S.L. Brown, C.M. Rayner, S. Perrier, Microwave-accelerated RAFT polymerization of polar monomers, Macromol. Rapid Commun. 28 (2007) 478-483. https://doi.org/10.1002/marc.200600755

[72] M. Leenen, F. Wiesbrock, R. Hoogenboom, U.S. Schubert, Microwave-assisted nitroxide-mediated polymerization of alkyl acrylates, E-Polymers 2005 (2005) 1-9. https://doi.org/10.1515/epoly.2005.5.1.741

[73] C. Zhang, H. Pan, X. Wang, S.K. Sun, Microwave-assisted ultrafast fabrication of high-performance polypyrrole nanoparticles for photothermal therapy of tumors: In vivo, Biomater. Sci. 6 (2018) 2750-2756. https://doi.org/10.1039/C8BM00653A

[74] A.A. Yadav, S.B. Kulkarni, C.D. Lokhande, Synthesis and characterization of polypyrrole thin film by MW-CBD method for NH3 gas sensor, Polym. Bull. 75 (2018) 4547-4553. https://doi.org/10.1007/s00289-018-2282-5

[75] D.B. Dupare, M.D. Shirsat, A.S. Aswar, Metal oxides doped PPY-PVA blend thin films based gas sensor, Sensors Transducers J. 101 (2009) 82-89.

[76] Z. Jiao, Y. Zhang, H. Fan, Ultrasonic-microwave method in preparation of polypyrrole-coated magnetic particles for vitamin D extraction in milk, J. Chromatogr. A 1457 (2016) 7-13. https://doi.org/10.1016/j.chroma.2016.06.041

[77] Y. Qi, Y. Cao, X. Meng, K. Yu, W. Si, W. Lei, Q. Hao, J. Li, F. Wang, Microwave-assisted synthesis of a polypyrrole/graphene composite using a pyrrole-induced graphene oxide hydrogel for the selective determination of dihydroxybenzenes, ChemistrySelect 3 (2018) 7713-7717. https://doi.org/10.1002/slct.201801306

[78] H. Mi, X. Zhang, Y. Xu, F. Xiao, Synthesis, characterization and electrochemical behavior of polypyrrole/carbon nanotube composites using organometallic-functionalized carbon nanotubes, Appl. Surf. Sci. 256 (2010) 2284-2288. https://doi.org/10.1016/j.apsusc.2009.10.053

[79] K.H. Kate, K. Singh, P.K. Khanna, Microwave formation of polypyrrole/Ag nano-composite based on interfacial polymerization by use of AgNO3, Synth. React. Inorganic, Met. Nano-Metal Chem. 41 (2011) 199-202. https://doi.org/10.1080/15533174.2010.538033

[80] M. Maruthapandi, K. Sharma, J.H.T. Luong, A. Gedanken, Antibacterial activities of microwave-assisted synthesized polypyrrole/chitosan and poly (pyrrole-N-(1-naphthyl) ethylenediamine) stimulated by C-dots, Carbohydr. Polym. 243 (2020) 116474. https://doi.org/10.1016/j.carbpol.2020.116474

[81] A. Belmokhtar, N. Sahli, A. Yahiaoui, M. Belbachir, Polycondensation of pyrrole and benzaldehyde catalyzed by Maghnite-H+, Express Polym. Lett. 1 (2007) 443-449. https://doi.org/10.3144/expresspolymlett.2007.62

[82] H. Liu, Q. Zhao, K. Wang, Z. Lu, F. Feng, Y. Guo, Facile synthesis of polypyrrole nanofiber (PPyNF)/NiO: X composites by a microwave method and application in supercapacitors, RSC Adv. 9 (2019) 6890-6897. https://doi.org/10.1039/C8RA09666J

[83] L. Xing, Q. Rong, Z. Ma, Non-enzymatic electrochemical sensing of hydrogen peroxide based on polypyrrole/platinum nanocomposites, Sensors Actuators, B Chem. 221 (2015) 242-247. https://doi.org/10.1016/j.snb.2015.06.078

[84] M. Josowicz, J. Janata, Electroactive polymers in chemical sensors, in: B. Scrosati (Ed.), Applications of Electroactive Polymers, Springer, Dordrecht, 1993, pp. 310-343. https://doi.org/10.1007/978-94-011-1568-1_10

[85] G.A. Snook, P. Kao, A.S. Best, Conducting-polymer-based supercapacitor devices and electrodes, J. Power Sources 196 (2011) 1-12. https://doi.org/10.1016/j.jpowsour.2010.06.084

[86] M. Yeganeh, M. Saremi, H. Rezaeyan, Corrosion inhibition of steel using mesoporous silica nanocontainers incorporated in the polypyrrole, Prog. Org. Coatings 77 (2014) 1428-1435. https://doi.org/10.1016/j.porgcoat.2014.05.007

[87] T.W. Lee, Investigation on the low luminous efficiency in a polymer light-emitting diode with a high work-function cathode by soft contact lamination, Adv. Funct. Mater. 17 (2007) 3128-3133. https://doi.org/10.1002/adfm.200700490

[88] C. Wang, W. Zheng, Z. Yue, C.O. Too, G.G. Wallace, Buckled, Stretchable polypyrrole electrodes for battery applications, Adv. Mater. 23 (2011) 3580-3584. https://doi.org/10.1002/adma.201101067

[89] R. Darabi, H. Karimi-Maleh, M. Akin, K. Arikan, Z. Zhang, R. Bayat, M. Bekmezci, F. Sen, Simultaneous determination of ascorbic acid, dopamine, and uric acid with a highly selective and sensitive reduced graphene oxide/polypyrrole-platinum nanocomposite modified electrochemical sensor, Electrochim. Acta 457 (2023) 142402. https://doi.org/10.1016/j.electacta.2023.142402

[90] W. Sun, X. Chen, Preparation and characterization of polypyrrole films for three-dimensional micro supercapacitor, J. Power Sources 193 (2009) 924-929. https://doi.org/10.1016/j.jpowsour.2009.04.063

[91] C. Shen, Y. Sun, W. Yao, Y. Lu, Facile synthesis of polypyrrole nanospheres and their carbonized products for potential application in high-performance supercapacitors, Polymer (Guildf). 55 (2014) 2817-2824. https://doi.org/10.1016/j.polymer.2014.04.042

[92] R. Oraon, A. De Adhikari, S.K. Tiwari, G.C. Nayak, Nanoclay-based hierarchical interconnected mesoporous CNT/PPy electrode with improved specific capacitance for high

performance supercapacitors, Dalt. Trans. 45 (2016) 9113-9126. https://doi.org/10.1039/C6DT00600K

[93] H. Liu, Z. Lu, J. Qin, K. Wang, F. Feng, Y. Guo, Microwave rapid synthesis of CuxO@polypyrrole nanofibre (PpyNF) composites for supercapacitors, Fullerenes Nanotub. Carbon Nanostructures 27 (2019) 947-952. https://doi.org/10.1080/1536383X.2019.1666366

[94] J. Wang, Y. Xu, X. Chen, X. Sun, Capacitance properties of single wall carbon nanotube/polypyrrole composite films, Compos. Sci. Technol. 67 (2007) 2981-2985. https://doi.org/10.1016/j.compscitech.2007.05.015

[95] K.H. An, K.K. Jeon, J.K. Heo, S.C. Lim, D.J. Bae, Y.H. Lee, High-capacitance supercapacitor using a nanocomposite electrode of single-walled carbon nanotube and polypyrrole, J. Electrochem. Soc. 149 (2002) A1058-A1062. https://doi.org/10.1149/1.1491235

[96] A. Davies, P. Audette, B. Farrow, F. Hassan, Z. Chen, J.Y. Choi, A. Yu, Graphene-based flexible supercapacitors: Pulse-electropolymerization of polypyrrole on free-standing graphene films, J. Phys. Chem. C 115 (2011) 17612-17620. https://doi.org/10.1021/jp205568v

[97] B. Zhang, P. Zhou, Y. Xu, J. Lin, H. Li, Y. Bai, J. Zhu, S. Mao, J. Wang, Gravity-assisted synthesis of micro/nano-structured polypyrrole for supercapacitors, Chem. Eng. J. 330 (2017) 1060-1067. https://doi.org/10.1016/j.cej.2017.07.183

[98] J. Li, H. Xie, Y. Li, Fabrication of graphene oxide/polypyrrole nanowire composite for high performance supercapacitor electrodes, J. Power Sources 241 (2013) 388-395. https://doi.org/10.1016/j.jpowsour.2013.04.144

[99] L. Cui, J. Shen, F. Cheng, Z. Tao, J. Chen, SnO2 nanoparticles@polypyrrole nanowires composite as anode materials for rechargeable lithium-ion batteries, J. Power Sources 196 (2011) 2195-2201. https://doi.org/10.1016/j.jpowsour.2010.09.075

[100] Y. Zhao, J. Li, N. Wang, C. Wu, G. Dong, L. Guan, Fully reversible conversion between SnO2 and Sn in SWNTs@SnO2@PPy coaxial nanocable as high performance anode material for lithium ion batteries, J. Phys. Chem. C 116 (2012) 18612-18617. https://doi.org/10.1021/jp304095y

[101] X. Du, T. Yang, J. Lin, T. Feng, J. Zhu, L. Lu, Y. Xu, J. Wang, Microwave-assisted synthesis of SnO2@polypyrrole nanotubes and their pyrolyzed composite as anode for lithium-ion batteries, ACS Appl. Mater. Interfaces 8 (2016) 15598-15606. https://doi.org/10.1021/acsami.6b03332

[102] J. Pokki, O. Ergeneman, K.M. Sivaraman, B. Özkale, M.A. Zeeshan, T. Lühmann, B.J. Nelson, S. Pané, Electroplated porous polypyrrole nanostructures patterned by colloidal lithography for drug-delivery applications, Nanoscale 4 (2012) 3083-3088. https://doi.org/10.1039/c2nr30192j

[103] S.H. Wu, H.T. Lu, H.T. Nguyen, P.Y. Lin, W.Y. Pan, A.E.Y. Chuang, Microwave-orchestrated phototherapeutic polypyrrole/soy extraction micro-complex for burn wound healing in a rodent model, J. Drug Deliv. Sci. Technol. 108 (2025) 106911. https://doi.org/10.1016/j.jddst.2025.106911

[104] F.A.G. da Silva Júnior, S.A. Vieira, S. de Avila Botton, M.M. da Costa, H.P. de Oliveira, Antibacterial activity of polypyrrole-based nanocomposites: A mini-review, Polimeros 30 (2020) 1-9. https://doi.org/10.1590/0104-1428.08020

[105] K. Kempe, C.R. Becer, U.S. Schubert, Microwave-assisted polymerizations: Recent status and future perspectives, Macromolecules 44 (2011) 5825-5842. https://doi.org/10.1021/ma2004794

[106] S.S. Jeon, C. Kim, J. Ko, S.S. Im, Pt nanoparticles supported on polypyrrole nanospheres as a catalytic counter electrode for dye-sensitized solar cells, J. Phys. Chem. C 115 (2011) 22035-22039. https://doi.org/10.1021/jp206535c

[107] M.E.I. Badawy, M.A.M. El-Nouby, P.K. Kimani, L.W. Lim, E.I. Rabea, A review of the modern principles and applications of solid-phase extraction techniques in chromatographic analysis, Anal. Sci. 38 (2022) 1457-1487. https://doi.org/10.1007/s44211-022-00190-8

Materials Research Foundations 189 (2026) 134-154 https://doi.org/10.21741/9781644904039-5

Chapter 5

Microwave-Assisted Synthesis of Spiro Heterocycles for Light Emitting Diodes

Priya, Rajni, Riyansha Verma, Sneh Lata[1*] and Anjaneyulu Bendi[2*]

[1]Central Synthetic Chemistry Lab, Department of Chemistry, Faculty of Applied and Basic Sciences, SGT University, Gurugram-122505, Haryana, India

[2]Department of Chemistry, Kalinga University, Kotni, Near Mantralaya, New Raipur-492 101, Chhattisgarh, India

snehlata_fosc@sgtuniversity.org, anjaneyulu@presidencyuniversity.in

Abstract

Spirocyclic molecules are defined as compounds with two rings connected via common atom. As they have structural complexity, distinct conformational characteristics and rigidity, spiro compounds promise to be intriguing skeletal systems in drug discovery. Heterocyclic compounds have incredible significance due to their diverse applications and biological activity. Microwave chemistry involves the application of microwave irradiation to chemical reactions which plays an important role in the synthesis of spiro heterocycles as well as other organic compounds. This chapter aims at highlighting the different approaches for the synthesis of spiro-heterocyclic compounds for pharmacological applications such as anticancer, antimicrobial and anti-inflammatory activity via microwave assisted techniques and in the development of optoelectronic devices.

Keywords

Microwave-Assisted Organic Synthesis, Spiro Heterocycles, Organic Light Emitting Diodes, Spiro-Oxindoles, Multicomponent

Contents

Materials Research Foundations 189 (2026) 134-154 https://doi.org/10.21741/9781644904039-5

1. Introduction

In organic chemistry, heterocyclic chemistry has become a key pillar, providing excellent tools for creating novel organic compounds that could potentially be used in materials chemistry. Heterocyclic moieties are useful building blocks for constructing a wide range of chemical structures [1-3]. Formerly referred to as bicyclic organic compounds, spiro compounds are made up of rings joined by a single atom and have several special properties, including 3D structural qualities associated with their intrinsic rigidity [4-7]. The material sciences, synthetic and medicinal chemistry, and other disciplines have shown a interest in the large family of fundamental structural motifs known as natural and synthetic heterocycles. Furthermore, spiro carbon has drawn special attention in medical chemistry due to its ability to give structural stiffness. Due to their considerable structural inflexibility and complexity, spiro compounds typically feature cyclic structures with one mutually sp^3 hybridized carbon atom encircled by two rings and intriguing synthetic encounters. The nitrogen, sulphur, and oxygen atoms that make up spiro heterocyclic compounds have distinctive characteristics of several natural and synthetic products that play important roles in biological processes and have demonstrated considerable pharmacological activity [8-12]. Applications of spiro heterocyclic compounds include the production of light-emitting diodes (OLEDs) as well as anti-inflammatory, antitumoral, anti-proliferative, anti-cancer, antibacterial, and anti-glaucoma agents.

The synthesis of spiro heterocycles can be done in a number of ways, but an effective and environmentally friendly method needs to be developed. A wide range of complex molecules have been synthesized using microwave-assisted MCRs, which have become a potent tool for the synthesis of heterocycles while offering a number of benefits over conventional synthetic techniques [13-16]. In microwave-assisted synthesis, microwave radiation is utilized to heat and accelerate chemical reactions. This causes molecules to vibrate or rotate since they are absorbing electromagnetic energy. Among the many benefits it offers are quicker reaction times, increased yields, and enhanced selectivity, all of which contribute to more effective procedures [17-21].

Despite the vast range of biological activities, the spiro-heterocyclic framework has been identified as a promising segment in several materials science fields, including organic light-emitting diodes (OLEDs). These diodes have sparked a great deal of research globally to create effective organic electroluminescent materials with a long lifespan and high efficiency. They are less expensive and lighter and can be treated using lithographic, inkjet, or simple casting techniques from a solution or molten state to create thin films or intricate patterns. Such versatility creates new possibilities for flexible electronics. Their consistent luminosity, low DC voltage operation, and the ability to adjust the emission colour through structural design are all advantages.

In this chapter, we outline the synthesis of spiro heterocyclic compounds for light-emitting diodes via microwave-assisted techniques [22-25]. We additionally go over the properties and applications of these compounds.

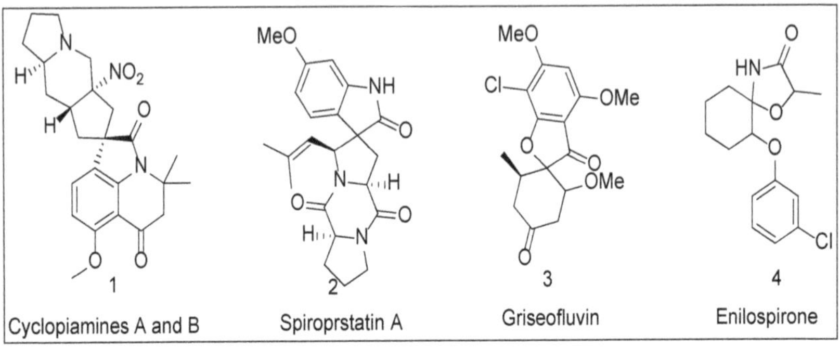

***Figure 1:** Some Naturally occurring Spiro heterocycles*

2. Chemistry of spiro heterocycles synthesized via microwave irradiations

Spiro heterocycles synthesised under microwave irradiations have emerged as promising candidates in the development of variety of scaffolds exhibiting pharmacological activities including anticancer, antimicrobial, anti-inflammatory and antiviral properties owing to their unique three-dimensional structures, excellent thermal stability, and higher yields. Beside serving as an important scaffold in the medicinal chemistry these compounds also demonstrate significant potential in development of organic light-emitting diodes (OLED) devices by reducing intermolecular interactions, enhancing charge transport properties and minimizing aggregation-caused quenching leading to OLED devices possessing superior luminescence efficiency, colour purity and improved device durability. Various synthetic approaches have been employed for the preparation of spiro heterocycles, including conventional thermal synthesis, metal-catalysed coupling reactions, multicomponent reactions, and photochemical methods. However, in this book chapter, we will primarily focus on the synthesis of spiro heterocycles via conventional and microwave (MW) irradiation method, which offers a rapid, energy-efficient, and environmentally friendly route for synthesizing spiro heterocycles with high yields and purity.

Bettina Rávai et al. reported a new, simple, and efficient microwave -assisted multicomponent synthetic method for the preparation of Spirooxindole dihydropyridine bisphosphonates (7) -a new class of compounds with potential pharmaceutical relevance. This one-pot reaction involves isatins (5), β-ketophosphonates (6), and primary amines, proceeding under catalyst and solvent-free conditions. The reaction parameters, including molar ratios, temperature, heating mode, and time, were carefully optimized to achieve moderate to good yields (20–70%) at 120 °C within 2 hours. The method showed broad substrate scope and was extended to various starting materials. Mechanistic investigations were supported by DFT calculations which suggested the formation of an iminophosphonate intermediate as a key step toward the desired spiro derivatives. The intermediate was successfully isolated and characterized. Altogether, 19 new spirooxindole dihydropyridine bisphosphonates were synthesized and structurally characterized by LC-MS measurements and ^{13}P NMR. This method offers a practical and sustainable approach for the synthesis of valuable spirocyclic compounds commonly found in bioactive molecules and pharmaceutical agents [26]. Scheme 1

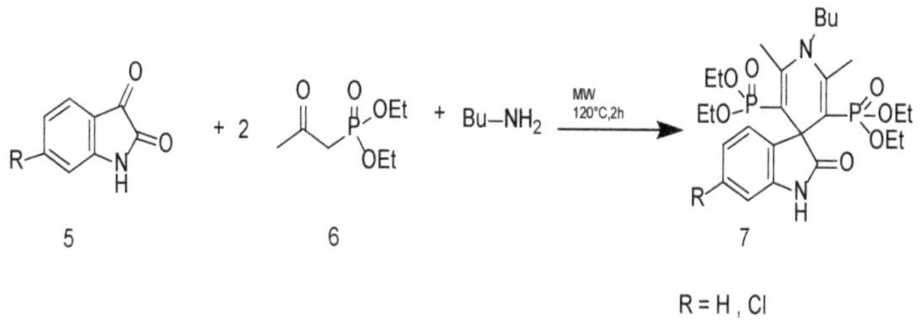

Scheme 1 *- Synthesis of spirooxindole dihydropyridine bisphosphonate.*

Pooja Paul et al. developed a sustainable and one pot efficient method using microwave assistance for the synthesis of 8,9-Dihydrospiro[chromeno[2,3-d]pyrimidine-5,3'-indoline derivative (11) in an aqueous environment, employing graphite oxide (GO) as a metal-free, heterogeneous carbocatalyst. This approach significantly reduces reaction times from hours, as observed with conventional heating, to just minutes under microwave irradiation. The use of GO not only prevents metal contamination and the use of toxic solvents but also improves the overall green chemistry aspect of the procedure. The catalyst, made from inexpensive and easily obtainable graphite powder, was extensively characterized through various analytical methods, including SEM, TEM, EDX, PXRD, FTIR, Raman spectroscopy, and TGA analysis. Notably, the catalyst exhibited remarkable recoverability and reusability for as many as eight consecutive cycles with negligible loss of catalytic performance. This approach yields high product quantities across a wide substrate range, with low catalyst amounts and straightforward work-up processes, emphasizing its practical relevance in the synthesis of biologically important spiro derivatives [27]. Scheme 2

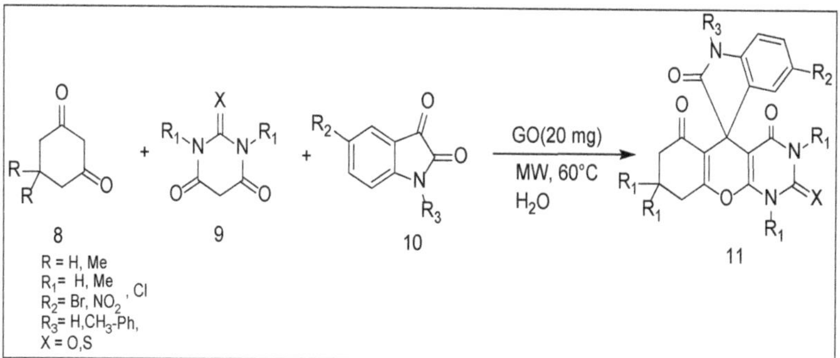

Scheme 2- *Synthesis of spiro derivatives of indole-pyrimidines (11).*

Emine Vildan Burgaz et al. formulated an efficient method for synthesis of 1,3,5,9-tetraone derivatives 14(a–j) through a microwave-assisted condensation reaction. The method involves the reaction of N,N-dimethyl barbituric acid (12a) and 1,3-diethylthiobarbituric acid (12b) with various penta-1,4-diene-3-one derivatives 13(a–f), resulting in the formation of structurally diverse spirocyclic compounds. Comparative optimization of conventional and microwave-assisted techniques demonstrated that microwave irradiation significantly enhanced the efficiency of the synthesis, offering shorter reaction times and excellent yields ranging from 78% to 92%. The structures of the synthesized compounds were confirmed through various spectral and analytical techniques, including ^{1}H NMR, ^{13}C NMR, LC-MS, and elemental analysis. This study underscores the effectiveness of microwave-assisted synthesis in generating novel spiro frameworks efficiently, with potential relevance in synthetic and medicinal chemistry [28]. Scheme 3

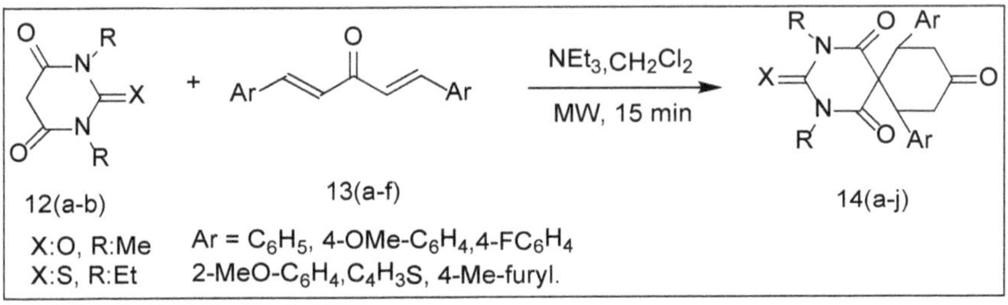

***Scheme 3-** Synthesis of 14(a-j) compounds.*

Rina Das et al. developed a series of spirofused heterocyclic compounds 18(K1-K8) using microwave irradiation method. The structure of these compounds was confirmed through FTIR, ^{1}HNMR, and Mass Spectroscopy. The synthesis was performed through an iodine-catalysed Biginelli-type condensation reaction under microwave irradiation, delivering a rapid, effective, and safe method for producing the desired spiroketal derivatives with high yields in considerably shorter reaction times. This approach exhibited the benefits of microwave-assisted techniques, presenting a simple, economical, and environmentally friendly pathway for the synthesis of spirofused heterocycles, rendering it a compelling alternative to traditional method [29]. Scheme 4

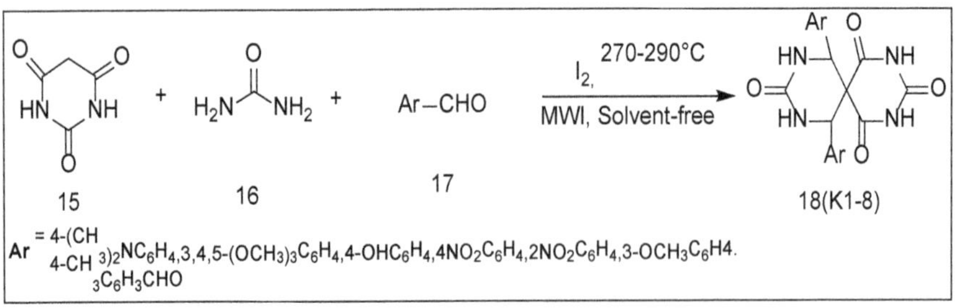

***Scheme 4-** Synthesis of novel Spirofused (Spiroketal) heterocyclic compounds.*

Nawal Kishore Sahu et al. described the development of a rapid and efficient microwave-assisted synthetic strategy for a novel series of functionalized spiro[indoline-3,2'-pyrrolidin]-2-one and spiro[indoline-3,3'-pyrrolizin]-2-one derivatives 22(a–f), 23(a–f), and 24(a–g), incorporating pharmaceutically relevant chalcones and amino acids in presence of methanol at 80 °C using a microwave-assisted approach. Compared to the traditional procedure, which took three hours, the microwave method drastically shortened the reaction time to just five minutes. Structure–activity relationship (SAR) was conducted on these derivatives, offering valuable information about how different functional groups affect their characteristics. Furthermore, molecular docking studies on selected derivatives (23a, 23e, 23f, and 24d) were performed to further validate the SAR findings [30]. Scheme 5

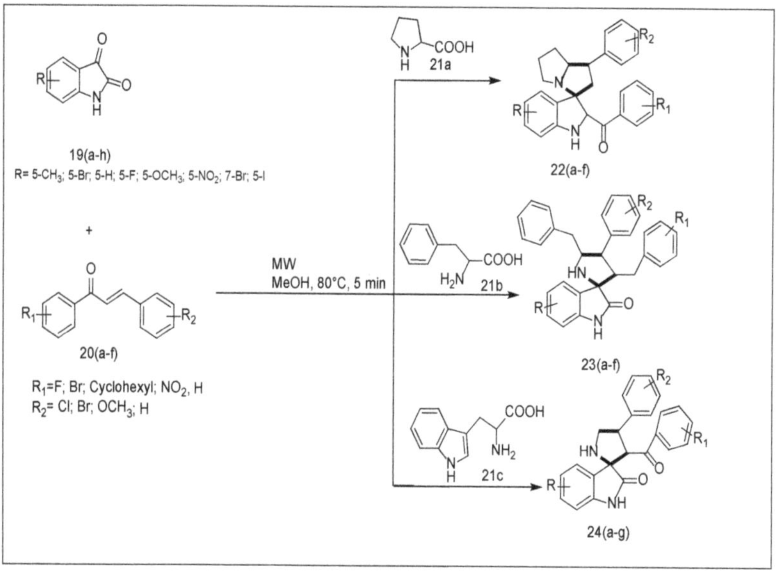

***Scheme 5** - Microwave-assisted synthesis of novel compounds 22(a–f), 23(a–f), and 24(a–g).*

Richa Sharma et al. developed a new series of nitrostyrene-based spiro[pyrrolidine-2,3'-oxindoles] 28(a–w) through a chemo- and regioselective [3+2] cycloaddition (Huisgen) reaction. The synthesis involved the reaction of substituted isatins 25(a–b), various α-amino acids 26(a–e), and (E)-2-aryl-1-nitroethenes 27(a–e) under microwave irradiation, providing an efficient and rapid synthetic route. Structural analysis of the synthesized spirooxindoles was conducted using ^{1}H and ^{13}C NMR techniques alongwith HRMS. Furthermore, the stereochemistry of the representative compound 28l was confirmed through single-crystal X-ray crystallographic analysis, establishing the spatial arrangement around the pyrrolidine ring in the spiro[pyrrolidine-2,3'-oxindole] framework. This methodology highlights the utility of microwave-assisted techniques in the rapid and selective construction of complex spiroheterocyclic structures [31]. Scheme 6

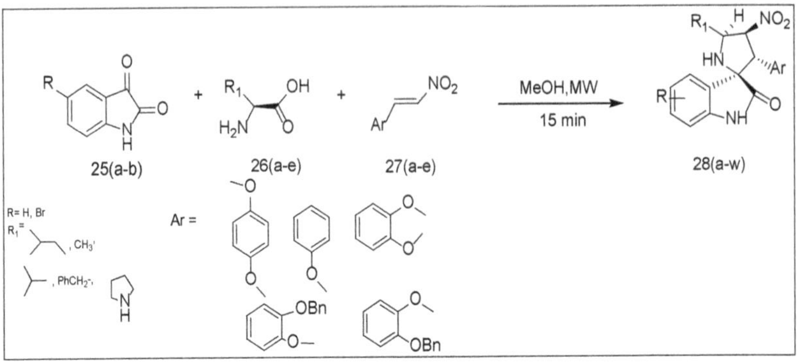

Scheme 6- *Synthesis of indole derivatives 28(a–w).*

F. H. Mahmoud Naglaa reported an environmentally benign and efficient microwave-assisted synthesis of spiro[indoline-3,7'-pyrazolo[3,4-b]pyridine] derivatives (33) under solvent- and catalyst-free one-pot multi-component reaction conditions. This approach, aligning with green chemistry principles, exhibited significant advantages such as rapid reaction progression, reduced reaction times, and simplified post-reaction work-up. Furthermore, the protocol consistently afforded excellent yields of the target molecules. Owing to their notable pharmacological profiles, these spiro-compounds are considered promising scaffolds for future drug design and development [32]. Scheme 7

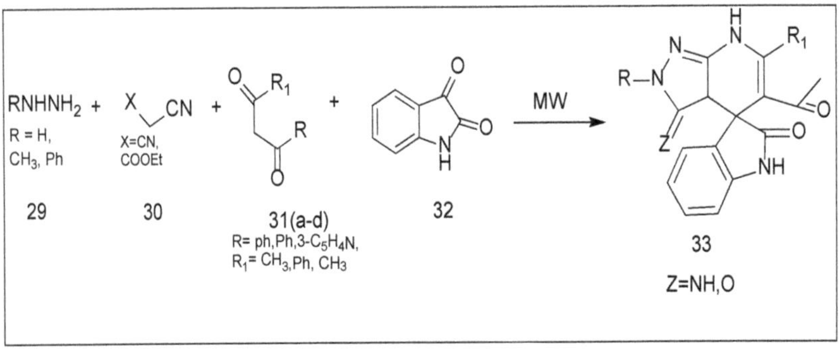

Scheme 7*-Synthesis of spiro derivatives 33*

Sabita Shroff et al. developed a fast and efficient microwave-assisted method for the synthesis of a novel series of spiro derivatives 38(a-f). These spiro diarylidene compounds, containing nitrogen atoms and α,β-unsaturated ketone groups, were synthesized through an aldol condensation reaction between compound (36) and various aryl aldehydes, followed by dehydration. Compared to traditional synthetic methods, the microwave-assisted approach offered the advantages of reduced reaction times and improved yields. The structures of these compounds were characterized using

Materials Research Foundations 189 (2026) 134-154 https://doi.org/10.21741/9781644904039-5

IR spectroscopy, ^1H and ^{13}C NMR spectroscopy, and mass spectrometry (MS). Density functional theory calculations performed with Gaussian 09 validated the structures. Notably, the 4-nitro-substituted derivative (compound 38a) and the 2,4,6-trimethoxy-substituted derivative (compound 38e) exhibited promising broad-spectrum biological activity, warranting further investigation into their therapeutic potential [33]. Scheme 8

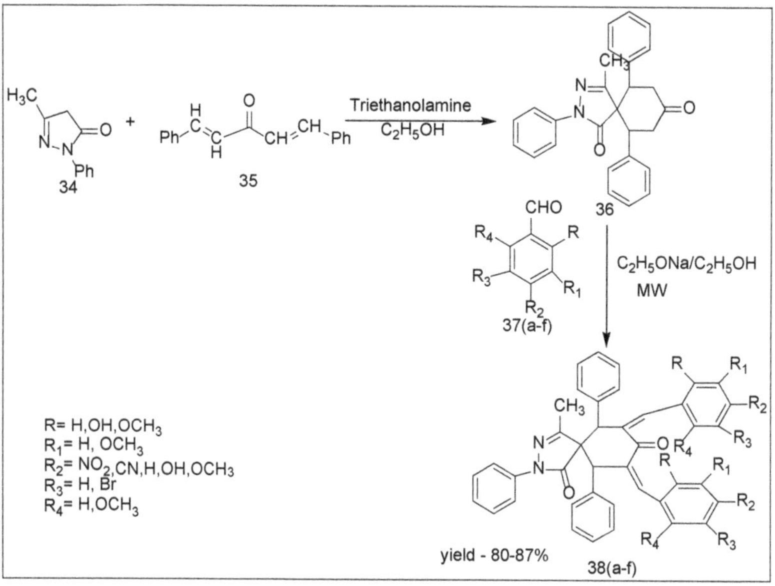

Scheme 8 *- Synthetic pathway of compounds 38(a–f)*

Aleff Castro et al. presented a simple alternative method for the synthesis of 1,3,4-thiadiazoline-based Spiro compounds, utilizing a combination of deep eutectic solvents and microwave irradiation in a one-pot procedure. Several DES systems, derived from different acids, were analysed as solvents for a three-component reaction involving isatin derivatives 39(a-l) (both monomeric and dimeric forms), thiosemicarbazide, (40) and acetic anhydride (41). Among them, the eutectic mixture of choline chloride and oxalic acid in a 1:1 molar ratio delivered superior results, providing yields of up to 79% across the series. For most of the products, the yields matched or outperformed those previously reported using traditional methods. These outcomes underscore the potential of one-pot synthesis as a quicker, more environmentally friendly, and highly effective alternative for producing thiadiazoline derivatives with notable cytotoxic properties [34]. Scheme 9

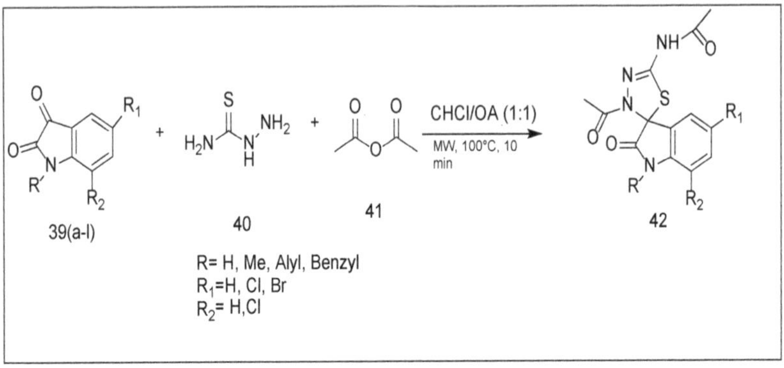

Scheme 9 *- Synthesis of spiro thiadiazolines.*

Dina F. Katowah et al. synthesised moieties of novel spiropyrrolizidine / pyrrolidineoxindole (46) using classical, ultrasonic, and microwave conditions. The ultrasonic method, performed in methanol, completed the reaction within 30 minutes at lower temperatures while under microwave irradiation, the reaction was efficiently completed in just 10 minutes under solvent-free conditions, resulting in high yields. Furthermore, molecular docking studies conducted on the synthesized compounds revealed promising interactions for several derivatives [35]. Scheme 10

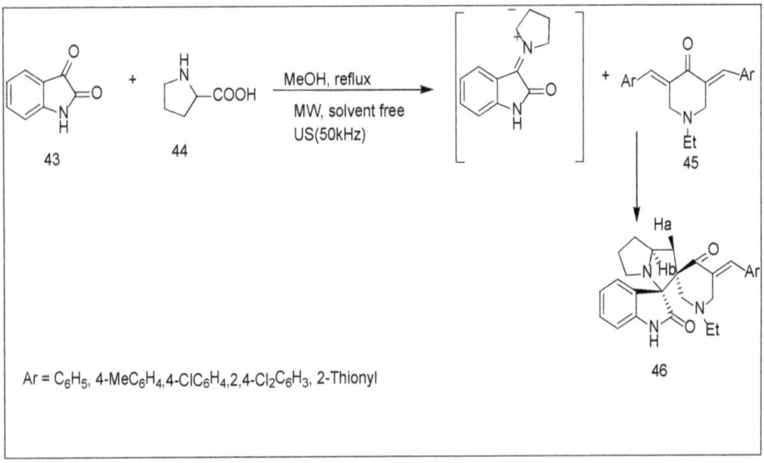

Scheme 10 *- Synthesis of dispiro-pyrrolizidine-oxindoles and spiropyrrolidine.*

Sonal Bhandari et al. introduced an efficient microwave-assisted, one-pot synthetic method for the preparation of pyrrolidine-fused bis-spirooxindoles. (51) The method involves the insitu generation of azomethine ylide (49) through the decarboxylative condensation of isatins (47) with

Materials Research Foundations 189 (2026) 134-154 https://doi.org/10.21741/9781644904039-5

primary α-amino acids(48), followed by a [3+2] cycloaddition with 3-alkenyl oxindoles. Utilizing ethanol as the solvent and avoiding any catalysts or additives, the reaction proceeded efficiently under microwave conditions and product was obtained in high yields. This approach highlights a green and practical route for building bis-spirooxindole frameworks featuring four stereocenters, offering a valuable platform for medicinal chemistry applications [36]. Scheme 11

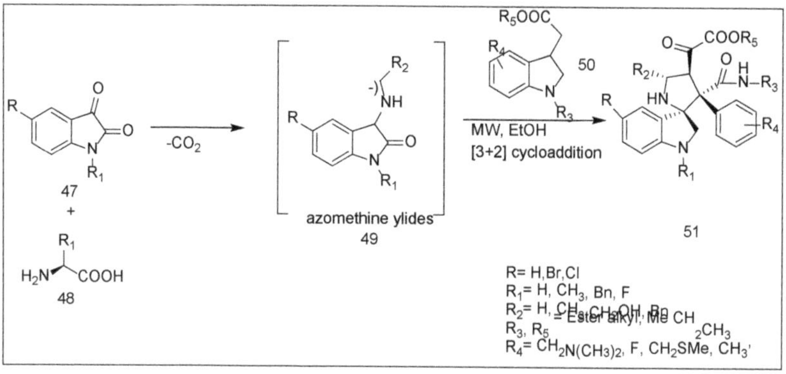

***Scheme 11** -Synthetic strategies for the construction of various bis-spirooxindoles.*

Sabita Nayak et al. described an efficient, straightforward, and adaptable approach for the synthesis of spiroindanone-pyrrolidine and spiroindanone-piperidine fused nitrochromene derivatives (55) via one-pot, three-component 1,3-dipolar cycloaddition. The azomethine ylides were generated in situ from the reaction of indane-1,3-dione with L-proline or pipecolic acid and then combined with 2-phenyl-nitrochromene dipolarophiles. The reactions proceeded smoothly under microwave-assisted conditions, resulting the desired cycloadducts in excellent yields (85–90%). The cycloaddition exhibited high regioselectivity and diastereoselectivity, with the regiochemical and stereochemical outcomes confirmed through detailed 2D NMR studies confirming the structural and stereochemical outcomes [37]. Scheme 12

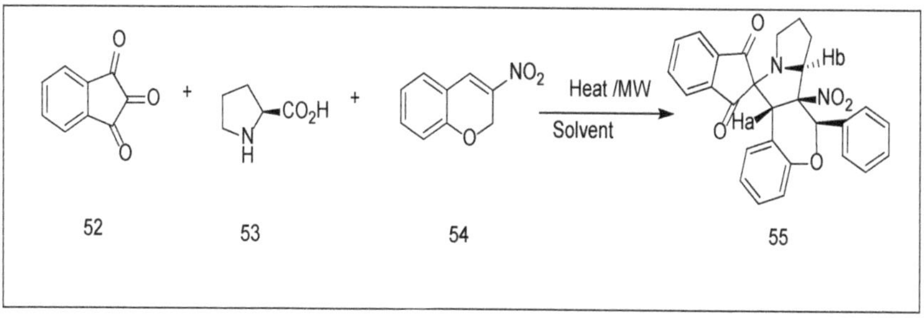

***Scheme 12** -Synthesis of compound 55.*

Materials Research Foundations 189 (2026) 134-154 https://doi.org/10.21741/9781644904039-5

Animesh Mondal et al. demonstrated a rapid and efficient methodology for the one-pot synthesis of diversified dihydrospiro[indeno[1,2-b]pyridine-4,3′-indoline] and pyridine analogues (60) under sustainable microwave-assisted conditions. The reactions were performed in a minimal amount of aqueous ethanol (1:1 v/v) without the use of any metal catalysts, aligning with green chemistry principles. The reaction conditions promote high efficiency, minimal waste production, and significant time savings, aligning with the core principles of sustainable synthetic chemistry. These methods offer promising pathways for the smooth construction of structurally fused spiro frameworks, reinforcing the growing importance of green technologies in heterocyclic synthesis [38]. Scheme 13

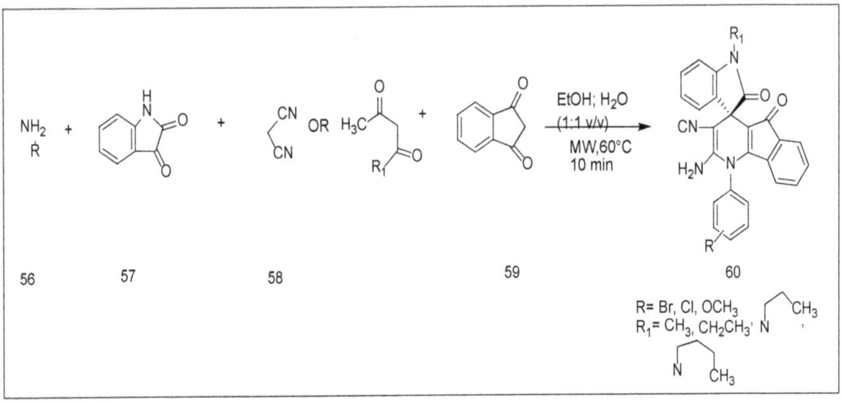

***Scheme 13**-Preparation of spiro-dihydropyridines chemo sensor to Zn^{2+} ion.*

Priyanka Nivrutti Shinde et al. described an efficient synthetic approach for spiro[indole-thiazolidine] derivatives 65(a-l) utilizing microwave-assisted reactions. Initially, isoniazid [61] reacted with isatin [62] to form pyridine derivative [63] through a microwave-assisted process. This intermediate [63] subsequently underwent cyclocondensation with mercaptoacetic acid, also under microwave irradiation, to yield the targeted spiro[indole-thiazolidine] compounds [64]. Further treatment of these spiro frameworks with various aromatic and heteroaromatic aldehydes resulted in the formation of a series of arylidene derivatives 65(a–l). The structures of all prepared molecules were thoroughly characterized and confirmed using spectroscopic analyses [39]. Scheme 14

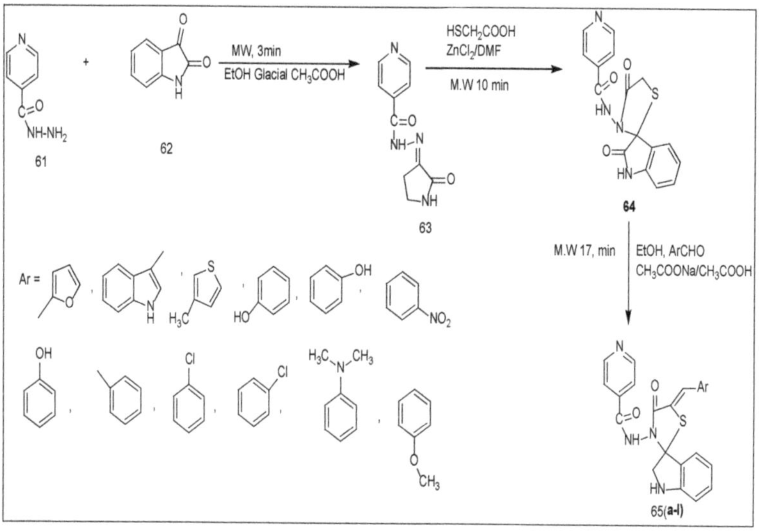

Scheme 14 *- Synthesis of Spiroindole-thiazolidine.*

Sheida Khojasteh-Khosro et al. reported an efficient and rapid microwave-assisted synthetic approach for the preparation of compound (70) These spirocyclic frameworks were synthesized using a four-component reaction involving phenylhydrazine or hydrazine hydrate (67), isatins (66), ketoesters (68), and either naphthylamine or 2-naphthol, (69) in the presence of nano-Co_3S_4 catalyst using microwave irradiation. The method offers a convenient, time-efficient, and high-yielding pathway for the construction of complex spiro heterocycles [40]. Scheme 15

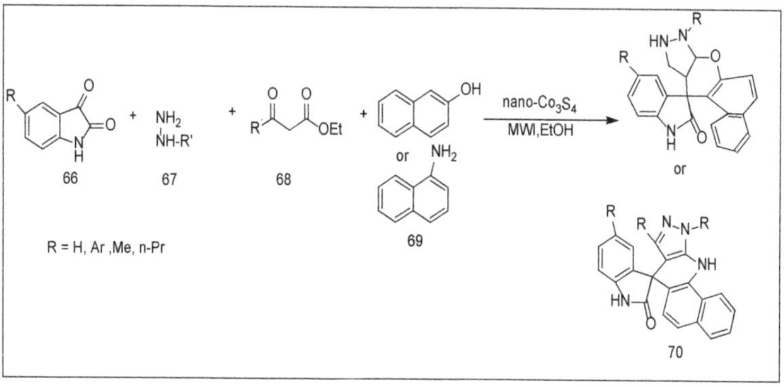

Scheme 15 *- Synthesis of spirooxindoles under microwave irradiations.*

Houda Gazzeh et al. synthesised a microwave-assisted approach for the 1,3-dipolar cycloaddition of compounds 71 with nitrilimines (72), which were generated in situ via dehydrohalogenation of hydrazonoyl chlorides. This method enabled the efficient synthesis of spiropyrazoline derivatives, achieving good to excellent yields. The reactions highlighted excellent chemo-, regio-, and diastereoselectivity. Structural confirmation of the products was achieved through spectroscopic techniques, elemental analysis, and single-crystal X-ray diffraction studies. Moreover, the luminescent behaviour of the synthesized heterocycles was assessed, revealing notable fluorescence properties in solution along with measured quantum yields [41]. Scheme 16

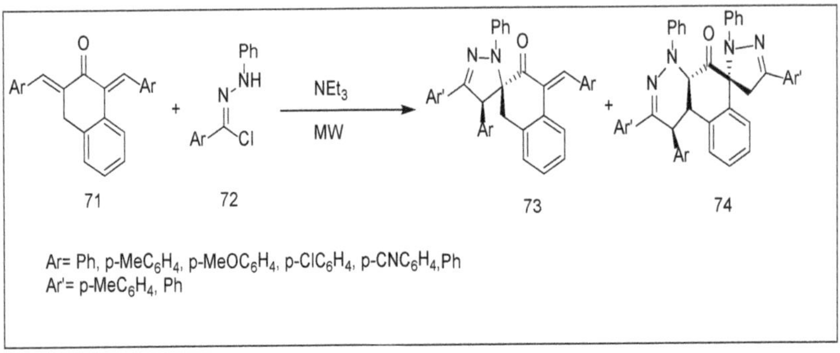

***Scheme 16** – Synthesis of spiro pyrazoline derivatives.*

In the above section of this chapter, it has been discussed that spiro heterocycles synthesised under microwave irradiations have demonstrated remarkable advantages in pharmacology including excellent thermal stability, higher yield and enhanced therapeutics but beyond the utility in medicinal chemistry these scaffolds due to their rigid spiro-architecture, electron tunability and durability are serving an important role in the designing of advanced OLEDS which will be highlighted in the following section.

Ye-Xin Zhang et al. created two host materials, SFCA (75) and SFCC (76), by incorporating either a carbazole or a diphenylamine unit to explore their potential uses as blue phosphorescent organic light-emitting diodes (PHOLEDs).Through meta conjugation, these diphenylamine and carbazole units were further connected to spiro-fused phenyl carbazole, which improved the triplet energies (2.72 eV and 2.84 eV) and matched the compound's compatibility with the blue phosphorescent material (FIrpic, E_T = 2.65). With high glass transition temperatures and high decomposition temperatures above 360 °C, these materials demonstrated exceptional thermal stability. Moreover, blue phosphorescent OLEDs made using these host materials and FIrpic as the dopant had maximum current efficiencies of 33.9 and 40.8 cd/A, respectively. These are some of the highest current efficiencies yet recorded for host materials based on fused N-phenylcarbazole [42]. Scheme 17

Materials Research Foundations 189 (2026) 134-154 https://doi.org/10.21741/9781644904039-5

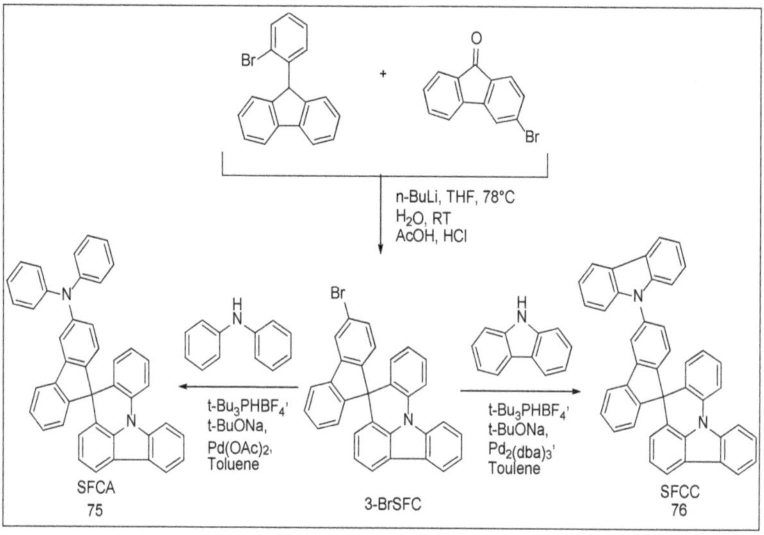

Scheme 17*-Synthesis of SFCA and SFCC.*

Yan Qian et al. developed a new spiro fluorene-based host material SFX-PF (77) to explore its application as light emitting diode through a simple two step synthesis without incorporating traditional hole and electron transport units. This host material was then characterised for photophysical properties which revealed that SFX-PF had a high decomposition temperature of around 311 °C and a high glass transition temperature with high triplet energy of 2.8 ev which was comparable to blue phosphorescent emitter FIrpic. Also, Devices doped with SFX-PF as the host material displayed low turn on voltage and maximum current efficiency of 29.3 cd/A and external quantum efficiency (EQE) 14.7% making the SFX-PF as a lead for designing blue PHOLED [43]. Scheme 18

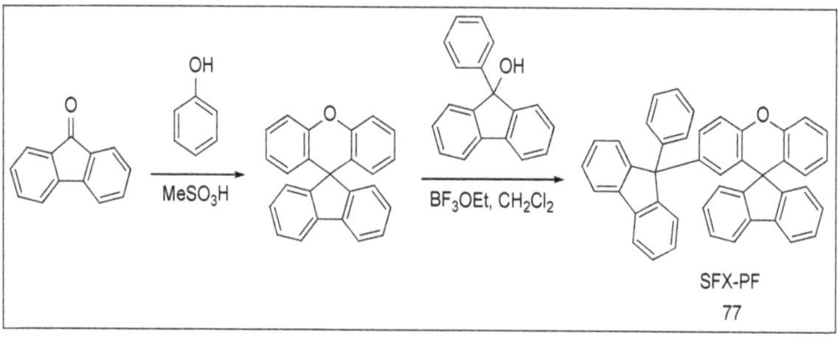

Scheme 18 *-Synthesis of SFX-PF.*

Xiang-Yang Liu et al. designed a series of spirofluorene-based host materials to explore its applications as green phosphorescent OLEDs. These materials were screened for the photophysical and electrochemical properties which revealed that different spiro-systems have minimal impact on the absorption, emission, and frontier molecular orbital properties of the compounds. However, when used as hosts in green PHOLEDs with the dopant Ir(ppy)2(acac), devices based on the spirobifluorene derivative (78) exhibited the highest electroluminescence performance with an external quantum efficiency (EQE) of over 18%. The superior device performance of 4 was attributed to its enhanced hole transport capability compared to the other hosts suggesting that even small structural variations in spirofluorene-based materials can lead to significant differences in their charge transport properties and OLED device performance [44]. Scheme 19

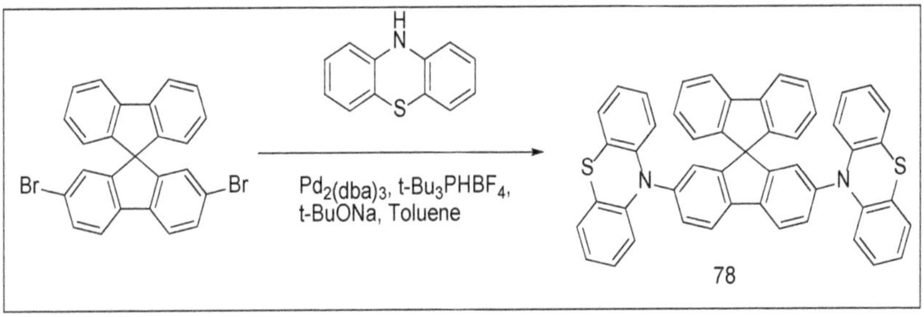

Scheme 19 – Synthesises of Spirofluorene-based derivative 78.

Lin He et al. described the synthesis of spiro-indenes (80) possessing carbon quaternary stereocenters through palladium-catalyzed asymmetric (4 + 2) dipolar cyclization by trapping π-allyl-Pd 1,4-dipoles with indene-involved ketenes generated in situ from 1-diazonaphthalene-2(1H)-ones (79) via visible light-induced Wolff rearrangement, where 6W blue light-emitting diodes (LEDs) were used as the light source. The reaction was optimized by utilising specialized chiral phosphoramidite ligands, achieving up to 86% yield, 97% enantiomeric excess, and high diastereoselectivity (up to 19:1 dr) [45]. Scheme 20

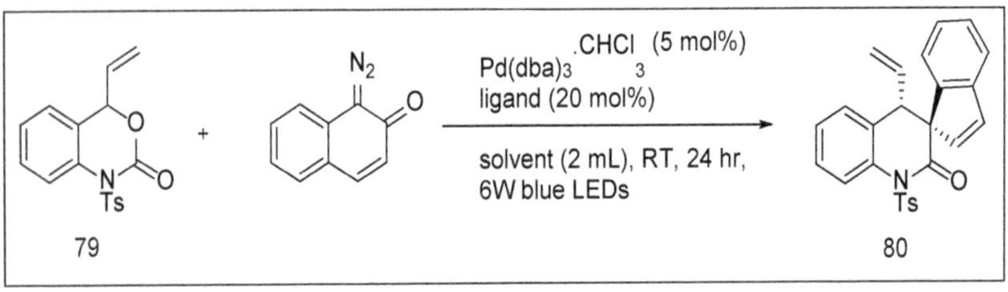

Scheme 20 - Synthesis of spiro-indenes.

Soon Ok Jeon et al. synthesized a pyridine-substituted spirofluorene-benzofluorene derivative – SPBP (81) and tested its suitability as an electron transport medium (ETM) in blue organic light-emitting diodes. This derivative was synthesized by Suzuki coupling, generating more than 98%

Materials Research Foundations 189 (2026) 134-154 https://doi.org/10.21741/9781644904039-5

pure product and exhibiting outstanding hole blocking properties due to electron-deficient pyridine units. The SPBP-based blue OLED outperformed the frequently used Alq3 ETM by 20% in quantum efficiency and 30% in power efficiency, with a high glass transition temperature of 122.6 °C and a large absorption peak at 298 nm. All of these studies point to SBP as a possible contender for improved OLED design [46]. Scheme 21

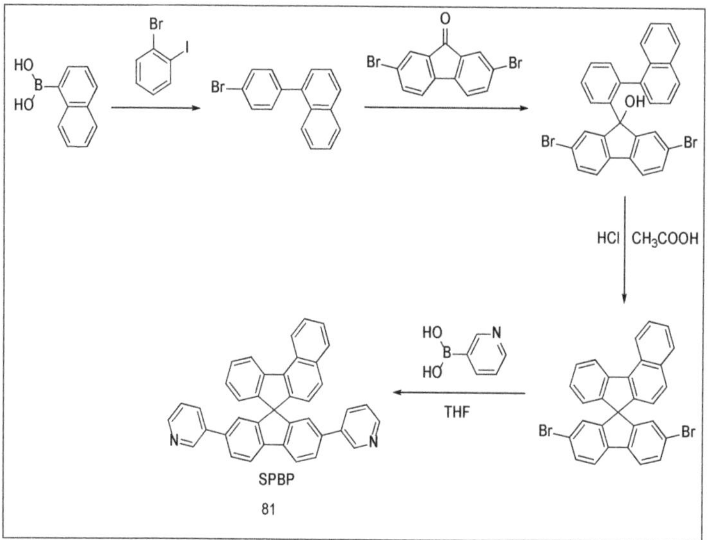

Scheme 21- *Synthesis of SPBP.*

List of abbreviations

OLED	- Organic light emitting diodes
MW	- Microwave irradiation
OEt	- Ethoxy
GO	- Graphite oxide
MeOH	- Methyl alcohol
DES	- Deep eutectic solvents
CNTs	- Carbon nanotubes
[Bmim]BF_4	- 1-Butyl-3-methyl-imidazolium tetrafluoroborate
TGA	- Thioglycolic acid
K_2CO_3	- Potassium carbonate
DMF	- Dimethylformamide
C_2H_5ONa	- Sodium ethoxide
US	- Ultrasound irradiation
NEt_3	- Triethylamine
HOMO	- Highest occupied molecular orbital

LUMO	- Lowest unoccupied molecular orbital
FMO	- Frontier molecular orbital
XRD	- X-ray diffraction
CH_3COOH	- Acetic acid
Ar	- Aryl/ aromatic
Ar-CHO	- aromatic aldehyde
SPBP	- (pyridin-3-yl)spiro[benzofluorene-7,9′-fluorene]
ETM	- Electron transport material
FIrpic	- Iridium(III) bis[(4,6-difluorophenyl)pyridinato-N,C^2] picolinate
EQE	- External Quantum Efficiency
SFX-PF	- 20-(9-phenyl-9H-fluoren-9-yl)spiro[fluorene-9,90-xanthene]
PHOLED	- blue phosphorescent organic light-emitting diodes

Conclusion and future perspective

In conclusion, spiro heterocycles have grown increasingly significant in medicinal, bioorganic, and pharmaceutical chemistry due to their biological activity and vast range of applications. The synthesis of spiro heterocyclic compounds is presented in this chapter utilizing a microwave-assisted approach that provides an effective yield, a shorter reaction time, and a simplified workup. Furthermore, these spiro compounds have been used to develop organic light-emitting diodes.

Significant efforts have been made to synthesize spiro heterocyclic compounds due to their importance in various fields of chemistry, including drug discovery, medicinal chemistry, natural products chemistry, materials science, and agrochemicals. Microwave multi-component reactions (MCRs) can be used to create this versatile class of compounds known as spiro heterocycles. Additionally, the application of microwave techniques enhances the reactivity and selectivity of reactions, reduces reaction times, and yields a higher amount of spiro compounds. However, despite the substantial progress made in producing these compounds for use in light-emitting diodes, several limitations still remain. Thus, in the coming years, a lot of focus must be placed on broadening the substrate scopes with high yields through the use of green pathways, which are safer, more cost-effective, atom-efficient, and offer scale-up synthesis.

References

[1] S.N. Afraj, A. Velusamy, M.C. Chen, M. Abd-Ellah, A.L. Abdelhady, Heterocyclic and heteropolycyclic moieties in organic hole transport materials for perovskite solar cells: Design, synthesis, and performance, Coord. Chem. Rev. 532 (2025) 216500. https://doi.org/10.1016/j.ccr.2025.216500

[2] A. Ding, M. Meazza, H. Guo, J.W. Yang, R. Rios, New development in the enantioselective synthesis of spiro compounds, Chem. Soc. Rev. 47 (2018) 5946-5996. https://doi.org/10.1039/C6CS00825A

[3] S. Dhadda, S. Sharma, P. Jakhar, H. Sharma, Contemporary progress in the green synthesis of spiro-thiazolidines and their medicinal significance: a review, RSC Adv. 13 (2023) 3723-3742. https://doi.org/10.1039/D2RA07474E

[4] X.-N. Zhang, X. Dong, Y. Wei, M. Shi, Access to 2′,3′-dihydro-1′H-spiro[indoline-3,4′-pyridin]-2-ones via amino acid derived phosphine-catalyzed asymmetric [4+2] annulation

with easily available oxindole-derived α,β-unsaturated imines, Tetrahedron 70 (2014) 2838-2846. https://doi.org/10.1016/j.tet.2014.02.052

[5] P. Saraswat, G. Jeyabalan, M.Z. Hassan, M.U. Rahman, N.K. Nyola, Review of synthesis and various biological activities of spiro heterocyclic compounds comprising oxindole and pyrrolidine moieties, Synth. Commun. 46 (2016) 1643-1664. https://doi.org/10.1080/00397911.2016.1211704

[6] R. Javahershenas, A. Makarem, K.D. Klika, Recent advances in microwave-assisted multicomponent synthesis of spiro heterocycles, RSC Adv. 14 (2024) 5547-5565. https://doi.org/10.1039/D4RA00056K

[7] M. Henary, C. Kananda, L. Rotolo, B. Savino, E.A. Owens, G. Cravotto, Benefits and applications of microwave-assisted synthesis of nitrogen-containing heterocycles in medicinal chemistry, RSC Adv. 10 (2020) 14170-14197. https://doi.org/10.1039/D0RA01378A

[8] S. Gulati, S.E. John, N. Shankaraiah, Microwave-assisted multicomponent reactions in heterocyclic chemistry and mechanistic aspects, Beilstein J. Org. Chem. 17 (2021) 819-865. https://doi.org/10.3762/bjoc.17.71

[9] M. Kamboj, S. Bajpai, G. Pandey, M. Yadav, B.K. Banik, Microwave-assisted synthesis of biologically relevant six-membered N-heterocycles, Curr. Microw. Chem. 10 (2023) 122-134. https://doi.org/10.2174/0122133356268693231114052121

[10] R. Singh, D. Bhardwaj, M.R. Saini, Recent advancement in the synthesis of diverse spiro-indeno[1,2-b]quinoxalines: A review, RSC Adv. 11 (2021) 4760-4804. https://doi.org/10.1039/D0RA09130H

[11] M.A. Borad, M.N. Bhoi, N.P. Prajapati, H.D. Patel, Review of synthesis of Spiro heterocyclic compounds from isatin, Synth. Commun. 44 (2014) 897-922. https://doi.org/10.1080/00397911.2013.843196

[12] N. Lashgari, G. Mohammadi Ziarani, R. Moradi, M. Zandiyeh, 4-Hydroxy-6-methyl-2-pyrone- A versatile synthon in the synthesis of heterocyclic scaffolds via multicomponent reactions, Heterocycles 96 (2018) 381. https://doi.org/10.3987/REV-17-872

[13] M. Mamaghani, R. Hossein Nia, Recent developments in the MCRs synthesis of pyridopyrimidines and Spiro-pyridopyrimidines: Recent developments in the MCRs synthesis of pyridopyrimidines and Spiro-pyridopyrimidines, J. Heterocycl. Chem. 54 (2017) 1700-1722. https://doi.org/10.1002/jhet.2783

[14] K. Kumar, Microwave-assisted diversified synthesis of pyrimidines: An overview, J. Heterocycl. Chem. 59 (2022) 205-238. https://doi.org/10.1002/jhet.4376

[15] F. Frecentese, I. Saccone, G. Caliendo, A. Corvino, F. Fiorino, E. Magli, E. Perissutti, B. Severino, V. Santagada, Microwave-assisted organic synthesis of heterocycles in aqueous media: Recent advances in medicinal chemistry, Med. Chem. 12 (2016) 720-732. https://doi.org/10.2174/1573406412666160502153553

[16] B.H. Rotstein, S. Zaretsky, V. Rai, A.K. Yudin, Small heterocycles in multicomponent reactions, Chem. Rev. 114 (2014) 8323-8359. https://doi.org/10.1021/cr400615v

[17] V. Nigam, S. Singh, S. Kasana, S. Kumar, B. Das Kurmi, G. Das Gupta, P. Patel, Revolutionizing indole synthesis: A microwave-powered approach, ChemistrySelect 9 (2024). https://doi.org/10.1002/slct.202402171

[18] P. Khanna, L. Khanna, S.J. Thomas, A.M. Asiri, S.S. Panda, Microwave assisted synthesis of Spiro heterocyclic systems: A review, Curr. Org. Chem. 22 (2018) 67-84. https://doi.org/10.2174/1385272821666170818161517

[19] I. Almas, A. Malik, N. Rasool, A. Kanwal, T. Khalid, H. Nawaz, Microwave-assisted protocol towards synthesis of heterocyclic molecules: A comparative analysis with conventional synthetic methodologies (years 2019-2023): A review, Mol. Divers. (2024). https://doi.org/10.1007/s11030-024-10981-y

[20] T.L. Lambat, P.K.P.G. Chopra, S.H. Mahmood, Microwave: A green contrivance for the synthesis of N-heterocyclic compounds, Curr. Org. Chem. 24 (2020) 2527-2554. https://doi.org/10.2174/1385272824999200622114919

[21] Q. Wu, H. Feng, D.-D. Guo, M.-S. Yi, X.-H. Wang, B. Jiang, S.-J. Tu, Microwave-assisted aqueous multicomponent reaction: Facile synthesis of polyfunctionalized indoline-spiro fused pyran derivatives: Synthesis of indoline-spiro fused pyran derivatives, J. Heterocycl. Chem. 50 (2013) 599-602. https://doi.org/10.1002/jhet.1537

[22] D. Luo, Q. Chen, B. Liu, Y. Qiu, Emergence of flexible white organic light-emitting diodes, Polymers (Basel) 11 (2019) 384. https://doi.org/10.3390/polym11020384

[23] X. Yang, X. Xu, G. Zhou, Recent advances of the emitters for high performance deep-blue organic light-emitting diodes, J. Mater. Chem. C Mater. Opt. Electron. Devices 3 (2015) 913-944. https://doi.org/10.1039/C4TC02474E

[24] L. Duan, L. Hou, T.-W. Lee, J. Qiao, D. Zhang, G. Dong, L. Wang, Y. Qiu, Solution processable small molecules for organic light-emitting diodes, J. Mater. Chem. 20 (2010) 6392. https://doi.org/10.1039/b926348a

[25] S. Janietz, H. Krueger, M. Thesen, B. Salert, A. Wedel, Concepts for the material development of phosphorescent organic materials processable from solution and their application in OLEDs, Organic Light Emitting Materials and Devices XVIII (2014) 918303. https://doi.org/10.1117/12.2061005

[26] B. Rávai, A.S. Németh, Z. Kelemen, Erika Bálint, Microwave-assisted multicomponent synthesis of spirooxindole dihydropyridine bisphosphonates, Eur. J. Org. Chem. 28 (2025) e202400873. https://doi.org/10.1002/ejoc.202400873

[27] P. Paul, M. Goswami, D. Chakravarty, S. Joshi, S.R Joshi, R. Nongkhlaw, Microwave-assisted synthesis of spiro heterocyclic scaffolds using graphite oxide as a heterogeneous carbocatalyst and study of their anti-microbial activities, ChemistrySelect 9 (2024) e202402637. https://doi.org/10.1002/slct.202402637

[28] E.V Burgaz, M. Yakut, i. Kunter, Efficient microwave synthesis and anti-cancer evaluation of new substituted 2,4- diazaspiro[5.5] undecane-1,5,9-trione (or 1,3,5,9-tetraone) derivatives, Arkivoc 8 (2024) 202412321. https://doi.org/10.24820/ark.5550190.p012.321

[29] R. Das, D.K Mehta, S. Gupta, S. Mujwar, V. Sharma, A. Goyal, S. Patel, A. Patel, Microwave-assisted synthesis, molecular docking study of spirofused heterocycles as anti-

microbial and anthelmintic potential, Letters in Organic Chemistry 20 (2023) 1182-1191. https://doi.org/10.2174/1570178620666230703111452

[30] N.K Sahu, R. Sharma, K.P Suhas, J. Joshi, K. Prakash, R. Sharma, R. Pratap, X. Hu, S. Kaur, M. Jain, C. Coluccini, P. Coghi, S. Chaudhary, Natural-product-inspired microwave-assisted synthesis of novel spirooxindoles as antileishmanial agents: Synthesis, stereochemical assignment, Bioevaluation, SAR, and Molecular Docking Studies, Molecules 28 (2023) 4817. https://doi.org/10.3390/molecules28124817

[31] R. Sharma, L. Yadav, A.A. Nasim, R.K Yadav, R.H. Chen, N. Kumari, F. Ruiqi, A. Sharon, N.K. Sahu, S.K. Ippagunta, P. Coghi, V.K.W. Wong, S. Chaudhary, Chemo-/regio-selective synthesis of novel functionalized spiro[pyrrolidine-2,30-oxindoles] under microwave irradiation and their anticancer activity, Molecules 28 (2023) 6503. https://doi.org/10.3390/molecules28186503

[32] F.H.M. Naglaa, S. A. Rizk, A.E. Galal, A.K. Ali, Expeditious microwavable one-pot synthesis and biological exploration of spiro[indoline-3,4′-pyrazolo[3,4-b] pyridine derivatives, Journal of the Iranian Chemical Society 19 (2022) 3711-3719. https://doi.org/10.1007/s13738-022-02568-x

[33] S. Shroff, P. P. Mohanta, I. Baitharu, B.P. Bag, A.K. Behera, Microwave-assisted synthesis of novel spiro diarylidenes and their antimicrobial assay, J. Serb. Chem. Soc. 87 (2022) 813-827. https://doi.org/10.2298/JSC210123031S

[34] A. Castro, I.M.G. Andrade, M.C. Coelho, D.P. da Costa, D. das N. Moreira, R.A. Maia, G. da S. Lima, G.F. dos Santos, B.G. Vaz, G.C.G. Militão, P.B.N. da Silva, M.L.A. de A. Vasconcellos, C.G. Lima-Junior, Multicomponent synthesis of spiro 1,3,4-thiadiazolines with anticancer activity by using deep eutectic solvent under microwave irradiation, J. Heterocyclic Chem. 60 (2023) 392-405. https://doi.org/10.1002/jhet.4591

[35] D.F. Katowah, H.M.E Hassaneen, T.A.Farghaly, Novel Spiro-pyrrolizidine-Oxindole and Spiropyrrolidine-Oxindoles: Green synthesis under classical, ultrasonic, and microwave conditions and molecular docking simulation for antitumor and type 2 diabetes, Arabian Journal of Chemistry 15 (2022) 103930. https://doi.org/10.1016/j.arabjc.2022.103930

[36] S. Bhandari, S. Sana, B. Sridhar, N. Shankaraiah, Microwave-assisted one-pot [3+2] cycloaddition of azomethine ylides and 3-alkenyl oxindoles: A facile approach to pyrrolidine-fused bis-spirooxindoles, ChemistrySelect 4 (2019) 1727-1730. https://doi.org/10.1002/slct.201802847

[37] S. Nayak, P. Panda, S. Mohapatra, B. Raiguru, N. Baral, Microwave-assisted one-pot, three-component regiospecific and sterospecific synthesis of Spiro indanone pyrrolidine/piperidine fused nitrochromene derivatives through 1,3-dipolar cycloaddition reactions, J. Heterocyclic Chem. 56 (2019) 1757-1770. https://doi.org/10.1002/jhet.3534

[38] A. Mondal, B. Naskar, S. Goswami, C. Prodhan, K. Chaudhuri, C. Mukhopadhyay, A quick accelerating microwave-assisted sustainable technique: permutated spiro-casing for imaging experiment, Mol. Divers. 24 (2020) 93-106. https://doi.org/10.1007/s11030-019-09934-7

[39] P.N. Shinde, M.A. Raskar, Synthesis and biological evaluation of Spiro (indole-thiazolidine) derivatives as antimicrobial agents, Int. J. Curr. Pharm. Res. 11 (2019) 71-74. https://doi.org/10.22159/ijcpr.2019v11i6.36346

[40] S. Khojasteh-Khosro, H. Shahbazi-Alavi, Preparation of spirooxindoles catalyzed by nano-Co3S4 under microwave irradiations, J. Chem. Res. 43 (2019) 107-111. https://doi.org/10.1177/1747519819841791

[41] H. Gazzeh, S. Boudriga, M. Askri, A. Khatyr, M. Knorr, C. Strohmann, C. Golz, Y. Rousselin, M.M. Kubicki, Stoichiometry-controlled cycloaddition of nitrilimines with unsymmetrical exocyclic dienones: Microwave-assisted synthesis of novel mono- and dispiropyrazoline derivatives, RSC Adv. 6 (2016) 49868-49875. https://doi.org/10.1039/C6RA09703K

[42] Y.X. Zhang, L. Ding, X.Y. Liu, Z.Q. Jiang, H. Chen, S.J. Ji, L.S. Liao, Spiro-fused N-phenylcarbazole-based host materials for blue phosphorescent organic light-emitting diodes, Organic Electronics. 20 (2015) 112-118. https://doi.org/10.1016/j.orgel.2015.02.014

[43] Y. Qian, G. Xie, S. Chen, Z. Liu, Y. Ni, X. Zhou, L. Xie, J. Liang, Y. Zhao, M. Yi, Y. Zhao, W. Wei, W. Huang, A new spiro[fluorene-9,90-xanthene]-based host material possessing no conventional hole- and electron-transporting units for efficient and low voltage blue PHOLED via simple two-step synthesis, Organic Electronics. 13 (2012) 2741-2746. https://doi.org/10.1016/j.orgel.2012.07.047

[44] X.Y. Liu, X. Tang, D. Zhao, B. Song, L. Ding, J. Fan, L.-S. Liao, A series of spirofluorene-based host materials for efficient phosphorescent organic light-emitting diodes, Organic Electronics. 61 (2018) 70-77. https://doi.org/10.1016/j.orgel.2018.06.046

[45] L. He, B.-L. Qu, M. Xiao, J.-Q. Yu, J. Yu, Y. Cheng, Z. Zhang, W.-J. Xiao, L.-Q. Lu, Synthesis of chiral spiro-indenes via Pd-catalyzed asymmetric (4 + 2) dipolar cyclization. Chem. Synth. 4 (2024). https://doi.org/10.20517/cs.2024.39

[46] S.O. Jeon, K.S. Yook, J.Y. Lee, Pyridine substituted spirofluorene derivative as an electron transport material for high efficiency in blue organic light-emitting diodes, Thin Solid Films 519 (2010) 890-893. https://doi.org/10.1016/j.tsf.2010.08.108

Materials Research Foundations 189 (2026) 155-208 https://doi.org/10.21741/9781644904039-6

Chapter 6

Zeolites via Microwave-Assisted Synthesis for Environmental Pollutant Adsorption

Iqra Urooj[1], Tooba Irfan[1], Zia Ur Rehman[2], Gulzar Muhammad[1*], Muhammad Ajaz Hussain[3], Aamna Majeed[1], Muhammad Arshad Raza[1]

[1]Department of Chemistry, Institute of Chemical Sciences, Government College University Lahore, 54000 Lahore, Pakistan

[2]Department of Chemistry, College of Physical Sciences & Engineering, University of Wyoming, Laramie

[3]Centre for Organic Chemistry, School of Chemistry, University of the Punjab, Lahore 54590, Pakistan

mgulzar@gcu.edu.pk

Abstract

This chapter provides a comprehensive overview of zeolites, with a focus on their microwave-assisted synthesis and diverse applications in controlling environmental pollution. It highlights the fundamental principles of microwave-assisted synthesis, emphasizing its role in enhancing the crystalline, structural uniformity, surface area, and adsorption capacity of zeolites. Key parameters such as crystallization temperature and time are discussed concerning optimizing synthesis efficiency. The chapter underscores the effectiveness of microwave-assisted synthesized zeolites in various environmental applications, including ion exchange, adsorption of volatile organic compounds (VOCs), nitrogen compounds, dyes and heavy metals, CO_2 capture, drying processes, hydrogen purification, and catalysis. The applications collectively demonstrate the versatility of zeolites as sustainable adsorbents for mitigating environmental pollution.

Keyword

Microwave-Assisted Synthesis, Environmental Pollutant, Adsorbent, Zeolites, Environmental Applications, Adsorption

Contents

Materials Research Foundations 189 (2026) 155-208 https://doi.org/10.21741/9781644904039-6

Materials Research Foundations 189 (2026) 155-208 https://doi.org/10.21741/9781644904039-6

List of Abbreviations

PBUs	Primary building units	SDAs	Structure-directing agents
SBUs	Secondary building units	ZSM-5	Zeolite Socony Mobil-5
ANA	Analcime	BET	Brunauer-Emmett-Teller
CHA	Chabazite	BEA	Beta zeolite
ERI	Erionite	ESA	Electric swing adsorption
FAU	Faujasite	RPSA	Rapid pressure swing adsorption
FER	Ferrierite	VSA	Vacuum swing adsorption
HEU	Heulandite	VPSA	Vacuum pressure swing adsorption
LAU	Laumontite	SMB	Simulated moving bed
MER	Merlinoite	TMB	True moving bed
MOR	Mordenite	ROG	Refinery off-gas
OFF	Offretite	SMROG	Steam methane reforming off-gas
PHI	Phillipsite	TSA	Temperature swing adsorption
YUG	Yugawaralite	PSA	Pressure swing adsorption
$AlPO_4$-5	Aluminophosphate-5	MZeo-HAP	Magnetic zeolite/hydroxyapatite
SAPO	Silicoaluminophosphate	MFI	Mobil-type five
LTA	Linde Type A	SOD	Sodalite
AFI	Framework type code of Aluminophosphate-5	VOCs	Volatile organic compounds

1. Introduction

1.1 Overview of Zeolites and Their Importance

Zeolites are aluminosilicates with a crystalline microporous structure frequently employed as "ion exchangers, adsorbents, and catalysts". Among them, some natural minerals are classified as tectosilicates. Si and Al oxide tetrahedra, known as primary building units (PBUs), are joined by oxygen ions to form secondary units (secondary building units (SBUs)) in two and three dimensions. Since the 1950s, their commercial availability, compositional tunability, thermal stability, and well-defined pore size have increased their use in both home and industrial applications, leveraging their unique properties [1,2].

Baron Axel Fredrik Cronstedt, a Swedish chemist and the pioneer of modern mineralogy, discovered zeolite materials in the eighteenth century when he noticed that moisture appeared on the surface of the mineral "stilbite" when heated [3]. The study of the chemical, physical, and mineralogical properties of natural zeolites was initiated after this discovery. Researchers laid the groundwork for the current use of natural and synthetic zeolites, even though structural studies on zeolites were limited at the time of discovery (in contrast to contemporary techniques). Based on Cronstedt's observations and research, the first synthetic zeolite-levinite was synthesized in the middle of the nineteenth century. A synthetic analogue zeolite-mordenite (MOR) [4] was synthesized in the first half of the twentieth century (following the first structural analysis of zeolite materials). Extensive research on the synthesis and the physical and chemical properties of zeolites (absorption, ion exchange, and dehydration) made it possible for zeolites to be used in industry.

1.2 Advancements in Synthesis Techniques

Zeolite synthesis has undergone significant advancements over time, resulting in substantial improvements in material properties by reducing synthesis time. Conventional hydrothermal techniques are somewhat effective but still require a prolonged time and high temperatures, which causes researchers to focus on exploring alternative approaches. Unlike conventional heating, microwave-assisted synthesis plays a transformative role in the production of zeolites. Traditional heating typically relies on thermal conduction, utilizing resistance heaters that convert the applied current into heat. It results in heating the reagent solely from the external side of the reactor via heat conduction, a time-consuming process that leads to energy loss. In microwave-assisted synthesis, the energy loss is reduced due to heat generation at the molecular level, which minimizes the time required for heat conduction and reaction, resulting in an enhanced reaction rate and a decrease in activation energy [5]. This also promotes rapid crystallization due to a higher heating rate. Unlike conventional heating, the microwave-assisted synthesis technique efficiently controls reaction parameters, performs selective heating when different microwave-absorbing characteristics are present within the reaction mixture, and reduces side reactions due to improved selectivity. It results in better reproducibility and reasonably controls the morphology, including shape and size [6].

The microwave-assisted synthesis technique has certain disadvantages associated with the vessel/reactor material, as non-metallic vessels, such as quartz, silica, or polytetrafluoroethylene, are required due to the blocking of microwaves by metals. As they interfere with the electromagnetic field, thermocouples and traditional thermometers are useless and can produce errors and sparks [7]. Standard optical fibers can measure temperatures up to 300 °C. Advanced doped or metal-free fibers may measure temperatures up to 1000-1550 °C [8,9]. Contactless solutions such as thermal cameras or pyrometers are also available. The devices make accurate temperature monitoring possible without interfering with microwave fields [10]. Unlike the slower and less flexible conventional methods, microwave heating offers fast and precise temperature control. It enables selective and localized heating, which is helpful for catalyst preparation and enhancing reaction rates. However, while hot spots can boost performance, they may also cause issues such as metal nanoparticle crystallization, which reduces efficiency [11]. Microwave synthesis remains practical and popular despite its drawbacks.

Materials Research Foundations 189 (2026) 155-208 https://doi.org/10.21741/9781644904039-6

1.3 Scope and Objective of Chapter

This chapter examines the growing importance of microwave-assisted synthesis in producing zeolites for removing contaminants from the environment. It aims to provide a comprehensive understanding of how microwave energy enhances the adsorption effectiveness of zeolites by modifying their physicochemical properties, accelerating the crystallization process, and reducing synthesis time. This chapter will cover the technical underpinnings of microwave-assisted synthesis, as well as its practical benefits over traditional techniques. The chapter also aims to examine recent developments, highlight essential applications in controlling environmental pollution, and discuss obstacles and potential paths forward in this developing discipline. To provide researchers and practitioners working on sustainable pollution mitigation strategies with a comprehensive reference, this chapter covers various topics, including applications involving actual environmental pollutants, structural and morphological characterization of zeolites, detailed analysis of synthesis techniques, and adsorption mechanisms.

2. Fundamentals of Zeolites

2.1 Structural Characteristics

The crystalline, micro-porous, hydrated aluminosilicates of zeolites typically form a three-dimensional network of $[SiO_4]^{4-}$ and $[AlO_4]^{5-}$ tetrahedra, which are linked together by oxygen atoms [12]. The zeolite structure is somehow considered as inorganic polymer due to formation of it through tetrahedral TO_4 units, where the T is Si^{4+} or Al^{3+} ion and each of the O atoms undergo to be linked by two T atoms (Fig. 1). The zeolite has structural formula $M_{x/n}[(AlO_2)_x(SiO_2)_y].wH_2O$, depending on the crystallographic unit cell, where M is an alkali or alkaline earth cation, n is the charge on the cation and w exhibits the number of water molecules per unit cell. The total number of tetrahedra per unit cell is depicted by x and y, while the ratio of y/x ranges from 1 to 5, especially in the silica zeolite case. Otherwise, it may range from 10 to 100 [13]. In the zeolite framework structure, the tetrahedra are the PBUs. In contrast, the geometric arrangements of tetrahedra are considered the SBUs, which may be simple " polyhedra, including cubes, hexagonal prisms, or cubo-octahedra" [14,15]. The SBUs undergo repetition to form the zeolite structure, which is then classified into eight groups.

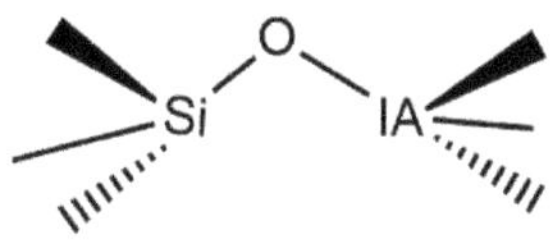

Figure 1. Silicon and aluminium tetrahedra within the zeolite framework.

The open-structure framework of zeolite features discrete channels or interlinked voids, which range in diameter from 0.3 to 20 Å and are surrounded by cations and water molecules. The tetrahedral AlO_4 within the zeolite framework contains a net negative charge, which is balanced by a cation that belongs to the IA and IIA groups of the periodic table. When an aqueous solution begins to pass through the channels and voids of zeolite, the cations interchange with other ions, containing the same sign charge, and narrow down the zeolite's channel pore diameter. The heat application is quite suitable for removing water, as the SBUs are tetrahedral arrangements

Materials Research Foundations 189 (2026) 155-208 https://doi.org/10.21741/9781644904039-6

containing four, six, ten, or more complex-membered rings. The cation's location, size, and coordination within the extra-framework can also affect the pore size. The SBUs for some zeolites are presented in Table 1.

Table 1. Classification of a few zeolite structures, SBUs, and structural properties [12]

Interlinked tetrahedra	SBUs	Structural Properties	Outcomes
4	4 oxygen rings	Four tetrahedrally coordinated Si or Al atoms form square-shaped rings, especially in small-pore zeolites	Single 4-membered rings contribute to framework rigidity.
5	5 oxygen rings	Five coordinated Si or Al atoms form a ring in medium-pore zeolites	Single 5-membered rings impart curvature to the framework.
6	6 oxygen rings	Six Si or Al atoms form a hexagonal ring	Single 6-membered rings contribute to the formation of large pores.
8	8 oxygen rings	Eight Si or Al atoms form rings in large-pore zeolites	Single 8-membered rings develop large pore openings.
8	4-4 oxygen rings	Two single 4-membered ring units form a cube-like ring	Double 4-membered rings are central to many zeolite frameworks and encapsulate fluoride ions.
12	6-6 oxygen rings	Two single 6-membered rings form large cages	Double 6-membered rings play a role in the formation of some zeolites.
16	8-8 oxygen rings	Two Single 8-membered ring units connected to form a ring, especially in extra-large pore zeolites	Double 8-membered rings develop a large pore opening.

2.2 Properties of Various Zeolite Frameworks

All zeolites share a common feature of containing a 3-dimensional structure, characterized by a 4-connected framework structure, which is formed by the sharing of TO_4 tetrahedra. The "T" in TO_4 refers to a tetrahedrally coordinated cation and forms a framework that contains channels and cavities. The general structure of the zeolite-containing framework is related to the pore opening size and the dimensionality of the channel system [14]. The ring size is crucial in defining the pore openings, called n-ring, where n is the number of T-atoms within the ring. Concerning pore size, the 8-ring is considered to have a small pore size opening, a 10-ring has a medium pore size, while a 12-ring has the highest pore size. The free diameters of the 8-, 10-, and 12-ring are determined using the calculation of the oxygen radius (1.35 Å), which are approximately 4.1, 5.5, and 7.4 Å, respectively. Still, the ring might be distorted, so the numbers are just a rough estimation [15].

Materials Research Foundations 189 (2026) 155-208 https://doi.org/10.21741/9781644904039-6

Structural features are pretty significant in different framework types of zeolites, including cages, channels, chains, and sheets. This variety of framework structural features might be designated as a cavity, cage, pentasil unit, crankshaft, double crankshaft chain, and even a 4.8^2 sheet or net, and is widely utilized and accepted for everyday use. For convenience, for example, a truncated octahedron (sodalite (SOD) unit) usually contains "six 4-rings, eight 6-rings", which is typically referred to as ($4^6 6^8$) cage. Irrespective of silicates, the Narsarsukite chain is commonly present in Aluminophosphate-5 ($AlPO_4$-5) structures, which have a framework type code (AFI). In high-silica zeolites such as Mobil-type five (MFI) and ZSM-11, the pentasil chain shares edge-sharing (5^8) cages [16]. The channel walls of various zeolites containing 1-dimensional pores usually contain 6-rings. Some zeolites contain only cages with ring sizes up to 6 without channels, such as the pure-silica clathrasils. Most zeolites contain 8-ring channels that intersect to develop 2- and 3-dimensional channel systems. This dimensional channel system plays a role in catalytic or sorption applications; for instance, coke formation may easily block the 1-dimensional-based channel features compared to the higher-dimensional channels, where alternative paths are allowed [15].

In the zeolite framework, the *ABC system* is crucial for describing the stacking sequence of layers, cages, or rings within zeolite frameworks. In metals or oxides, the ABC system is used to describe the stacking of layers of closest-packed atoms, a crystal chemistry term used to describe the various structures of zeolite types. Among the 133 zeolites, the 15 framework types are known as the ABC-6 family of zeolite frameworks, explained by the stacking of hexagonal arrays of 6-rings, such as SOD and CHA. While other zeolites are described by this crystal term involving sheets of SOD cages, such as FAU and EMT. The stacking sequence is an elegant way to describe the framework of zeolites and expresses the natural pattern as it builds the real material along with this framework [16]. Different patterns are used to classify zeolite structural frameworks, including *pore opening, structural subunit, channel system, framework density, loop-based configuration, and coordination arrangement*. The pattern of framework identifications, including coordination arrangements and loop-based configurations related to each of the T-atoms within the zeolite framework, is noteworthy in determining the classification of zeolites [15].

2.3 Classification of Zeolites

The general zeolite formula displays the monovalent and divalent cations by M and D, respectively. These two parts are easily distinguished as being mutually dependent on each other, although they are different; yet, they form a homogeneous complex, exhibiting both chemical and physical characteristics. In square brackets, the portion shows that the tetrahedral framework undergoes characterization via an overall negative charge, which increases as the ratio of Si/Al decreases. The exchangeable extra framework cations are referred to as the portion outside the square brackets and facilitate neutralizing the negative charge of the framework. The additional framework cations are coordinated via the water molecules outside the square bracket. The formula $M_xD_y[Al_{(x+2y)}Si_{(x+2y)}O_{2n}].mH_2O$ plays a role when extra-framework cations or rare earth ions are present within the zeolite structure and provides detailed analysis for advanced or synthesized zeolite structures.

The structural determination is the primary method for identifying the zeolite family, due to the significant differences in chemical composition. TO_4 tetrahedra are PBUs that form a three-dimensional network by sharing oxygens with two tetrahedra and are usually present within the silicates (tectosilicates) framework [17]. The sharing coefficient is somewhat two or smaller than

Materials Research Foundations 189 (2026) 155-208 https://doi.org/10.21741/9781644904039-6

the silicate frameworks. Theoretical networks can be estimated as a small number of complex SBUs, typically developed via a finite number of tetrahedra (up to 16), which can be easily calculated within the silicate framework. The frameworks typically differ when SBUs are arranged in various patterns and are generally coded according to their structure type, following the name of the material type used. The topological symmetry is the framework's symmetry when it is considered in its regular shape, concerning movements, and the topological relationship between tetrahedral centres (nodes) remaining intact plays a role in the characterization of each of the structure types. The zeolite species up to 52 are classified into different structure types (up to 38). At the same time, the *topological symmetry is reduced to absolute symmetry* due to the ordered distribution of tetrahedral cations and extra-framework ions (cations, holes, and water molecules), the squeezing of the framework, and the repulsion of extra-framework cations [18]. The classification and structure type code for some crucial zeolites are provided, along with their chemical formulae (Table 2).

Table 2. Classification of a few zeolites, their structure type code with schematic chemical formulae, Si/Al (mol/mol), and dominant extra-framework cations [18,19]

Code	Zeolite Name	Discovery year	Schematic Chemical Formula	Si/Al (mol/mol)	Dominant cations
ANA	Analcime, Wairakite	1797	$Na_{16}[Al_{16}Si_{32}O_{96}].16H_2O$ $Ca_8[Al_{16}Si_{32}O_{96}].16H_2O$	1.5-2.8	Na^+, Ca^{2+}
CHA	Chabazite, Willhendersonite	1792	$(Ca_{0.5}, Na, K)_4[Al_4Si_8O_{24}].12H_2O$ or $Ca_2(Ca_{0.5},K)_2[Al_6Si_6O_{24}].10H_2O$	1.4-4.0	Ca^{2+}, Na^+, K^+ Ca^{2+}
ERI	Erionite	1898	$K_2(Ca_{0.5},Na)_7\,[Al_9Si_{27}O_{72}].28H_2O$	2.6-3.8	Ca^{2+}, Na^+, K^+
FAU	Faujasite	1842	$(Na, K, Ca_{0.5}, Mg_{0.5})_{56}[Al_{56}Si_{136}O_{384}].235H_2O$	2.1-2.8	Na^+, Ca^+
FER	Ferrierite	1918	$(Mg_{0.5}, Na, K)_6[Al_6Si_{30}O_{72}].20H_2O$	4.9-5.7	Mg^{2+}, Na^+, K^+
HEU	Heulandite, Clinoptilolite	1923	$(Na, K, Ca_{0.5})_7[Al_7Si_{29}O_{72}].22H_2O$	4.0-5.7	Ca^{2+}, Na^+, K^+
LAU	Laumontite	1805	$Ca_4[Al_8Si_{16}O_{48}].16H_2O$	1.9-2.4	Ca^{2+}
MER	Merlinoite	1977	$K_6Ca_2Na\,[Al_{11}Si_{21}O_{64}].22H_2O$	-	K^+
MOR	Mordenite/ Maricopaite	1864	$Na_3Ca_2K[Al_8Si_{40}O_{96}].28H_2O$ or $Pb_7Ca_2[Al_{12}Si_{36}(O, OH)_{100}]. n\,(H_2O, OH)$	4.0-5.7	Na^+, Ca^{2+}, K^+ Pb^{2+}
OFF	Offretite	1890	$Ca. Mg. K\,[Al_5Si_{13}O_{36}].16H_2O$	-	Ca^{2+}, Mg^{2+}, K^+
PHI	Phillipsite/ Harmotome	1825	$K_2(Na, Ca_{0.5})_3[Al_5Si_{11}O_{32}].12H_2O$ or $Ba_2(Na, Ca_{0.5})\,[Al_5Si_{11}O_{32}].12H_2O$	1.1-3.3	K^+, Na^+, Ca^{2+} Ba^{2+}
YUG	Yugawaralite	1952	$Ca_2[Al_4Si_2O_{32}].8H_2O$	-	Ca^{2+}

2.4 Natural and Synthetic Zeolites

2.4.1 Natural Zeolites

Natural zeolites, microporous aluminosilicate minerals usually of volcanic and hydrothermal origin, occur as *crystalline forms or fine grains* in igneous, metamorphic, and sedimentary rocks [20]. The marine deposits remain largely inaccessible due to economic and technological

Materials Research Foundations 189 (2026) 155-208 https://doi.org/10.21741/9781644904039-6

constraints, despite the presence of zeolites in significant quantities in ocean floor sediments. Through surface mining methods like opencast extraction, terrestrial deposits, which are frequently contained in tuffs and clays, can be easily exploited, allowing for their wide industrial and environmental use. Clinoptilolite, MOR, and CHA are the most operationally significant zeolites in nature, and their special crystalline frameworks provide them with exceptional ion-exchange and selective adsorption capabilities.

Natural zeolites are particularly useful in environmental engineering and agricultural applications due to their excellent selectivity for ammonium [21] and heavy metal ions [22,23], which are among their primary advantages. Zeolite-filled sorption columns are listed in environmental applications as an effective and economical remediation method for wastewater containing radioactive isotopes and heavy metals [24,25]. This is comparable to the effective removal of ammonium ions from industrial, municipal, and agricultural effluents using natural zeolite filtration systems [26,27]. To assist in regulating the spread of pesticides and fertilizers, the agriculture sector uses natural zeolites as agrochemical transporters [28]. To improve crop yields, zeolites are also added to soil amendments to enhance the availability of nutrients and improve moisture retention. Even in aquaculture, they absorb ammonium ions, maintaining the water in fish ponds clean. Due to their ion-exchange properties, they are also utilized as feed additives to enhance animal productivity and health. Natural zeolites are also used in home environments because of their potent cation-exchange capabilities. In pet litters, they work exceptionally well as absorbents to help regulate moisture and odors [29].

Despite their numerous benefits, natural zeolites have drawbacks that prevent them from being used in certain industrial settings. Their comparatively tiny channel diameters constitute a significant limitation; for example, the most common natural zeolite, clinoptilolite, has channel openings between 0.30 and 0.40 nm [30]. This limits their use in procedures that call for separating or transforming larger gas molecules and organic chemicals by preventing their adsorption. Furthermore, zeolite deposits in nature are limited and non-renewable. The increasing global demand for specialized molecular sieves and adsorbents with particular properties has highlighted the need for synthetic alternatives. Laboratory-synthesized zeolites have been the focus of much of the current materials science research because they overcome many of the shortcomings of natural zeolites by having pore diameters, framework compositions, and thermal properties that can be precisely controlled.

2.4.2 Synthetic Zeolites

Although zeolites were first identified as naturally occurring minerals, more than a hundred structures have since been created in laboratory settings. Zeolites are naturally formed over thousands of years by the interplay of basic lake waters and volcanic ash. On the other hand, synthetic zeolites undergo development by simulating hydrothermal conditions in carefully regulated laboratory settings, which involve high pressures and temperatures. The synthetic approach is somewhat expensive, requiring clean substrates, specialized equipment, and a significant energy input. Current research trends focus on identifying more sustainable and economical raw materials, often derived from natural or industrial waste products, to reduce the cost of zeolite synthesis and its environmental impact [30].

Numerous *natural and waste-derived materials* can be used to create synthetic zeolites. Natural sources of silica are commonly used, including diatomite [31], volcanic glasses (like perlite [32] and pumice [33]), and clay minerals (such as kaolinite and halloysite [32]). Furthermore, fly ash [34], rice husk ash [35], and expanded perlite [36] waste are examples of aluminosilicate waste products that have drawn interest as affordable feedstocks for the zeolite manufacturing process (Fig. 2). However, the purity needed for premium commercial applications is sometimes lacking in zeolites made from such complicated source ingredients. Notwithstanding this *drawback*, there are substantial financial benefits to using inexpensive natural and waste substrates instead of high-purity synthesized compounds.

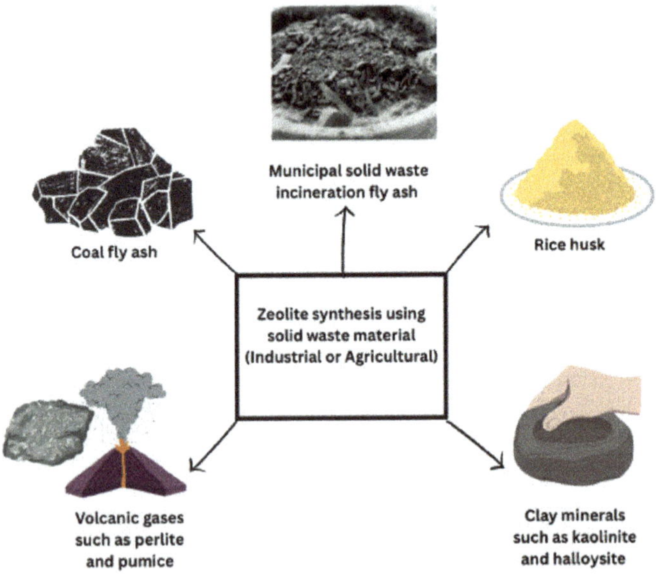

Figure 1. Different solid waste materials are used in the synthesis of zeolites to control environmental pollution.

In the 1950s, aluminosilicates were first synthesized hydrothermally. Depending on the circumstances of the procedure, zeolites crystallized in a few hours or days when aluminosilicate materials were heated in the presence of alkaline solutions at regulated temperatures and pressures [37]. In modern landscape, there are a variety of ways to produce synthetic zeolite, such as "hydrothermal synthesis (at standard or high pressure), microwave-assisted synthesis [38], method with molten salt [39], the fusion approach [40], alkali activation [41], and dialysis-based synthesis [42]". The hydrothermal synthesis replicates the conditions of natural zeolite formation and remains one of the most effective methods for producing zeolites. Typically, the process is operated in an alkaline medium (pH > 8.5) at temperatures ranging from 80 to 350 °C. Autoclaves have a series of "*dissolving, condensation, gelatinization, and crystallization stages*" while conducting the reaction at high pressure [43]. The desired zeolite phases are generated selectively by carefully regulating the synthesis conditions.

Materials Research Foundations 189 (2026) 155-208 https://doi.org/10.21741/9781644904039-6

On the other hand, *microwave-assisted synthesis* is a modern technique that speeds up the crystallization of zeolites by rapidly and uniformly heating the reaction mixture using microwave energy. Additionally, this reaction uses less energy than traditional hydrothermal methods, and even nucleation rates are increased by the regulated heating, producing finer and more consistent crystallite sizes. While regulating with low-temperature systems, microwave synthesis is somewhat beneficial as it facilitates the creation of Na-P1 and zeolite X (a type of zeolite) and provides an economical and environmentally friendly alternative for industrial-scale production. Techniques like the fusion process and the molten salt approach enable the creation of distinctive zeolite types, such as SOD and cancrinite. Still, they are less frequently employed due to their higher costs and complexity. The overall cost of synthetic zeolites produced using these methods is typically in the middle of that of natural zeolites and pure synthetic products [38].

Synthetic zeolites perform better than the natural zeolites during removal of dangerous radioactive isotopes like strontium and cesium [44,45] and synthetic zeolites also disclose a notable adsorption capability for heavy metal ions such as cadmium (Cd^{2+}), lead (Pb^{2+}), zinc (Zn^{2+}), copper (Cu^{2+}), nickel (Ni^{2+}), and chromium (Cr^{3+}) [46,47]. Synthetic zeolites play a specific role in catalytic applications due to their wider, linked channels, which remain active for extended periods. At the same time, those with smaller pores often deactivate quickly due to pore blockage [48]. Even polarity and surface properties of zeolites can be modified for various catalytic applications [49] by causing variation in the synthesis parameters, especially the ratio of silicon-to-aluminium (Si/Al). For instance, high-silica zeolites are more hydrophobic and can be used in aqueous environments as they do not cause water poisoning during acid-catalyzed processes [50]. Several disadvantages of synthesized zeolites include that they are typically made as fine crystalline powders, with individual crystals measuring only a few microns in size. Despite efforts to create synthetic zeolite granulates suitable for real-world applications, these materials remain commercially inaccessible, and the handling of synthetic zeolites may be hindered by their delicate particle nature [30]. Still, synthetic zeolites are attractive options for environmental cleanup technologies since they have been demonstrated to absorb up to twice as much oil as natural clinoptilolite [51].

2.5 Products for Zolitization

In the synthesis of zeolites, the type of zeolite produced is influenced by the chemical makeup of the feedstock, along with reaction factors including time, temperature, and pressure [52]. To make zeolites such as "hydroxysodalite, ANA, and Na-P1", higher pressures and temperatures are required. Some zeolite syntheses, including zeolite X, zeolite Na-P1, hydroxysodalite, and zeolite A, require lower temperatures (below 100 °C). In addition, the synthesis of zeolite A occurs by increasing aluminium concentration (Si/Al ratio < 1). According to microstructural analyses, the synthesized zeolites at lower temperatures typically contain smaller crystallites with less developed morphologies compared to zeolites synthesized at high temperatures. In real-world applications, zeolites with larger pore sizes, such as zeolite X and Na-P1, are preferred due to their enhanced sorption capabilities. Although using *organic templates* during zeolite synthesis increases the production costs, it still encourages the creation of particular zeolite phases. For modern synthesis processes, the *key objective* is to minimize the generation of highly alkaline effluent, aligning with sustainable and eco-friendly production practices [30].

Materials Research Foundations 189 (2026) 155-208 https://doi.org/10.21741/9781644904039-6

3. Microwave-assisted Synthesis of Zeolites

3.1 Principle of Microwave Heating

An electromagnetic radiation generator [53,54] is typically present within a laboratory microwave oven, generating wavelengths ranging from 0.01 to 1 m or frequencies from 30 to 0.3 GHz. In microwave-assisted synthesis, ionic conduction and dipole rotation convert the microwave energy into heat [55,56]. Within the reaction mixture, the microwaves are absorbed by dipoles just like water [57]. The molecule dipole moments rotate in unison and realign themselves as the microwave electric field changes [58,59]. The liquid's temperature rises due to the heat produced during this process. In the reaction mixture, the ionic conduction is facilitated by charged particles such as metal ions. The microwave field causes charged particles to oscillate, allowing them to move back and forth and collide with nearby objects, thereby producing heat. At a specific temperature and frequency, the "loss tangent" represents a solvent's capacity to transfer heat from microwave energy, which combines the two effects. Therefore, microwave irradiation speeds up the chemical reactions and their rate, enhances the product's uniformity, and allows the use of a wide range of integrated "metals and silicon to aluminium ratios" [60-62]. Meanwhile, the "hot spot" undergoes development via microwave heating, which influences particular reaction's working mechanisms, such as "heterogeneous Suzuki-Miyaura coupling and how toluene vapour adsorbs and releases from the granular activated carbon" [63,64].

In microwave synthesis, the product size is typically controlled by adjusting the rate of new particle formation and the rate at which these particles undergo crystallization. Microwave heating accelerates the dissolution of reactants, including precursors or gels, resulting in a high number of nuclei that crystallize immediately from the reactant mixture. It forms small particles with uniform sizes and a repeating growth process, which is repeated several times, resulting in large particle formation with consistent sizes [65]. The rate enhancement is considered a thermal or kinetic effect in microwave irradiation. Since irradiating polar molecules speeds up the energy flow, the mixture's temperature would rise quickly and consistently.

Additionally, the reaction is accelerated by superheating the mixture, producing hot patches [66,67]. Rapid microwave irradiation can create this hot zone, thereby accelerating the development of nanocrystals and ensuring uniform seeding in the precursor solution [68]. Microwave synthesis affects the product's crystal structure by regulating the temperature and reaction time. Due to their rapid crystallization rate, products are likely to produce a less stable, kinetically advantageous phase, such as AlPOs with high pore sizes [69,70]. By accelerating the rate of crystallization and generating kinetically overdriven growth conditions that are far from equilibrium, microwave irradiation regulates the product's shape. Anisotropic morphologies, including nanorods, cylinders, rings, and plates, are the outcome of this reaction, disrupting the isotropic development of crystals [71,72]. The processes that result in different morphologies are divided into the following categories: "micelle growth, direct attachment, development catalyzation by seeds, and shape," and are usually facilitated by stabilizing molecule's surface adherence. When paired with modified seeded growth, the processes produce products with hetero-oligomer orientation or core-shell structure [73,74].

3.2 Different Methods for Microwave-assisted Synthesis

In this modern era, advanced microwave technology has a more powerful impact for utilization in the synthesis of zeolites, including different synthesis methods assisted by microwave, including "microwave irradiation, microwave-assisted hydrothermal synthesis, microwave-assisted ageing synthesis, microwave-assisted ionothermal synthesis, along with different microwave-related methods" for synthesizing zeolites.

3.2.1 Microwave Irradiation

In zeolite synthesis, the commonly utilized synthesis involves heating/microwave irradiation. This synthesis prepares the uniform mixture by varying "aluminium to silica ratios and adding structure-directing agents (SDAS)". In general synthesis of zeolite via microwave-assisted synthesis, the dissolution of raw material, stirring to obtain homogeneous mixture and pass-through microwave irradiation resulting in nucleation and crystal growth (Fig. 26.3). Even, the morphology and structure during zeolite synthesis can be controlled by adjusting the reaction conditions including design and size of autoclaves, microwave intensity and temperature, and duration of reaction [75].

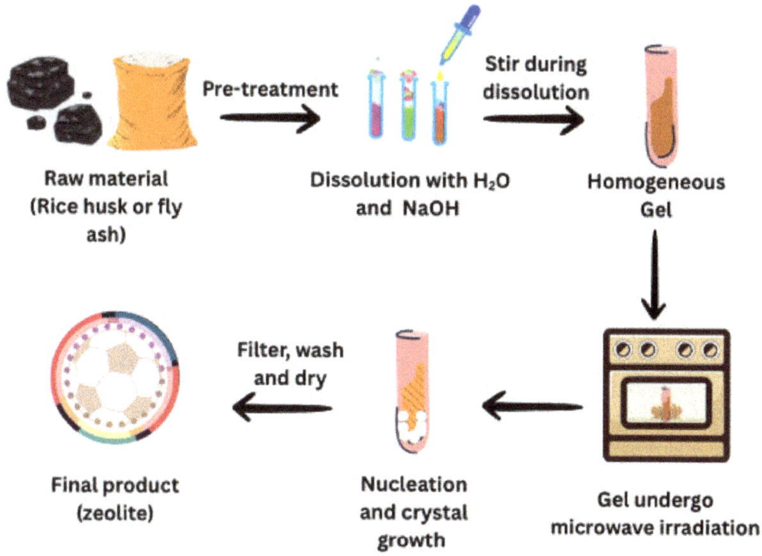

Figure 2. Zeolite synthesis mechanism via microwave irradiation.

3.2.2 Zeolite Particles

Microwave irradiation can be used to prepare zeolite particles due to its significant advantages, including short reaction time and formation of uniform crystallinity. In zeolite synthesis, the shape, porosity, and particle size can be controlled precisely by adjusting the pH, template concentration, and Si/Al ratio. Microwave methods are used to synthesize zeolites such as AFI and CHA, having various morphologies, including aggregated spheres, plates, rods, prisms, and barrels [76]. The carbon-templated mesoporous Zeolite Socony Mobil-5 (ZSM-5) zeolites are also synthesized by adjusting the amount of carbon supplied during microwave irradiation, and even meso-porosity and acid site distribution can be adjusted during synthesis [77]. In Silicalite-1 zeolites, particle size control can be obtained during microwave synthesis by varying the fluoride media, mineralizing agents, and sodium hydroxide content [78]. By microwave irradiation, the nanosized FAU-type zeolite synthesized within water-in-oil microemulsions, exhibiting improved characteristics and enhancing BET (Brunauer-Emmett-Teller) surface area [79].

Zeolite W, presenting prismatic morphology, is prepared via microwave heating, and ethylene glycol is added within it for homogeneity [80]. Compared to traditional heating, microwave heating forms MOR zeolite crystals with a pure, oval plate shape, exhibiting a greater surface area [81]. Silicoaluminophosphate (SAPO-16) zeolite is also formed using a template based on hexamethyleneimine, usually within a non-aqueous medium, under microwave radiation [82]. Irrespective of conventional heating, microwave irradiation impacts the synthesis of zeolite when the same precursor is used, resulting in the formation of phase-pure NaP rather than MCM-22 [83]. Using environmentally friendly silica sources, including rice husk ash, K-F nano-zeolites are also synthesized via microwave irradiation without the use of organo-templates [84]. Na-P and Linde Type A (LTA) type zeolites are prepared using coal fly ash within 30 min or less and play a specific role in efficiently adsorbing methyl orange dye [85]. Combining coal fly ash with biomass fly (rich in potassium) forms potassium-type zeolites with a quicker ammonium adsorption saturation, approximately 16 h under microwave, compared to 24 h under oil-bath heating [86]. Additionally, diatomite, a naturally occurring clay, was utilized to produce SOD zeolites in a microwave setting [87].

3.2.3 Zeolite Membrane

In zeolite membrane synthesis, microwave irradiation is a valuable method for creating various sizes, shapes, and pore configurations. This synthesis is beneficial as it enhances the membrane permeance and perm-selectivity for different gas combinations. The synthesis of zeolite membrane via microwave irradiation exhibits a short synthesis time compared with conventional methods (Fig. 4). NaA zeolite membranes on ZrO_2 substrates provide strong perm-selectivity for H_2/C_3H_8 (8.13) and H_2/N_2 (4.01), even surpassing Knudsen diffusion limits [88]. FAU-type zeolite membranes, porous α-Al_2O_3 substrates are also synthesized via this method, exhibiting single-gas permeation and single-liquid vapor permeation characteristics of zeolite along with perm-selectivity of H_2 and N_2 to be as high as 4.25 [89].

Materials Research Foundations 189 (2026) 155-208 https://doi.org/10.21741/9781644904039-6

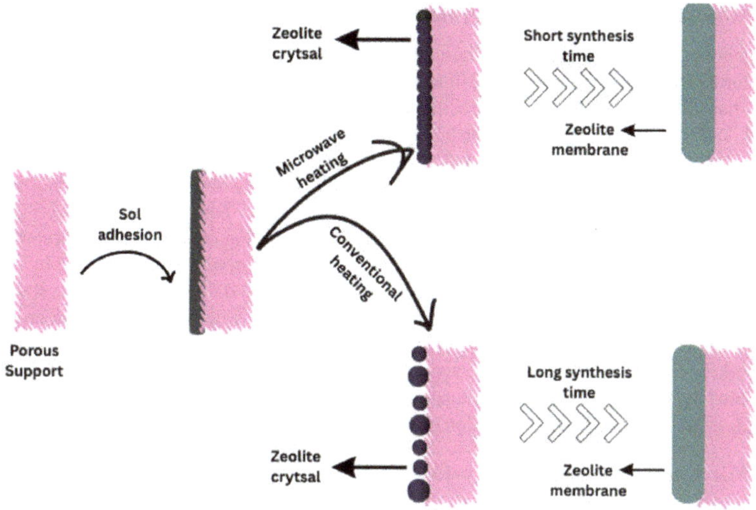

Figure 3. Zeolite membrane synthesis via microwave irradiation compared with conventional methods.

On ceramic tubes, zeolite membranes related to "deca-dodecasil 3 rhombohedral (DDR)" are produced, representing CO_2/CH_4 selectivity (190 at 0.14 MPa) and decreased thickness (1.5 μm). When 60 min pass, a fully intergrown DDR membrane develops, along with the formation of sub-micron crystals in 10 min [90]. SAPO-5 membranes are synthesized via microwave-assisted synthesis by in situ crystallization on α-Al_2O_3. On a continuous SAPO-5 membrane, small non-zeolitic pores also developed [91]. Zeolite T membranes are formed using microwave-assisted hydrothermal techniques in esterification reactions, thereby exceeding the limitations of chemical equilibrium. It plays a role in improving separation factors for water/n-butyl alcohol mixtures [92]. By microwave irradiation, the microstructures of zeolites can also be controlled, which influences the thickness, size, shape, and orientation of the crystals. Even slight modifications to microwave settings, CHA-type membranes with small crystal sizes, and significant separation factors have also been developed. During synthesis, the temperature and time of microwave heating must be controlled, resulting in CHA membranes that exhibit a high separation factor and good reproducibility [93]. In a free SDA environment, LTA zeolite layers without defects are also synthesized by controlling the membrane thickness to 2–3 μm, along with the crystal size, which is also up to 1–3 μm [94]. Furthermore, continuous *b*-oriented MFI films and pure (h0h)-oriented silicalite-1 membranes are also formed, exhibiting enhanced separation and permeation capabilities. The continuity of the membranes can also be improved by inducing specific enhancements in the temperature and time of crystallization, as well as the density of nuclei [95,96].

3.2.4 Microwave-assisted Hydrothermal Synthesis

In this synthesis, conventional hydrothermal heating is followed by microwave irradiation for zeolite formation, resulting in better zeolites with improved surface area, crystallinity, and higher quality than those obtained through traditional heating. EMT zeolites are formed by the seed-directed, organic-template-free approach, resulting in high crystallinity (162%). The formed zeolites, obtained by microwave-assisted hydrothermal synthesis, exhibited good crystallinity, improved morphology, and smoother surfaces [97]. HZSM-5 zeolites are also synthesized via microwave-assisted hydrothermal techniques, which outperform conventional HZSM-5 zeolites produced by long-term hydrothermal heating in terms of improving crystallinity, surface area, and aromatic selectivity [98]. The size of crystals and the zeolite's morphology are adjusted by altering the conditions regarding hydrothermal and microwave. The crystalline SAPO-34 crystals were prepared for the first time via microwave irradiation, where the irradiation power had an impact on both the structure and size. In contrast, the time of microwave influenced the structure and framework based on spherical or cubic aggregates [99]. Even the NaA nanoparticles doped with Ag (60 nm) tend to form via microwave-assisted hydrothermal synthesis using silica regenerated from industrial waste without the addition of organic compounds [100].

3.2.5 Microwave-assisted Ageing Synthesis

To enhance seeding and precursor uniformity, this ageing synthesis is combined with microwave synthesis before hydrothermal synthesis and is utilized instead of traditional stirring. Inexpensive HMI templates are employed in the development of tiny, highly porous ZSM-5 (BET surface area up to 333 m^2/g), which requires microwave ageing at 60 °C for 1.0 h [101]. The provided temperature causes an impact on nucleation, crystallinity, and ageing conditions that need to be managed carefully. The microspheres of zeolite-P modified by bismuth, containing micropores, are also formed using microwave-assisted ageing for 1.0 h, resulting in the uniform formation of bismuth oxychloride nanoplate precursors with a thickness of 45 nm. This synthesis method results in a decrease in reaction time of at least 83% [102]. Spherical MCM-22 zeolite with central pores and a particle size of 8 μm can also be synthesized by this microwave-assisted ageing synthesis, resulting in shortening the crystallization duration to 4 days and increasing the ratio of Si/Al, following 1.0 h of ageing at a specific temperature (60 °C) [103]. Even the morphology of MCM-22 zeolite can also be adjusted by altering the temperature and ageing time during synthesis. The synthesis of pure ANA zeolite, with an average particle size of 9.0 μm, is performed using this method, which involves 0.5 h of microwave ageing followed by 48 h of conventional heating at 160 °C [104]. This synthesis methodology offers the benefits of reducing the synthesis duration for zeolite P and enhancing its zeolite surface area, along with high crystallinity (97%) [105].

3.2.6 Microwave-assisted Ionothermal Synthesis

Microwave-assisted ionothermal synthesis is an alternative approach to the conventional method of dissolving zeolite in water, enabling synthesis via microwave. In microwave-assisted ionothermal synthesis, "ionic liquids" serve as solvents and SDAs for the crystal-based products. By combining ionothermal-dependent techniques with microwaves, the "nano-sized containing NaY zeolite" is synthesized by utilizing metakaolin, improving the pore volume (0.37 cm^3/g) and surface area (778 m^2/g). The formed nano-sized zeolites disclosed a decrease in micropore volume (0.29 cm^3/g) compared with conventionally synthesized zeolites [106]. The microwave irradiation-assisted ionothermal synthesis plays a crucial role in developing hierarchical structured FeAlPO-

5, thereby enhancing the mesoporous volume and catalytic performance by facilitating the formation of tiny mesopores [107]. This synthesis also plays a key role in forming a high-iron concentration-based hierarchical Fe-LEV zeolite structure from a eutectic mixture, with structure-directing effects that increase the surface area [108]. Pure AEL-type aluminophosphate zeolites ($AlPO_4$-11 and SAPO-11) are also formed by using this method. The formed pure AEL-type zeolite structures, characterized by cuboid-like crystallites, are weakened by optimal directed growth, which typically occurs through fast crystallization during microwave irradiation, resulting in an impact on the zeolite morphology [109]. This synthesis approach controls the morphology, surface area, and structure of the zeolite.

3.2.7 Other Methods for Zeolite's Microwave-assisted Synthesis

Microwave induction typically complements other methods in zeolite formation and modifies its properties to enhance production. By combining microwave-assisted micro-explosion synthesis with H_2O_2 decomposition from interwoven mesoporous networks in different zeolites such as "beta-P, silicate-1, H-form ZSM-5, zeolite Y, SAPO-34, and MCM-22-P". The formed zeolites enhance the superior acidic sites by maintaining microporosity and catalytic performance [110]. By de-templating ZSM-5 and combining it with cetyltrimethylammonium bromide and NaOH, followed by microwaving for 1.0 h at 150 °C, a hierarchical porous ZSM-5@MCM-41 is formed. The formed porous zeolite exhibited an improvement in pore volume (0.76 cm^3/g) and surface area (787.8 m^2/g) [111]. In the metathetic synthesis of Ag-MWO_4 on zeolite A, where M represents metals such as Fe, Ni, and Sr, microwave irradiation is used to immobilize small Ag particles (3-5 μm) within a porous MWO_4/zeolite matrix [112]. Due to microwave-assisted seed secondary growth, MFI-type zeolite is developed, resulting in decreased synthesis time and improved gas perm-selectivity [113]. Microwave irradiation, combined with the ozone or calcination process, speeds up the template removal process in MFI membrane production. The process results in minimizing the degradation time for MFI membranes (up to 1.2 μm thick) to 44 h [114]. The different zeolites prepared via microwave-assisted synthesis are summarized in Table 3.

Table 3. Important zeolites prepared via microwave-assisted synthesis

Type of Zeolite	Substrate/ Conditions	Morphology	Size [μm]	Applications	Synthesis Method	Reference
Zeolite T	373 K, 30-48 h	Rod-like	3	-	Microwave irradiation	[115]
MFI	438 K, 1.0 h	Puck-like	0.25	-	Microwave irradiation	[116]
Zeolite A	0.33 h	Puck-like crystal	0.04-0.08	-	Microwave irradiation	[117]
Silicalite-1	400 K, 3.0 h	Puck-like crystal	0.1	-	Microwave irradiation	[118]
ZSM-5	438 K, 1 h	-	0.8-1	-	Microwave irradiation	[77]
FAU	Porous α-Al_2O_3, 363 K, 20 min	-	-	Selectivity between mixtures of water-ethanol and water-isopropanol	Microwave membrane	[89]
NaA Zeolite	ZrO_2, 363 K, 15-30 min	-	-	H_2/C_3H_8 and H_2/N_2 selectivity	Microwave membrane	[88]

NaA Zeolite	α-Al_2O_3, 393 K, 30 min	-	-	Desalination	Microwave membrane	[119]
Zeolite CHA	Stainless steel, 373-448 K, 4-30 h	-	-	Selectivity between mixtures of water-ethanol and water-isopropanol	Microwave membrane	[93]
Silicalite-1	α-alumina tube, 443 K, 1-4 h	-	-	n-butane/i-butane selectivity	Microwave membrane	[96]
NaA	Microwave 60 min, 313 K for 3 h	Clustered spherical/cubic	0.04-0.09	-	Microwave-assisted hydrothermal	[100]
SAPO-34	Microwave 140-210 W, then 473 K up to 3 h	Spherical and cubic	0.4-0.5	-	Synthesis using microwave-assisted hydrothermal	[99]
MFI	Hydro-thermal treatment, along with heating for 1 h at 443 K + Microwave for 1 h at 443 K	Cubic	0.14	-	Synthesis using microwave-assisted hydrothermal	[120]
BZSM-5	Microwave 30 min, heat at 453 K for 1-3 days	Hexagonal	3-6	-	Microwave-assisted hydrothermal synthesis	[121]
SAPO-34	Microwave at 200 W for 1 h and then heat at 473 K for 18 h	Cubic	0.126	-	Microwave-assisted hydrothermal synthesis	[122]

3.3 Microwave Techniques in Modification of Zeolite Catalysts

In significant molecule transformation, the mesoporous aluminosilicates of the M41S family play an important role in hosting structural heteroatoms and anchored functional groups. They act as selective catalysts due to the non-availability of firm acidity in crystalline zeolites, which have a pore size of 20-100 Å [13,123]. Si, Al-MCM-41 has been formed via aged precursor gels (mixed silica source), having good thermal stability, and usually takes 1.0 h to synthesize at optimum temperature (150 °C). The condensation reaction of the silicate network is increased via microwave heating. A variety of methods have been in use to prepare "Si and Al-MCM-41" that usually require microwave irradiation power up to 650 W in a microwave oven for 1-30 min [126], and photoluminescence spectroscopy [127] plays a role in providing a detailed investigation regarding the zeolite formation for this reaction.

Materials Research Foundations 189 (2026) 155-208 https://doi.org/10.21741/9781644904039-6

3.4 Microwave Activation of Catalytic Reactions

To modify the zeolite-based catalysts, ion exchange or impregnation is used, along with microwave heating, to provide the necessary heat input [124]. The Na^+ in SOD and FAU have been replaced by Co^{2+} quickly via microwaves [125]. Even some La-H-ZSM-5 zeolites, formed by varying levels of La^{3+} exchange, are used in microwave synthesis and exhibit acidity typically associated with catalytic cracking activity [126]. In zeolite Y, the palladium clusters are usually encapsulated through microwave-mediated ion exchange between $[Pd(NH_3)_4](NO_3)_2 \cdot H_2O$ and Na-Y, resulting in the formation of catalysts that exhibit exceptional activity for complete CO oxidation at room temperature [127].

The zeolite-based catalysts of Cu/ZSM-5 are prepared by dissolving $CuCl_2$ in Na-ZSM-5, then heating the combined solids in a microwave, which plays an active role in CO activation and NO_x reduction [128,129]. The X-ray diffraction patterns exhibit the characteristic peaks of the zeolite. At the same time, Differential Thermal Analysis curves do not reveal any peak related to the melting point of $CuCl_2$ up to a critical weight ratio. The infrared technology has shown higher CO adsorption by the Cu-ZSM-5 formed using microwave irradiation compared to the sample prepared by conventional ion exchange. In Na-Y zeolite, the copper chloride and other metal salts (Li, Au, Ni, and Ru) are dispersed using this technique. They are also helpful in forming intrazeolite Au (I) carbonyl and nitrosyl complexes [130].

Several catalysts required for the conversion of syngas to methanol have also been prepared by H-ZSM-5 by three different techniques, such as "solution ion exchange with Cu (II) acetate, thermal solid-state exchange with $CuCl_2$ at ≥ 400 °C, and microwave solid-state ion exchange with $CuCl_2$". In thermal solid-state exchange, the Cu^{2+} is produced primarily. In contrast, in solution, ion exchange and microwave solid-state ion exchange methods typically form products that are usually dominated by Cu^+ species. The definitive results are unavailable, as no microwave temperature record is available. The microwave-treated materials of BET at 145-158 °C, having surface areas approximately 900 m^2/g, are somewhat similar to thermally treated catalysts at temperatures up to 850 °C [131]. The microwave-assisted dispersion is also expanding to different systems, including MgO/KL [132], $ZnCl_2$/Na-Y [133], and [MoO_3, $CuCl_2$, $NiX_2 \cdot 6H_2O$]Na-ZSM-5 (X = Cl and NO_3), Na-Y, and MCM-41 [134]. Even during $AlPO_4$-5 crystallization [139], unstable dye molecules have also been encapsulated by rapid microwave synthesis to develop zeolite photocatalysis.

3.5 Factors of Microwave-assisted Synthesis Impact on Zeolite

Microwave-assisted synthesis is a rapid alternative in zeolite synthesis, as it addresses factors such as prolonged crystallization times and high energy consumption. The properties of zeolites undergo to be improved, exhibiting high crystallinity and adsorption capacity (Fig. 5). The crystallization time for zeolites prepared via microwave-assisted synthesis is accelerated as the example of beta-type zeolites displayed an improved specific surface area (463 m^2/g) in just 4.0 h compared with conventional hydrothermal techniques (take 48 h) for attaining particular surface area up to 483 m^2/g. The properties improve due to constant microwave energy, increasing nucleation and crystal growth rates [135]. The microwave-assisted synthesis prefers pure-phase formation; for instance, the zeolite NaP [83] is obtained without any impurity compared with conventional hydrothermal methods, resulting in the formation of MCM-22. Even the zeolites obtained via microwave irradiation result in higher crystallinity with fewer impurity phases, such as in zeolite A (formed from metakaolinite, a highly crystalline product, which is received in 4 h with minimal impurities compared with conventional methods (long durations) [136].

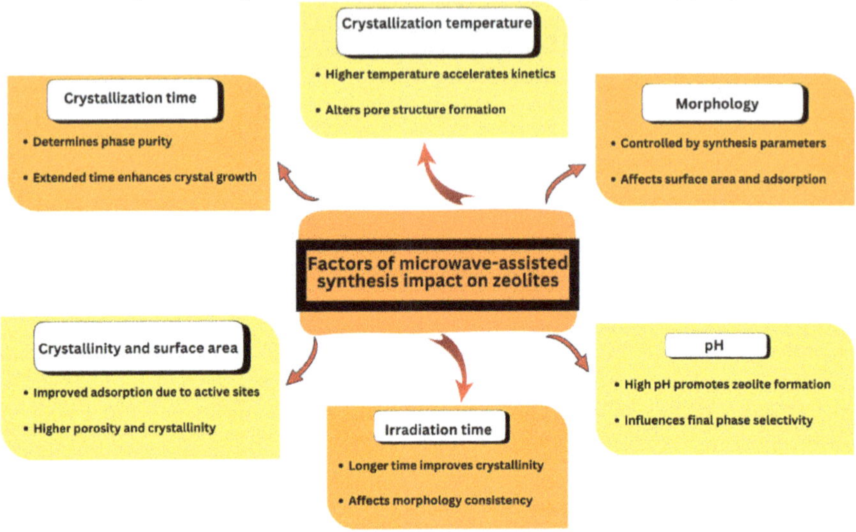

Figure 4. Different microwave-assisted synthesis factors impact on the zeolite.

The synthesized zeolites possessed a higher surface area with well-defined morphology and specific pore structures. For example, zeolite A (formed by coal fly ash) results in well-defined cubic crystals suitable for use as an adsorbent in environmental applications [137]. Even the nanosized high-silica beta zeolites (BEA) are also prepared via microwave irradiation at temperatures up to 100 °C (for 20 min), resulting in metastable nanoparticle formation at first. After subsequent hydrothermal treatment, uniform nanometer-sized BEA was synthesized [138]. The process of dealumination and desilication is enhanced by microwave-assisted synthesis, which enhances pore structure and acidity. For instance, the BEA zeolite synthesized by the microwave-assisted method portrays efficient dealumination [139]. In ZSM-5, microwave-assisted synthesis facilitates the generation of an ordered and mesoporous surface. At the same time, BEA and MOR zeolites are also prepared via microwave-assisted synthesis, resulting in enhanced catalytic activity due to moderate acidity. Even the synthesis time is a key factor influencing zeolite formation, resulting in a 50% decrease and an increase in cation exchange capacity of up to 85% [140].

3.6 Selectivity

In the microwave-assisted synthesis of zeolites, selectivity usually occurs through two mechanisms. At first, one reaction pathway is favoured due to the rapid heating rates over others. For instance, in naphthalene sulphonation, as the heat rate increases, the product ratio for the two isomers shifts from 55:45 to 95:5. In that case, two selectivity patterns are predicted under Arrhenius kinetic models. At the same time, the first path is generally observed experimentally [141]. Selectivity is based on reaction temperature conditions rather than heating rates, and

Materials Research Foundations 189 (2026) 155-208 https://doi.org/10.21741/9781644904039-6

microwave irradiation helps achieve rapid heating rates. Secondly, in a heterogeneous system, the selectivity is induced by differentia, heating effects. For instance, pure zeolite Y is formed during synthesis under microwave heating at a temperature of 150 °C within 20 min. In conventional heating, pure zeolite Y is formed, resulting in only 10% conversion during synthesis after 45 min. Traditional heating causes Na-Y to disappear entirely after 180 min, resulting in the domination of P-type zeolite [142,143]. The kinetic control is even seen due to the suppression of undesired phases by microwave heating. Selectivity phenomena are typically estimated within microwave-activated catalytic reactions and may also be observed in the cryptomelane/bixbyite system. However, there is still a need to observe keenly the positive differential heating effects within microporous material synthesis [144].

4. Basics of Adsorption Process

4.1 Adsorption

The term "adsorption" means the accumulation of molecules, atoms, or ions from a liquid phase at the solid surface. The nanoporous material used in adsorption typically has a high specific surface area, which is achieved through its porosity. To categorize nanoporous material, it usually depends on pore size, such as microporous (dp < 20 Å), mesoporous (20 Å < dp < 500 Å), and macroporous (dp > 500 Å) structures. Microporous zeolites have excellent adsorption capabilities [145].

4.2 Adsorption Mechanism

The adsorption mechanism plays a crucial role in optimizing the features of adsorbents, including functional groups, porosity, and structure, to remove water pollutants efficiently [146]. In the adsorption mechanism, the adsorbent and adsorbate typically interact by moving across the water-adsorbent interface, thereby minimizing the free energy in accordance with thermodynamic principles [147]. The interaction between organic and inorganic pollutants and zeolite-based adsorbents typically occurs through several mechanisms, including electrostatic attraction, π-π stacking, hydrogen bonding, and pore filling, which act simultaneously [148]. In the adsorption of pollutants, two phenomena, such as trapdoor phenomena and guest-induced framework deformation, significantly influence the adsorption process [149]. In trapdoor phenomena, cation relocation is triggered by adsorbate interaction, initially obstructing the pore and allowing the cation to undergo displacement, thereby enabling a molecule to enter the pore. In guest-induced framework deformation, framework deformation is induced by temperature or guest species and is associated with the flexibility of the framework [150,151]. Due to its high selectivity, the trapdoor process in zeolite adsorption is gaining increasing interest and is induced by H_2O- and CO_2-based molecules. The molecules interact strongly with the additional cation framework of the zeolite, exhibiting sensitivity to external stimuli such as temperature [149]. For diffuse, close-fitting molecules, such as ethene adsorption on molecular sieve ITQ-55 and CO_2 adsorption on RHO, zeolite flexibility is crucial for adsorption [152,153].

4.3 Thermodynamics in Adsorption Process

Adsorption isotherms usually study adsorption thermodynamics. The adsorption isotherms at constant temperature and different pressures generally measure the equilibrium amount of adsorption at the surface. The Type I isotherm is typically observed in microporous solids, such as zeolites, which exhibit steep adsorption due to strong interactions between the adsorbate and adsorbent [145,149]. In Type IV isotherms, hysteresis loops are observed due to capillary condensation and are common among mesoporous solids. While the Langmuir model applies to monolayer adsorption on homogeneous surfaces, it has limitations for heterogeneous materials, such as zeolites [154]. The surface heterogeneity is determined using accurate models, such as the Toth, Sips, and Unilan models [149]. By introducing a parameter n, the Toth model is used for adjusting surface heterogeneity [155]. The Sips model plays a role in determining a Gaussian energy distribution while undergoing deviation from Henry's law at low pressures [156]. For condensable vapours, models based on the Dubinin–Radushkevich and Dubinin-Astakhov equations are suitable for describing micropore filling mechanisms. Adsorption may involve two ways, including physisorption, which exhibits weak forces ($\leq$ 50 kJ/mol), or chemisorption, which depicts chemical bonding ($\geq$ 50 kJ/mol), and isosteric heat trends estimate the strength and nature of adsorbate-adsorbent interactions, influencing the process design and energy requirements for desorption [149].

4.4 Diffusion-controlled Process

In a diffusion-controlled process, adsorption involves the transfer of external mass, which is governed by the flow pattern and the shape of the adsorbent. This process is typically modelled via a linear driving force. In microporous material like zeolites, the diffusion-controlled process usually occurs through successive jumps between adsorption sites [162] and is modelled by Crank's solution for spheres [149]. The Knudsen diffusion is somehow dominant in mesoporous material, while Fick's law is applied in macropores. When adsorbents are bonded with different binders, including clays or silica, slow diffusion occurs because the pores in the adsorbents are blocked [157,158]. In molecular sieving, kinetic selectivity relies on a narrow pore size distribution [1], resulting in the characterization of kinetics on elementary, microscopic, and macroscopic scales [159].

4.5 Computational Methods

Computational methods are employed when adsorption on zeolites is complex and requires experimental determination. Even Monte Carlo simulations play a specific role in determining thermodynamic properties, such as adsorption isotherms and heats of adsorption [160]. The simulations are based on statistical mechanics, which determines pore volume, size distribution, and surface area of adsorbents, primarily when the structure of the adsorbent is known. Molecular dynamics simulations are used at a wide level to compute transport properties, including the determination of diffusivities and mean square displacements. By computational study, these studies typically approach the screening of real or theoretical adsorbents for the separation of alkane isomers [161], the capture of CO_2 [162], the separation of alkenes from alkenes [163], and the study of other mixture adsorption [164].

4.6 Commercial Zeolite-based Adsorbent Manufacturing

The goal of commercial manufacturing of adsorbents is to develop high-quality products with minimal material loss. The manufacturing process typically begins with the zeolite synthesis stage, where zeolite powder is efficiently formed with high purity and yield. Different ambient pressure reactors are used to synthesize low-silica zeolites, while hydrothermal synthesis is preferable for high-silica zeolites, such as Beta and ZSM-5, under high-pressure conditions due to their structural properties. Commercial synthesis reactors are typically differentiated in manufacturing by two methods: heating and agitation methods, which impact crystallization and purity. The zeolite powder is washed after synthesis to remove any residual by-products and then dried until it reaches a friable state, ready for use in the next agglomeration stage. The zeolite powder may undergo ion-exchange and other modifications before being subjected to the agglomeration step during the manufacturing process. This agglomeration approach has certain potential for products in which the need for ion-exchange, albeit at high levels (50-70%), is not required, and they exhibit facile modification, favourable ion-exchange behaviour, and are non-toxic [165].

The following manufacturing step is agglomeration, which includes extrusion and bead forming, standard methods. In both extrusion and bead-forming processes, mixing is the first step in incorporating binders and additives, such as extrusion aids, which act as a glue to strengthen the final product. In extrusion, the mixture forms a wet paste that is passed through an extruder to form rod-shaped particles. Extrusion aids facilitate the abrasive nature of zeolites, making for easy processing and improved plasticity. Standard dies, which contain hole sizes ranging from 1/8 to 1/16 inch, play a role in determining particle diameter. At the same time, extrudate length is usually controlled by natural breakage or a rotating chopper linked with the extruder. In the bead-forming process, rotational motion is used to create spherical particles in an apparatus that includes pan-granulators, which shape products in a rotating, inclined pan. Even Nauta forming, a mixing method that employs dual-axis rotation in a conical bowl, facilitates achieving similar results. The agglomeration approach aims to produce target-size particles efficiently with suitable yield and throughput [166].

The agglomerated zeolite particles typically possess sufficient intrinsic strength, enabling them to withstand handling and conveyance to the drying and calcination stages. It is essential for the manufacturing process. During this manufacturing phase, the temperature increases steadily to eliminate moisture and other volatiles and to set the binding agent thermally [165]. In many products, binders such as clay ensure compatibility with zeolites due to their aluminosilicate structure and high surface hydroxyl density. The binders enhance physical strength, usually measured by crush strength and attrition values. The particle strength must be suitable for use in adsorption vessels with service lives exceeding 10 years [166]. The bonded network encapsulates zeolite crystallites through binder setting, resulting in enhanced durability and requiring a temperature above 600 °C, which is near the zeolite's thermal limit. Hydrothermal damage is prevented through controlled temperature ramping and the use of adequate purge gas, ensuring micropore accessibility while maintaining residual moisture below 1 wt% [165,167].

In forming zeolite adsorbents in any shape, the emerging technique is 3D printing, offering design flexibility; however, it is currently confined to laboratory-scale applications, although some commercial providers exist [165]. The product's final shape typically contains a binder content that ranges between 5 and 30%, requiring a balance. As the binder level increases, it improves abrasion resistance but minimizes porosity, resulting in diffusional limitations and diminished adsorption performance [149]. To address this concern, binderless zeolitic adsorbents are now available on the market that utilise an intermediate binder to shape the material, and are typically followed by hydrothermal or thermal processes. It reduces the use of binder and creates a zeolite-based adsorbent entirely composed of zeolite, exhibiting improved adsorption capacity and kinetics, and suppressing binder-related diffusion barriers [168].

4.7 Adsorption Processes

4.7.1 Swing Adsorption Processes

Swing adsorption is a crucial technique, also known as temperature swing adsorption (TSA) or pressure swing adsorption (PSA), in commercial gas adsorption operations. The mechanism of this technique works by cycling either temperature or pressure to promote separation [169]. In this process, two beds must be arranged in parallel to facilitate a basic swing adsorption configuration. The one bed must be regenerated via inert purging, pressure or temperature adjustments, or adsorption [149]. The best performance indicators in this process involve product purity (the molar fraction of the intended product), recovery (the ratio of the required component in the product to the feed), and adsorbent productivity (the amount of adsorbent and the amount of feed processed per time) [170]. When selecting an adsorbent, the working capacity is crucial in determining the difference in loading between the adsorption and desorption processes, which are influenced by both the isotherm shape and adsorption capacity. TSA is suitable for removing pollutants at < 10%, typically < 2%, while PSA is usually employed for bulk separations in which the target component can surpass 10% of the stream [169]. Different variants of the swing adsorption process have also been proposed, which could be used frequently in hybrid processes, including electric swing adsorption (ESA), rapid pressure swing adsorption (RPSA), vacuum swing adsorption (VSA), and vacuum pressure swing adsorption (VPSA) [149,171].

4.7.2 Simulated Moving Bed Adsorption

Simulated moving bed (SMB) adsorption is an adsorptive separation process performed in the liquid phase, also known as true moving bed (TMB). The purpose of developing the SMB process is to overcome the TMB limitations by eliminating the physical movement of adsorbent particles [172]. This process does not utilise a patented rotary valve system that requires switching inlet and outlet ports; Broughton and Gerhold first introduced this innovation in 1957 [173]. The SMB system has several sections or zones that manage the flow of components throughout the system, allowing different processes to co-occur, including adsorption, desorption, and displacement, which occur cyclically. This process consists of four interconnected sections or zones, along with an eluent or diluent (D), which facilitates the separation and desorption. D can displace stronger adsorbates, which can then be removed via distillation and recycled [149]. Section 1 (desorption of component A) is divided into four sections, using D to displace adsorbent A, resulting in a rich A extract. In section II (desorption of component B), adsorbent B is displaced by adsorbent A and diluent (D). In section III (adsorption of A and displacement of B and D), adsorbent A is adsorbed, and adsorbents B and D are removed as raffinate. Section IV (desorption of D and final separation) tends to recycle the D, and the liquid phase and inlet or outlet ports flow from section I towards IV. In industrial applications, zeolites are used as adsorbents to separate normal paraffins, notably for separating normal from iso-paraffins, and for the separation of xylenes [172].

4.8 Post-synthetic Modification of Zeolites

To achieve the desired properties, zeolites often require post-synthetic treatments after synthesis, with tailored compositions, to enhance their catalytic or adsorptive performance [149]. The most common modification treatment for zeolites is ion-exchange. This treatment is conducted in an aqueous solution containing a high concentration of the desired cation, resulting in modifications to the material's acid-base, redox, and textural properties, including pore sizes and interactions

Materials Research Foundations 189 (2026) 155-208 https://doi.org/10.21741/9781644904039-6

with adsorbates. The zeolite undergoes filtering, washing, and drying through repeated treatment, and is favourable for small cations, including metals or ammonium. The large organic SDAs are unsuitable for this process due to size limitations [174].

The other treatment includes surface functionalization, such as impregnation or ion exchange of transition and noble metals (Ag and Cu), which enhances catalytic and adsorption characteristics for gases, mainly including olefins, H_2, and CO [175]. This treatment requires chemical vapor deposition and transfers metal reactivity to the zeolite surface [174]. Even non-metallic functional group grafting, such as amine grafting, is crucial in enhancing the CO_2 capture and minimizing the hydrophilicity by sustaining CO_2 affinity. Even amine-grafted mesoporous silicas are also suitable for this type of modification [176].

Silanization is a modification treatment that results in deactivating the external surface acidity and tuning the pore properties using chemical vapor deposition. Alkoxysilanes, such as tetramethylorthosilicate or tetraethylorthosilicate, are used in this treatment and impact the catalytic behaviour [177]. In preparation of catalyst or adsorbent from zeolites, activation can be done through calcination [178] to remove organic SDAs by combustion in air and transform ammonium forms into acidic H, resulting in annealing of silanol defects. In para-containing organic SDAs, the phosphorus is removed by hydrogenation followed by calcination [179]. By promoting the hydrolysis of T-O-T bonds, steaming is another modification treatment that alters the acidic and structural characteristics, causing dealumination, desilication, and the formation of meso-porosity. This treatment results in a loss of crystallinity, which ultimately depends on the severity of the treatment [180]. Different methods regarding post-synthetic treatment of zeolites are mentioned in Table 4.

Table 4. Post-synthetic modification methods for zeolite-based adsorbents [165]

Modification Methods	Framework Impact	Influence on Adsorption	Example
Ion-exchange	Alters the non-framework cations composition	Influences adsorption selectivity and equilibrium isotherms	LiX, CaX, and 3A, 4A, 5A
Impregnation	Non-Framework Composition	Change equilibrium isotherms and pore size tuning	Cu (I) Y for a selective CO adsorbent
Dealumination	Eliminates framework Al atoms, increasing Si/Al ratio	Produces high-silica, shifts the equilibrium isotherms, and uses low-polarity adsorbents	High silica for hydrophobic separations
Desilication	Framework Composition by extracting Si atoms	Modifies equilibrium isotherms and enhances meso-porosity and diffusion	ZSM-5 (Mesoporous)
Alumination/Zincation	Compositional framework by insertion of Al or Zn atoms	Improves adsorbent polarity and active sites, and influences equilibrium isotherms	Increasing adsorbent polarity
Framework Substitution	Replaces Si or Al in the framework with heteroatoms	Alters electronic properties, shifts the adsorption strength, and changes the kinetics	Ga-ZSM-5, Ge-modified zeolites
Binder Conversion	Macropore/Mesopore	Improves adsorption kinetics and pore accessibility	Binderless X and A zeolites

Materials Research Foundations 189 (2026) 155-208 https://doi.org/10.21741/9781644904039-6

4.9 Impact of pH Variation on Adsorption

The pH significantly affects the adsorption efficiency and alters the surface charge of the adsorbent and the chemical behaviour of solutes. The surface of the adsorbent is typically influenced by the pH, undergoing protonation and deprotonation, resulting in a range of charges from positively charged to neutral to negatively charged. The alterations impact electrostatic interaction with the solute [181]. At different pH levels, the solute speciation, particularly for metal ions, undergoes transformations due to polymerization, complexation, or precipitation [182]. At low pH, the functionalized zeolites having thiol and dithizone groups exhibit reduced metal ion adsorption due to H_3O^+ competition. While at high pH values, adsorption increases due to surface deprotonation. Metal hydroxide precipitation limits further adsorption at pH above 7 [183].

At a pH of 4-5, the natural zeolite polypyrrole composites facilitate vanadium adsorption through the deprotonated amino group; however, at higher pH levels, their efficiency decreases due to vanadium polymerization [184]. Zeolite Y-chitosan magnetic composite at pH 2 exhibits the most effective removal of Cr (VI) due to a strong electrostatic attraction between protonated sites and $HCrO_4^-$ anions [185]. At pH 5 and 3, the amino-functionalized MCM-41 exhibits maximum removal of Pb (II) and MnO_4^-, respectively, and is influenced by surface charge and solute protonation. At pH 2, clinoptilolite-activated carbon macro-composite tends to remove Acid Orange 7 up to 99% due to the positively charged surface [186]. The dye adsorption by MnO_2-modified MCM-41 is typically enhanced at pH 2.9 due to the negative surface charge, resulting in approximately 85% removal at pH 10 [187].

5. Applications of Microwave-assisted Synthesized Zeolites

Zeolites are chemically and thermally stable; even conventional aluminosilicates do not lose their crystallinity when steamed at high temperatures due to their stability up to 700 °C. They are readily soluble in strong bases and acids [1,188]. To reduce framework-based on aluminium in the synthesis of FCC catalyst Ultra Stabilized Y zeolite, steaming is frequently employed post-synthetically, improving stability and customizing defect structures [189]. High- and pure-silica zeolites can tolerate high temperatures (up to 1300 °C), nearly similar to quartz's melting point (1713 °C). Even the high- and pure-silica zeolites resist chemicals and do not dissolve in concentrated bases and hydrofluoric acid [190]. The crystallinity of zeolite can be reduced or diminished using excessive mechanical stress, such as grinding [191].

The porosity within the zeolite's structure is a key characteristic in ion exchange and other applications involving the adsorption of environmental pollutants. In contrast, in some applications, exact size and shape selectivity is an essential requirement, and their consistent and narrow pore size distribution offers better outcomes. To achieve the best results, zeolites undergo modification via synthetic or post-synthetic methods to incorporate specific properties within the zeolite pores by adding extra-framework components with redox and acid-base characteristics. Different LTA zeolite-based types (3A, 4A, and 5A, and FAU), which are primarily used for zeolites (mostly X and Y), are the most commonly employed in applications, including the removal of environmental pollutants, largely due to their early commercial availability [149]. In environmental pollution control, zeolites play a crucial role in reducing contaminants by efficiently adsorbing metal ions and dyes, as well as absorbing gases such as ammonia, VOCs, and other harmful pollutants to mitigate environmental pollution (air, soil, and water) (Fig. 6).

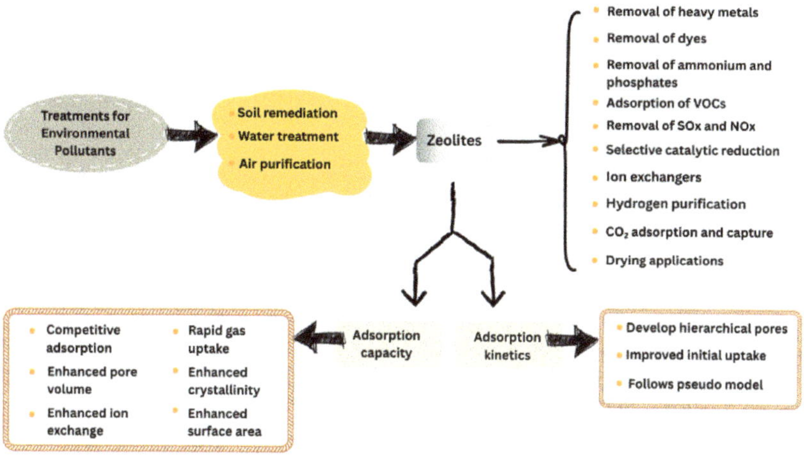

Figure 5. Applications of zeolites in mitigating environmental pollution.

5.1 Zeolites as Ion Exchangers

To control environmental and wastewater pollution, zeolite-based adsorbents are used to soften water in laundry detergents. The initial application of various zeolite-based adsorbents, including A, X, CHA, and MOR, was made at an industrial level as ion exchangers, with properties estimated in the 1950s [192]. The pore size of the zeolite-based adsorbents is small, resulting in ion-exchange with the limitation that bigger cations, especially those with hydration shells, cannot pass through the small pores. The size exclusion is responsible for the possible ion exchange and sieving by zeolites, along with the different affinities of the ions. In cobalt exchange, the sodium version of zeolite A, such as zeolite 4A, is used for separating different cations, including Ni^{2+} and Co^{2+} [193].

Different structures of zeolites (natural and synthetic), including "ANA, CHA, HEU, ERI, FAU, FER, LTA, MER, MFI, and MOR," play a crucial role as ion exchangers due to the variety of ion-exchange isotherms and selectivity. High-silica zeolites usually prefer large, monovalent cations, while low-silica zeolites generally prefer small, multivalent cations to provide the best possible results as adsorbents. The cations exhibit delayed exchange kinetics due to their considerable hydration energies, including Li^+ and Mg^{2+}. Transition metal cations are generally less desirable for this application based on the substance, and ion-exchanger measurements are challenging to assess in diluted solutions using low-silica zeolites. For instance, in sodium-exchanged zeolites, the sluggish exchange of sodium for the hydronium ion causes alkalinization of the water, making it challenging to determine the ion-exchange properties [194]. In wastewater treatment, zeolites are used, even in laundry detergents, to reduce eutrophication and serve as a phosphate substitute [195]. Among different zeolites, zeolite 4A plays a role in the extraction of strontium radioactive ions and heavy metals, while clinoptilolite is used for ammonium removal [193,196]. The pollutants, such as organic waste and anions like F^-, are also removed by zeolite adsorbents through filtration and adsorption. Synthetic zeolites, such as CHA, FAU, and LTA, play a crucial role in mitigating radioactive decay, particularly in removing caesium-based radioactive waste [149,193].

Materials Research Foundations 189 (2026) 155-208 https://doi.org/10.21741/9781644904039-6

5.2 Selective Adsorption and Molecular Sieving of Zeolites

Zeolite adsorbents are suitable for selective adsorption and molecular sieving due to their microporosity and uniform pore diameters, which result in the segregation of various molecules. Damour and Friedel conducted early research on zeolite adsorption in 1840 and 1896, respectively. McBain coined the term "molecular sieves," which elucidates the nature of highly selective adsorption in zeolites [149]. Zeolites play a wide role in separation procedures, as the ideal adsorbent should be selective with high adsorption capacity, be able to be regenerated, and exhibit long-lasting, stable, and economical characteristics [169]. The selective adsorption by zeolites demands kinetic and thermodynamic selectivity based on pore size, shape, and chemical composition.

Zeolites typically utilize this kinetic selectivity to retain molecules of a larger size compared to the pore diameters, a phenomenon known as molecular sieving. The noble characteristics of zeolite, including temperature stability and post-synthetic modifications, make it a worthwhile material for separation-based applications. In industrial separation, zeolites (3A, 4A, and 5A) have been used as the 4A zeolite (Na form), having a pore size of 4 Å, undergo alteration by ion exchange with Ca^{2+} and K^+ and produce the zeolite 5A (5 Å) and zeolite 3A (3 Å), respectively [165]. Due to the modifications resulting in a change in pore size, they are suitable for adsorbing specific compounds through selective adsorption. For instance, the zeolite 5A plays a crucial role in adsorbing linear hydrocarbons over branched ones, and zeolite 3A is used in water adsorption [149,197].

The zeolite adsorption properties may be affected due to structural alterations, including cation translocation and framework deformation, referred to as gating effects, which can result in improved adsorption selectivity. For instance, the CHA is often modified due to trapdoor phenomena, in which molecules enter the zeolite pores through cation translocation, playing a role in CO_2/CH_4 separations [198]. In CO_2 adsorption on RHO zeolites and ethene adsorption on ITQ-55 [152], the flexibility of the zeolite plays a crucial role in adsorption and diffusion processes, highlighting the potential use of zeolites as adaptable adsorbents in separation procedures.

5.3 Purification of Hydrogen by Zeolites

Refinery off-gas (ROG) and steam methane reforming off-gas (SMROG) are typically utilized in refineries, accounting for approximately 95% of the hydrogen produced. SMROG is composed of syngas (CO and H_2 mixture) and maximizes the H_2 output and minimizes the CO levels using the water-gas shift reaction. After the shift process, SMROG comprises 70-80% H_2, 15-25% CO_2, 3-6% CH_4, 1-3% CO, and trace amounts of N_2 at a pressure of around 30 bar. While ROG undergoes saturation with H_2O, it has C_3H_8 and C_{4+} hydrocarbons in trace amounts along with 65-90% H_2, 2-3% CH_4, and 4-8% C_2H_6. Hydrogen purification is crucial for obtaining a highly pure (>98%) H_2 product, which can be further used in the production of ammonia synthesis or as a secondary steam with >99% CO_2 for sequestration or consumption [199].

The two main techniques, including PSA, facilitate 85% of hydrogen production (provide pure products (99.9%) and membrane-based separation is highly used in hydrogen purification. With adsorbents based on noble metals, H_2 interacts weakly, while PSA adsorbs contaminants, and H_2 interacts easily with the inorganic materials [149]. The PSA method employs stacked beds of adsorbents for the best separation of zeolite 5A and activated carbon, or activated carbon and zeolite X and Y. Even, CO_2 collecting instruments are placed before or after the primary hydrogen PSA. The RPSA development, along with monoliths and sheets, tends to increase the mass transfer

Materials Research Foundations 189 (2026) 155-208 https://doi.org/10.21741/9781644904039-6

coefficients and advances the sorption of steam methane reforming using high-temperature adsorbents for CO_2 separation simultaneously [200].

5.4 Drying Applications

Due to their reversible water-adsorption capability, zeolites are used in both laboratory and industrial drying applications. The natural and synthetic low-silica zeolite types, including A (3A, 4A, and 5A) and X (13X), typically contain negatively charged framework and extra-framework cations, which make them hydrophilic and enable their strong interactions with water molecules [1]. In PSA, the zeolite type, such as 13X, is used, and even in TSA, these take part in drying gas and liquid streams. In industrial drying applications, NaA zeolite membranes outperform conventional adsorbents, including active alumina and silica. At high temperatures, the zeolite membranes play a role in selectively adsorbing water from the mixture [201].

During the TSA process, the first zeolite used is zeolite 4A for drying natural gas. Zeolites 5A and 13X are used simultaneously for drying and sweetening natural gas [202]. Zeolite 3A has been used to dry olefins, preventing oligomerization within the adsorbent pores by acting as a molecular filter. Additional polar substances within drying mixture, including alcohols or ethanol/water azeotropes, are another application for this molecular sieving action [149]. High-silica MOR and CHA exhibit high chemical stability and are utilized in dry streams that typically contain extremely acidic gases. Zeolite beads are used non-regeneratively to dry organic solvents on a laboratory scale [12,203].

5.5 Nitrogen Removal from Natural Gas

During landfill gas, natural gas and biogas processing, separating nitrogen from methane is crucial to enhance the heating value of the gas mixture. The separation must comply with pipeline (<3%) and liquefied natural gas (<1%) specifications [204]. Cryogenic distillation is the most common method for nitrogen removal and is highly energy-intensive. Adsorption and membrane-based techniques are also used for separation, and they must be selective toward N_2 or CH_4. The well-known commercial zeolites play a crucial role in adsorbing methane due to their high polarizability [205]. The PSA-based methane-selective process is not very favourable as the methane is usually present in greater amounts and requires large bed volumes. The nitrogen-selective separation can be achieved through kinetic control or molecular sieving due to the smaller size of nitrogen (3.64 Å) compared to methane (3.76 Å).

In nitrogen separation, zeolite 4A could be used as suggested and patented by Habgood [206]. The limitation of its use is due to its hydrophilicity and high activation temperature, which restricts its practicality. In separating N_2 from CH_4, natural clinoptilolites and their calcium-exchanged forms are used. For the PSA process, the magnesium-exchanged clinoptilolite is used for nitrogen removal [207]. Different titanosilicate materials, including "ETS-4, CTS-1, Ba-ETS-4, and Sr-ETS-4", are usually developed by Engelhard Corporation and exhibit excellent performance in nitrogen removal applications at a trim level [205,208]. The materials typically have 8-ring openings and a mixed octahedral-tetrahedral framework, which is generally tuned by purification techniques (ion exchange and thermal treatment), and are commercialized as a molecular gate. The thermal stability increases and even reuses after ion exchange, resulting in pore openings (3-4 Å) and kinetic differentiation between nitrogen and methane [149,209].

5.6 Zeolites in Carbon Dioxide Removal and Capture

The primary sources of methane (CH_4) are considered to be fossil fuels, such as natural gas, and renewable sources, including biogas or landfill gas. It is a key fuel that might be used in global power generation (23%) and petrochemical processes, including methane reforming for hydrogen or syngas production. Oil and coal are the primary sources of natural gas, while landfill gas and biogas are obtained from the anaerobic digestion of waste in sewage plants and landfills [210]. There is a need to remove CO_2, H_2O, and N_2 from the methane-rich mixtures, as they reduce the fuel value and can cause corrosion and plugging. The hydrocarbons from C_2 to C_3 enhance the heating value, and the heavier hydrocarbons must be eliminated from methane-rich mixtures to avoid condensation problems [211].

The removal of CO_2 is quite essential due to its capability to dilute, act as a solvent, and be a corrosive "sour" gas. The methods employed for its removal include absorption via aqueous alkanolamines, adsorption, membrane separation, cryogenic distillation, and methanation [212]. The absorption process is suitable for large-scale operations, but its high energy cost restricts its use in small-scale applications. To surpass the limitations, PSA and membrane methods are an alternative way to remove CO_2 economically and energy-efficiently. PSA utilizes adsorbents including zeolites, metal-organic frameworks, carbon molecular sieves, and titanosilicates for CO_2/CH_4 separations.

Among these, high commercial potential is achieved through traditional zeolites, such as A, X-type, Y-type, silicalite, MOR, clinoptilolite, and ETS-4. The selectivity of CO_2 usually arises from strong interactions with cations in polar zeolites or molecular sieving mechanisms. The regeneration might undergo complications in high-polarity zeolites because they adsorb water strongly [213]. The selectivity of zeolites, including RHO, CHA, AEI, and AlPO/SAPO frameworks, is enhanced by trapdoor and guest-induced effects, leading to improved performance [149,214]. The commercial viability might be affected when costly organic SDAs are required. At the same time, Zeolite 13X and certain low-polarity FAU, LTA, and CHA zeolites ensure balanced selectivity, adsorption capacity, and regenerability. The pure silica and $AlPO_4$-based adsorbents exhibit high adsorbent capacity with a lower heat of adsorption during CO_2 removal from methane-rich mixtures [213]. The microwave-assisted synthesis of zeolite A using natural kaolinite via meta-kaolinization at 100 °C for 4.0 h has demonstrated a high CO_2 capacity (2.18 mmol/g) at an optimum temperature and pressure of 25 °C and 1 bar, respectively. The processing time and temperature are reduced due to the formation of high-crystalline and pure zeolite, which provides more active sites, making it a promising cost-effective method for carbon capture applications [136].

From anthropogenic sources, including transportation, industry, and energy production, CO_2 emissions have been increasing rapidly over the past 50 years and are now exceeding the biosphere's reabsorption capacity. Among the greenhouse gases, CO_2 is the primary one associated with climate change, so mitigation strategies are urgently needed. The most effective approach to reducing CO_2 emissions is carbon capture and sequestration, given the continued reliance on fossil fuels throughout this century. Carbon capture and sequestration are necessary in various industries, including natural gas processing, hydrogen production, fertiliser production, and power generation [215]. The dominant CO_2 removal method is chemical absorption with aqueous amines, primarily used for post-combustion streams. Other methods, such as swing adsorption, membranes, and chemical looping, are alternative methods for carbon capture and sequestration. Due to high energy requirements for CO_2 capturing in ultra-dilute sources, the direct air capture process is also in development. At the same time, zeolites have limited use in this approach due to their ability to adsorb water [216]. Even using zeolite in the direct air capture approach signifies a certain drop in CO_2 capacity under humid conditions for Li-LSX, K-LSX, and 13X [217]. The Zn-CHA has potential in carbon capture and sequestration [218]. Zeolite 13X is a benchmark material used in post-combustion capture, where flue gases contain 3-30% CO_2. Zeolite 13X is utilized in post-combustion at low pressure due to its greater capacity and selectivity. The hydrophilicity of the zeolite's framework necessitates process adaptations [219,220]. High-silica zeolites, AlPOs, and SAPOs are actually low-polarity adsorbents, resulting in reduced water uptake with promising performance. High selectivity and adsorbent capacity are estimated in SAPO-RHO, NaTEA-PST-20, and SAPO-56; however, scalability remains a concern that needs to be addressed [149].

5.7 Removal of Miscellaneous Pollutants

The other pollutants in industrial streams include various gases, such as CO, NH_3, NO_2, NO, H_2S, and SO_2, which must be controlled as they are produced during combustion and chemical production. Environmental pollutants, such as carbon monoxide (CO), are highly toxic and are produced by steam reforming processes and incomplete combustion of fuels. In hydrogen production, zeolites like 5A are used for CO removal, and Ag^+-exchanged zeolites are utilized for other industrial applications such as the separation of CO from air or nitrogen [221]. Another toxic gas, ammonia, is highly reactive and is widely produced in fertilizer and chemical manufacturing. In addition to water scrubbing and biological filtration, ammonia can be removed through adsorption on zeolites. The conventional aluminosilicate zeolite is less competitive due to its higher heat of adsorption compared to high-silica-containing MFI and pure silica-based zeolites (LTA and FAU) for efficient ammonia removal using swing adsorption [222].

In fossil fuel processing and organic decomposition, the hydrogen sulfide (H_2S) is toxic and corrosive and can be removed by zeolites such as 4A and 5A due to competitive adsorption with CO_2 and H_2O. The best zeolites are pure silica CHA and MFI, having a hydrophobic nature and a lower adsorption heat value, resulting in easier regeneration. The harmful by-products of combustion, such as nitrogen oxides (NO_x) and sulfur oxides (SO_2), are significant environmental pollutants that are removed using zeolites, including NaY and Na-ZSM-5, for flue gas treatment [212]. In SO_2 removal, zeolite 5A plays an outperforming role. In contrast, zeolite types, including 13X and 4A, are suitable for removing trace amounts of flue gas, highlighting the potential of zeolites in adsorbing capacity and contributing to the control of environmental pollution [223].

5.8 Adsorption for Volatile Organic Compounds and Dyes

The zeolite X has a microporous structure that limits its adsorbing nature toward VOCs and synthetic dyes. The zeolite X has been modified with the sequential dealumination-desilication with microwave assistance, converting it into mesoporous zeolites and enhancing its adsorption properties. The surface area of mesoporous zeolites becomes 136-155 m^2/g, which is an advancement compared with the original zeolite X's surface area (18 m^2/g). This modified zeolite X has demonstrated increased adsorption of methylene blue, reaching a maximum of 88.94 mg/g. The adsorption isotherm of methylene blue on this modified zeolite follows the Langmuir model, and the kinetic data fit the pseudo-second-order model. Even mesoporous zeolite X can be regenerated by ethanol immersion and ultrasound after saturation with methylene blue, sustaining the adsorption capacity at up to 75% after four cycles. To remove toluene, the mesoporous zeolite

Materials Research Foundations 189 (2026) 155-208 https://doi.org/10.21741/9781644904039-6

X undergoes a reaction with TiO_2, followed by pass-through adsorption and photocatalytic oxidation. The results revealed a 3.3-fold improvement compared to treating TiO_2 alone. The factors influencing toluene photocatalytic oxidation efficiency include mixing ratio, flow rate, and humidity. In a humid environment, the UV treatment enhances the TiO_2/mesoporous zeolite X photocatalyst's toluene oxidation performance within the first 30 minutes. Thus, the efficiency of zeolites with low Si/Al ratios can be improved for treating synthetic dyes and VOCs by a novel microwave-assisted dealumination-desilication method [224].

The rice husk, containing a high amount of silica, is used in the synthesis of zeolite Na-Y via a microwave-assisted method. High surface area (266.7 m^2/g) of synthesized zeolite Na-Y is obtained by extracting silica with 2 M HCl and undergoing microwave heating at 100-200 W. The crystalline structure of the synthesized zeolite, with particle sizes ranging from 0.1 to 1.4 μm, plays a crucial role in adsorbing textile dyes with high adsorption efficiencies, as demonstrated by the results for methylene blue (94.2%), methyl red (94.7%), and methyl orange (93.89%). The synthesis of zeolite from agricultural waste demonstrates its eco-friendly nature in removing dyes from wastewater, confirming a cost-effective synthesis and suitability for reducing industrial and factory wastewater pollution [225]. The magnetic zeolite/hydroxyapatite (MZeo-HAP) nanocomposite is prepared to remove anionic dyes from aqueous solutions. The synthesized nanocomposite of zeolite has shown improvement in surface area (103.56 m^2/g) and resulting adsorption of different anionic dyes, including Reactive Orange 5, Reactive Orange 16, and Congo Red, at pH 2 with contact time up to 30 min. The adsorption by this nanocomposite is exhibited as endothermic and favours adsorption efficiently at a temperature near 40 °C. The Reactive Orange 5, Reactive Orange 16, and Congo Red have been adsorbed onto a nanocomposite zeolite with adsorption capacities of up to 92.45, 88.31, and 104.05 mg/g, respectively. The dominant adsorption mechanism in this synthesized zeolite-based nanocomposite is chemisorption and follows the Freundlich isotherm and pseudo-second-order kinetics. This synthesized nanocomposite serves as a potential adsorbent in reducing wastewater concerns and environmental pollution due to overcoming the challenges of poor stability [226].

For VOCs such as acetone, benzene, toluene, and xylene, high-silica nano-sized BEA is used as an adsorbent. Zeolites exhibit high hydrothermal stability and hydrophobicity, which can be prepared using microwave-assisted hydrothermal methods, thereby reducing the need for fluoride and seeds. The formed nano-sized BEA zeolites have size ranges from 25 to 180 nm, along with a Si/Al ratio from 9 to infinity. The synthesis of BEA generates hydroxyl free radicals. The generated hydroxyl free radicals enhance the Si-O-Si bonds and improve the zeolite's properties. A significant improvement has been shown by the formed BEA in VOCs adsorption capacity, especially for toluene, compared with conventionally synthesized BEA. The Sips model is followed by the toluene adsorption behaviour, indicating that micropores dominate the adsorption process. The highest toluene adsorption capacity is estimated to be up to 124.76 cm^3/g, surpassing other conventionally formed Beta zeolites due to its large pore volume and strong hydrophobicity. Due to its dynamic adsorption capacity, this formed zeolite can be used for environmental pollutants [138].

5.9 Leaching Characteristics and Adsorption of Cations and Dyes

To address environmental concerns, the Na-P1 zeolites are synthesized from municipal solid waste incineration fly ash via microwave-assisted hydrothermal treatment, achieving high crystallinity (51.68%). The synthesis utilizes magnetic stirring, exhibiting a cation exchange capacity of up to

2.58 meq/g, which is 133 times greater than that of municipal solid waste incineration fly ash. The Si/Al ratio during zeolite NaP1 synthesis is adjusted by varying the Na_2SiO_3 dosage (30 wt%) and the 1 M NaOH concentration, which depicts the reaction feasibility. In synthetic products, the leachability of heavy metals usually meets pollution control standards and exhibits a BET surface area and total pore volume of zeolite Na-P1 up to 61.42 m^2/g and 0.44 cm^3/g, respectively. The improved surface area has demonstrated the potential of this modified zeolite as an adsorbent for cations and dyes, with adsorption capacities for Cu^{2+} ions and methylene blue of 84.65 and 84.55 mg/g, respectively. This successful adsorption capacity is achieved by a modified zeolite, which has shown potential in treating wastewater and water softening to control environmental pollutants [227].

Coal fly ash can also be used to synthesise zeolite A via microwave-assisted synthesis, providing an alternative, energy-efficient, and sustainable approach to mitigating environmental pollution. The optimized microwave power and Si/Al ratio yield a high-purity zeolite A with reduced impurities and a high surface area (29.54 m^2/g). The synthesized zeolite A has a cation exchange capacity of up to 3.10 meq/g, making it suitable for use as an adsorbent and ion exchanger. This method must mitigate environmental pollution (hazardous fly ash transforms into a valuable product), and it is ideal for industrial and environmental applications in cation and dye adsorption [137]. Different forces govern the basic mechanism of an adsorbate interacting with an adsorbent, helping to understand how pollutants bind to the active site of the adsorbent surface (Fig. 7).

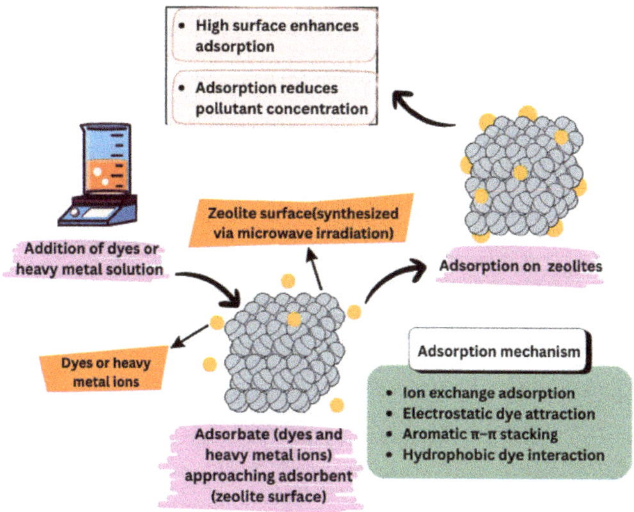

Figure 6. General mechanism for dyes or heavy metal ions on the zeolite surface.

5.10 Zeolites in Heavy Metal Removal

The microwave-assisted synthesis of zeolite A and ANA enhances the surface area and crystallinity peaks. It eliminates 100 mg/L of heavy metal ions such as Co, V, As, Pb, and Ni from

Materials Research Foundations 189 (2026) 155-208 https://doi.org/10.21741/9781644904039-6

aqueous solutions. The optimized conditions for the reaction include a pH of 6, a dosage of 1.0 g/L, and a contact time of up to 90 min. Zeolite A has the highest efficiency in metal removal due to its greater surface area than ANA. The adsorption kinetics follow pseudo-first-order kinetics and are fitted with various models, including the Langmuir, Freundlich, Dubinin-Radushkevich, and Temkin isotherm models. In both zeolite A and ANA, the surface physisorption is used for the removal mechanism, and the adsorption capacity for five metal ions follows the order as Co > V > As > Pb > Ni. Both zeolites can be reused, demonstrating effective performance over eight cycles, which results in cost-effective alternatives for removing heavy metals from wastewater and mitigating environmental concerns [228]. The zeolites prepared via microwave-assisted synthesis have a wide range of roles as eco-friendly adsorbents due to their enhanced surface area and active sites. The properties enhance the adsorption capacity and efficiency of synthesized zeolites, which mitigate environmental pollution and remove various heavy metals, dyes, and other organic pollutants, as described in Table 5.

Table 5. Different microwave-assisted synthesized zeolites and their role as adsorbents to mitigate environmental pollution

Zeolite Type	Precursor	Pollutant Type	Outcomes	Reference
Zeolite A	Metakaolinite	CO_2	2.2 mmol/g adsorption capacity	[136]
Zeolite A	Coal fly ash	Cation adsorption (such as metal ions)	3.10 mequiv/g cation exchange capacity	[137]
Na-Y Zeolite	Rice husk ash	Organic dyes (such as methylene blue, methyl orange, and methyl red)	Methylene blue, methyl orange, and methyl red show 94.2, 93.89, and 94.7% adsorption efficiency, respectively	[225]
MZeo-HAP nanocomposite	Zeolite + Hydroxyapatite	Anionic dyes (such as Reactive Orange 5, Reactive Orange 16, and Congo Red)	Reactive Orange 5, Reactive Orange 16, and Congo Red show adsorption capacity of 92.45, 88.31, and 104.05 mg/g, respectively	[226]
Zeolite Na-P1	Municipal solid waste incineration fly ash	Metal ions and Dyes (such as Cu^{2+} ion, methylene blue)	Cu^{2+} and methylene blue show 84.65 and 84.55 mg/g adsorption capacity, respectively	[227]
Zeolite A	Low-grade kaolin	Organic dye pollutant (such as methylene blue)	Methylene blue shows 2.44 mg/g adsorption capacity	[229]
Rare earth-doped Zeolite Y	Aluminosilicate sources	Organosulfur pollutant (such as thiophene)	75% desulfurization efficiency	[230]
Zeolite X (Zn-exchanged)	Aluminosilicate sources	H_2S (found in biogas)	Desulfurization capacity up to 13.8 mgS/g	[231]
BEA	Tetraethylammonium hydroxide	CO_2	2.16 mmol/g	[135]

5.11 Zeolites in Selective Catalytic Reduction

In selective catalytic reduction of NO_x, the zeolite catalyst has been used widely since the 1980s and somehow the utilization of metal-exchanged zeolites, including Cu-LTA [232], Cu-CHA

[233], Cu-ZSM-5 [234], Cu-SSZ-13 [235], Cu-KFI [236], Fe-LTA [237], Fe-ZSM-5 [238], and Fe-SSZ-13 [239]. The Cu-SSZ-13 is a small-pore zeolite that plays a crucial role in nitrogen selectivity and exhibits high activity over a range of temperatures (160-550 °C) [235]. At high temperatures (250-400 °C), the Cu (II) species are usually determined to be active within Cu-CHA-based zeolites [233]. Due to hydrocarbon-assisted NO_x reduction, the Cu^{2+} exchanged zeolites, including SSZ-13 and ZSM-5, enhanced NO_x activity [240]. Even the selective catalytic reduction-based activity is also improved by Cu-KFI and exhibits better hydrothermal stability [236]. Due to the stable Fe(III) dimers, Fe-SSZ-13 can retain its activity, based on selective catalytic reduction, at temperatures above 500 °C and does not undergo ageing [239].

In CO_2 reduction, zeolites are crucial in improving catalytic reduction as hierarchical ZSM-5 hosts Re (1) photocatalysts and exhibit high activity [241]. The reduction of CO_2 into CO and oxidation of H_2O to O_2 occur by the Ti and Fe-modified HZSM-5, forming photoactive centres. Even higher selectivity and structural stability are shown by Rh@HZSM-5, while the hydrogen efficiency of CO_2 is improved using NaA zeolite membranes [201]. Fe-based zeolites are utilized in Fenton-based reactions, resulting in the degradation of organic compounds, including methylene blue, phenol, and paracetamol [242]. Other zeolites, including Cu-Y, FAU, and clinoptilolite, enhance the Fenton catalysis. Iron leaching can be reduced using zeolites such as Fe-ZSM-5, Fe-Beta, Fe-MFI, and Fe-Y, which sustain the reaction's activity [75,243].

6. Recommendations and Conclusion

In zeolite research, the primary key challenge is the *availability* of *raw materials*, especially silica sources. Commercial silica is typically derived from sand and exhibits considerable variability in reactivity and selectivity. This results in a challenge in synthesizing synthetic zeolites from silica and alumina, as their raw materials are expensive. There is a need to explore alternative raw materials, including clay minerals, natural zeolites, coal ash, and industrial waste, for the economic synthesis of zeolites. The waste material from industries, factories, and vehicles is a big concern for environmental pollution. This issue needs to be mitigated when using it to produce valuable products. Efforts should be made to develop methods for utilizing waste materials as alternative raw materials in specific applications, such as water purification, heavy metal removal, and ammonium treatment. For instance, *kaolin* [244] is a clay mineral that could be used as an aluminosilicate source during the synthesis of zeolites. Kaolin transforms reactive metakaolin after calcination and is further processed to form zeolite. Sometimes, certain impurities, such as iron, titanium, and calcium, can cause contamination within the zeolites, affecting their catalytic activity and hardness characteristics. Optimizing calcination conditions is crucial for achieving a more reactive metakaolin phase, thereby enhancing the properties and efficiency of zeolites. Even fly ash [245] could also be used as a promising material in synthesizing zeolites, which are typically obtained from coal combustion, another source of environmental pollution. In various applications of water treatment, catalysis, and other industries, fly ash remains a topic of ongoing research and is a focus for further innovations in this area. The waste materials must be reused and recycled within zeolite synthesis to mitigate concerns about waste disposal and may offer *cost-effective alternatives* in producing zeolites. This approach is highly concerned with *green synthesis* and exhibits sustainable zeolite synthesis by broadening its applications in rising environmental concerns.

Materials Research Foundations 189 (2026) 155-208 https://doi.org/10.21741/9781644904039-6

In conclusion, zeolites possess versatile characteristics that enable their use in various industrial and environmental applications, including agriculture, water treatment, biomedicine, oil refining, and biotechnology. In the modern era, humanity has faced numerous challenges, including drought, freshwater scarcity, climate change, and the pandemic. Environmental concerns are on the rise, and zeolites are playing a crucial role in mitigating carbon dioxide emissions and sequestration. The microporous crystalline surface of zeolite is used as an adsorbent, ion-exchanger, purifier of hydrogen (using zeolite 5A), for drying industrial streams (using zeolites 3A and 4A), and for the separation of nitrogen from methane. As the focus shifts to an eco-friendly environment and green chemistry, the primary objective is to adopt low-energy synthesis routes to mitigate water and environmental pollution and reduce carbon footprints [165]. The global challenges need to be addressed through the promising potential of zeolites, such as their applications in reducing CO_2 emissions and the development of sustainable energy solutions. As trends emerge daily, advancements in computational modelling and data mining are also improving to accelerate discoveries tailored to industrial needs. Despite certain limitations in cost and synthesis obstacles, the applications of zeolites using efficient synthesis methods have demonstrated that zeolites prepared via microwave-assisted synthesis are a key material for *sustainable* future innovations.

References

[1] E.M. Flanigen, Molecular sieve zeolite technology-the first twenty-five years, Pure Appl. Chem. 52 (1980) 2191-2211. https://doi.org/10.1351/pac198052092191

[2] R.M. Milton, Molecular sieve adsorbents, U.S. Patent 2,882,243, 1959.

[3] K. Margeta, A. Farkaš, Introductory chapter: Zeolites-from discovery to new applications on the global market, in: K. Margeta, A. Farkaš (Eds.), Zeolites-New Challenges, IntechOpen, London, 2020, pp. 1-10. https://doi.org/10.5772/intechopen.92907

[4] P.K. Bajpai, M.S. Rao, K. Gokhale, Synthesis of mordenite type zeolites, Ind. Eng. Chem. Prod. Res. Dev. 17 (1978) 223-227. https://doi.org/10.1021/i360067a009

[5] I. Bilecka, M. Niederberger, Microwave chemistry for inorganic nanomaterials synthesis, Nanoscale 2 (2010) 1358-1374. https://doi.org/10.1039/b9nr00377k

[6] J.A. Dahl, B.L. Maddux, J.E. Hutchison, Toward greener nanosynthesis, Chem. Rev. 107 (2007) 2228-2269. https://doi.org/10.1021/cr050943k

[7] L.S. Gangurde, G.S. Sturm, T.J. Devadiga, A.I. Stankiewicz, G.D. Stefanidis, Complexity and challenges in non-contact high temperature measurements in microwave-assisted catalytic reactors, Ind. Eng. Chem. Res. 56 (2017) 13379-13391. https://doi.org/10.1021/acs.iecr.7b02091

[8] P. Xu, Y. Dong, D. Zhou, C. Fu, J. Zhang, H. Zhang, Z. Lu, L. Chen, X. Bao, 1200 °C high-temperature distributed optical fiber sensing using Brillouin optical time domain analysis, Appl. Opt. 55 (2016) 5471-5478. https://doi.org/10.1364/AO.55.005471

[9] X. Lin, Y. Liang, L. Jin, L. Wang, Dual-polarized fiber laser sensor for photoacoustic microscopy, Sensors 19 (2019) 4632. https://doi.org/10.3390/s19214632

[10] J.L. Hueso, R. Mallada, J. Santamaria, Gas-solid contactors and catalytic reactors with direct microwave heating: current status and perspectives, Catal. Today 423 (2023) 113927. https://doi.org/10.1016/j.cattod.2022.10.009

[11] Y. Suttisawat, S. Horikoshi, H. Sakai, P. Rangsunvigit, M. Abe, Enhanced conversion of tetralin dehydrogenation under microwave heating: effects of temperature variation, Fuel Process. Technol. 95 (2012) 27-32. https://doi.org/10.1016/j.fuproc.2011.11.006

[12] D.W. Breck, Zeolite Molecular Sieves: Structure, Chemistry, and Use, First ed., John Wiley & Sons, New York, 1974.

[13] H. van Bekkum, J. Jansen, E.M. Flanigen, Introduction to Zeolite Science and Practice, First ed., Elsevier, Amsterdam, 1991.

[14] L.B. McCusker, C. Baerlocher, Zeolite structures, in: H. van Bekkum, E.M. Flanigen, P.A. Jacobs, J.C. Jansen (Eds.), Introduction to Zeolite Science and Practice, Elsevier, Amsterdam, 2007, pp. 13-37.

[15] L.B. McCusker, C. Baerlocher, Zeolite structures, in: H. van Bekkum, E.M. Flanigen, P.A. Jacobs, J.C. Jansen (Eds.), Introduction to Zeolite Science and Practice, Elsevier, Amsterdam, 2001, pp. 37-67.

[16] C. Baerlocher, L.B. McCusker, D.H. Olson, Atlas of Zeolite Framework Types, Sixth rev. ed., Elsevier, Amsterdam, 2007.

[17] T. Zoltai, Classification of silicates and other minerals with tetrahedral structures, Am. Mineral. 45 (1960) 960-973.

[18] E. Passaglia, R.A. Sheppard, The crystal chemistry of zeolites, Rev. Mineral. Geochem. 45 (2001) 69-116. https://doi.org/10.2138/rmg.2001.45.2

[19] C. Colella, Natural zeolites, in: J. Čejka, H. van Bekkum (Eds.), Zeolites and Ordered Mesoporous Materials: Progress and Prospects, Elsevier, Amsterdam, 2005, pp. 13-40. https://doi.org/10.1016/S0167-2991(05)80004-7

[20] G. Gottardi, E. Galli, Natural Zeolites, First ed., Springer, Berlin, Heidelberg, 2012.

[21] P. Vassileva, D. Voikova, Investigation on natural and pretreated Bulgarian clinoptilolite for ammonium ions removal from aqueous solutions, J. Hazard. Mater. 170 (2009) 948-953. https://doi.org/10.1016/j.jhazmat.2009.05.062

[22] M. Sprynskyy, B. Buszewski, A.P. Terzyk, J. Namieśnik, Study of the selection mechanism of heavy metal (Pb2+, Cu2+, Ni2+, and Cd2+) adsorption on clinoptilolite, J. Colloid Interface Sci. 304 (2006) 21-28. https://doi.org/10.1016/j.jcis.2006.07.068

[23] J. Gorimbo, B. Taenzana, A.A. Muleja, A.T. Kuvarega, L.L. Jewell, Adsorption of cadmium, nickel and lead ions: equilibrium, kinetic and selectivity studies on modified clinoptilolites from the USA and RSA, Environ. Sci. Pollut. Res. 25 (2018) 30962-30978. https://doi.org/10.1007/s11356-018-2992-0

[24] T. Belova, Adsorption of heavy metal ions (Cu2+, Ni2+, Co2+ and Fe2+) from aqueous solutions by natural zeolite, Heliyon 5 (2019) e02320. https://doi.org/10.1016/j.heliyon.2019.e02320

[25] A.L. Ciosek, G.K. Luk, An innovative dual-column system for heavy metallic ion sorption by natural zeolite, Appl. Sci. 7 (2017) 795. https://doi.org/10.3390/app7080795

[26] P. Belousov, A. Semenkova, T. Egorova, A. Romanchuk, S. Zakusin, O. Dorzhieva, E. Tyupina, Y. Izosimova, I. Tolpeshta, M. Chernov, Cesium sorption and desorption on glauconite, bentonite, zeolite, and diatomite, Minerals 9 (2019) 625. https://doi.org/10.3390/min9100625

[27] E. Ivanova, M. Karsheva, B. Koumanova, Adsorption of ammonium ions onto natural zeolite, J. Univ. Chem. Technol. Metall. 45 (2010) 295-302.

[28] C. Sangeetha, P. Baskar, Zeolite and its potential uses in agriculture: A critical review, Agric. Rev. 37 (2016) 101-108. https://doi.org/10.18805/ar.v0iof.9627

[29] M. Rehakova, S. Čuvanová, M. Dzivak, J. Rimár, Z. Gavalová, Agricultural and agrochemical uses of natural zeolite of the clinoptilolite type, Curr. Opin. Solid State Mater. Sci. 8 (2004) 397-404. https://doi.org/10.1016/j.cossms.2005.04.004

[30] M. Król, Natural vs. synthetic zeolites, Crystals 10 (2020) 620-622. https://doi.org/10.3390/cryst10070622

[31] G. Garcia, E. Cardenas, S. Cabrera, J. Hedlund, J. Mouzon, Synthesis of zeolite Y from diatomite as silica source, Microporous Mesoporous Mater. 219 (2016) 29-37. https://doi.org/10.1016/j.micromeso.2015.07.015

[32] O. Biel, P. Rożek, P. Florek, W. Mozgawa, M. Król, Alkaline activation of kaolin group minerals, Crystals 10 (2020) 264-268. https://doi.org/10.3390/cryst10040268

[33] N. Burriesci, M. Crisafulli, L. Saija, G. Polizzotti, Hydrothermal synthesis of zeolites from rhyolitic pumice of different geological origins, Mater. Lett. 2 (1983) 74-78. https://doi.org/10.1016/0167-577X(83)90038-1

[34] P. Kunecki, R. Panek, M. Wdowin, T. Bień, W. Franus, Influence of the fly ash fraction after grinding process on the hydrothermal synthesis efficiency of Na-A, Na-P1, Na-X and Sodalite zeolite types, Int. J. Coal Sci. Technol. 8 (2021) 291-311. https://doi.org/10.1007/s40789-020-00332-1

[35] D. Prasetyoko, Z. Ramli, S. Endud, H. Hamdan, B. Sulikowski, Conversion of rice husk ash to zeolite beta, Waste Manag. 26 (2006) 1173-1179. https://doi.org/10.1016/j.wasman.2005.09.009

[36] M. Król, W. Mozgawa, J. Morawska, W. Pichór, Spectroscopic investigation of hydrothermally synthesized zeolites from expanded perlite, Microporous Mesoporous Mater. 196 (2014) 216-222. https://doi.org/10.1016/j.micromeso.2014.05.017

[37] R. Barrer, E. White, The hydrothermal chemistry of silicates. Part II. Synthetic crystalline sodium aluminosilicates, J. Chem. Soc. (1952) 1561-1571. https://doi.org/10.1039/jr9520001561

[38] X. Querol, A. Alastuey, A. López-Soler, F. Plana, J.M. Andrés, R. Juan, P. Ferrer, C.R. Ruiz, A fast method for recycling fly ash: Microwave-assisted zeolite synthesis, Environ. Sci. Technol. 31 (1997) 2527-2533. https://doi.org/10.1021/es960937t

[39] M. Park, C.L. Choi, W.T. Lim, M.C. Kim, J. Choi, N.H. Heo, Molten-salt method for the synthesis of zeolitic materials: I. Zeolite formation in alkaline molten-salt system, Microporous Mesoporous Mater. 37 (2000) 81-89. https://doi.org/10.1016/S1387-1811(99)00196-1

[40] M. Zhang, H. Zhang, D. Xu, L. Han, D. Niu, B. Tian, J. Zhang, L. Zhang, W. Wu, Removal of ammonium from aqueous solutions using zeolite synthesized from fly ash by a fusion method, Desalination 271 (2011) 111-121. https://doi.org/10.1016/j.desal.2010.12.021

[41] P. Rożek, M. Król, W. Mozgawa, Geopolymer-zeolite composites: a review, J. Clean. Prod. 230 (2019) 557-579. https://doi.org/10.1016/j.jclepro.2019.05.152

[42] H. Tanaka, H. Eguchi, S. Fujimoto, R. Hino, Two-step process for synthesis of a single phase Na-A zeolite from coal fly ash by dialysis, Fuel 85 (2006) 1329-1334. https://doi.org/10.1016/j.fuel.2005.12.022

[43] C.S. Cundy, P.A. Cox, The hydrothermal synthesis of zeolites: precursors, intermediates and reaction mechanism, Microporous Mesoporous Mater. 82 (2005) 1-78. https://doi.org/10.1016/j.micromeso.2005.02.016

[44] A.Y. Lonin, V. Levenets, I. Neklyudov, A. Shchur, The usage of zeolites for dynamic sorption of cesium from waste waters of nuclear power plants, J. Radioanal. Nucl. Chem. 303 (2015) 831-836. https://doi.org/10.1007/s10967-014-3597-9

[45] O.A. Moamen, I. Ismail, N. Abdelmonem, R.A. Rahman, Factorial design analysis for optimizing the removal of cesium and strontium ions on synthetic nano-sized zeolite, J. Taiwan Inst. Chem. Eng. 55 (2015) 133-144. https://doi.org/10.1016/j.jtice.2015.04.007

[46] K. He, Y. Chen, Z. Tang, Y. Hu, Removal of heavy metal ions from aqueous solution by zeolite synthesized from fly ash, Environ. Sci. Pollut. Res. 23 (2016) 2778-2788. https://doi.org/10.1007/s11356-015-5422-6

[47] B. Kozera-Sucharda, B. Gworek, I. Kondzielski, The simultaneous removal of zinc and cadmium from multicomponent aqueous solutions by their sorption onto selected natural and synthetic zeolites, Minerals 10 (2020) 343. https://doi.org/10.3390/min10040343

[48] A. Nizami, O. Ouda, M. Rehan, A. El-Maghraby, J. Gardy, A. Hassanpour, S. Kumar, I. Ismail, The potential of Saudi Arabian natural zeolites in energy recovery technologies, Energy 108 (2016) 162-171. https://doi.org/10.1016/j.energy.2015.07.030

[49] J.W. Ward, The nature of active sites on zeolites: I. The decationated Y zeolite, J. Catal. 9 (1967) 225-236. https://doi.org/10.1016/0021-9517(67)90248-5

[50] A. Corma, State of the art and future challenges of zeolites as catalysts, J. Catal. 216 (2003) 298-312. https://doi.org/10.1016/S0021-9517(02)00132-X

[51] L. Bandura, M. Franus, G. Józefaciuk, W. Franus, Synthetic zeolites from fly ash as effective mineral sorbents for land-based petroleum spills cleanup, Fuel 147 (2015) 100-107. https://doi.org/10.1016/j.fuel.2015.01.067

[52] S. Gonthier, L. Gora, I. Güray, R.W. Thompson, Further comments on the role of autocatalytic nucleation in hydrothermal zeolite syntheses, Zeolites 13 (1993) 414-418. https://doi.org/10.1016/0144-2449(93)90113-H

[53] R.N. Baig, R.S. Varma, Alternative energy input: Mechanochemical, microwave and ultrasound-assisted organic synthesis, Chem. Soc. Rev. 41 (2012) 1559-1584. https://doi.org/10.1039/C1CS15204A

[54] P. Lidström, J. Tierney, B. Wathey, J. Westman, Microwave assisted organic synthesis-a review, Tetrahedron 57 (2001) 9225-9283. https://doi.org/10.1016/S0040-4020(01)00906-1

[55] M.A. Surati, S. Jauhari, K. Desai, A brief review: microwave assisted organic reaction, Arch. Appl. Sci. Res. 4 (2012) 645-661.

[56] B.L. Hayes, Recent advances in microwave-assisted synthesis, Aldrichimica Acta 37 (2004) 66-77.

[57] N.A. Khan, S.H. Jhung, Synthesis of metal-organic frameworks (MOFs) with microwave or ultrasound: rapid reaction, phase-selectivity, and size reduction, Coord. Chem. Rev. 285 (2015) 11-23. https://doi.org/10.1016/j.ccr.2014.10.008

[58] J. Klinowski, F.A.A. Paz, P. Silva, J. Rocha, Microwave-assisted synthesis of metal-organic frameworks, Dalton Trans. 40 (2011) 321-330. https://doi.org/10.1039/C0DT00708K

[59] V. Polshettiwar, M.N. Nadagouda, R.S. Varma, Microwave-assisted chemistry: a rapid and sustainable route to synthesis of organics and nanomaterials, Aust. J. Chem. 62 (2009) 16-26. https://doi.org/10.1071/CH08404

[60] H. Wang, J.R. Zhang, J.J. Zhu, A microwave assisted heating method for the rapid synthesis of sphalrite-type mercury sulfide nanocrystals with different sizes, J. Cryst. Growth 233 (2001) 829-836. https://doi.org/10.1016/S0022-0248(01)01629-3

[61] X. Xu, W. Yang, J. Liu, L. Lin, Synthesis of NaA zeolite membrane by microwave heating, Sep. Purif. Technol. 25 (2001) 241-249. https://doi.org/10.1016/S1383-5866(01)00108-3

[62] J. Cai, J. Liu, Z. Gao, A. Navrotsky, S.L. Suib, Synthesis and anion exchange of tunnel structure akaganeite, Chem. Mater. 13 (2001) 4595-4602. https://doi.org/10.1021/cm010310w

[63] S. Horikoshi, A. Osawa, M. Abe, N. Serpone, On the generation of hot-spots by microwave electric and magnetic fields and their impact on a microwave-assisted heterogeneous reaction in the presence of metallic Pd nanoparticles on an activated carbon support, J. Phys. Chem. C 115 (2011) 23030-23035. https://doi.org/10.1021/jp2076269

[64] Z. Yang, H. Yi, X. Tang, S. Zhao, Q. Yu, F. Gao, Y. Zhou, J. Wang, Y. Huang, K. Yang, Potential demonstrations of "hot spots" presence by adsorption-desorption of toluene vapor onto granular activated carbon under microwave radiation, Chem. Eng. J. 319 (2017) 191-199. https://doi.org/10.1016/j.cej.2017.02.157

[65] M. Baghbanzadeh, L. Carbone, P.D. Cozzoli, C.O. Kappe, Microwave-assisted synthesis of colloidal inorganic nanocrystals, Angew. Chem. Int. Ed. 50 (2011) 11312-11359. https://doi.org/10.1002/anie.201101274

[66] E. Haque, C.M. Kim, S.H. Jhung, Facile synthesis of cuprous oxide using ultrasound, microwave and electric heating: effect of heating methods on synthesis kinetics, morphology and yield, CrystEngComm 13 (2011) 4060-4068. https://doi.org/10.1039/c0ce00920b

[67] W.C. Conner, G. Tompsett, K.H. Lee, K.S. Yngvesson, Microwave synthesis of zeolites: 1. Reactor engineering, J. Phys. Chem. B 108 (2004) 13913-13920. https://doi.org/10.1021/jp037358c

[68] C. Li, Y. Wei, A. Liivat, Y. Zhu, J. Zhu, Microwave-solvothermal synthesis of Fe3O4 magnetic nanoparticles, Mater. Lett. 107 (2013) 23-26. https://doi.org/10.1016/j.matlet.2013.05.117

[69] N.A. Khan, J.H. Park, S.H. Jhung, Phase-selective synthesis of a silicoaluminophosphate molecular sieve from dry gels, Mater. Res. Bull. 45 (2010) 377-381. https://doi.org/10.1016/j.materresbull.2010.01.003

[70] S.H. Jhung, T. Jin, Y.H. Kim, J.S. Chang, Phase-selective crystallization of cobalt-incorporated aluminophosphate molecular sieves with large pore by microwave irradiation, Microporous Mesoporous Mater. 109 (2008) 58-65. https://doi.org/10.1016/j.micromeso.2007.04.031

[71] B. Viswanath, P. Kundu, A. Halder, N. Ravishankar, Mechanistic aspects of shape selection and symmetry breaking during nanostructure growth by wet chemical methods, J. Phys. Chem. C 113 (2009) 16866-16883. https://doi.org/10.1021/jp903370f

[72] Y. Xia, Y. Xiong, B. Lim, S.E. Skrabalak, Shape-controlled synthesis of metal nanocrystals: simple chemistry meets complex physics?, Angew. Chem. Int. Ed. 48 (2009) 60-103. https://doi.org/10.1002/anie.200802248

[73] L. Carbone, P.D. Cozzoli, Colloidal heterostructured nanocrystals: synthesis and growth mechanisms, Nano Today 5 (2010) 449-493. https://doi.org/10.1016/j.nantod.2010.08.006

[74] R. Buonsanti, M. Casavola, G. Caputo, P.D. Cozzoli, Advances in the chemical fabrication of complex multimaterial nanocrystals, Recent Pat. Nanotechnol. 1 (2007) 224-232. https://doi.org/10.2174/187221007782360420

[75] X. Zeng, X. Hu, H. Song, G. Xia, Z.Y. Shen, R. Yu, M. Moskovits, Microwave synthesis of zeolites and their related applications, Microporous Mesoporous Mater. 323 (2021) 111262. https://doi.org/10.1016/j.micromeso.2021.111262

[76] S.H. Jhung, J.S. Chang, Y.K. Hwang, S.E. Park, Crystal morphology control of AFI type molecular sieves with microwave irradiation, J. Mater. Chem. 14 (2004) 280-285. https://doi.org/10.1039/b309142b

[77] J.B. Koo, N. Jiang, S. Saravanamurugan, M. Bejblová, Z. Musilová, J. Čejka, S.E. Park, Direct synthesis of carbon-templating mesoporous ZSM-5 using microwave heating, J. Catal. 276 (2010) 327-334. https://doi.org/10.1016/j.jcat.2010.09.024

[78] M.A. Sanhoob, O. Muraza, Synthesis of silicalite-1 using fluoride media under microwave irradiation, Microporous Mesoporous Mater. 233 (2016) 140-147. https://doi.org/10.1016/j.micromeso.2016.01.004

[79] Z. Cheng, S. Han, W. Sun, Q. Qin, Microwave-assisted synthesis of nanosized FAU-type zeolite in water-in-oil microemulsion, Mater. Lett. 95 (2013) 193-196. https://doi.org/10.1016/j.matlet.2013.01.002

[80] Y.H. Seo, E.A. Prasetyanto, N. Jiang, S.M. Oh, S.E. Park, Catalytic dehydration of methanol over synthetic zeolite W, Microporous Mesoporous Mater. 128 (2010) 108-114. https://doi.org/10.1016/j.micromeso.2009.08.011

[81] G. Li, H.M. Hou, R.S. Lin, Rapid synthesis of mordenite crystals by microwave heating, Solid State Sci. 13 (2011) 662-664. https://doi.org/10.1016/j.solidstatesciences.2010.12.040

[82] S. Suresh, I.A.K. Reddy, N. Venkatathri, Synthesis of SAPO-16 molecular sieve in non-aqueous medium by microwave method using hexamethyleneimine as a template, Microporous Mesoporous Mater. 263 (2018) 275-281. https://doi.org/10.1016/j.micromeso.2017.12.008

[83] H.L. Zubowa, H. Kosslick, D. Müller, M. Richter, L. Wilde, R. Fricke, Crystallization of phase-pure zeolite NaP from MCM-22-type gel compositions under microwave radiation, Microporous Mesoporous Mater. 109 (2008) 542-548. https://doi.org/10.1016/j.micromeso.2007.06.002

[84] S.F. Wong, K. Deekomwong, J. Wittayakun, T.C. Ling, O. Muraza, F. Adam, E.P. Ng, Crystal growth study of KF nanozeolite and its catalytic behavior in aldol condensation of benzaldehyde and heptanal enhanced by microwave heating, Mater. Chem. Phys. 196 (2017) 295-301. https://doi.org/10.1016/j.matchemphys.2017.04.061

[85] T. Al-Dahri, A.A. Abdul Razak, I.H. Khalaf, S. Rohani, Response surface modeling of the removal of methyl orange dye from its aqueous solution using two types of zeolite synthesized from coal fly ash, Mater. Express 8 (2018) 234-244. https://doi.org/10.1166/mex.2018.1433

[86] T. Fukasawa, A. Horigome, A.D. Karisma, N. Maeda, A.N. Huang, K. Fukui, Utilization of incineration fly ash from biomass power plants for zeolite synthesis from coal fly ash by microwave hydrothermal treatment, Adv. Powder Technol. 29 (2018) 450-456. https://doi.org/10.1016/j.apt.2017.10.022

[87] S. Zeng, R. Wang, Z. Zhang, S. Qiu, Solventless green synthesis of sodalite zeolite using diatomite as silica source by a microwave heating technique, Inorg. Chem. Commun. 70 (2016) 168-171. https://doi.org/10.1016/j.inoche.2016.06.013

[88] Z.L. Cheng, X.X. Qin, Microwave-assisted secondary synthesis of NaA zeolite membranes grown on ZrO2 substrate, Appl. Mech. Mater. 477 (2014) 1481-1484. https://doi.org/10.4028/www.scientific.net/AMM.477-478.1481

[89] G. Zhu, Y. Li, H. Zhou, J. Liu, W. Yang, FAU-type zeolite membranes synthesized by microwave assisted in situ crystallization, Mater. Lett. 62 (2008) 4357-4359. https://doi.org/10.1016/j.matlet.2008.07.026

[90] M. Wang, L. Bai, M. Li, L. Gao, M. Wang, P. Rao, Y. Zhang, Ultrafast synthesis of thin all-silica DDR zeolite membranes by microwave heating, J. Membr. Sci. 572 (2019) 567-579. https://doi.org/10.1016/j.memsci.2018.11.049

[91] W. Yang, B. Zhang, X. Liu, Synthesis and characterization of SAPO-5 membranes on porous α-Al2O3 substrates, Microporous Mesoporous Mater. 117 (2009) 391-394. https://doi.org/10.1016/j.micromeso.2008.07.015

[92] Z. Han, L. Yanshuo, Z. Guangqi, L. Jie, L. Liwu, Y. Weishen, Microwave synthesis of a- & b-oriented zeolite T membranes and their application in pervaporation-assisted esterification, Chin. J. Catal. 29 (2008) 592-594. https://doi.org/10.1016/S1872-2067(08)60059-5

[93] N. Hu, Y. Li, S. Zhong, B. Wang, F. Zhang, T. Wu, Z. Yang, R. Zhou, X. Chen, Microwave synthesis of zeolite CHA (chabazite) membranes with high pervaporation performance in absence of organic structure directing agents, Microporous Mesoporous Mater. 228 (2016) 22-29. https://doi.org/10.1016/j.micromeso.2016.03.018

[94] M.A. Baig, F. Patel, K. Alhooshani, O. Muraza, E.N. Wang, T. Laoui, In-situ aging microwave heating synthesis of LTA zeolite layer on mesoporous TiO2 coated porous alumina support, J. Cryst. Growth 432 (2015) 123-128. https://doi.org/10.1016/j.jcrysgro.2015.09.012

[95] X. Li, Y. Yan, Z. Wang, Continuity control of b-oriented MFI zeolite films by microwave synthesis, Ind. Eng. Chem. Res. 49 (2010) 5933-5938. https://doi.org/10.1021/ie1000136

[96] K. Sun, B. Liu, S. Zhong, A. Wu, B. Wang, R. Zhou, H. Kita, Fast preparation of oriented silicalite-1 membranes by microwave heating for butane isomer separation, Sep. Purif. Technol. 219 (2019) 90-99. https://doi.org/10.1016/j.seppur.2019.03.018

[97] S. Xia, Y. Chen, H. Xu, D. Lv, J. Yu, P. Wang, Synthesis EMT-type zeolite by microwave and hydrothermal heating, Microporous Mesoporous Mater. 278 (2019) 54-63. https://doi.org/10.1016/j.micromeso.2018.11.012

[98] S. Vichaphund, V. Sricharoenchaikul, D. Atong, Selective aromatic formation from catalytic fast pyrolysis of Jatropha residues using ZSM-5 prepared by microwave-assisted synthesis, J. Anal. Appl. Pyrolysis 141 (2019) 104628. https://doi.org/10.1016/j.jaap.2019.104628

[99] F.M. Shalmani, R. Halladj, S. Askari, Effect of contributing factors on microwave-assisted hydrothermal synthesis of nanosized SAPO-34 molecular sieves, Powder Technol. 221 (2012) 395-402. https://doi.org/10.1016/j.powtec.2012.01.036

[100] S.N. Azizi, S. Ghasemi, M. Mikhchian, Microwave-assisted synthesis of NaA nanozeolite from slag and performance of Ag-doped nanozeolite as an efficient material for determination of hydrogen peroxide, RSC Adv. 6 (2016) 52058-52066. https://doi.org/10.1039/C6RA06724G

[101] J.W. Jun, I. Ahmed, C.U. Kim, K.E. Jeong, S.Y. Jeong, S.H. Jhung, Synthesis of ZSM-5 zeolites using hexamethylene imine as a template: effect of microwave aging, Catal. Today 232 (2014) 108-113. https://doi.org/10.1016/j.cattod.2013.08.017

[102] A. Samadi-Maybodi, S.M. Pourali, Microwave-assisted aging synthesis of bismuth modified zeolite-P microspheres via BiOCl nanoflake transformation, Microporous Mesoporous Mater. 167 (2013) 127-132. https://doi.org/10.1016/j.micromeso.2012.02.012

[103] Y. Wu, X. Ren, J. Wang, Effect of microwave-assisted aging on the static hydrothermal synthesis of zeolite MCM-22, Microporous Mesoporous Mater. 116 (2008) 386-393. https://doi.org/10.1016/j.micromeso.2008.04.027

[104] S.N. Azizi, M. Yousefpour, Static and ultrasonic-assisted aging effects on the synthesis of Analcime zeolite, Z. Anorg. Allg. Chem. 636 (2010) 886-890. https://doi.org/10.1002/zaac.200900415

[105] S.N. Azizi, N. Asemi, The effect of ultrasonic and microwave-assisted aging on the synthesis of zeolite P from Iranian perlite using Box-Behnken experimental design, Chem. Eng. Commun. 201 (2014) 909-925. https://doi.org/10.1080/00986445.2013.793675

[106] X. Wang, C. Yan, Synthesis of nano-sized NaY zeolite composite from metakaolin by ionothermal method with microwave assisted, Inorg. Mater. 46 (2010) 517-521. https://doi.org/10.1134/S0020168510050146

[107] X. Zhao, J. Chen, Z. Sun, A. Li, G. Li, X. Wang, Formation mechanism and catalytic application of hierarchical structured FeAlPO-5 molecular sieve by microwave-assisted ionothermal synthesis, Microporous Mesoporous Mater. 182 (2013) 8-15. https://doi.org/10.1016/j.micromeso.2013.08.005

[108] X. Zhao, W. Duan, Q. Wang, D. Ji, Y. Zhao, G. Li, Microwave-assisted ionothermal synthesis of Fe-LEV molecular sieve with high iron content in low-dosage of eutectic mixture, Microporous Mesoporous Mater. 275 (2019) 253-262. https://doi.org/10.1016/j.micromeso.2018.09.005

[109] Y.P. Xu, Z.J. Tian, S.J. Wang, Y. Hu, L. Wang, B.C. Wang, Y.C. Ma, L. Hou, J.Y. Yu, L.W. Lin, Microwave-enhanced ionothermal synthesis of aluminophosphate molecular sieves, Angew. Chem. Int. Ed. 45 (2006) 3965-3970. https://doi.org/10.1002/anie.200600054

[110] B. Zhang, Y. Zhang, Y. Hu, Z. Shi, A. Azhati, S. Xie, H. He, Y. Tang, Microexplosion under microwave irradiation: a facile approach to create mesopores in zeolites, Chem. Mater. 28 (2016) 2757-2767. https://doi.org/10.1021/acs.chemmater.6b00503

[111] X.L. Luo, F. Pei, W. Wang, H.M. Qian, K.K. Miao, Z. Pan, Y.S. Chen, G.D. Feng, Microwave synthesis of hierarchical porous materials with various structures by controllable desilication and recrystallization, Microporous Mesoporous Mater. 262 (2018) 148-153. https://doi.org/10.1016/j.micromeso.2017.11.037

[112] C.S. Lim, Preparation and characterization of Ag-MWO4/zeolite (M = Fe, Ni) composites by cyclic microwave-assisted metathetic method, Asian J. Chem. 26 (2014) 1853. https://doi.org/10.14233/ajchem.2014.15831B

[113] Z. Tang, S.J. Kim, X. Gu, J. Dong, Microwave synthesis of MFI-type zeolite membranes by seeded secondary growth without the use of organic structure directing agents, Microporous Mesoporous Mater. 118 (2009) 224-231. https://doi.org/10.1016/j.micromeso.2008.08.029

[114] J. Motuzas, S. Heng, P.Z. Lau, K. Yeung, Z. Beresnevicius, A. Julbe, Ultra-rapid production of MFI membranes by coupling microwave-assisted synthesis with either ozone or calcination treatment, Microporous Mesoporous Mater. 99 (2007) 197-205. https://doi.org/10.1016/j.micromeso.2006.06.042

[115] R. Zhou, S. Zhong, X. Lin, N. Xu, Synthesis of zeolite T by microwave and conventional heating, Microporous Mesoporous Mater. 124 (2009) 117-122. https://doi.org/10.1016/j.micromeso.2009.05.001

[116] Y.K. Hwang, J.S. Chang, S.E. Park, D.S. Kim, Y.U. Kwon, S.H. Jhung, J.S. Hwang, M.S. Park, Microwave fabrication of MFI zeolite crystals with a fibrous morphology and their applications, Angew. Chem. Int. Ed. 44 (2005) 556-560. https://doi.org/10.1002/anie.200461403

[117] Z. Chen, S. Li, Y. Yan, Synthesis of template-free zeolite nanocrystals by reverse microemulsion-microwave method, Chem. Mater. 17 (2005) 2262-2266. https://doi.org/10.1021/cm048039g

[118] S.H. Jhung, T. Jin, Y.K. Hwang, J.S. Chang, Microwave effect in the fast synthesis of microporous materials which stage between nucleation and crystal growth is accelerated by microwave irradiation?, Chem. Eur. J. 13 (2007) 4410-4417. https://doi.org/10.1002/chem.200700098

[119] L. Wang, J. Yang, J. Wang, W. Raza, G. Liu, J. Lu, Y. Zhang, Microwave synthesis of NaA zeolite membranes on coarse macroporous α-Al2O3 tubes for desalination, Microporous Mesoporous Mater. 306 (2020) 110360. https://doi.org/10.1016/j.micromeso.2020.110360

[120] W. Xu, T. Zhang, R. Bai, P. Zhang, J. Yu, A one-step rapid synthesis of TS-1 zeolites with highly catalytically active mononuclear TiO6 species, J. Mater. Chem. A 8 (2020) 9677-9683. https://doi.org/10.1039/C9TA13851J

[121] M. Abrishamkar, S.N. Azizi, H. Kazemian, The effects of various sources and templates on the microwave-assisted synthesis of BZSM-5 zeolite, Z. Anorg. Allg. Chem. 637 (2011) 312-316. https://doi.org/10.1002/zaac.201000309

[122] H. Akhoundzadeh, M. Taghizadeh, H.S. Pajaie, Synthesis of highly selective and stable mesoporous Ni-Ce/SAPO-34 nanocatalyst for methanol-to-olefin reaction: role of polar aprotic N, N-dimethylformamide solvent, Particuology 40 (2018) 113-122. https://doi.org/10.1016/j.partic.2017.11.004

[123] A.C. Kresge, M.E. Leonowicz, W.J. Roth, J. Vartuli, J. Beck, Ordered mesoporous molecular sieves synthesized by a liquid-crystal template mechanism, Nature 359 (1992) 710-712. https://doi.org/10.1038/359710a0

[124] G. Bond, R. Moyes, D. Whan, Recent applications of microwave heating in catalysis, Catal. Today 17 (1993) 427-437. https://doi.org/10.1016/0920-5861(93)80046-4

[125] R.P. Zerger, K.C. McMahon, M.D. Seltzer, R. Michel, S.L. Suib, Preparation of highly dispersed cobalt clusters in zeolites via microwave discharge methods, J. Catal. 99 (1986) 498-505. https://doi.org/10.1016/0021-9517(86)90375-1

[126] Y.F. Li, J.Q. Zhu, H. Liu, P. Wang, H.P. Tian, Increase in the hydrothermal stability of the La-modified ZSM-5 zeolite, Acta Phys.-Chim. Sin. 27 (2011) 52-58. https://doi.org/10.3866/PKU.WHXB20110130

[127] T.H. Ji, X.P. Meng, N.Z. Wu, S.J. Li, Y.J. Liu, The synthesis and catalytic performance of Pd cluster encaged in Y-type zeolite, Chem. J. Chin. Univ. 18 (1997) 6-10.

[128] F.S. Xiao, W. Xu, S. Qiu, R. Xu, The high dispersion of CuCl2 in NaZSM-5 by using microwave technique, Catal. Lett. 26 (1994) 209-215. https://doi.org/10.1007/BF00824046

[129] F.S. Xiao, W. Xu, S. Qiu, R. Xu, New route for dispersion of inorganic salts onto the channel surfaces of microporous crystals: high dispersion of CuCl2 in zeolites using a microwave technique, J. Mater. Chem. 4 (1994) 735-739. https://doi.org/10.1039/jm9940400735

[130] H. Karge, H. Pfeifer, W. Hölderich, Zeolites and Related Microporous Materials: State of the Art 1994, First ed., Elsevier, Amsterdam, 1994.

[131] S. Deng, Y. Lin, Microwave heating synthesis of supported sorbents, Chem. Eng. Sci. 52 (1997) 1563-1575. https://doi.org/10.1016/S0009-2509(97)00495-8

[132] J.H. Zhu, Y. Wang, Y. Chun, Z. Xing, Q.H. Xu, MgO/KL: strong basic zeolite prepared by microwave radiation, Mater. Lett. 35 (1998) 177-182. https://doi.org/10.1016/S0167-577X(97)00249-8

[133] D.H. Yin, D.L. Yin, Study on the solid-state reaction of ZnCl2 and Y zeolites under microwave irradiation, Acta Phys.-Chim. Sin. 14 (1998) 448-452. https://doi.org/10.3866/PKU.WHXB19980513

[134] F.S. Xiao, S. Zheng, J. Sun, R. Yu, S. Qiu, R. Xu, Dispersion of inorganic salts into zeolites and their pore modification, J. Catal. 176 (1998) 474-487. https://doi.org/10.1006/jcat.1998.2054

[135] H.S. You, H. Jin, Y.H. Mo, S.E. Park, CO2 adsorption behavior of microwave synthesized zeolite beta, Mater. Lett. 108 (2013) 106-109. https://doi.org/10.1016/j.matlet.2013.06.088

[136] M.R. Oliveira, J.A. Cecilia, D. Ballesteros-Plata, I. Barroso-Martín, P. Núñez, A. Infantes-Molina, E. Rodríguez-Castellón, Microwave-assisted synthesis of zeolite A from metakaolinite for CO2 adsorption, Int. J. Mol. Sci. 24 (2023) 14040. https://doi.org/10.3390/ijms241814040

[137] B. Makgabutlane, L.N. Nthunya, E.N. Nxumalo, N.M. Musyoka, S.D. Mhlanga, Microwave irradiation-assisted synthesis of zeolites from coal fly ash: an optimization study for a sustainable and efficient production process, ACS Omega 5 (2020) 25000-25008. https://doi.org/10.1021/acsomega.0c00931

[138] Z. Ma, H. Deng, L. Li, Q. Zhang, G. Chen, C. Sun, H. He, J. Yu, Fluoride-free and seed-free microwave-assisted hydrothermal synthesis of nanosized high-silica beta zeolites for effective VOCs adsorption, Chem. Sci. 14 (2023) 2131-2138. https://doi.org/10.1039/D2SC06389A

[139] B. Chandra Shekara, B. Jai Prakash, Y. Bhat, Dealumination of zeolite BEA under microwave irradiation, ACS Catal. 1 (2011) 193-199. https://doi.org/10.1021/cs1000448

[140] N. Salahudeen, A review on zeolite: application, synthesis and effect of synthesis parameters on product properties, Chem. Afr. 5 (2022) 1889-1906. https://doi.org/10.1007/s42250-022-00471-9

[141] D. Stuerga, K. Gonon, M. Lallemant, Microwave heating as a new way to induce selectivity between competitive reactions. Application to isomeric ratio control in sulfonation of naphthalene, Tetrahedron 49 (1993) 6229-6234. https://doi.org/10.1016/S0040-4020(01)87961-8

[142] H. Chon, S.K. Ihm, Y.S. Uh, Progress in Zeolite and Microporous Materials, First ed., Elsevier, Amsterdam, 1996.

[143] A. Arafat, J. Jansen, A. Ebaid, H. van Bekkum, Microwave preparation of zeolite Y and ZSM-5, Zeolites 13 (1993) 162-165. https://doi.org/10.1016/S0144-2449(05)80272-6

[144] Q. Zhang, J. Luo, E. Vileno, S.L. Suib, Synthesis of cryptomelane-type manganese oxides by microwave heating, Chem. Mater. 9 (1997) 2090-2095. https://doi.org/10.1021/cm970129g

[145] M. Thommes, K. Kaneko, A.V. Neimark, J.P. Olivier, F. Rodriguez-Reinoso, J. Rouquerol, K.S. Sing, Physisorption of gases, with special reference to the evaluation of surface area and pore size distribution (IUPAC technical report), Pure Appl. Chem. 87 (2015) 1051-1069. https://doi.org/10.1515/pac-2014-1117

[146] Q. Ye, Q. Li, X. Li, Removal of heavy metals from wastewater using biochars: adsorption and mechanisms, Environ. Pollut. Bioavailab. 34 (2022) 385-394. https://doi.org/10.1080/26395940.2022.2120542

[147] N. Jiang, R. Shang, S.G. Heijman, L.C. Rietveld, Adsorption of triclosan, trichlorophenol and phenol by high-silica zeolites: adsorption efficiencies and mechanisms, Sep. Purif. Technol. 235 (2020) 116152. https://doi.org/10.1016/j.seppur.2019.116152

[148] S. Kumari, J. Chowdhry, M. Kumar, M.C. Garg, Zeolites in wastewater treatment: a comprehensive review on scientometric analysis, adsorption mechanisms, and future prospects, Environ. Res. 260 (2024) 119782. https://doi.org/10.1016/j.envres.2024.119782

[149] E. Pérez-Botella, S. Valencia, F. Rey, Zeolites in adsorption processes: state of the art and future prospects, Chem. Rev. 122 (2022) 17647-17695. https://doi.org/10.1021/acs.chemrev.2c00140

[150] S. Ghojavand, B. Coasne, E.B. Clatworthy, R. Guillet-Nicolas, P. Bazin, M. Desmurs, L. Aguilera, V. Ruaux, S. Mintova, Alkali metal cations influence the CO2 adsorption capacity of nanosized chabazite: modeling vs experiment, ACS Appl. Nano Mater. 5 (2022) 5578-5588. https://doi.org/10.1021/acsanm.2c00537

[151] X. Wang, N. Yan, M. Xie, P. Liu, P. Bai, H. Su, B. Wang, Y. Wang, L. Li, T. Cheng, The inorganic cation-tailored "trapdoor" effect of silicoaluminophosphate zeolite for highly selective CO2 separation, Chem. Sci. 12 (2021) 8803-8810. https://doi.org/10.1039/D1SC00619C

[152] P.J. Bereciartua, Á. Cantín, A. Corma, J.L. Jordá, M. Palomino, F. Rey, S. Valencia, E.W. Corcoran, P. Kortunov, P.I. Ravikovitch, Control of zeolite framework flexibility and pore topology for separation of ethane and ethylene, Science 358 (2017) 1068-1071. https://doi.org/10.1126/science.aao0092

[153] J.G. Min, K.C. Kemp, H. Lee, S.B. Hong, CO2 adsorption in the RHO family of embedded isoreticular zeolites, J. Phys. Chem. C 122 (2018) 28815-28824. https://doi.org/10.1021/acs.jpcc.8b09996

[154] I. Langmuir, The adsorption of gases on plane surfaces of glass, mica and platinum, J. Am. Chem. Soc. 40 (1918) 1361-1403. https://doi.org/10.1021/ja02242a004

[155] J. Tóth, Uniform interpretation of gas/solid adsorption, Adv. Colloid Interface Sci. 55 (1995) 1-239. https://doi.org/10.1016/0001-8686(94)00226-3

[156] R. Sips, On the structure of a catalyst surface, J. Chem. Phys. 16 (1948) 490-495. https://doi.org/10.1063/1.1746922

[157] A. Charkhi, M. Kazemeini, S.J. Ahmadi, S. Ammari, Effect of bentonite binder on adsorption and cation exchange properties of granulated nano nay zeolite, Adv. Mater. Res. 335 (2011) 423-428. https://doi.org/10.4028/www.scientific.net/AMR.335-336.423

[158] J. Zhang, Y. Mao, J. Li, X. Wang, J. Xie, Y. Zhou, J. Wang, Ultrahigh mechanically stable hierarchical mordenite zeolite monolith: direct binder-/template-free hydrothermal synthesis, Chem. Eng. Sci. 138 (2015) 473-481. https://doi.org/10.1016/j.ces.2015.08.016

[159] J. Kärger, D.M. Ruthven, D.N. Theodorou, Diffusion in Nanoporous Materials, First ed., John Wiley & Sons, Hoboken, New Jersey, 2012. https://doi.org/10.1002/9783527651276

[160] E. Haldoupis, S. Nair, D.S. Sholl, Pore size analysis of > 250000 hypothetical zeolites, Phys. Chem. Chem. Phys. 13 (2011) 5053-5060. https://doi.org/10.1039/c0cp02766a

[161] D. Dubbeldam, R. Krishna, S. Calero, A.O. Yazaydın, Computer-assisted screening of ordered crystalline nanoporous adsorbents for separation of alkane isomers, Angew. Chem. Int. Ed. 51 (2012) 11867-11871. https://doi.org/10.1002/anie.201205040

[162] A.H. Farmahini, S. Krishnamurthy, D. Friedrich, S. Brandani, L. Sarkisov, Performance-based screening of porous materials for carbon capture, Chem. Rev. 121 (2021) 10666-10741. https://doi.org/10.1021/acs.chemrev.0c01266

[163] J. Kim, L.C. Lin, R.L. Martin, J.A. Swisher, M. Haranczyk, B. Smit, Large-scale computational screening of zeolites for ethane/ethene separation, Langmuir 28 (2012) 11914-11919. https://doi.org/10.1021/la302230z

[164] J.J. Gutiérrez-Sevillano, S. Calero, Computational approaches to zeolite-based adsorption processes, in: S. Valencia, F. Rey (Eds.), New Developments in Adsorption/Separation of Small Molecules by Zeolites, Springer, Berlin, 2020, pp. 57-83. https://doi.org/10.1007/430_2020_66

[165] C. Martínez, J. Pérez-Pariente, Zeolites and Ordered Porous Solids: Fundamentals and Applications, First ed., Editorial Universitat Politècnica de València, Valencia, 2011.

[166] A. Pfenninger, Manufacture and use of zeolites for adsorption processes, in: C. Baerlocher, J.M. Bennett, W. Depmeier, A.N. Fitch, H. Jobic, H. Koningsveld, W.M. Meier, A. Pfenninger, O. Terasaki (Eds.), Structures and Structure Determination, Springer, Berlin, 2001, pp. 163-198.

[167] C.G. Coe, S.M. Kuznicki, Polyvalent ion exchanged adsorbent for air separation, U.S. Patent 4,481,018, 1984.

[168] K. Gleichmann, B. Unger, A. Brandt, Manufacturing of industrial zeolite molecular sieves, Chem. Ing. Tech. 89 (2017) 851-862. https://doi.org/10.1002/cite.201600164

[169] R.T. Yang, Adsorbents: Fundamentals and Applications, First ed., John Wiley & Sons, Hoboken, New Jersey, 2003.

[170] R.T. Yang, Gas Separation by Adsorption Processes, First ed., World Scientific, Singapore, 1997. https://doi.org/10.1142/p037

[171] R. Ribeiro, C. Grande, A.E. Rodrigues, Electric swing adsorption for gas separation and purification: a review, Sep. Sci. Technol. 49 (2014) 1985-2002. https://doi.org/10.1080/01496395.2014.915854

[172] D. Ernest, J. Henley, J.D. Seader, Separation Process Principles: Chemical and Biochemical Operations, Third ed., John Wiley & Sons, Hoboken, New Jersey, 2011.

[173] D.B. Broughton, C.G. Gerhold, Continuous sorption process employing fixed bed of sorbent and moving inlets and outlets, U.S. Patent 2,985,589A, 1961.

[174] V. Valtchev, G. Majano, S. Mintova, J. Pérez-Ramírez, Tailored crystalline microporous materials by post-synthesis modification, Chem. Soc. Rev. 42 (2013) 263-290. https://doi.org/10.1039/C2CS35196J

[175] A. Jayaraman, R.T. Yang, C.L. Munson, D. Chinn, Deactivation of π-complexation adsorbents by hydrogen and rejuvenation by oxidation, Ind. Eng. Chem. Res. 40 (2001) 4370-4376. https://doi.org/10.1021/ie0102753

[176] A.M. Varghese, G.N. Karanikolos, CO2 capture adsorbents functionalized by amine-bearing polymers: a review, Int. J. Greenh. Gas Control 96 (2020) 103005. https://doi.org/10.1016/j.ijggc.2020.103005

[177] C. Connor, K. Möller, H. Manstein, The effect of silanization on the catalytic and sorption properties of zeolites, Cattech 5 (2001) 172-182. https://doi.org/10.1023/A:1014068622252

[178] I.C. Medeiros-Costa, E. Dib, N. Nesterenko, J.P. Dath, J.P. Gilson, S. Mintova, Silanol defect engineering and healing in zeolites: opportunities to fine-tune their properties and performances, Chem. Soc. Rev. 50 (2021) 11156-11179. https://doi.org/10.1039/D1CS00395J

[179] L. Gómez-Hortigüela, B. Bernardo-Maestro, Insights into the Chemistry of Organic Structure-Directing Agents in the Synthesis of Zeolitic Materials, First ed., Springer, Cham, 2018. https://doi.org/10.1007/978-3-319-74289-2

[180] C.J. Heard, L. Grajciar, F. Uhlík, M. Shamzhy, M. Opanasenko, J. Čejka, P. Nachtigall, Zeolite (in) stability under aqueous or steaming conditions, Adv. Mater. 32 (2020) 2003264. https://doi.org/10.1002/adma.202003264

[181] N. Muhamad, N. Abdullah, M.A. Rahman, K.H. Abas, A.A. Aziz, M.H.D. Othman, J. Jaafar, A.F. Ismail, Removal of nickel from aqueous solution using supported zeolite-Y hollow fiber membranes, Environ. Sci. Pollut. Res. 25 (2018) 19054-19064. https://doi.org/10.1007/s11356-018-2074-3

[182] S. Abello, A. Bonilla, J. Perez-Ramirez, Mesoporous ZSM-5 zeolite catalysts prepared by desilication with organic hydroxides and comparison with NaOH leaching, Appl. Catal. Gen. 364 (2009) 191-198. https://doi.org/10.1016/j.apcata.2009.05.055

[183] H. Javadian, M. Taghavi, Application of novel polypyrrole/thiol-functionalized zeolite Beta/MCM-41 type mesoporous silica nanocomposite for adsorption of Hg2+ from aqueous solution and industrial wastewater: kinetic, isotherm and thermodynamic studies, Appl. Surf. Sci. 289 (2014) 487-494. https://doi.org/10.1016/j.apsusc.2013.11.020

[184] N.H. Mthombeni, S. Mbakop, A. Ochieng, M.S. Onyango, Vanadium (V) adsorption isotherms and kinetics using polypyrrole coated magnetized natural zeolite, J. Taiwan Inst. Chem. Eng. 66 (2016) 172-180. https://doi.org/10.1016/j.jtice.2016.06.016

[185] A. Gaffer, A.A. Al-Kahlawy, D. Aman, Magnetic zeolite-natural polymer composite for adsorption of chromium (VI), Egypt. J. Pet. 26 (2017) 995-999. https://doi.org/10.1016/j.ejpe.2016.12.001

[186] C.K. Lim, H.H. Bay, C.H. Neoh, A. Aris, Z. Abdul Majid, Z. Ibrahim, Application of zeolite-activated carbon macrocomposite for the adsorption of Acid Orange 7: isotherm, kinetic and thermodynamic studies, Environ. Sci. Pollut. Res. 20 (2013) 7243-7255. https://doi.org/10.1007/s11356-013-1725-7

[187] B. Nanda, A.C. Pradhan, K. Parida, A comparative study on adsorption and photocatalytic dye degradation under visible light irradiation by mesoporous MnO2 modified MCM-41 nanocomposite, Microporous Mesoporous Mater. 226 (2016) 229-242. https://doi.org/10.1016/j.micromeso.2015.12.027

[188] G.H. Kühl, Modification of zeolites, in: J. Weitkamp, L. Puppe (Eds.), Catalysis and Zeolites: Fundamentals and Applications, Springer, Berlin, 1999, pp. 81-197. https://doi.org/10.1007/978-3-662-03764-5_3

[189] R. Szostak, Secondary synthesis methods, in: H. van Bekkum, E.M. Flanigen, P.A. Jacobs, J.C. Jansen (Eds.), Introduction to Zeolite Science and Practice, Elsevier, Amsterdam, 2001, pp. 261-297.

[190] R. Barrer, Factors governing the growth of crystalline silicates, Discuss. Faraday Soc. 5 (1949) 326-337. https://doi.org/10.1039/df9490500326

[191] A. Buzimov, S. Kulkov, W. Eckl, S. Pappert, L. Gömze, E. Kurovics, I. Kocserha, R. Géber, Effect of mechanical treatment on properties of zeolites with chabazite structure, J. Phys.-Conf. Ser. 790 (2017) 012004. https://doi.org/10.1088/1742-6596/790/1/012004

[192] R.P. Townsend, E.N. Coker, Ion exchange in zeolites, in: H. van Bekkum, E.M. Flanigen, P.A. Jacobs, J.C. Jansen. (Eds.), Introduction to Zeolite Science and Practice, Elsevier, Amsterdam, 2001, pp. 467-524. https://doi.org/10.1016/S0167-2991(01)80253-6

[193] D. Breck, Crystalline molecular sieves, J. Chem. Educ. 41 (1964) 678-684. https://doi.org/10.1021/ed041p678

[194] R. Harjula, A. Dyer, S.D. Pearson, R.P. Townsend, Ion exchange in zeolites. Part 1. Prediction of Ca2+-Na+ equilibria in zeolites X and Y, J. Chem. Soc., Faraday Trans. 88 (1992) 1591-1597. https://doi.org/10.1039/FT9928801591

[195] H.G. Karge, J. Weitkamp, Zeolites as Catalysts, Sorbents and Detergent Builders: Applications and Innovations, First ed., Elsevier, Amsterdam, 1989.

[196] L. Ames, Characterization of a strontium-selective zeolite, Am. Mineral. 47 (1962) 1317-1326.

[197] J.P. Ausikaitis, D.R. Garg, Adsorption separation cycle, U.S. Patent 4,373,935, 1983.

[198] J. Shang, G. Li, R. Singh, Q. Gu, K.M. Nairn, T.J. Bastow, N. Medhekar, C.M. Doherty, A.J. Hill, J.Z. Liu, Discriminative separation of gases by a "molecular trapdoor" mechanism

in chabazite zeolites, J. Am. Chem. Soc. 134 (2012) 19246-19253. https://doi.org/10.1021/ja309274y

[199] K. Liu, C. Song, V. Subramani, Hydrogen and Syngas Production and Purification Technologies, First ed., John Wiley & Sons, Hoboken, New Jersey, 2010. https://doi.org/10.1002/9780470561256

[200] B.P. Russell, K.M. Patel, H. Rastelli, Pressure swing adsorption process and apparatus for purifying a hydrogen-containing gas stream, U.S. Patent 10,399,032 B2, 2019.

[201] H. Li, C. Qiu, S. Ren, Q. Dong, S. Zhang, F. Zhou, X. Liang, J. Wang, S. Li, M. Yu, Na+-gated water-conducting nanochannels for boosting CO2 conversion to liquid fuels, Science 367 (2020) 667-671. https://doi.org/10.1126/science.aaz6053

[202] R.M. Milton, Sweetening and drying of natural gas, U.S. Patent 3,078,634, 1963.

[203] D.R. Burfield, K.H. Lee, R.H. Smithers, Desiccant efficiency in solvent drying. A reappraisal by application of a novel method for solvent water assay, J. Org. Chem. 42 (1977) 3060-3065. https://doi.org/10.1021/jo00438a024

[204] T.E. Rufford, S. Smart, G.C. Watson, B. Graham, J. Boxall, J.D. Da Costa, E. May, The removal of CO2 and N2 from natural gas: a review of conventional and emerging process technologies, J. Pet. Sci. Eng. 94 (2012) 123-154. https://doi.org/10.1016/j.petrol.2012.06.016

[205] M. Tagliabue, D. Farrusseng, S. Valencia, S. Aguado, U. Ravon, C. Rizzo, A. Corma, C. Mirodatos, Natural gas treating by selective adsorption: material science and chemical engineering interplay, Chem. Eng. J. 155 (2009) 553-566. https://doi.org/10.1016/j.cej.2009.09.010

[206] H.W. Habgood, Removal of nitrogen from natural gas, U.S. Patent 2,843,219A, 1958.

[207] C.C. Chao, Selective adsorption on magnesium-containing clinoptilolites, U.S. Patent 4,964,889, 1990.

[208] S.M. Kuznicki, V.A. Bell, S. Nair, H.W. Hillhouse, R.M. Jacubinas, C.M. Braunbarth, B.H. Toby, M. Tsapatsis, A titanosilicate molecular sieve with adjustable pores for size-selective adsorption of molecules, Nature 412 (2001) 720-724. https://doi.org/10.1038/35089052

[209] S.U. Nandanwar, D.R. Corbin, M.B. Shiflett, A review of porous adsorbents for the separation of nitrogen from natural gas, Ind. Eng. Chem. Res. 59 (2020) 13355-13369. https://doi.org/10.1021/acs.iecr.0c02730

[210] T.L. Valley, A.R. Richard, M. Fan, The progress in water gas shift and steam reforming hydrogen production technologies-a review, Int. J. Hydrogen Energy 39 (2014) 16983-17000. https://doi.org/10.1016/j.ijhydene.2014.08.041

[211] Y. Yang, N. Burke, S. Ali, S. Huang, S. Lim, Y. Zhu, Experimental studies of hydrocarbon separation on zeolites, activated carbons and MOFs for applications in natural gas processing, RSC Adv. 7 (2017) 12629-12638. https://doi.org/10.1039/C6RA25509D

[212] A.J. Kidnay, W.R. Parrish, D.G. McCartney, Fundamentals of Natural Gas Processing, Second ed., CRC Press, Boca Raton, 2019. https://doi.org/10.1201/9780429464942

[213] E.J. Garcia, J. Perez-Pellitero, G.D. Pirngruber, C. Jallut, M. Palomino, F. Rey, S. Valencia, Tuning the adsorption properties of zeolites as adsorbents for CO2 separation: best compromise between the working capacity and selectivity, Ind. Eng. Chem. Res. 53 (2014) 9860-9874. https://doi.org/10.1021/ie500207s

[214] E. Pérez-Botella, R. Martínez-Franco, N. González-Camuñas, Á. Cantín, M. Palomino, M. Moliner, S. Valencia, F. Rey, Unusually low heat of adsorption of CO2 on AlPO and SAPO molecular sieves, Front. Chem. 8 (2020) 588712. https://doi.org/10.3389/fchem.2020.588712

[215] M. Bui, C.S. Adjiman, A. Bardow, E.J. Anthony, A. Boston, S. Brown, P.S. Fennell, S. Fuss, A. Galindo, L.A. Hackett, Carbon capture and storage (CCS): the way forward, Energy Environ. Sci. 11 (2018) 1062-1176. https://doi.org/10.1039/C7EE02342A

[216] A. Kumar, D.G. Madden, M. Lusi, K.J. Chen, E.A. Daniels, T. Curtin, J.J. Perry, M.J. Zaworotko, Direct air capture of CO2 by physisorbent materials, Angew. Chem. Int. Ed. 54 (2015) 14372-14377. https://doi.org/10.1002/anie.201506952

[217] A.N. Stuckert, R.T. Yang, CO2 capture from the atmosphere and simultaneous concentration using zeolites and amine-grafted SBA-15, Environ. Sci. Technol. 45 (2011) 10257-10264. https://doi.org/10.1021/es202647a

[218] D. Fu, Y. Park, M.E. Davis, Zinc containing small-pore zeolites for capture of low concentration carbon dioxide, Angew. Chem. Int. Ed. 61 (2022) e202112916. https://doi.org/10.1002/anie.202112916

[219] F. Chang, J. Zhou, P. Chen, Y. Chen, H. Jia, S.M. Saad, Y. Gao, X. Cao, T. Zheng, Microporous and mesoporous materials for gas storage and separation: a review, Asia-Pac. J. Chem. Eng. 8 (2013) 618-626. https://doi.org/10.1002/apj.1717

[220] M.E. Boot-Handford, J.C. Abanades, E.J. Anthony, M.J. Blunt, S. Brandani, N. Mac Dowell, J.R. Fernández, M.C. Ferrari, R. Gross, J.P. Hallett, Carbon capture and storage update, Energy Environ. Sci. 7 (2014) 130-189. https://doi.org/10.1039/C3EE42350F

[221] F.G. Kerry, Industrial Gas Handbook: Gas Separation and Purification, Second ed., CRC Press, Boca Raton, 2007. https://doi.org/10.1201/9781420008265

[222] I. Matito-Martos, A. Martin-Calvo, C. Ania, J. Parra, J.M. Vicent-Luna, S. Calero, Role of hydrogen bonding in the capture and storage of ammonia in zeolites, Chem. Eng. J. 387 (2020) 124062. https://doi.org/10.1016/j.cej.2020.124062

[223] M. Ozekmekci, G. Salkic, M.F. Fellah, Use of zeolites for the removal of H2S: a mini-review, Fuel Process. Technol. 139 (2015) 49-60. https://doi.org/10.1016/j.fuproc.2015.08.015

[224] L. Anh, T. Nhu, L. Hien, N. Thong, N. Phuong, N. Dung, N. Duong, N. Long, Highly mesoporous X zeolites synthesized by microwave-assisted sequential dealumination-desilication for methylene blue adsorption and toluene photocatalytic removal, Int. J. Environ. Sci. Technol. 22 (2025) 1-18. https://doi.org/10.1007/s13762-025-06436-y

[225] A.D. Karisma, S. Altway, E. Agustiani, E.O. Ningrum, D.R. Zuchrillah, Synthesis of zeolite as a textile dye waste adsorbent from rice husk ash using microwave heating method, J. Phys.-Conf. Ser. 2344 (2022) 012014. https://doi.org/10.1088/1742-6596/2344/1/012014

[226] F. Piri, A. Mollahosseini, M.M. Hosseini, Enhanced adsorption of dyes on microwave-assisted synthesized magnetic zeolite-hydroxyapatite nanocomposite, J. Environ. Chem. Eng. 7 (2019) 103338. https://doi.org/10.1016/j.jece.2019.103338

[227] Q. Zhou, X. Jiang, Q. Qiu, Y. Zhao, L. Long, Synthesis of high-quality NaP1 zeolite from municipal solid waste incineration fly ash by microwave-assisted hydrothermal method and its adsorption capacity, Sci. Total Environ. 855 (2023) 158741. https://doi.org/10.1016/j.scitotenv.2022.158741

[228] T.S. Jamil, H. Youssef, Microwave synthesis of zeolites from Egyptian kaolin: evaluation of heavy metals removal, Sep. Sci. Technol. 51 (2016) 2876-2886. https://doi.org/10.1080/01496395.2016.1229337

[229] T. Abdullahi, Development and optimization of zeolite synthesis route from natural koalin for adsorption of dyes, PhD Thesis, Universiti Tun Hussein Onn Malaysia, 2019.

[230] N. Salahudeen, A. Ahmed, A. Al-Muhtaseb, B. Jibril, R. Al-Hajri, S. Waziri, M. Dauda, J. Al-Sabahi, Microwave-assisted adsorptive desulfurization of model diesel fuel using synthesized microporous rare earth metal-doped zeolite Y, J. Eng. Res. 12 (2015) 44-52. https://doi.org/10.24200/tjer.vol12iss1pp44-52

[231] Q.L. Nguyen, T.T.B. Hoang, Selective adsorption of H2S in biogas using zeolite prepared by microwave-assisted method, CTU J. Innov. Sustain. Dev. (2016) 52-56. https://doi.org/10.22144/ctu.jsi.2016.008

[232] T. Ryu, N.H. Ahn, S. Seo, J. Cho, H. Kim, D. Jo, G.T. Park, P.S. Kim, C.H. Kim, E.L. Bruce, Fully copper-exchanged high-silica LTA zeolites as unrivaled hydrothermally stable NH3-SCR catalysts, Angew. Chem. 129 (2017) 3304-3308. https://doi.org/10.1002/ange.201610547

[233] K.A. Lomachenko, E. Borfecchia, C. Negri, G. Berlier, C. Lamberti, P. Beato, H. Falsig, S. Bordiga, The Cu-CHA deNOx catalyst in action: temperature-dependent NH3-assisted selective catalytic reduction monitored by operando XAS and XES, J. Am. Chem. Soc. 138 (2016) 12025-12028. https://doi.org/10.1021/jacs.6b06809

[234] J.H. Baik, S.D. Yim, I.S. Nam, Y.S. Mok, J.H. Lee, B.K. Cho, S.H. Oh, Modeling of monolith reactor washcoated with CuZSM-5 catalyst for removing NO from diesel engine by urea, Ind. Eng. Chem. Res. 45 (2006) 5258-5267. https://doi.org/10.1021/ie060199+

[235] Y.J. Kim, J.K. Lee, K.M. Min, S.B. Hong, I.S. Nam, B.K. Cho, Hydrothermal stability of CuSSZ13 for reducing NOx by NH3, J. Catal. 311 (2014) 447-457. https://doi.org/10.1016/j.jcat.2013.12.012

[236] J. Kim, S.J. Cho, D.H. Kim, Facile synthesis of KFI-type zeolite and its application to selective catalytic reduction of NOx with NH3, ACS Catal. 7 (2017) 6070-6081. https://doi.org/10.1021/acscatal.7b00697

[237] D. Jo, T. Ryu, G.T. Park, P.S. Kim, C.H. Kim, I.S. Nam, S.B. Hong, Synthesis of high-silica LTA and UFI zeolites and NH3-SCR performance of their copper-exchanged form, ACS Catal. 6 (2016) 2443-2447. https://doi.org/10.1021/acscatal.6b00489

[238] J.H. Park, J.H. Choung, I.S. Nam, S.W. Ham, N2O decomposition over wet-and solid-exchanged Fe-ZSM-5 catalysts, Appl. Catal. B Environ. 78 (2008) 342-354. https://doi.org/10.1016/j.apcatb.2007.09.020

[239] L. Kovarik, N.M. Washton, R. Kukkadapu, A. Devaraj, A. Wang, Y. Wang, J. Szanyi, C.H. Peden, F. Gao, Transformation of active sites in Fe/SSZ-13 SCR catalysts during hydrothermal aging: a spectroscopic, microscopic, and kinetics study, ACS Catal. 7 (2017) 2458-2470. https://doi.org/10.1021/acscatal.6b03679

[240] I. Heo, S. Sung, M.B. Park, T.S. Chang, Y.J. Kim, B.K. Cho, S.B. Hong, J.W. Choung, I.S. Nam, Effect of hydrocarbon on DeNOx performance of selective catalytic reduction by a combined reductant over Cu-containing zeolite catalysts, ACS Catal. 9 (2019) 9800-9812. https://doi.org/10.1021/acscatal.9b02763

[241] K.D. Dubois, A. Petushkov, E. Cardona, S.C. Larsen, G. Li, Adsorption and photochemical properties of a molecular CO2 reduction catalyst in hierarchical mesoporous ZSM-5: an in situ FTIR study, J. Phys. Chem. Lett. 3 (2012) 486-492. https://doi.org/10.1021/jz201701y

[242] Y. Wu, M. Yang, S. Hu, L. Wang, H. Yao, Characteristics and mechanisms of 4A zeolite supported nanoparticulate zero-valent iron as Fenton-like catalyst to degrade methylene blue, Toxicol. Environ. Chem. 96 (2014) 227-242. https://doi.org/10.1080/02772248.2014.931960

[243] R. Gonzalez-Olmos, M.J. Martin, A. Georgi, F.D. Kopinke, I. Oller, S. Malato, Fe-zeolites as heterogeneous catalysts in solar Fenton-like reactions at neutral pH, Appl. Catal. B Environ. 125 (2012) 51-58. https://doi.org/10.1016/j.apcatb.2012.05.022

[244] C.A. Rios, C.D. Williams, M.A. Fullen, Nucleation and growth history of zeolite LTA synthesized from kaolinite by two different methods, Appl. Clay Sci. 42 (2009) 446-454. https://doi.org/10.1016/j.clay.2008.05.006

[245] M. Inada, Y. Eguchi, N. Enomoto, J. Hojo, Synthesis of zeolite from coal fly ashes with different silica-alumina composition, Fuel 84 (2005) 299-304. https://doi.org/10.1016/j.fuel.2004.08.012

Materials Research Foundations 189 (2026) 209-236
https://doi.org/10.21741/9781644904039-7

Chapter 7

Microwave-Assisted Zeolite Synthesis for Biomedical Applications

Muhammad Saria[1], Zeeshan Taj[1], Sumreen Hayat[1], Muhammad Saqalein[1], Razia Noreen[2], Muhammad Faisal[3], Saima Muzammil[1,*]

[1]Institute of Microbiology, Government College University, Faisalabad, Pakistan

[2]Department of Biochemistry, Government College University, Faisalabad, Pakistan

[3]Institute of Plant Breeding and Biotechnology, MNS-University of Agriculture, Multan, Pakistan

saimamuzammil83@gmail.com

Abstract

Zeolites, microporous crystalline aluminosilicates featuring interconnected tetrahedra, are exceptionally versatile in scientific and industrial applications. This chapter provides a detailed examination of their characteristics, natural and synthetic varieties, and production techniques (including hydrothermal and sol-gel methods) to achieve specific functionalities. Zeolites are essential in catalysis, adsorption, environmental cleanup (water/soil), agriculture (soil improvement/nutrient release), and developing biomedical applications like drug delivery and tissue engineering. Furthermore, the chapter covers their utility in improving polymer membranes, facilitating polymer cracking, and tackling climate change. It concludes by outlining current obstacles (like affordable synthesis) and prospective novel uses, emphasizing zeolites' vital contribution to contemporary progress.

Keywords

Zeolite, Zeolite Imidazolate Frameworks, Metal Organic Frameworks, Natural Zeolites, Drug Delivery System

Contents

1. Introduction

Zeolites are microporous crystalline substances with a framework structure comprised of four-connected TO_4 units. The 'T' atom, located in the tetrahedron's center, is predominantly silicon or aluminum, but can include other appropriate elements. These central atoms (Al, Si, or P) form tetrahedra by bonding with oxygen atoms (primary building blocks). These tetrahedra then link via oxygen bridges, creating secondary building blocks such as rings and prisms of different sizes [1]. Axel Fredrik Cronstedt, in 1756, named "zeolite" based on the steam produced by a mineral upon rapid heating. That mineral was later classified as stilbite, a microporous crystalline aluminosilicate. Barrer's 1940s research initiated major zeolite advancements. The 1950s and 60s, considered a "golden age," saw the commercialization of ultrastable Y zeolite (USY), which significantly improved oil refining via fluid catalytic cracking (FCC) [2].

Zeolites are valuable catalysts and adsorbents due to their ion-exchange characteristics and cage-like porous structure. The efficiency of zeolites is directly related to pore size, and thus they are microporous (<2 nm), mesoporous (2-50 nm), or macroporous (>50 nm) in nature [3]. Zeolites consist of central atom tetrahedral building units. The building units then link through the oxygen bridges to form varied rings and cage-like structures as illustrated in Fig. 1 [1].

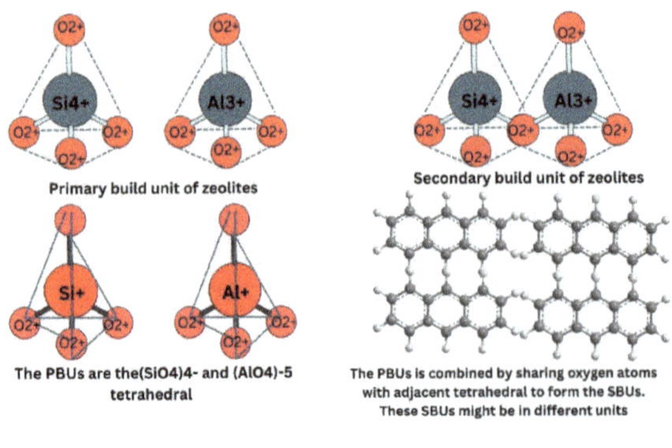

Figure 1. Primary and secondary building units of zeolites. [1]

Materials Research Foundations 189 (2026) 209-236 https://doi.org/10.21741/9781644904039-7

Zeolites, aluminosilicates by nature and hydrated, are utilized as low-cost source material for ceramic membranes. The structures of membranes are being extensively studied due to their mechanical and chemical stability, which is superior compared to that of polymeric and inorganic membranes [4]. Zeolites are useful for polymer cracking because they contain homogeneous micropores that impart shape selectivity and improve product specificity. They are also solid acid catalysts, which adds to their versatility [5].

Zeolite activity is regulated by channel dimensionality, pore size, pore volume, and extra-framework species. Although typically aluminosilicates (Al, Si, O), homopolar (pure silica) and heteroatom-substituted (Ge, B, etc.) varieties exist, meaning one may have specialized functions. Exchangeable cations are used to give charge balance. Low-pore-density and high-pore-size zeolites are required for a variety of applications, hence new zeolite structure discovery continues [6]. Zeolites are highly versatile materials used in numerous technologies, including catalysis, separations, drug delivery, and sensing. Choosing the right zeolite for a specific application depends on factors like pore structure, chemical makeup, water affinity, acidity, and stability. Zeolites have a particularly significant impact on industrial catalysis. Natural zeolites form through hydrothermal processes, primarily associated with volcanic activity. They can be found as crystals in igneous and metamorphic rocks, or as finer grains accumulated in sedimentary formations. Because of their affordability, natural zeolites are primarily utilized in large-scale mineral applications [7].

The discovery of natural zeolites marked a significant advancement in mineralogy, as their unique surface and structural properties have proven valuable across diverse fields, including agriculture, industry, animal husbandry, cosmetics, and biotechnology. Initially found in basalt rock vugs and cavities, zeolites were later discovered in sedimentary formations [8]. They also have potential for catalytic and antibacterial applications and are widely used as pet litter. The increased amount of wastewater from industries and the failure of traditional treatment technologies to provide for discharge needs have resulted in the severe degradation of water quality. Water pollution, a cause for life danger, is notably added to by textile industry dye effluent disposal. These toxic dyes, which were discharged in large quantities into water bodies, pose severe environmental and health risks. The addition of zeolites improves polymer membranes' mechanical strength, properties, and adsorption capacity. These composite materials, e.g., zeolite-polysulfone or polyvinyl alcohol/ZSM-5, are effective in the removal of heavy metal cations and dyes. Zeolite membrane-based systems offer benefits in the guise of low cost, mechanical rigidity, high selectivity, and porosity [4].

Zeolites are vital catalysts in the chemical industry, which produces 85% of daily necessities. They boost efficiency and cut energy use in chemical manufacturing, and are increasingly used to convert greenhouse gases and biomass into valuable products. Zeolites are key to sustainable development by enabling resource efficiency, renewable energy solutions, environmental improvement, and applications in medicine [2]. Zeolites, due to their nanoporous structure, offer selective adsorption and molecular sieving, making them versatile in various fields, including agriculture. In agriculture, they're used as soil amendments, slow-release carriers, crop protectants, and feed additives. Soil application of zeolites enhances crop yield by improving mineral and water nutrition and photosynthetic activity, particularly in herbaceous crops. Recent studies explore foliar application for pest and disease control, leveraging zeolite dehydration and particle film formation. This approach, mainly in tree crops, impacts leaf physiology and optics, product quality, and shelf-life, suggesting zeolites as a potential tool to mitigate climate change effects like water and heat stress [9].

2. Types of Zeolites

Zeolites are crystalline materials with intricate, three-dimensional frameworks that contain tiny, precisely sized pores. They exist both naturally and synthetically, and are categorized by their specific framework structure and type. Framework structure is vital for understanding zeolite chemistry. Zeolite classification, however, depends on framework type, which characterizes the symmetrical connection of the tetrahedral atoms, distinct from the framework structure itself.

2.1 Natural Zeolite

Natural zeolites, found globally with diverse mineral compositions, lack precise reserve estimates. Over 70 natural zeolite types exist, with major sources in Slovakia, Bulgaria, Greece, Italy, and the USA. Formed from volcanic glass under various conditions, their structure consists of a 3D network of Si and Al tetrahedra, creating channels and cages that house water molecules and cations like Na^+, K^+, Ca^{2+}, and Mg^+. The Si/Al ratio significantly influences their properties, and the presence of Al imparts a negative charge to the framework [10]. Among the most prevalent natural zeolites are clinoptilolite, chabazite, mordenite, stilbite, erionite, heulandite, and phillipsite [9].

The growing challenge of clean water access is exacerbated by industrial and population growth, leading to increased pollution from heavy metals, inorganic and organic pollutants, and microorganisms. Heavy metals are especially concerning because of their toxicity and severe health impacts. While various water treatment methods exist, adsorption stands out for its cost-effectiveness and operational simplicity. Finding a suitable adsorbent with high capacity, selectivity, and sustainability is crucial. Naturally occurring clays like zeolites are promising due to their abundance and favorable adsorption properties. Zeolites, especially Italian chabazite, significantly improve soil properties, leading to reduced water and fertilizer needs and increased crop yields. Their high drainage and water retention capacity minimize water loss in various soil types. Additionally, their reversible water properties stabilize root-level humidity and temperature [11]. Adsorption is a versatile and evolving separation technology used in various applications like water treatment and gas purification. Its development is closely linked to zeolites, microporous aluminosilicates with precise pore sizes. Zeolites enable selective separation of molecules like CO_2 and hydrocarbons, and their tunable properties make them effective for diverse industrial separations, including drying, air separation, and isomer separation [12].

Soil contamination by herbicides like metsulfuron-methyl (MSM) reduces fertility. While antidotes exist, adsorbents offer a promising solution for soil decontamination. Materials such as activated carbon, organic clays, and organo-zeolites can detoxify herbicide residues. Natural minerals like silica, alumina, and clays also accelerate herbicide degradation. Studies have shown that natural zeolites, especially when acid-activated, effectively adsorb herbicides like atrazine, making them potential tools for removing herbicides from water and soil [13]. High-moisture sludge disposal is a challenge, with traditional methods being energy-intensive. Smoldering offers a lower-cost alternative, but it produces harmful flue gases (CO, NOx, SO_2, VOCs). This necessitates costly purification, hindering its wider adoption. To address this, adding catalysts, particularly abundant and cost-effective natural zeolites, directly to the sludge during smoldering is proposed. Natural zeolites' unique structure and catalytic/adsorbent properties can potentially reduce emissions and heavy metal pollutants, offering a more sustainable solution [14].

Natural zeolites aid drug identification. Paracetamol (PAR), a common analgesic, is metabolized in the liver to a toxic metabolite (NAPQI). Existing PAR detection methods (HPLC, spectroscopy) are often slow, expensive, and lack sensitivity. Electrochemical methods offer a simpler, cheaper, and more accurate alternative with less sample prep. To improve bare electrode performance for PAR detection, modifications with nanomaterials like graphene oxide (GO) and Ni^{2+} zeolite (NiZ) in a carbon paste electrode (CPE/NiZ/GO sensor) are being explored as demonstrated in Fig. 2 [15].

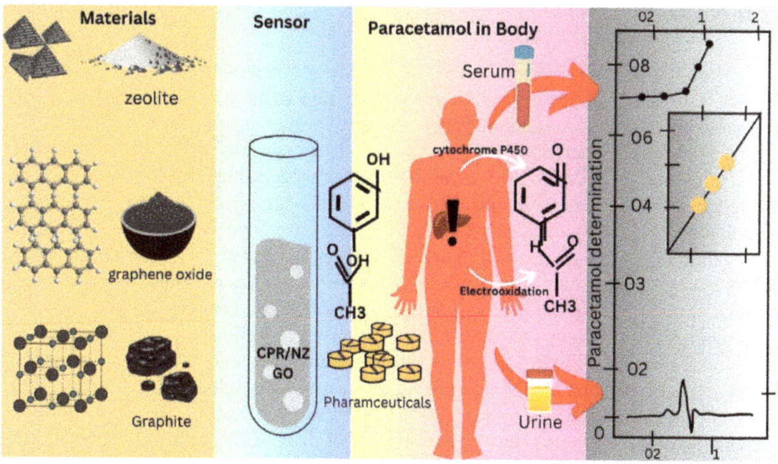

Figure 2. Zeolite modified CPE for paracetamol detection [15]

Natural zeolite (NZ) is cheap and useful, especially for cleaning up the environment. But it's not great at removing dyes like methylene blue (MB) from wastewater. So, scientists are making better materials called activated carbon/natural zeolite (AC/NZ) nanocomposites to grab more of that dye [16]. Natural zeolites, despite their utility, have restricted industrial applications due to their fixed crystal structures. Primarily, their small pore sizes, exemplified by clinoptilolite's 0.30-4 nm channels, limit the adsorption of larger molecules. Additionally, these deposits are non-renewable. Natural zeolites, limited by impurities and inconsistent composition, are often replaced by synthetic zeolites for industrial use. To reduce costs and utilize abundant resources, several solid wastes, including coal fly ash, coal gangue, rice husk, sugarcane bagasse, and coal

Materials Research Foundations 189 (2026) 209-236 https://doi.org/10.21741/9781644904039-7

gasification slag, are being explored as cost-effective silica-aluminum sources for synthetic zeolite production [17].

2.2 Synthetic Zeolite

The International Zeolite Association recognizes 46 natural zeolite minerals, but over 250 synthetic zeolites have been created, with more being developed. Synthetic zeolites offer customizable properties, making them highly versatile. Their key advantages include: extremely high surface areas (up to 1000 m^2/g), controllable water absorption and hydrophobicity through silica-alumina ratio and surface hydroxyl content adjustments, tunable active sites for specific applications, and pore sizes designed for targeted molecular interactions [18]. Synthetic zeolites, including A, X, Y, and P, are created by heating materials like China clay, feldspar, and soda ash. Zeolites A, X, and Y feature cubo-octahedral cages of SiO_4 and AlO_4 tetrahedra. Zeolites X and Y, known for their stability and large pores, are crucial catalysts in gasoline production through gas oil cracking. Zeolite A (NaA or 4A), synthesized from sodium aluminosilicate ($Na_{12}(AlO_2)_{12}(SiO_2)_{12}\cdot 27H_2O$), is a synthetic zeolite used in detergents to soften hard water and replace environmentally harmful phosphates. Its high ion exchange capacity (4-10 μm particle size) prevents fabric discoloration during washing [1].

Ammonium pollution is a serious problem. Adsorption using synthetic zeolite, especially from coal fly ash (CFA), is a promising solution due to its high capacity. Studies show CFA-derived zeolite effectively removes ammonium from wastewater, with removal rates ranging from 60-82% using various synthesis methods [19]. Metal-organic frameworks (MOFs), like zeolites, are crystalline microporous materials with extensive surface areas and uniform pores, sharing similar structural and physical characteristics. However, MOFs are synthetic hybrid metal-organic materials that are differentiated by metal ion and organic ligand connections, whereas zeolites are inorganic aluminosilicates. Because of their vast surface area and adaptable architectures, MOFs particularly zeolitic imidazolate frameworks, or ZIFs are versatile porous materials that are perfect for drug delivery. Zinc-based MOFs are preferred for their safety. Pore size dictates their use, and MOFs surpass zeolites in flexibility and stability, allowing for tailored applications [20]. Zeolite A (NaA or 4A), a synthetic sodium aluminosilicate, is commonly used in detergents to soften hard water and replace environmentally harmful sodium tripolyphosphate. Its high ion exchange capacity and 4-10 μm particle size prevent fabric graying and dye transfer during washing. Zeolites offer diverse topologies: varying pore sizes and channel dimensions, including internal cages for reactions or nanoparticle encapsulation. Nanoparticle dispersion enhances catalytic activity. Zeolite encapsulation prevents fouling and stabilizes nanoparticles. Close proximity of catalytic and acid sites is crucial for hydrogenation, and zeolite micropores provide catalytic selectivity [21]. Metal nanoparticles can be trapped in these channels as depicted in Fig. 3 [22].

Materials Research Foundations 189 (2026) 209-236 https://doi.org/10.21741/9781644904039-7

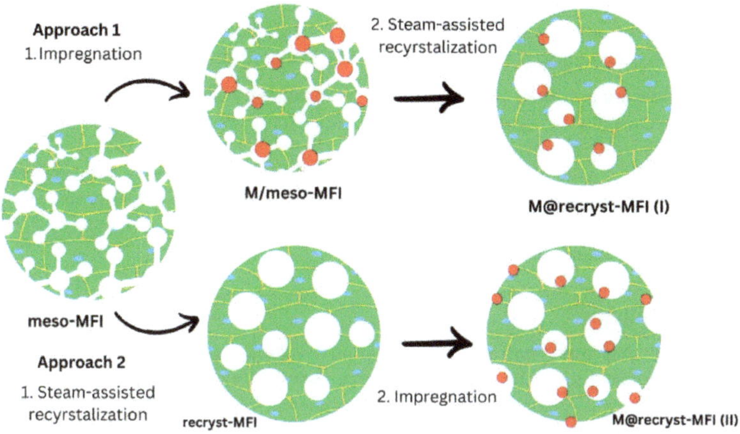

Figure 3. Metal nanoparticles encapsulated within an MFI framework [22]

Zeolites, with their structured pores and stability, are excellent supports for metal catalysts, enhancing activity and selectivity. Recent advances in synthesis and characterization have boosted the development of metal zeolite composites for heterogeneous catalysis [23]. Zeolites function as either solid catalysts themselves or as supports for metal nanoparticle catalysts. For diesel vehicle NOx emission control, NH_3-SCR is a leading technology. While V_2O_5-based catalysts were initially used, their toxicity and limitations led to their replacement. Zeolitic catalysts, particularly Cu-exchanged zeolites like Cu-ZSM-5, are favored for their high activity and stability. However, Cu-ZSM-5 suffers from poor water resistance, low-temperature activity, and structural instability at high temperatures due to dealumination, active site shifts, and hydrocarbon adsorption, ultimately leading to catalyst deactivation [24]. Iron (Fe) readily integrates into zeolite frameworks, especially Fe^{3+}, mimicking aluminum. This substitution enhances zeolite hydrophobicity and oxidation capabilities, making it suitable for VOC removal in humid conditions. Fe-modified zeolites also show improved yields in biomass pyrolysis and phenol hydroxylation. This is because Fe's easy incorporation leads to zeolites with favorable hydrophobic, acidic, and redox properties, resulting in high performance across various reactions [25].

Heavy metal soil pollution (lead, cadmium, zinc, etc.) from industrialization and fertilizer overuse is a significant agricultural problem, exacerbated by low pH which increases metal solubility and uptake into the food chain. Zeolites, with their high cation-exchange capacity, effectively sequester these positively charged heavy metals. Studies show zeolites can retain substantial amounts of metals like lead and reduce plant uptake of radioactive strontium. They also remove heavy metals from sewage sludge and mitigate copper pollution in viticulture by adsorbing Cu+ ions [8]. Synthetic zeolites find widespread use across various sectors. In agriculture, they enhance soil properties, act as fertilizer carriers, and support soilless cultivation. Industrially, they serve as catalysts, adsorbents, and ion exchangers in processes like petroleum refining and environmental remediation. In food, zeolites are used in packaging, drying, and to improve product quality, including wine. Notably, in winemaking, zeolites offer versatile applications from stabilization to sensor enhancement, addressing wine defects through their unique properties like sorption and ion exchange, making them promising alternatives to traditional fining agents [26].

Synthetic zeolites are preferred for their controlled, rapid synthesis, enabling tailored properties and high purity, leading to superior adsorption. Heteroatom zeolites, created by incorporating diverse elements, further enhance performance in chemical production, pollution control, and biomass conversion. Synthetic zeolites, while highly effective, are expensive, driving the search for cost-effective alternatives. Research is focused on understanding their performance, how they bind pollutants, and their environmental effects (like leaching and salinity) to assess their practical use in cleaning up contaminated soil [27].

3. Zeolite Synthesis

A swift progress has been seen due to the recent developments in the synthesis process and characterization of zeolites in the form of synergistic benefits by combining active metal sites with the inherited features of zeolites. Synthetic zeolite interest is growing due to diverse industrial applications. Modern synthesis allows tailoring zeolites using silica and clay, or even waste materials. Hydrothermal synthesis (using water) is common, while solvothermal (using organic solvents) and ionothermal (using ionic liquids) methods offer alternatives. Essentially, hydrothermal and ionothermal methods are subsets of solvothermal synthesis.

3.1 Hydrothermal Synthesis

Typically, zeolites are created in a controlled, heated environment using autoclaves or digestors. This process, known as hydrothermal synthesis, involves reacting gels in alkaline solutions at temperatures ranging from 80 to 200 °C, replicating the conditions found during natural zeolite formation. Traditional zeolite synthesis involves a lengthy (24-96 hours), hydrothermal process (90-150°C, 1-15 bar, alkaline) converting aluminosilicate gels (from aluminum and silicon sources, or cheaper alternatives like clays and waste) into crystalline zeolites. Crystal formation peaks early, yielding 0.1-10 μm crystals. This method produces various zeolite types (chabazite, Na-P1, etc.) depending on process parameters [28]. Indeed, the hydrothermal synthesis of zeolites has gained a lot of interest due to their potential traits, which have led to the subsequent increase in their demand [29].

The intricate hydrothermal process by which zeolites crystallize makes it difficult to fully understand how their pores form. Consequently, zeolite synthesis largely depends on empirical, 'trial-and-error' approaches, which hinders progress in the field. Steaming and acid leaching, rather than simple thermal treatment, are preferred for creating zeolite defect domains. This process enhances diffusion by [30].

- Making silica-aluminum debris more labile, increasing pore space.
- Generating extra-framework species (EFAL, EFSI) and amorphous silica-alumina within pores.

A secondary process, desilication, further refines the zeolite structure by selectively removing silicon under alkaline conditions [30].

3.2 Sol-Gel Synthesis

Throughout the 20th century, considerable improvements have been achieved both in the laboratory and commercial zeolites synthesis. The sol-gel process involves creating an oxide network through the polycondensation of precursors in a liquid, transitioning from a sol (solid in liquid) to a gel (polymer network with liquid). This versatile process, while widely used for silica, is also applicable to other metal oxides (like TiO_2, SnO_2, WO_3, ZnO) and carbon, allowing for diverse material design based on experimental conditions [31]. Sol-gel methods are increasingly used for MOF synthesis due to their simplicity and ability to control pore size. Research shows sol-gels enable MOF functionalization, composite formation, and creation of structures like thin films and monoliths. Factors like metal source, solvent, particle size, and drying conditions significantly impact sol-gel formation. While monolithic MOFs often have lower surface areas than powders, they offer improved volumetric efficiency due to higher density. Ligand modulation is explored for ZIF-8 gels, though their use in monolith production remains largely uninvestigated [32].

Aluminum alloy is prone to corrosion, necessitating effective protection. Sol-gel coatings are a promising solution, but directly adding corrosion inhibitors can cause defects. Nanocontainers, especially MOFs like ZIF-8, offer controlled inhibitor release and self-healing capabilities. ZIF-8's pH-responsive nature and stability make it ideal for encapsulating inhibitors, enhancing corrosion protection in sol-gel coatings [31]. Zeolite membranes are promising for gas separation, but defects hinder performance. SSZ-13 membrane defects were effectively patched using silica sol with varying ethanol/TEOS ratios. This method, confirmed by various characterization techniques, preserved the zeolite structure while significantly improving CO_2/CH_4 separation. Specifically, an ethanol/TEOS ratio of 95.5 boosted single gas selectivity from 33 to 176, with only a 15% reduction in CO_2 permeance, and achieved a remarkable mixed gas selectivity of 660 [33]. Silica-supported catalysts, often made via base-catalyzed sol-gel, find broad use in reactions like DRM and oxidation. These methods yield varied catalyst shapes. Bio-oxidation is a new, mild approach. Silica encapsulation boosts catalyst performance and reuse in industries from syngas to biosensors [31].

3.3 Mechano-chemical Synthesis of Zeolites

Mechanochemical reactions have played a critical role in the solution-based chemistry, allowing the reactions to take place at room temperature without any need for bulk solutions, and provide important parameters to synthesize substances that are otherwise challenging to produce. Mechanochemical synthesis of MOFs provides an alternate strategy to produce both pure MOFs and MOFs hybrids by lowering the usage of solvents [34]. An important direction for utilizing the mechanochemical approach in zeolites synthesis involves the preparation of reactants through mechanical activation before the crystallization process starts. This approach is easy, quick, and straight forward provides more production, and often requires solely a preliminary milling phase before heat processing. Furthermore, it generates a reduced volume of liquid waste compared to the hydrothermal (HT) approach [35].

Mechanical force, specifically grinding, forms the basis of the mechanochemical method, disrupting the ordered arrangement of atoms within solid reactants. During this grinding, either introduced or liberated from hydrated compounds, minimal quantities of liquid may be present. These liquid components can promote interactions among the initial substances, resulting in the creation of transitional compounds and the desired final products. Milling force in the mechanochemical approach enables uniform mixture formation, reduces particle size, and makes the surface more chemically active [36]. Moreover, this approach is becoming more popular in zeolites synthesis as it refines crystals into nano-powders for enhanced properties, speeds up synthesis for various zeolite and MOF structures, and efficiently introduces heteroatoms like Sn, outperforming manual grinding [37].

3.4 Microwave Assisted Synthesis

Various methods synthesize advanced zeolites, including hydrothermal, ion-exchange, and CVD. Microwave heating stands out for its simplicity, speed, and improved uniformity, significantly accelerating zeolite formation compared to conventional methods. Experiments demonstrate microwave heating can increase the rates of ZSM-5 formation fourfold and improve gel dissolution during synthesis of zeolite T [38]. Microwave heating is a rapid, efficient, and cost-effective method of synthesizing nanoporous materials with greater efficiency compared to electrical heating. Microwave heating provides high surface area products such as silicas, carbons, MOFs, and metal oxides for applications in drug delivery, water filtration, energy storage, gas adsorption, and catalysis. Silica's stability and biocompatibility make it widely used. Mesoporous silica, like MCM-41 and SBA-15, offers high surface areas and tunable pores, but conventional synthesis is time-consuming. Microwave heating provides a quicker, more uniform option, decreasing synthesis durations and enhancing material quality [39].

Mesoporous silica, versatile due to its properties, is often synthesized via hydrothermal methods. Microwave-assisted hydrothermal synthesis accelerates this process. These materials are effective in waste removal because of their stability, large surface area, and adjustable pores. Both conventional and non-conventional silica sources, including industrial by-products, are used to create sorbents for pollutant extraction [39]. Mesoporous silicas, crucial for various applications, suffer from stability and acidity issues. Traditional synthesis is complex and polluting. Microwave-assisted chelation of Y zeolite creates MZMs, a promising alternative with enhanced properties and reduced environmental impact [40]. Porous materials with adjustable characteristics, known as metal-organic frameworks (MOFs), show promise for energy storage. Their synthesis methods, including microwave (MW)-assisted synthesis, significantly impact their structure and performance. Traditional methods like solvothermal and hydrothermal synthesis are time-consuming. MW synthesis offers rapid, uniform heating, reducing synthesis time from days to minutes and improving MOF properties. This method minimizes toxic solvent use and enhances control over MOF morphology, crucial for advanced applications [41].

Microwave-assisted synthesis is gaining popularity for metal-organic framework (MOF) production due to its speed and morphology control. Utilizing microwave radiation (300 MHz-300 GHz) accelerates nucleation and shortens crystallization. Microwave-assisted flow chemistry significantly advances green synthesis by leveraging small-diameter flow reactors to overcome microwave penetration limitations in large-scale reactions. Advanced synthesizers enable precise control, while reactor designs utilize either microwave-absorbing or transparent materials, complemented by catalysts that facilitate internal heating. This combination enhances the synthesis of fine chemicals, nanoparticles, and novel material [42]. Microwave-assisted organic synthesis accelerates discovering and developing drugs through rapid, uniform heating and solvent-free reactions. It's crucial for screening drug formulas, peptide synthesis, and DNA amplification. This technology enables faster, more sustainable pharmaceutical synthesis, impacting areas like target identification, pharmacokinetics, and chemical library production [43].

Materials Research Foundations 189 (2026) 209-236 https://doi.org/10.21741/9781644904039-7

4. Biomedical Applications of Zeolites

Zeolites, both natural and synthetic, offer diverse applications due to their unique properties, with particle size being a critical factor. Nano-zeolites and green synthesis methods are driving advancements, promising sustainable and expanded applications across various fields. Zeolites and ZIFs, owing to their interesting properties, can be a possible candidate for biomedical applications such as tissue engineering, regenerative medicine, and drug delivery. Zeolites and zeolite imidazolate frameworks (ZIFs) are good biomaterials, especially when blended with biodegradable polymers. Although conventional ceramic biomaterials such as calcium phosphates are used for bone regeneration, zeolites and ZIFs provide versatile properties for next-generation biomedical applications [44].

Table 1 Exploring the Role of Zeolites in Biomedical Mechanisms and Their Outcomes

Application Area	Zeolite/ZIF Type	Mechanism/ Function	Key Outcome	References
Tissue engineering / Implant-associated infection	ZSM-5 zeolite (mordenite framework inverted-type)	High surface area and micropores improved drug loading and sustained release; biocompatibility supports bone marrow stromal cell proliferation; bactericidal effect via gentamicin release	Greater drug loading and sustained release vs. hydroxyapatite; reduced bacterial adhesion; inhibited biofilm formation (*S. epidermidis*); excellent biocompatibility with bone marrow stromal cells	[45]
Wastewater treatment (removal of heavy metals, pharmaceuticals, and dyes)	ZIF-8 and ZIF-67 nanocomposites	High surface area and tunable porosity facilitate adsorption via electrostatic interactions, hydrogen bonding, π–π stacking, and ion exchange; photocatalysis occurs through electron–hole generation and ROS production, enabling degradation of pollutants; composite formation improves stability and performance compared to pristine ZIFs	Demonstrated efficient removal of diverse contaminants (heavy metals, pharmaceuticals, dyes); exhibited outstanding chemical/thermal stability; composites showed enhanced adsorption and photocatalytic activity; discussed performance factors, challenges, and future perspectives for real wastewater application	[46]
CO_2 capture and wastewater treatment	Zeolite-Y	Microwave-assisted hydrothermal synthesis improved crystallinity, surface area, and pore volume; high Si/Al ratio provided enhanced	Zeolite-Y synthesized from Clay 1 showed CO_2 adsorption capacity of 5.23 mmol/g at 298 K and 1 bar, methylene blue removal of 94.3%,	[47]

		thermal stability; adsorption driven by large surface area and microporous structure	surface area of 485 m^2/g, and pore volume of 0.24 cm^3/g; demonstrated dual potential for CO_2 capture and dye removal, while converting waste clay into high-value material	
Groundwater treatment for fluoride removal	Hydroxyapatite (HAp)-modified zeolite (ZH8) composite	HAp incorporation enhanced fluoride affinity via ion exchange and chemical adsorption; adsorption process followed Langmuir isotherm (monolayer adsorption) and pseudo-first-order kinetics; column studies indicated Thomas model behavior	Achieved up to 92.23% fluoride removal under optimal batch conditions; maximum adsorption capacity of 39.38 mg/g; equilibrium reached within 4 h; reusability maintained >85% efficiency for three cycles; column studies showed breakthrough times up to 23 h and 99% yield at 5 ppm F^-; demonstrated strong selectivity and practical applicability for drinking water treatment	[48]
Targeted and pH-responsive drug delivery	Zeolite/Chitosan (ZA/CS) composite	High drug loading and entrapment efficiency due to porous zeolite structure and polymeric chitosan matrix; pH-responsive release following Ritger–Peppas kinetic model with anomalous/non-Fickian diffusion behavior; biocompatibility confirmed by MTT assay	Achieved drug loading of 1989.13 mg/g and entrapment efficiency of 99.88%; sustained ibuprofen release with 55% released at pH 1.2 after 2h; diffusion exponent values (n = 0.863 at pH 1.2 and n = 0.981 at pH 7.4) indicated mixed diffusion mechanisms; demonstrated excellent biocompatibility, highlighting suitability for therapeutic delivery systems	[49]
Antimicrobial applications	metal-modified zeolite A	Porous framework facilitates ion exchange and controlled release of antimicrobial metal	Unmodified zeolites showed limited antibacterial effects at high doses; metal-modified zeolites	[50]

		ions (Ag^+, Cu^{2+}, Zn^{2+}); enhanced antibacterial activity at lower concentrations compared to unmodified zeolites	exhibited strong antimicrobial activity at low concentrations; Ag-modified zeolite A demonstrated MIC of 16 µg/mL against multiple bacterial strains due to high ion-exchange capacity and sustained ion release; identified as promising long-lasting antimicrobial material	
Antibacterial	ZIF-8 synthesized	mechanism of action included cell membrane disruption and protein binding (molecular docking with *E. coli* protein 5AZC)	Ethanol-, methanol-, and water-based ZIF-8 showed yields of 66.77%, 73.14%, and 68.12%, respectively; ethanol-based ZIF-8 had the smallest crystallite size (24.76 nm) and highest zeta potential (+37 mV); water-based ZIF-8 exhibited the strongest antibacterial activity (inhibition zone 17.30 mm, MIC = 50 µg/100 µL); thermal stability up to 800 °C; strong correlation found between crystallinity, surface charge, and antibacterial activity	[51]
Dental biomaterials and bone tissue engineering	Hydroxyapatite (HA) composite reinforced with perlite-derived zeolite	Zeolite reinforcement improved surface roughness and porosity of HA composites; hydrothermal synthesis ensured stable structure; apatite layer formation promoted by Ca/P ratio; in vitro tests in Ringer's solution confirmed progressive bioactivity and mineralization over time		[52]

Wound healing	Silver ion-exchanged zeolite (Ag-Z)	Zeolite enhanced surface roughness and water absorption, aiding wound exudate uptake; Ag-loaded zeolite provided antibacterial activity via ion release and improved cell adhesion; composite matrix offered mechanical stability and biocompatibility	Composite films showed improved water uptake, >85% cell viability, strong antibacterial activity against *E. coli* and *S. aureus*; incorporation of Ag-Z improved tensile strength and modulus of elasticity; demonstrated potential as an effective wound dressing material	[53]
Bone regeneration and orthopedic biomaterials	Nanometricalumin ophosphate zeolites (SAPO-34, SAPO-5)	Nanoporous structure and nanosheet-like morphology facilitated apatite formation and ion exchange; incorporation of Ca^{2+} and Li^{+} cations enhanced bioactivity; degradation under physiological conditions enabled sustained release of bioactive ions; nanoporous topology and surface chemistry promoted osteogenic differentiation	Promoted apatite mineralization within 14 days; showed controlled degradation (5–22 wt. %); released Li^{+} and Ca^{2+} ions sustainably; maintained cytocompatibility up to 250 µg/mL; stimulated osteoblast adhesion, viability, and differentiation without osteogenic supplements; identified as promising osteoinductive biomaterials for bone repair and regeneration	[54],[55]

4.1 Zeolites in Drug Delivery

Zeolites, being modifiable materials, are potential MDDS materials for controlled drug delivery and enhanced bioavailability and have certain benefits over common excipients and even clays. Zeolites, due to their unique structure, biocompatibility, and controlled release characteristics, are potential candidates for site-specific drug delivery and have certain advantages over other delivery systems like liposomes and nanoparticles. They can be made into polymers for multiple delivery routes and are efficient detoxification scavengers. While abundant and affordable, zeolites' potential carcinogenic properties must be considered, though this characteristic could also be leveraged for cancer treatment [56]. MOFs, especially ZIFs are attracting attention for drug delivery owing to their high porosity, variable pore size, and biocompatibility. ZIFs, with structures similar to zeolites, offer advantages like greater surface area and chemical stability. Zinc-based MOFs, in particular, are promising due to zinc's low toxicity and therapeutic properties. The ability to tailor pore size with different metals and linkers allows for controlled drug loading and release, making ZIFs versatile for various biomedical applicationsas depicted in Fig. 4 [20].

Materials Research Foundations 189 (2026) 209-236 https://doi.org/10.21741/9781644904039-7

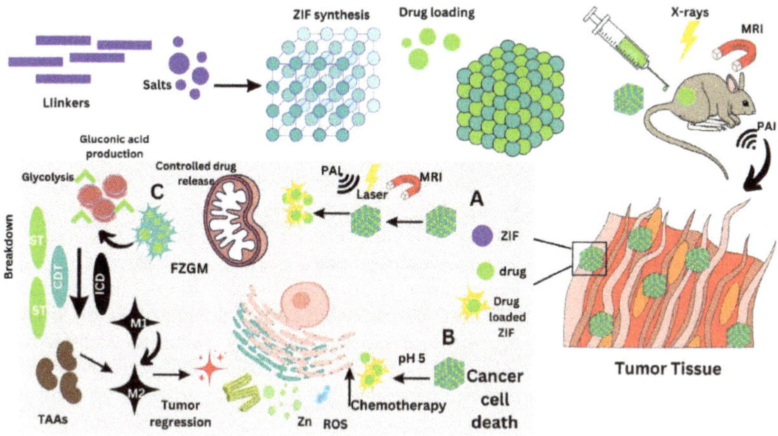

Figure 4. ZIFs: Stimuli-responsive nanomaterials for targeted drug delivery and cancer therapy, showcasing tunable properties and biomedical potential. [20]

Nanoparticle drug delivery overcomes drug limitations with enhanced solubility and targeted release. ZMOFs, a type of MOF with zeolite-like properties, offer tunable pores for optimized drug loading and release. Nanoscale zeolites further improve drug delivery, particularly for cancer treatment, by enhancing macromolecule interactions. Colon cancer, a leading cause of death, is treated with chemotherapy, but toxicity limits effectiveness.Zeolites, biocompatible and porous, offer sustained drug release. Their structure allows encapsulation of drugs like 5-fluorouracil (5-Fu), a common chemotherapy agent. Studies showed zeolites are promising drug carriers, but research often neglects the impact of varying gastrointestinal fluid composition beyond pH [57]. Microwave irradiation was used to synthesize nanoscale, adenine-incorporated zinc-based MOFs (b-MOFs) for paclitaxel delivery. These nanoparticles, around 200nm, achieved high paclitaxel loading (637 mg/g), increased drug solubility, and showed enhanced cancer cell toxicity compared to free paclitaxel, demonstrating their potential as effective anticancer nanocarrier [58].

Zinc zeolites (X and Y types) were developed as novel carriers for 6-mercaptopurine (MERC), a poorly bioavailable chemotherapeutic drug. MERC was successfully loaded onto the zeolites via coordination with zinc ions, resulting in a controlled, non-burst release profile. Approximately 30% of the drug was released within the first 10 hours, with a total release of 78-88% over 20 hours. Cytotoxicity tests on MCF-7 cells confirmed effective MERC release from the carriers, while the zeolites themselves showed biocompatibility [59]. Microporous aluminosilicates, known as zeolites, can improve drug effectiveness when used as drug delivery systems (DDS) due to their catalytic properties and strong ability to adsorb substances. Zeolites, specifically ZMT, modified with Zn(II), offer a biocompatible drug delivery system to combat cancer treatment challenges like drug leakage and toxicity, utilizing ibuprofen's anti-cancer properties [60]. Nanocarriers improve drug delivery; understanding endocytosis and surface modification is key. Zeolite-based 5-FU delivery, especially LTL, showed strong anticancer effects, particularly in breast cancer. LTL

Materials Research Foundations 189 (2026) 209-236 https://doi.org/10.21741/9781644904039-7

internalization was caveolin-mediated, and functionalization enhanced uptake and reduced tumors without toxicity. Faujasite-Y zeolite was studied as an isoniazid (INH) carrier for controlled release. Adsorption was optimal at pH 3, with molecular modeling detailing the mechanism.The zeolite stabilized INH in acidic environments without changing its release, indicating potential for improved tuberculosis treatments [61].

Zeolites and mesoporous silica effectively remove pharmaceuticals from water. Adsorption depends on sorbent hydrophobicity, pore structure, and drug properties. Scopoletin (SC), a medicinal coumarin with various bioactivities, was encapsulated in ZIF-8 for controlled release. Characterization confirmed SC loading. The SC:ZIF-8 system revealed 22.5% encapsulation efficiency and a pH-dependent release, with different release kinetics compared to pure SC. SC exhibited expectorant activity at higher doses, but the ZIF-8 system did not at low doses [62]. Magnetic clinoptilolite zeolite (CZ) nano/micro-particles were developed as multifunctional carriers for drug delivery, imaging, and localized heating. They demonstrated effective drug adsorption and release, magnetic heating capabilities, and anti-amyloidogenic properties. This makes CZ particles a promising platform for imaging, therapy, and controlled drug delivery, with potential applications in cancer hyperthermia and amyloid disease treatment [63]. ZIF-8 effectively encapsulates hydrophobic, hydrophilic, and amphiphilic drugs, like doxorubicin (DOX) and caffeine, offering controlled release. ZIF-8's pH-triggered activation at tumor sites and high loading capacity enhance drug delivery. Surface modifications, such as PEGylation, reduce toxicity. ZIF-8 also improves the properties of materials like hyaluronic acid for wound healing. Despite its potential, toxicity considerations are important for in vivo applications. To improve drug release in chitosan bone scaffolds, calcium zeolite was added and compared to hydroxyapatite [64].

4.2 Zeolites for Antimicrobial Uses

Antibiotic resistance necessitates new antimicrobial strategies. Zeolites and MOFs, with their microporous structures, offer controlled release of antimicrobial metal ions. Despite their potential in coatings and medical devices, standardized evaluation methods are lacking. Nanoparticles (NPs), especially AgZnO, offer a solution through direct cell wall disruption. Zeolite-supported AgZnO nanocomposites, leveraging zeolite's porous structure, aim to enhance antibacterial effect on *S. aureus* and *E. coli.* Zinc species on LTA zeolite (NaA), prepared via ultrasound, showed strong antimicrobial activity. ZnO_2-NaA, when dried, completely inhibited *E. faecalis* growth and showed significant antifungal activity against *C. albicans*, while also reducing fibroblast toxicity as shown in Fig. 5 [65].

Materials Research Foundations 189 (2026) 209-236 https://doi.org/10.21741/9781644904039-7

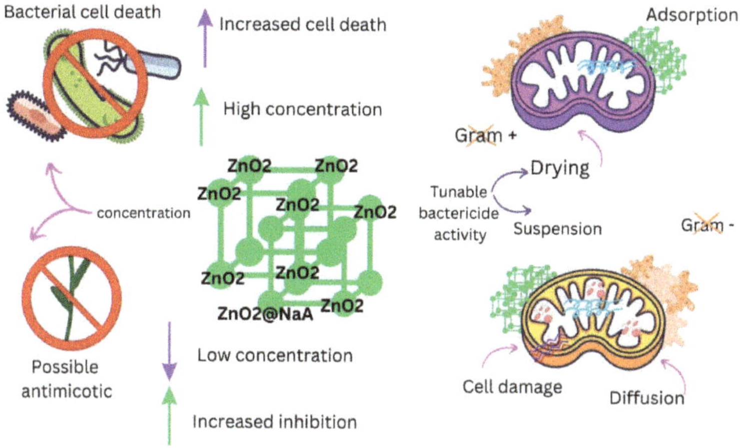

Figure 5. LTA zeolite with zinc: Effective antimicrobial properties. [65]

Metal-loaded Mobil-twenty-two zeolite added to EVA resin improved antimicrobial properties and mechanical strength. Silver, copper, and zinc were effective, with copper/zinc being cost-effective alternatives, creating antimicrobial EVA films. Zeolite-modified polylactic acid (PLA) or polypropylene (PP) composites, copper-saturated zeolites (LZ)/LZ-nCu at 5% showed antibacterial and biocompatible properties, suitable for biomedical applications, while 10% concentrations exhibited toxicity and reduced mechanical strength. Zeolite improved PP's thermal stability and modulus, but slightly decreased PLA's [66]. Microbially induced concrete corrosion (MICC) degrades concrete via microbial acid production, particularly from sulfur-oxidizing bacteria. Geopolymers, with their acid-resistant aluminosilicate structure, offer a durable alternative. Zeolite incorporation further enhances geopolymer properties by improving microstructure and reducing permeability through a "seeding effect." Loading zeolite with potassium sorbate, a safe antimicrobial, provides sustained protection against MICC by slowly releasing the agent, creating a highly resistant geopolymer composite [67].

Zeolites are promising dental materials due to their biocompatibility, ion-exchange, and antibacterial potential (via embedded ions). However, their impact on dental material strength must be assessed. ZIF-8, a metal-organic framework, offers both odor adsorption and antimicrobial properties. Synthesized via a water-based method, ZIF-8 NPs are applied with a binder, aiming to create a durable and effective anti-odor textile without compromising aesthetics [68]. Contaminated produce poses food safety risks; alternatives to chlorine are sought. Strawberries especially need effective preservation. ZnONPs and zeolites are valuable due to their safe antimicrobial properties. Combining them into nanocomposites (ZnO/ZeoNC) enhances their effectiveness. *Aloe vera* gel (AG), with its antimicrobial and antioxidant properties, further offers a natural approach to extend shelf life and protect against microbial decay in fruits like strawberries [69].

Trauma care requires hemostatic dressings with antimicrobial properties. While existing dressings have limitations, research focuses on improved materials like nanocomposites and nanofibers. Cotton-zeolite dressings, supplemented with silver and ascorbic acid, demonstrate improved hemostasis and antibacterial activity for extended care. Antibiotic-laden hospital wastewater fuels resistance and environmental harm. Zeolite, a stable, porous mineral, offers potential for treatment, especially when modified with antibacterial metals like silver. Amoxicillin (AMX), a common and persistent antibiotic, is a key target for removal from hospital wastewater due to its environmental impact [70]. Bacterial water pollution demands effective treatment. Zeolite, modified with antibacterial metals via impregnation, offers an effective approach. Water is at risk from industrial toxins too; MNPs are a potential solution, requiring stabilization. Zeolite or polymer supports, especially self-assembled nanocomposites, address this. These materials effectively remove dyes, nitro compounds, and antibiotics from wastewater [71].

To combat bacterial resistance, stable antimicrobial carriers are needed. Zeolite, using a dry process, effectively delivered thymol, enhancing its stability and antimicrobial activity, offering a simpler alternative to essential oils encapsulation. Microbial growth in construction materials causes damage and health problems. Aqueous paints, replacing VOCs, increase this risk. Metal-modified zeolites offer a safe, antimicrobial solution for gypsum composites [72]. Zeolite (Y-H) facilitated rapid, eco-friendly synthesis of antimicrobial benzamides. These compounds demonstrated efficacy against *S. pyogenes* and fungi, with their mechanisms clarified by molecular docking.Zeo/Ag-GO nanocomposite, produced via semi-powder metallurgy, exhibited superior antibacterial and biocompatible properties over Zeo/Ag and Zeo, making it an innovative material for infection-resistant bone implants. Silver-MFI zeolite exhibited strong antibacterial activity, exceeding antibiotics, against *E. coli*, *S. aureus*, and *MRSA*, linked to zeolite acidity. This makes it a strong contender for combating resistant bacterial infections [73].

Zeolite stability and Ag NPs' surface plasmon resonance (SPR) are leveraged for improved antimicrobial efficacy, offering a simpler, and broader-application alternative to current complex methods [74]. Eradicating *H. pylori* is challenging due to antibiotic resistance, necessitating susceptibility testing and exploring complementary therapies like probiotics. Ammonium removal, crucial for *H. pylori*'s survival, is a promising strategy. Clinoptilolite, a natural zeolite, effectively inhibits *H. pylori* growth by exchanging sodium for ammonium ions, offering a potential alternative treatment. Moisture-activated, silver-zeolite PUF releases nanoparticles, demonstrating antibacterial activity against *E. coli* and especially *M. luteus* [75].

4.3 Zeolite-Based Scaffolds forTissue Engineering

Tissue regeneration faces challenges due to aging, accidents, and disease, necessitating advanced biomaterials. Advanced biomaterials, particularly ZIF-8 MOFs, are being explored for tissue regeneration due to their biocompatibility, porosity, and drug-delivery capabilities, addressing limitations of traditional transplants.Tissue engineering uses drug-loaded mats, especially with glucosamine and ZIF-8 nanoparticles, to address cartilage repair challenges by enabling controlled drug release [76]. Covalently bonded gelatin-FAU zeolite composites, with controlled aggregation and improved thermal stability, demonstrated biocompatibility and could be beneficial for tissue engineering and wound dressings [77]. Polyhedral OligomericSilsesquioxane-Polycaprolactone-Zeolite (POSS-PCL-Zeo) nanofibers offer a promising bone tissue engineering scaffold, combining biocompatibility and mechanical strength to enhance bone regeneration. GEN-loaded

Materials Research Foundations 189 (2026) 209-236 https://doi.org/10.21741/9781644904039-7

ZSM-5 scaffolds, created via a space holder technique, demonstrated high porosity (70-78%), good mechanical strength, and apatite formation, demonstrating potential for bone tissue engineering [78].

Zeolites are viable biomaterials for bone tissue engineering because they are osteo-inductive, porous, and biocompatible nature, serving as effective scaffolds, implant coatings, and therapeutic agents for bone disorders. Polycaprolactone-zeolite scaffolds, created via solvent casting and freeze-drying, demonstrated encouraging properties for cancellous bone tissue engineering, including porosity, uniform zeolite distribution, increased mechanical strength, and enhanced bioactivity [79]. To improve drug release in chitosan bone scaffolds, calcium zeolite was added and compared to hydroxyapatite. Zeolite-chitosan showed enhanced drug retention and controlled release of risedronate, suggesting improved potential for bone tissue engineering [80]. A vancomycin-alginate coating on ZSM-5 zeolite scaffolds effectively enhanced mechanical strength, provided robust antibacterial activity against *S. aureus*, and maintained favorable porosity, cell viability, and calcium deposition, making it a promising candidate for clinical bone repair. Addressing the complex demands of large bone defect repair, which requires both osteogenesis and angiogenesis, ZIF-8 MOFs emerge as a potential vector for miRNA delivery. Their ability to overcome barriers like instability and poor cellular uptake, via pH-responsive degradation and enhanced endosomal escape, offers a significant advancement in gene therapy [81].

Zeolite coatings on titanium implants offer improved osteointegration through controlled calcium release, addressing limitations of traditional methods. They also provide antibacterial and corrosion-resistant properties. Zeolites, with their controlled ion release, are potential biomaterials for bone tissue engineering, offering advantages over traditional materials by enhancing osteogenesis through tunable frameworks, coatings, and drug delivery [82]. For skin regeneration, a novel hydrogel was created using gelatin, agarose, zeolite, and pomegranate peel extract. Zeolite facilitated controlled release, while the extract's antioxidants and the polymers' biocompatibility aimed to enhance tissue repair. Fe-Ca-SAPO-34-enhanced chitosan structure promotes h-DPSC differentiation for bone and tooth regeneration, showing promise in hard tissue engineering [83].

4.4 Role of Zeolites in Biosensors and Diagnostics

Biosensors use biomolecules (DNA, enzymes, etc.) and transducers for sensitive analyte detection. ZIF nanomaterials, with high stability and porosity, enable efficient biomolecule immobilization, boosting biosensor performance. Bioactive ZIFs, created via synthesis incorporation, maintain biomolecule activity [84]. Alzheimer's Disease (AD), impacting the elderly, requires early detection beyond current slow, complex diagnostics. Biosensors offer rapid, accessible POC diagnosis of AD biomarkers, enabling real-time detection. Zeolite nanomaterials in an impedance biosensor for early Alzheimer's detection by targeting AβO. Zeolites enhance biomolecule interaction, facilitating aptamer-antibody immobilization for sensitive detection. Effective probe design and immobilization are essential for optimal performance [85]. A ZIF-8 MOF fluorescent biosensor offers rapid, sensitive COVID-19 RNA detection, overcoming current diagnostic limitations. Its properties ensure efficient and highly selective RNA detection. ZIF-MOFs provide a promising solution for cancer theranostics by enabling efficient nucleic acid delivery, overcoming challenges in uptake and degradation, because of their high loading capacity and tunable properties [86].

Materials Research Foundations 189 (2026) 209-236 https://doi.org/10.21741/9781644904039-7

Zeolites and ZIFs, exhibit high surface area and biocompatibility, enhance biosensor performance by providing stable, cost-effective biomolecule immobilization and offering environmentally friendly alternatives. MOFs show promise against infectious diseases due to their tunable porosity. They're explored for treatment as antimicrobials, drug carriers, therapeutic composites, nanozymes, and in phototherapy. For diagnosis, MOF-based biosensors detect pathogens via electrochemical, fluorometric, and colorimetric methods. Overcoming current limitations is crucial to maximize MOFs' impact [87]. MOFs are explored for diagnosing and treating cancer, diabetes, neurological, and ocular diseases, with research focusing on their synthesis, sensor development, and drug delivery applications. MOFs and zeolites combat viruses through: selective biosensing, virus-removing barriers, and protective biocomposites for drug delivery and vaccine development [88].

ZEO-IO, a nanocomposite of fly ash, enables sensitive NT-ProBNP detection of early myocardial infarction. It is capable of picomolar detection limits with aptamer-based sensing with stability and selectivity high in serum. A label-free biosensor based on ZIF-8 makes sensitive detection of cyto protein interactions possible with superior performance compared to conventional assays in speed and sensitivity [89]. MOF-derived ZIFs present stable, porous architectures for chemical sensing. Their composites with improved properties allow sensitive detection through electrochemical, luminescence, colorimetric, and SERS processes [90].

Conclusion

Through their wide range of structural features, zeolites have become profoundly versatile compounds for use in numerous and varied applications. From their original significance in mineralogy and industrial catalysis to current applications in biomedical processes and detoxification of the environment, zeolites continue to lead the way. Their distinctive microporous structure and labile chemical composition allow them to function as selective adsorbents, catalysts, and drug-delivery agents.

Evolution of zeolite synthesis from natural synthesis to newer methods like hydrothermal, sol-gel, mechanochemical, and microwave methods has allowed the synthesis of zeolites with particular properties. Such properties to regulate pore size, form, and chemical functionality have extended their scope of applications, solving water purification, soil decontamination, and drug delivery problems.

Natural zeolites provide cost-effective solutions for large-scale applications, while synthetic zeolites offer greater precision and purity, making them essential for high-performance uses. The development of nanocomposites, such as zeolite-polymer membranes and metal-zeolite catalysts, further enhances their performance by improving mechanical stability, selectivity, and catalytic activity.

In the biomedical field, zeolites and zeoliticimidazolate frameworks (ZIFs) are transforming drug delivery systems with their ability to enable controlled release and targeted therapies. Their biocompatibility and adjustable pore sizes make them ideal for encapsulating and delivering drugs, offering significant advantages over conventional delivery methods. Despite their many benefits, challenges persist, including the need for more cost-effective synthesis methods, a deeper understanding of long-term environmental impacts, and addressing potential toxicity concerns. Ongoing research into heteroatom-substituted zeolites, nanoparticle encapsulation, and advanced

Materials Research Foundations 189 (2026) 209-236 https://doi.org/10.21741/9781644904039-7

characterization techniques is expected to further expand the potential applications of these remarkable materials.

In summary, zeolites are a cornerstone of materials science, providing sustainable solutions and driving technological progress across various industries. As research continues to uncover their full potential, zeolites are set to play an increasingly critical role in tackling global challenges and enhancing quality of life.

References

[1] T. Derbe, S. Temesgen, M. Bitew, A short review on synthesis, Char. Appl. Zeol. Adv. In Mater. Sci. Eng. 2021 (2021) 6637898. https://doi.org/10.1155/2021/6637898

[2] W. Yan, Y. Li, F.-S. Xiao, Z. Liu, J. Yu, The future of zeolites, ACS Publ. (2024) 7103-7105. https://doi.org/10.1021/acs.chemmater.4c01675

[3] K. Jacob, J. Clement, Zeolite atn: Topological characterization and predictive analysis on potential energies using entropy measures, J. Mol. Struct. 1299 (2024) 137101. https://doi.org/10.1016/j.molstruc.2023.137101

[4] M.K. Tufail, M. Ifrahim, M. Rashid, I.U. Haq, R. Asghar, U. Uthappa, M. Selvaraj, M. Kurkuri, Chemistry of zeolites and zeolite-based composite membranes as a cutting-edge candidate for removal of organic dyes & heavy metal ions: Progress and future directions, Sep. Purif. Technol. (2024) 128739. https://doi.org/10.1016/j.seppur.2024.128739

[5] S. Tsubota, S. Kokuryo, K. Tamura, K. Miyake, Y. Uchida, A. Mizusawa, T. Kubo, N. Nishiyama, Exploring the effect of brønsted acidity of MFI-type zeolites on catalytic cracking temperature of low-density polyethylene, Catal. Sci. Technol. 14 (2024) 1369-1374. https://doi.org/10.1039/D3CY01622F

[6] G. Raman, Forecasting low framework density zeolites from synthesis descriptors using machine learning, J. Solid State Chem. 327 (2023) 124290. https://doi.org/10.1016/j.jssc.2023.124290

[7] E. Kianfar, A. Mahler, Zeolites: Properties, applications, modification and selectivity, 2020.

[8] E. Cataldo, L. Salvi, F. Paoli, M. Fucile, G. Masciandaro, D. Manzi, C.M. Masini, G.B. Mattii, Application of zeolites in agriculture and other potential uses: A review, Agronomy 11 (2021) 1547. https://doi.org/10.3390/agronomy11081547

[9] G. Conversa, S. Pacifico, P. La Rotonda, C. Lazzizera, A. Bonasia, A. Elia, Foliar application of natural zeolites affects the growth and productivity of processing tomato, Eur. J. Agron. 154 (2024) 127100. https://doi.org/10.1016/j.eja.2024.127100

[10] M. Senila, O. Cadar, Modification of natural zeolites and their applications for heavy metal removal from polluted environments: Challenges, Rece. Adv. Pers. Heli. 10 (2024) e25303. https://doi.org/10.1016/j.heliyon.2024.e25303

[11] D. Prisa, Study and evaluation of natural zeolite and dried zeolite for the cultivation of friggitello pepper, World J. Adv. Res. Rev. 19 (2023) 632-641. https://doi.org/10.30574/wjarr.2023.19.2.1626

[12] E. Pérez-Botella, S. Valencia, F. Rey, Zeolites in adsorption processes: State of the art and future prospects, Chem. Rev. 122 (2022) 17647-17695. https://doi.org/10.1021/acs.chemrev.2c00140

[13] A.V. Pastukhov, M.M. Ilyin, N.D. Chkanikov, Acid-activated natural zeolite and montmorillonite as adsorbents for decomposing metsulfuron-methyl herbicide, Inorg. Chem. Commun. 158 (2023) 111615. https://doi.org/10.1016/j.inoche.2023.111615

[14] L. Ma, Y. Zhao, C. Zhang, X. Su, Y. Qiao, Q. Fang, J. Huang, D. Zhang, Reducing co/no and absorbing heavy metals in self-sustained smouldering of high-moisture sludge by regulating inert media with low-cost natural zeolite, Environ. Pollut. 337 (2023) 122556. https://doi.org/10.1016/j.envpol.2023.122556

[15] R. Porada, N. Wenninger, C. Bernhart, K. Fendrych, J. Kochana, B. Baś, K. Kalcher, A. Ortner, Targeted modification of the carbon paste electrode by natural zeolite and graphene oxide for the enhanced analysis of paracetamol, Microchem. J. 187 (2023) 108455. https://doi.org/10.1016/j.microc.2023.108455

[16] F. Mohamed, M. Shaban, S.K. Zaki, M.S. Abd-Elsamie, R. Sayed, M. Zayed, N. Khalid, S. Saad, S. Omar, A.M. Ahmed, Activated carbon derived from sugarcane and modified with natural zeolite for efficient adsorption of methylene blue dye: Experimentally and theoretically approaches, Sci. Rep. 12 (2022) 18031. https://doi.org/10.1038/s41598-022-22421-8

[17] C. Cao, W. Xuan, S. Yan, Q. Wang, Zeolites synthesized from industrial and agricultural solid waste and their applications: A review, J. Environ. Chem. Eng. 11 (2023) 110898. https://doi.org/10.1016/j.jece.2023.110898

[18] W. Wang, L. He, Q. Luo, L. Wang, J. Wang, H. Chen, Z. Miao, Q. Yao, M. Sun, Synthesis and application of core-shell, Holl. yolk-shell Mult. Stru. Zeol. Microporous Mesoporous Mater. 362 (2023) 112766. https://doi.org/10.1016/j.micromeso.2023.112766

[19] H.-D. Zhou, C.-Y. Wang, Q. Wang, B.-X. Xu, G. Zhu, A sodium-acetate-modified Synt. Zeol. RSC Adv. 14 (2024) 17843-17854. https://doi.org/10.1039/D4RA01547A

[20] H. Akhtar, U. Amara, K. Mahmood, M. Hanif, M. Khalid, S. Qadir, Q. Peng, M. Safdar, M. Amjad, A. Tahir, Drug carrier wonders: Synthetic strategies of zeolitic imidazolates frameworks (ZIFs) and their applications in drug delivery and anti-cancer activity, Adv. In Coll. Inte. Sci. 329 (2024) 103184. https://doi.org/10.1016/j.cis.2024.103184

[21] J. Si, R. Guo, Y. Zhang, W. Ning, Y. Sun, W. Li, S. Miao, Synthesis of Linde A-type zeolite from ball clay with incorporated ruthenium and application in hydrogenation catalysis, Appl. Clay Sci. 239 (2023) 106897. https://doi.org/10.1016/j.clay.2023.106897

[22] A.A. Vasconcelos, T. Len, A.d.N. de Oliveira, A.A.F.d. Costa, A.R.d.S. Souza, C.E.F.d. Costa, R. Luque, G.N.d. Rocha Filho, R.C.R. Noronha, L.A.S.d. Nascimento, Zeolites: A theoretical and practical approach with uses in (bio) chemical processes, Appl. Sci. 13 (2023) 1897. https://doi.org/10.3390/app13031897

[23] Q. Zhang, S. Gao, J. Yu, Metal sites in zeolites: Synthesis, Char. Cata. Chem. Rev. 123 (2022) 6039-6106. https://doi.org/10.1021/acs.chemrev.2c00315

[24] J. Zhang, Z. Li, H. Yi, X. Tang, H. Cheng, Q. Yu, Synthesis and application prospect of small-pore zeolites in vehicle exhaust purification, Fuel 348 (2023) 128577. https://doi.org/10.1016/j.fuel.2023.128577

[25] J. Zhang, X. Tang, H. Yi, Q. Yu, Y. Zhang, J. Wei, Y. Yuan, Synthesis, Char. Appl. Fe-zeolite A Review, Appl. Cata. A Gene. 630 (2022) 118467. https://doi.org/10.1016/j.apcata.2021.118467

[26] M. Noviello, C.E. Gattullo, M. Faccia, V.M. Paradiso, G. Gambacorta, Application of natural and synthetic zeolites in the oenological field, Food Res. Int. 150 (2021) 110737. https://doi.org/10.1016/j.foodres.2021.110737

[27] B. Hudcová, M. Osacký, M. Vítková, A. Mitzia, M. Komárek, Investigation of zinc binding properties onto natural and synthetic zeolites: Implications for soil remediation, Microporous Mesoporous Mater. 317 (2021) 111022. https://doi.org/10.1016/j.micromeso.2021.111022

[28] N. Kordala, M. Wyszkowski, Zeolite properties, Meth. Synt. Sele. Appl. Mole. 29 (2024) 1069. https://doi.org/10.3390/molecules29051069

[29] T.-L. Wu, Y.-H. Chen, W.-D. Hsu, Phase transition pathway of hydrothermal zeolite synthesis, Phys. Chem. Mine. 48 (2021) 1-10. https://doi.org/10.1007/s00269-020-01125-3

[30] A. Intang, P. Susmanto, M.D. Bustan, S. Haryati, Determination of swelling operation parameters to improve the hierarchy of natural zeolite lampung after synthesis, Sout. Afri. J. Chem. Eng. 50 (2024) 125-134. https://doi.org/10.1016/j.sajce.2024.08.004

[31] Y. Huang, C. Zhao, Y. Li, C. Wang, T. Shen, D. Cheng, H. Yang, Development of self-healing sol-gel anticorrosion coating with pH-responsive 1H-benzotriazole-inbuilt zeolitic imidazolate framework decorated with silica shell, Surf. Coat. Technol. 466 (2023) 129622. https://doi.org/10.1016/j.surfcoat.2023.129622

[32] E. Hunter-Sellars, P.A. Saenz-Cavazos, A.R. Houghton, S.R. McIntyre, I.P. Parkin, D.R. Williams, Sol-gel synthesis of high-density zeolitic imidazolate framework monoliths via ligand assisted methods: Exceptional porosity, Hydr. Appl. In Vapo. Adso. Adva. Func. Mater. 31 (2021) 2008357. https://doi.org/10.1002/adfm.202102716

[33] A. Aydani, H. Maghsoudi, A. Brunetti, G. Barbieri, Silica sol gel assisted defect patching of ssz-13 zeolite membranes for CO2/CH4 separation, Sep. Purif. Technol. 277 (2021) 119518. https://doi.org/10.1016/j.seppur.2021.119518

[34] M. Ahmadi, S.M. Ayyoubzadeh, F. Ghorbani-Bidkorbeh, S. Shahhosseini, S. Dadashzadeh, E. Asadian, M. Mosayebnia, S. Siavashy, An investigation of affecting factors on mof characteristics for biomedical applications: A systematic review, Heliyon 7 (2021). https://doi.org/10.1016/j.heliyon.2021.e06914

[35] S. Al-Nahari, D. Laurencin, B. Alonso, Solvent-free synthesis of zeolites: new insights into the mechanism and non-mechanochemical route, Microporous Mesoporous Mater. 350 (2023) 112445. https://doi.org/10.1016/j.micromeso.2023.112445

[36] Q. Yao, Y. Peng, M. Chen, Y. Wang, J. Ding, B. Ma, Q. Wang, S. Lu, One-step high efficiency synthesis of zeolite from fly ash by mechanochemical method as a low-cost adsorbent for cadmium removal, J. Environ. Chem. Eng. 12 (2024) 111877. https://doi.org/10.1016/j.jece.2024.111877

[37] A. Kornas, J.E. Olszówka, P. Klein, V. Pashkova, zeolites.A. Milling activation for the solvent-free synthesis of practical guide, Catalysts 11(2) (2021) 246, . https://doi.org/10.3390/catal11020246

[38] X. Zeng, X. Hu, H. Song, G. Xia, Z.-Y. Shen, R. Yu, M. Moskovits, Microwave synthesis of zeolites and their related applications, Microporous Mesoporous Mater. 323 (2021) 111262. https://doi.org/10.1016/j.micromeso.2021.111262

[39] B. Díaz de Greñu, R. de Los Reyes, A.M. Costero, P. Amorós, J.V. Ros-Lis, Recent progress of microwave-assisted synthesis of silica materials, Nano. 10 (2020) 1092. https://doi.org/10.3390/nano10061092

[40] S. Abdulridha, Y. Jiao, S. Xu, R. Zhang, A.A. Garforth, X. Fan, Mesoporous zeolitic materials (mzms) derived from zeolite y using a microwave method for catalysis, Fron. In Chem. 8 (2020) 482. https://doi.org/10.3389/fchem.2020.00482

[41] P.T. Phan, J. Hong, N. Tran, T.H. Le, The properties of microwave-assisted synthesis of metal-organic frameworks and their applications, Nano. 13 (2023) 352. https://doi.org/10.3390/nano13020352

[42] T. Mooney, M. Ilamanova, B. Török, Microwave-assisted flow chemistry for green synthesis and other applications, Bent. Sci. Publ. Dire. 9 (2022) 65-69. https://doi.org/10.2174/2213335610666221208163107

[43] B. Jadhav, P. Ige, J. Kokat, R. Mokate, Microwave-assisted synthesis: A new technology in drug discovery, Mini Rev. Med. Chem. 9 (2022) 340-58. https://doi.org/10.2174/138955751090903034 0

[44] J. Matusiak, A. Przekora, W. Franus, Zeolites and zeolite imidazolate frameworks on a quest to obtain the ideal biomaterial for biomedical applications: A review, Mater. Today 67 (2023) 495-517. https://doi.org/10.1016/j.mattod.2023.06.008

[45] Y.P. Guo, T. Long, Z.F. Song, Z.A. Zhu, Hydrothermal fabrication of ZSM-5 zeolites: Biocompatibility, drug delivery property, and bactericidal property, J. Biom. Mater. Res. Part. B Appl. Biom. 102 (2014) 583-591. https://doi.org/10.1002/jbm.b.33037

[46] M.A. Nazir, S. Ullah, M.U. Shahid, I. Hossain, T. Najam, M.A. Ismail, A. Rehman, M.R. Karim, S.S.A. Shah, Zeolitic imidazolate frameworks (zif-8 & zif-67): Synthesis and application for wastewater treatment, Sep. Purif. Technol. 356 (2025) 129828. https://doi.org/10.1016/j.seppur.2024.129828

[47] M. Abdalqadir, S.R. Gomari, D. Hughes, Novel synthesis and application of zeolite-y from waste clay for efficient CO2 capture and water purification, Colloids Surf. A: Physicochem. Eng. Asp. 704 (2025) 135460. https://doi.org/10.1016/j.colsurfa.2024.135460

[48] R. Boteju, L. Zheng, H.M. Wasana, Q. Wu, Y. Wei, H. Zhong, Y. Wang, A. De Alwis, Hydroxyapatite-modified zeolite for fluoride removal from drinking water: Adsorption mechanism investigation and column study, Water 17 (2025) 908. https://doi.org/10.3390/w17060908

[49] A.M. Eldeeb, E. Serag, M. Elmowafy, M.E. El-Khouly, pH-responsive zeolite-a/chitosan nanocarrier for enhanced ibuprofen drug delivery in gastrointestinal systems, Int. J. Biol. Macromol. 289 (2025) 138879. https://doi.org/10.1016/j.ijbiomac.2024.138879

[50] H. Tabesh, S. Kharrazi, M. Bashiri Barazandeh, P. Ebadoulah Poursafa, A. Poorkhalil, A comprehensive study on enhancing microbicidal activity of pure and ion-exchanged zeolites through structural and chemical determinants, IET Nanobiotechnol. 2025 (2025) 7012728. https://doi.org/10.1049/nbt2/7012728

[51] A.I.O. Ibrahim, D. Moodley, E. Maboza, A. Olivier, L. Petrik, Zeolite-y-loaded chitosan nanoparticles as endodontic antimicrobial agent: An in vitro study, Eur. J. Dent. (2025). https://doi.org/10.1055/s-0045-1802947

[52] S.A. Amur, B.P. Sharma, N.A. Soomro, Q. Khuhro, M. Tariq, H. Liang, M. Kazi, M. Nur-e-Alam, Synthesis, Char. Dens. Func. Theo. Stud. Anti. Acti. Mole. Dock. Zeol. Imid. Framework8 Appl. Orga. Chem. 39 (2025) 7826. https://doi.org/10.1002/aoc.7826

[53] B. Tunç, M. Muratoğlu, Synthesis of hydroxyapatite/zeolite/perlit hybrid coatings on human tooth by hydrothermal method and their in vitro behaviour, Proc. The Inst. Mech. Engi. Part. H J. Eng. In Medi. 239 (2025) 548-559. https://doi.org/10.1177/09544119251342007

[54] N. Sarrafan, S. Hassanzadeh-Tabrizi, N. Koupaei, Synthesis and characterization of novel polycaprolactone/chitosan/ag-exchange zeolite hydrogel as a wound dressing, J. Polym. Res. 32 (2025) 230. https://doi.org/10.1007/s10965-025-04454-5

[55] C. Torres, E.C. Whitehead, B.P. Le Monnier, D. Peña, A. Palomino-Calderón, M. Neira, P. Romero-Hasler, W.L. Grayson, M. Tsapatsis, C. Covarrubias, Silicoaluminophosphate (sapo) zeolites with nanometric dimensions: Structural tuning for osteoinductive applications in bone regeneration, Mate. 40 (2025) 102401. https://doi.org/10.1016/j.mtla.2025.102401

[56] T. Pandya, S. Patel, M. Kulkarni, Y.R. Singh, A. Khodakiya, S. Bhattacharya, B.G. Prajapati, Zeolite-based nanoparticles drug delivery systems in modern pharmaceutical research and environmental remediation, Heliyon 10 (2024) e36417. https://doi.org/10.1016/j.heliyon.2024.e36417

[57] A. Abd-Elsatar, M. Farag, H. Youssef, S. Salih, M. Mounier, E. El-Meliegy, Different zeolite systems for colon cancer therapy: monitoring of ion release, Cyto. Drug. Rele. Beha. Prog. In Biom. 8 (2019) 101-113. https://doi.org/10.1007/s40204-019-0115-8

[58] T.T.T. Nguyen, B.Q.G. Le, V.T.H. Nguyen, J.H. Lee, N.X.D. Mai, L.H.T. Nguyen, T.L.H. Doan, Microwave-assisted synthesis of porous biomolecule-incorporated metal-organic frameworks as efficient nanocarriers for anti-cancer drugs, Chem. 8 (2023) 202301543. https://doi.org/10.1002/slct.202301543

[59] M. Jakubowski, M. Kucinska, M. Ratajczak, M. Pokora, M. Murias, A. Voelkel, M. Sandomierski, Zinc forms of faujasite zeolites as a drug delivery system for 6-mercaptopurine, Microporous Mesoporous Mater. 343 (2022) 112194. https://doi.org/10.1016/j.micromeso.2022.112194

[60] Y.A. Neolaka, Y. Lawa, M. Riwu, H. Darmokoesoemo, H. Setyawati, J. Naat, B.A. Widyaningrum, U.O. Aigbe, K.E. Ukhurebor, R.B. Onyancha, Synthesis of zinc (ii)-natural zeolite mordenite type as a drug carrier for ibuprofen: drug release kinetic modeling and cytotoxicity study, Results Chem. 4 (2022) 100578. https://doi.org/10.1016/j.rechem.2022.100578

[61] I.M. Souza, A. Borrego-Sánchez, C.I. Sainz-Díaz, C. Viseras, S.B. Pergher, Study of faujasite zeolite as a modified delivery carrier for isoniazid, Mater. Sci. Eng. C 118 (2021) 111365. https://doi.org/10.1016/j.msec.2020.111365

[62] P.A. Sampaio, S.C. Serafim, P.M.N. Menezes, E.C.V. Pereira, P.G.S. de Sá, J.M.T. de Alencar Filho, R.G. de Oliveira Júnior, P.J.R. Neto, J.A. da Silva, L.A. Rolim, Development and characterization of the zeolite imidazolate framework for a modified release of the drug scopoletin, J. Drug Deliv. Sci. Technol. 61 (2021) 102131. https://doi.org/10.1016/j.jddst.2020.102131

[63] V. Hovhannisyan, K. Siposova, A. Musatov, S.-J. Chen, Development of multifunctional nanocomposites for controlled drug delivery and hyperthermia, Sci. Rep. 11 (2021) 5528. https://doi.org/10.1038/s41598-021-84927-x

[64] S. Feng, X. Zhang, D. Shi, Z. Wang, Zeolitic imidazolate framework-8 (zif-8) for drug delivery: a critical review, Front. Chem. Sci. Eng. 15 (2021) 221-237. https://doi.org/10.1007/s11705-020-1927-8

[65] J.I.D.L. Ramirez, V.A.R. Villegas, R.D. Cadena-Nava, E. Loredo-Garcia, F. Chávez-Rivas, V. González-Torres, V. Petranovskii, Antimicrobial activity of the lta zeolite modified by zinc species, Microporous Mesoporous Mater. 380 (2024) 113295. https://doi.org/10.1016/j.micromeso.2024.113295

[66] L.M. Romero, D.A. Palacio, S. Esquivel, G.A. Sánchez-Sanhueza, M. Montaño, D. Rojas, A. Jaramillo, C. Medina, C. Montalba, M.F. Meléndrez, Contact antibacterial and biocompatible polymeric, Comp. With. Copp. Zeol. Fill. Copp. Oxid. Nano. A Step. Towa. New Raw Mater. For The Biom. Indu. Poly. 315 (2024) 127795. https://doi.org/10.1016/j.polymer.2024.127795

[67] X. Kang, H. Ye, Geopolymers functionalised by antibacterial zeolite against biocorrosion, Cem. Concr. Compos. 150 (2024) 105569. https://doi.org/10.1016/j.cemconcomp.2024.105569

[68] H.S. Jhinjer, M. Jassal, A.K. Agrawal, Nanosized ZIF-8 based odor adsorbing and antimicrobial finish for polyester fabrics, Appl. Surf. Sci. 639 (2023) 158153. https://doi.org/10.1016/j.apsusc.2023.158153

[69] A. Partoazar, S. Afrasiabi, F. Ghasemzadeh, M. Ghazi-Khansari, M.M.S. Dallal, spp. Antimicrobial activity of zeolite/zinc oxide nanocomposite containing aloe vera gel against shigella and its durability effects on strawberries, Bionanoscience 13 (2023) 2096-2102. https://doi.org/10.1007/s12668-023-01165-3

[70] N. Torkian, A. Bahrami, A. Hosseini-Abari, M.M. Momeni, M. Abdolkarimi-Mahabadi, A. Bayat, P. Hajipour, H.A. Rourani, M.S. Abbasi, S. Torkian, Synthesis and characterization of ag-ion-exchanged zeolite/TiO2 nanocomposites for antibacterial applications and photocatalytic degradation of antibiotics, Environ. Res. 207 (2022) 112157. https://doi.org/10.1016/j.envres.2021.112157

[71] M. Riaz, N. Khan, S.A. Khan, Z. Ahmad, M.A. Khan, M. Iqbal, H.A. Hemeg, E.M. Bakhsh, S.B. Khan, Enhanced catalytic reduction/degradation of organic pollutants and antimicrobial activity with metallic nanoparticles immobilized on copolymer modified with nay zeolite films, J. Mol. Liq. 359 (2022) 119246. https://doi.org/10.1016/j.molliq.2022.119246

[72] M. Król, J. Syguła-Cholewińska, T. Sawoszczuk, Zeolite-supported aggregate as potential antimicrobial agents in gypsum composites, Materials 15 (2022) 3305. https://doi.org/10.3390/ma15093305

[73] P. Peixoto, J.F. Guedes, E. Rombi, A.M. Fonseca, C.A. Aguiar, I.C. Neves, Metal ion-zeolite materials against resistant bacteria, MRSA Indu. & Eng. Chem. Res. 60 (2021) 12883-12892. https://doi.org/10.1021/acs.iecr.1c01736

[74] E. Torres-Flores, N. Flores-López, C. Martínez-Núñez, J. Tánori-Córdova, M. Flores-Acosta, M. Cortez-Valadez, Silver nanoparticles in natural zeolites incorporated into commercial coating: Antibacterial study, Appl. Phys. A 127 (2021) 1-11. https://doi.org/10.1007/s00339-020-04227-5

[75] G. Cerri, M. Farina, A. Brundu, E. Gavini, A. Salis, W. Dathe, Antibacterial activity of Zn-loaded Cuban zeolite against Helicobacter pylori in comparison to its Na-loaded and unmodified counterparts, Environ. Geoc. Heal. 43 (2021) 2037-2048. https://doi.org/10.1007/s10653-020-00781-2

[76] N. Ranjbar, M. Kolahdoozan, H. Ebadi-Dehaghani, Electrospun polycaprolactone/gelatin mat incorporated with glucosamine-loaded zeolite imidazolate framework-8 nanoparticles for cartilage tissue engineering, J. Polym. Res. 31 (2024) 9. https://doi.org/10.1007/s10965-023-03847-8

[77] T. Menard, B. Zarbaliyev, A. Echalard, E. Bullier-Marchandin, F. Gens, G. Ladam, N. Guliyeva, B. Louis, G. Lutzweiler, Synthesis and characterization of carboxylic amine-grafted fau zeolites as inorganic fillers to design biocompatible composites, Microporous Mesoporous Mater. 363 (2024) 112834. https://doi.org/10.1016/j.micromeso.2023.112834

[78] N. Nazemi, N. Rajabi, Z. Aslani, M. Kharaziha, M. Kasiri-Asgarani, H. Bakhsheshi-Rad, A. Najafinezhad, A. Ismail, S. Sharif, F. Berto, Synthesis and characterization of gentamicin-loaded ZSM-5 scaffold: Cytocompatibility and antibacterial activity, J. Biom. Appl. 37 (2023) 979-991. https://doi.org/10.1177/08853282221140672

[79] N. Zakeri, H. Rezaie, J. Javadpour, M. Kharaziha, Fabrication and characterization of polycaprolactone â zeolite nanocomposite for bone tissue engineering, J. Adv. Mater. In Eng. Este. 39 (2022) 77-94.

[80] M. Sandomierski, K. Adamska, M. Ratajczak, A. Voelkel, Chitosan-zeolite scaffold as a potential biomaterial in the controlled release of drugs for osteoporosis, Int. J. Biol. Macromol. 223 (2022) 812-820. https://doi.org/10.1016/j.ijbiomac.2022.11.071

[81] H. Feng, Z. Li, W. Xie, Q. Wan, Y. Guo, J. Chen, J. Wang, X. Pei, Delivery of therapeutic miRNAs using nanoscale zeolitic imidazolate framework for accelerating vascularized bone regeneration, Chem. Eng. J. 430 (2022) 132867. https://doi.org/10.1016/j.cej.2021.132867

[82] G. Lutzweiler, Y. Zhang, F. Gens, A. Echalard, G. Ladam, J. Hochart, T. Janicot, N. Mofaddel, B. Louis, Deciphering the role of faujasite-type zeolites as a cation delivery platform to sustain the functions of mc3t3-e1 pre-osteoblastic cells, Mater. Adv. 3 (2022) 8616-8628. https://doi.org/10.1039/D2MA00768A

[83] G. Navidi, M. Allahvirdinesbat, S.M.M. Al-Molki, S. Davaran, P.N. Panahi, M. Aghazadeh, A. Akbarzadeh, A. Eftekhari, K.D. Safa, Design and fabrication of m-sapo-34/chitosan

scaffolds and evaluation of their effects on dental tissue engineering, Int. J. Biol. Macromol. 187 (2021) 281-295. https://doi.org/10.1016/j.ijbiomac.2021.07.104

[84] S.M. Mousavi, F.F. Nezhad, M.H. Akmal, R.H. Althomali, N. Sharma, V. Rahmanian, R. Azhdari, A. Gholami, M.M. Rahman, W.-H. Chiang, Recent advances and synergistic effect of bioactive zeolite imidazolate frameworks (ZIFs) for biosensing applications, Talanta 275 (2024) 126097. https://doi.org/10.1016/j.talanta.2024.126097

[85] Z. Qiu, X. Zhang, N. Jia, X. Li, R. Li, S.C. Gopinath, M. Jiao, Zeolite nanomaterial-modified dielectrode oxide surface for diagnosing Alzheimer's disease by dual molecular probed impedance sensor, Turk. J. Biochem. 48 (2024) 668-674. https://doi.org/10.1515/tjb-2023-0079

[86] L. Wang, Z. Liang, D. Li, Y. Chen, G. Ma, Y. Liang, Synthesis and application of zeolitic imidazolate frameworks-based nucleic acid delivery system in tumor diagnosis and therapy: A review, Biomed. Mater. 18 (2023) 052002. https://doi.org/10.1088/1748-605X/ace39d

[87] C.R. Quijia, R.C. Alves, G. Hanck-Silva, R.C. Galvão Frem, G. Arroyos, M. Chorilli, Metal-organic frameworks for diagnosis and therapy of infectious diseases, Crit. Rev. Microbiol. 48 (2022) 161-196. https://doi.org/10.1080/1040841X.2021.1950120

[88] R. Aguilar-Palma, M. Malankowska, J. Coronas, Applications of metal-organic frameworks and zeolites to virus detection and control: Biosensors, Barr. Bioc. Zeit. Für Anor. Und Allg. Chem. 647 (2021) 1532-1541. https://doi.org/10.1002/zaac.202000453

[89] L.D. Trino, L.G. Albano, D.C. Granato, A.G. Santana, D.H. De Camargo, C.C. Correa, C.C. Bof Bufon, A.F. Paes Leme, Zif-8 metal-organic framework electrochemical biosensor for the detection of protein-protein interaction, Chem. Mater. 33 (2021) 1293-1306. https://doi.org/10.1021/acs.chemmater.0c04201

[90] J. Zhang, Y. Tan, W.-J. Song, Zeolitic imidazolate frameworks for use in electrochemical and optical chemical sensing and biosensing: A review, Microchim. Acta 187 (2020) 1-23. https://doi.org/10.1007/s00604-020-4173-3

Chapter 8

Microwave-Assisted Synthesis of Ionic Liquids: Green Solvents for Modern Chemistry

Guocai Tian[1]*

[1]State Key Laboratory of Complex Non-Ferrous Metal Resource Clean Utilization, Faculty of Metallurgical and Energy Engineering, Kunming University of Science and Technology, Kunming 650093, China

tiangc@kust.edu.cn, tiangc01@gmail.com

Abstract

Ionic liquids, as one of the new green solvents, which have important application potential in catalysis, electro-chemistry, biomass conversion and other fields since their high thermal stability and small volatility, and controllable physico-chemical properties. However, traditional synthesis methods have problems such as long reaction time, high-energy consumption, and many by-products. Microwave assisted synthesis technology, with its efficient heating, precise temperature control, and energy-saving advantages, has become a key breakthrough point for the large-scale production of ionic liquids. This chapter introduces the structure, physical and chemical properties, and applications of ionic liquids. It provides a systematic review of the current research status, reaction mechanisms, existing problems, optimization strategies, industrial application cases of microwave-assisted synthesis of the ionic liquids, and explores their future development directions.

Keywords

Ionic Liquids, Microwave Assisted Synthesis, Green Chemistry, Green Solvents, Process Intensification

Contents

Materials Research Foundations 189 (2026) 237-267 https://doi.org/10.21741/9781644904039-8

1. Introduction

Ionic liquids (ILs) are ionic systems composed of the structurally asymmetric, relatively larger volume of organic cations and the small volume organic or inorganic anions that remain liquid at room temperature [1-11]. Its melting point is usually below 100 ℃. Since ethyl ammonium nitrate was first reported in 1914, it experienced explosive development in the 1990s due to the increase of green chemistry and ILs. Compared with the traditional organic solvents such as acetone and dichloromethane, ionic liquids have the following advantages [8-11]. ILs have low vapor pressure and near zero volatility (vapor pressure < 10^{-9} Pa), reducing volatile organic compounds pollution; Wide liquid temperature range (-50~400 ℃), suitable for various applications; The design-ability of structural properties is regulated by changing the combination of the cations or anions to control solubility, polarity, and catalytic activity. Special solubility, good conductivity, design-ability, good stability, and environmental friendliness are widely used in fields such as catalysis, organic synthesis, electro-chemistry, extraction and separation, and materials [12-14].

Ionic liquids not only provide a new reaction environment for reaction systems, but also play a role in promoting reactions, achieving the goal of environmental friendliness. They are expected to bring breakthrough progress to modern industries facing important issues such as safety and pollution, and have a very broad application prospect. ILs, as reaction media, have unique properties that not only replace some organic solvents in catalytic reactions, organic synthesis, and separation and purification, but also solve the problems of high pollution, low yield, and difficult separation in traditional chemical reactions, achieving green chemistry; Moreover, it can cross to integrate with existing technologies to form integrated innovation. To further expand the development space and improve the functionality of existing technologies, such as the combined application of ILs and supercritical fluids, electro-chemistry, nano-material preparation, microwave radiation, and other technologies. However, traditional ILs synthesis relies on long-term heating (6-24 hours) and various organic solvents, which does not adhere to green chemistry principles. The introduction of microwave-assisted synthesis technology [15-20] has significantly improved reaction efficiency and sustainability.

Microwaves (typically at a frequency of 2.45 GHz) achieve molecular-level heating through dipole polarization and ion conduction mechanisms microwave technology is a new kind of green chemistry method that has been widely applied in the field of synthesis. The heating method of microwave is different from traditional one. Traditional heating is carried out from the surface of the materials to its inside through radiation, convection, and conduction. The thermal effect of microwave heating is caused by the magnetic loss or dielectric loss of materials in electromagnetic fields, which has the characteristics of fast synthesis, mild reaction, and heating process accompanied by "non thermal effects" [15, 16]. The heating effect of microwave is closely related to the molecular polarization of substances. Experiments have shown that polar solvents strongly absorb microwaves and heat up quickly, while non-polar solvents absorb very little or no microwaves. From the perspective of microwave chemistry, ILs have high polarizability and can serve as excellent microwave absorbents, resulting in fast heating rates [17]. Due to the fact that most ILs typically contain weakly coordinated ions or their preparation materials have high polarity, high polarization potential rather than coordination ability, and high microwave absorption rate, they are suitable for microwave heating technology. In addition, ILs have strong microwave absorption capabilities, so researchers have developed microwave-assisted methods for synthesizing functionalized ILs[18-19]. The core advantages of microwave synthesis include: selective heating; polar molecules (such as ILs precursors) preferentially absorb energy; reaction rate improvement; microwave heating can shorten reaction time by more than 90%; energy consumption reduction, compared to oil bath heating, it saves 50-80% energy. After decades of development, the microwave method for preparing ILs has become relatively mature. In recent years, BASF's Basel™ The process adopts microwave synthesis of 1-butyl-3-methylimidazolium chloride salt ([Bmim]Cl), compressing the time of reaction from 8 hours to 15 minutes and achieving kilogram scale amplification. This chapter will summarize the research progress of microwave technology in the synthesis of ILs, hoping to provide reference for the microwave synthesis and application of ILs.

2. Composition and synthesis method of ionic liquids

2.1 Common combinations of anions and cations

ILs are often composed of the organic cations, with larger and asymmetric cations and the smaller inorganic or organic anions [6-11]. At present, research and application of ILs have been performed for many years, resulting in a rapid increase in the types of ILs. The reported types of ILs have exceeded 10^{18}. Cuing to the different types of cations, they can be classified into holiness, triangle, guanine, ammonium alcohol, Quaternary ammonium, shithole, Quaternary phosphorus, pyrimidine, prole, formidable and other cations. Common anions include lactic acid, acetic acid, sulfuric acid, trifluoroacetic acid, propionic acid, halogen, tetrafluoroboric acid, nitric acid, bis(trifluoromethanesulfonicacid)imide, hexafluoro- phosphate, tetrazolium, and azole anions. Figure 1 shows the common anions and cations for various ILs.

Materials Research Foundations 189 (2026) 237-267 https://doi.org/10.21741/9781644904039-8

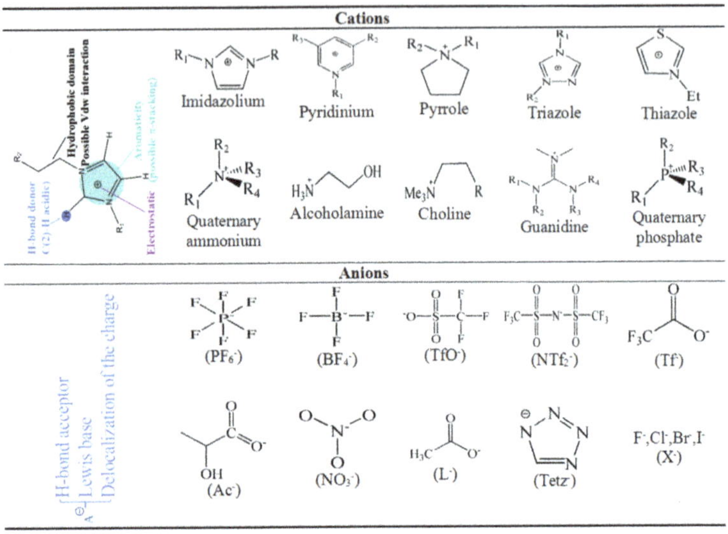

Figure 1 Structure of common cations and anions for ionic liquid

2.2 Synthesis of ionic liquids

Traditionally, there are two basic ways to synthesize ILs, that is the direct (one-step) synthesis and the indirect (two-steps) synthesis [6-11].

2.2.1 Direct synthesis (one-step method)

The one-step or direct synthesis method of ILs is through the reaction of acid-base neutralization or quaternization, which is operating simple, economic, no by-products, and the products are easy to purify [6-19]. For example, nitroethylamine ILs are prepared by neutralizing an ethylamine aqueous solution with the nitric acid [20]. The specific preparation process is as follows: after neutralization reaction, water is removed under a vacuum. To keep the ILs purity, it always dissolved it in organic solvents such as tetrahydrofuranor acetonitrile, treated by activated carbon, and finally removing the organic solvent with vacuum to obtain the product ionic liquid. Various of tetrafluoroborate base ILs were synthesized using this method with different cations [21]. In addition, various ILs can also be synthesized in one step through the reactions of quaternization [22,23], such as 1-butyl-3-methylimidazolium chloride ([Bmim]Cl), 1-buthyl-3-methylimidazolium trifluoromethanesulfonate ([Bmim]CF_3SO_3), etc. The one-step synthesis has the advantages of high yield, simple operation, few by-products, and easy purification, but it is generally difficult to obtain the target complex ILs.

2.2.2 Two-Step Method

When the one-step synthesis is difficult to obtain the specific ILs, most two-step methods are used. The two-step route for synthesizing ILs includes two steps [6-11]. Firstly, a halide salt with the

Materials Research Foundations 189 (2026) 237-267 https://doi.org/10.21741/9781644904039-8

target cation $[Cat]^+$ is synthesized, such as [Cat]X (X=Cl, Br, I). Then, ion exchange, metathesis reaction, complexation reaction, or electrolysis [24] are carried out to exchange the halide X^- ion into the Y^- anion of the target ILs or add Lewis's acid to obtain the target ionic liquid, as shown in Figure 2. In the second step, when using the metal salts such as MY (commonly NH_4Yor AgY), the AgY precipitate or NH_4, HX gases generated are easily removed. The strong proton acid HY addition requires the reaction to be carried out under low-temperature stirring conditions, followed by multiple water washes until neutral, extraction of the organic solvents with ILs. Finally, the organic solvent is removed by vacuum to obtain pure ILs [23]. Most ILs such as tetrafluoroboric acid, hydrogen sulfate, and hexafluorophosphate, such as N-butyl-pyridinium tetrafluoroborate ($[BPy]BF_4$), 1-buthyl-3-methylimidazolium tetrafluoroborate($[Bmim]BF_4$), and 1-butyl-3-methyl-imidazolium hydrogen sulfate ($[Bmim]HSO_4$), are synthesized using a two-step method. Overall, this method has good adaptability and high yield, but it is necessary to ensure complete anion displacement. Otherwise, even a small amount of residual halide anions will have a great impact on the physical and chemical properties and applications of the final product. Moreover, the synthesis cycle is long. The purification difficulty is high, and a large amount of organic solvents are required, which not only wastes resources but also pollutes the environment, which does not abide by the green chemistry principles.

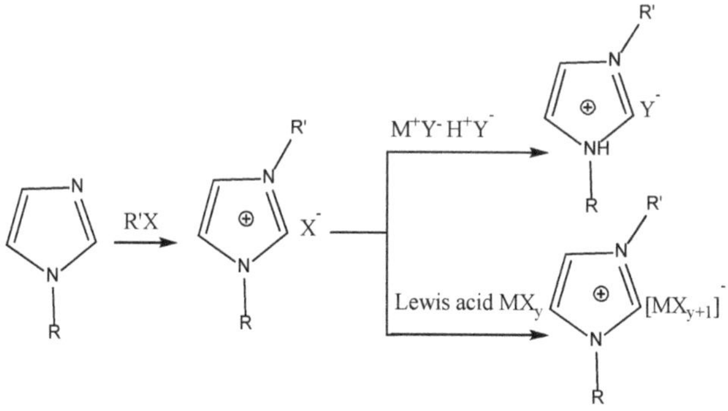

Figure 2 Two-step Synthesis Scheme of ionic liquids

3. Properties and applications of ionic liquids

3.1 Physical and chemical properties

(1) Melting point. Melting point is one of the key indicators for evaluating the properties of ILs [6-11]. ILs have excellent kinetic controllability in many reactions, with some ILs having a melting point as low as170 K and a maximum liquid path of 400 K. For the widely reported imidazole ILs, it has shown that as the volume of imidazole cations increases, the molecular charge becomes more dispersed, leading to a decrease in molecular symmetry and a decrease of the melting point for ILs. In addition, it was found that increasing the anions' size can effectively decrease the melting point of ILs. ILs have a lower melting point than other solvents. The melting point of the ILs is affected by the chain length of the alkyl on the ring of cations. The ILs with longer side alkyl chains have a lower melting point. When the carbon number on the alkyl chain is greater than 6,

Materials Research Foundations 189 (2026) 237-267 https://doi.org/10.21741/9781644904039-8

the melting point increases with the increase of chain length [8-11]. Most ionic liquids have good solubility in aqueous solutions, which mainly is determined by the hydrophobic groups of cations and the hydrogen bonds of anions. By designing a reasonable combination of the anions and cations, the solubility of ionic liquid systems can be changed.

(2) Density. The density of ILs is less influenced by external factors such as pressure and temperature [7-9]. Except for some pyrrole and guanidine salts with densities ranging from 0.9 to 0.97 g/cm^3, which are higher than water and generally above 1.1 g/cm^3, the density of most ILs is roughly between 1.1 and 1.6 g/cm^3. The density of ILs is mainly determined by the types of cations and anions, with anions having a more significant impact on density. Usually, the larger volume of anions, the higher ILs density is, while the larger volume of the organic cations, the lower ILs density. Even small changes in the structure of the cation can finely adjust the density of the ILs.

(3) Viscosity. ILs always have the higher viscosity than traditional organic solvents [8-10] about 1-3 times. Due to the effect of the structure of anions and cations, the viscosity changes significantly. High viscosity is a disadvantageous factor in the ILs application. Higher viscosity can affect the movement and transport of molecules or ions in ILs, reducing rates of reactions. In most applications, ILs can be mixed with other low viscosity compounds. Highly sensitive to temperature changes and the presence of pollutants, even a slight increase in temperature or the presence of impurities can result in a significant lower in the viscosity of ILs.

(4) Vapor pressure. The vapor pressure of ILs is extremely low. Compared with traditional solvents, ILs have very low vapor pressure and are extremely difficult or generally non-volatile near room temperature or at room temperature, saving solvents and reducing environmental pollution [7-9].

(5) Thermal stability. ILs have large thermal stability [8-10]. They are the thermal decomposition temperatures of ionic liquids, reflecting the upper limit of their operating temperature and serving as a measure of their thermal stability. ILs have excellent thermal stability, with a thermal decomposition temperature typically above 573 K. ILs have non-volatile properties and almost no vapor pressure. It has been found that 70% of ILs can be kept for a long time at a low temperature or room temperature, which gives ILs a longer lifespan.

(6) Electro-chemical stability. The electro-chemical window is wide and has good conductivity [6-11]. The electro-chemical window is greater than 4 V, which is higher than that of general solvents. This allows ILs to exist stably in solution and is not easily degraded by acidic or alkaline environments, which can prevent hydrogen evolution reactions. The conductivity of ILs is greatly affected by the cation type. The conductivity can be adjusted by designing cations.

(7) Good structural and property adjustability. Different functional groups can be introduced onto anions and cations according to specific needs, or different cations and anions can be combined to develop the ILs with specific needs [7-10]. Generally, carbon chains can be extended or substituent can be introduced to enable them to possess both organic and some inorganic properties, thereby altering their characteristics.

(8) Polarity. The polarity performance directly affects the difficulty of the reaction [8-11]. For example, ILs with 1-alkyl-3-methylimidazole as the cation have a polarization ability equivalent to short-chain alcohols, which weakens with the increase of the alkyl chain length. Changing the anion types has a little effect on the polarization ability.

3.2 Application of ionic liquids

Compared with traditional organic electrolytes and solvents, ILs have many excellent physical and chemical properties [6-11]. Extremely low vapor pressure, can be operated near room temperature or at room temperature, is not easy to evaporate, and reduces solvent loss. Low melting point, ILs exhibit a liquid state at or near room temperature (with a wide range of liquid states). Higher thermal and chemical stability makes it less prone to decomposition during reactions. The wide electro-chemical window and strong conductivity of ILs make them widely used in the field of electro-chemistry. Excellent structural adjust-ability, in order to meet certain or specific needs, can usually be obtained by changing the type or structure of anions and cations to change their physical and chemical properties. High gas solubility can serve as a collector for specific gases. Operating at or near room temperature results in low energy consumption and can be reemployed multiple times. Because of their excellent properties, ILs are widely used in various fields such as catalysis, material preparation, extraction and separation, organic synthesis, secondary batteries, and solar cells [6-11,14,25-31]. ILs, as a new kind of green solvents, have received widespread attention from numerous researchers in the word. Figure 3 shows the properties and main applications of ILs, which are introduced below.

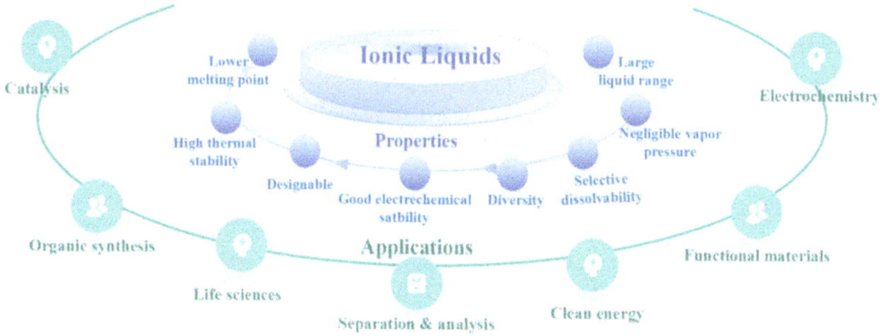

Figure 3 The properties and application of Ionic liquids

(1) Catalyst. The prominent characteristics of ILs as catalysts are high selectivity and good low-temperature activity. ILs as a catalyst can alter the mechanism of the chemical reactions [6], improve reaction rates and selectivity, and increase yields. ILs can serve as both catalysts and solvents, making reaction conditions milder and greatly improving conversion and selectivity. Additionally, ILs can be recycled and reused.

(2) Electro-chemistry. ILs are very suitable as electrolytes and can be used as media in devices composed of conductive polymers because of their high thermal stability, low vapor pressure, broad electro-chemical stability, in-combustibility, wide potential window, high ionic conductivity, and easy formation of gel with polymers. ILs have applications in batteries, electro-deposition, electro-polishing, electro-chemical synthesis, capacitors, sensors, capillary electrophoresis, and other fields [6-8].

(3) Organic synthesis. ILs have significant excellent properties as a new type of solvent. Widely used in Friedel Crafts and hydrogenation, Heck reactions, redox and Diels-Alder reactions, etc. [7-9]. In ILs, the overall electrostatic field still dominates, and there is strong interaction between

Materials Research Foundations 189 (2026) 237-267 https://doi.org/10.21741/9781644904039-8

cations and anions. Compared with volatile and flammable molecular liquids, it has almost no vapor pressure, so it can replace organic solvents that cause environmental pollution and safety accidents due to their volatility and flammability.

(4) Separation and purification. Because of their unique physico-chemical properties, ILs are highly suitable as solvents for separation and purification. There have been numerous reports in this area, such as using ILs to extract and recover butanol from fermentation broth. Extracting the non-volatile organic compounds in ILs using supercritical CO_2 as flotation, leaching, and extraction agents for efficient separation of minerals and valuable metal elements [8-11].

(5) Material preparation. ILs, as a structurally design-able green solvent, exhibit unique advantages in material preparation. The tun-ability of its anions and cations endows it with excellent solubility, thermal stability, and conductivity, making it widely used in the synthesis of nano-materials, polymer modification, and development of energy materials. For example, imidazole based ILs can assist in the preparation of highly catalytic metal nano-particles or serve as templates to regulate the pore size of molecular sieves. Functionalized ILs, such as amino modified ones, can improve the dispersibility of graphene in composite materials and enhance electrode performance. In addition, ILs have also shown excellent performance in cellulose dissolution, PVDF membrane modification, and other fields. Their low volatility and environmentally friendly characteristics provide new ideas for the preparation of green materials [8-11].

(6) Life sciences. ILs have demonstrated unique application potential in the field of life sciences, with their low toxicity, high bio-compatibility, and designability making them ideal biological media. For example, choline based ILs can be applied for protein stabilization and drug delivery, improving the bioavailability of insoluble drugs. Amino acid based ILs exhibit selective bactericidal activity in antibacterial materials. Imidazole based ILs can be used as efficient solvents for extraction of DNA/RNA, replacing traditional organic reagents. In addition, the good conductivity is also suitable for the development of biosensors. The current research focuses on optimizing the biodegradability and targeting of ILs to expand their application prospects in biomedicine, diagnosis, and treatment [8].

(7) Clean energy. ILs have become key materials since their good thermal stability, wide electro-chemical window, and controllable physico-chemical properties. As efficient electrolytes, they can significantly enhance the cycle life and energy density of lithium/sodium ion batteries and super-capacitors. In fuel cells, proton based ILs can replace traditional acidic media, improve proton conductivity, and reduce corrosiveness [6-8]. In addition, the excellent CO_2 solubility can be used for carbon capture technology, and as a catalyst carrier, it can optimize biomass energy conversion efficiency. The current research is dedicated to developing low-cost and high stability functionalized ILs to promote their large-scale applications in energy storage, catalysis, and carbon neutrality technologies.

4. Basic principles of microwave-assisted synthesis

Microwaves belong to a type of electromagnetic wave. In the international classification of electromagnetic wave frequency bands, the types of electromagnetic waves are arranged according to the length of the wavelength. Generally, electromagnetic waves with a wavelength of 1mm-1m or a frequency of 300 MHz -300 GHz is referred to as microwaves [32-42]. Microwave has various applications. To avoid crossing, it has specific applications in specific frequencies and frequency bands, such as broadcasting and television (470-90 MHz), civilian communication (890-940 MHz), wireless and mobile phones (940-1350 MHz). The commonly used microwave frequency range is shown in Table 1. The frequency of industrial and household microwave heating in most countries is 2450 MHz ±50 MHz. The commonly used frequencies for scientific and medical research and industrial production are 433 MHz, 2450 MHz, 915 MHz, and 5800 MHz. Microwaves with wavelengths of 1-25 cm are dedicated to radar, while most of the remaining bands are used for telecommunications transmission. At present, the commonly used frequencies for heating in China are 915 MHz ± 50 MHz and 2450 MHz ± 50 MHz [39-42].

Table 1 Common Microwave Frequency Range

Frequency range (MHz)	Common main frequency (MHz)	Band (m)	Center wavelength (m)
890-940	915	L	0.330
2400-2500	2450	S	0.122
5725-5875	5800	C	0.052
22000-22250	22125	K	0.014

The temperature is usually used to indicate the degree of heating of materials, which essentially reflects the average kinetic energy of molecules within the material. When the temperature rises, the speed of molecular motion increases, and the thermal motion becomes more intense. So, in order to raise the temperature of an object, it is necessary to make the molecules inside the object vibrate, increasing its average kinetic energy [24,25]. The traditional heating method is to transfer energy from the zone of high-temperature to the zone with low temperature through contact, which is commonly referred to as thermal conduction. In this process of energy transfer, the molecules in the zone of high-temperature and the molecules in the zone of low temperature undergo elastic collisions when in contact, which can be understood as the transfer of kinetic energy. The molecules in the zone of high-temperature transfer its kinetic energy to the molecules in the zone with low temperature, causing the molecular vibrations in the zone of low temperature to accelerate and the temperature to naturally rise. The common way to use fire to boil water in daily life is to use heat conduction, but the heating time of heat conduction is relatively slow because the kinetic energy transfer of a molecule is from the outside to the inside, requiring a longer heating time, which relies on the material thermal conductivity. At the same time, molecules in the high-temperature zone will rapidly dissipate energy to the surrounding areas during the process of energy transfer, so the heating efficiency is not high.

Microwave heating is the utilization of microwaves as an energy source, which produces thermal effects when they interact with material molecules and are absorbed. When microwaves act on matter, they may generate electronic polarization, atomic polarization, interface polarization, and dipole turning polarization, among which dipole turning polarization plays a major role in heating the matter [39-42]. For microwave heating, it is worth noting that the object being heated needs to contain polar molecules in order to be heated, while non-polar molecules do not have permanent dipole moments and have low absorption of microwave energy, making them unable to be heated. In the process of microwave heating, polar molecules in the heated material are affected by the electromagnetic field and generate dipoles. The two ends of polar molecules carry positive and negative charges, and they rotate under the action of rotational torque in the electric field. Positive charges turn into negative charges. Negative charges turn into positive charges, or existing dipoles in the material are rearranged, and the arrangement of dipoles becomes orderly. The electric field inside the reaction chamber is constantly changing, and dipoles will undergo relaxation phenomena following the changes in the electric field. The frequency of electric field transformation is the same as that of microwave frequency. For example, the commonly used microwave heating frequency is 2.45 GHz, which means that the electric field undergoes 2.45 billion changes in one second. The polar molecules in the heated material also undergo a transformation at a frequency of 2.45 billion per second. At such a high frequency, the thermal motion of the molecules becomes very intense [24-25]. As molecules rearrange in the direction of the constantly changing high-frequency electric field, they must overcome the interference and obstacles of their original thermal motion and inter-molecular interactions, producing effects similar to rapid collisions and friction, achieving molecular level "stirring" and generating a large amount of heat. A large amount of thermal energy is generated inside the material, causing the temperature of the material to rapidly rise, achieving the conversion between microwave energy and thermal energy. Microwave heating is a type of "cold and heat source" that generates and comes into contact with an object, producing electromagnetic energy instead of hot air. Microwave heating is a new heating method that has a series of unique advantages that traditional heating methods do not possess, such as fast and uniform heating speed, and the ability to selectively heat substances.

Microwave heating systems can generally be divided into microwave generators, circulators, water loads, waveguides, and resonant cavities [25-35]. The microwave generator is responsible for converting electrical energy into microwave energy, and the magnetron, as a typical microwave generator, has a high cost-effectiveness. At the same time, solid-state microwave sources continue to develop and have advantages in parameter adjustment, gradually replacing the position of magnetrons. The ring resonator and water load serve to protect the magnetron and prevent it from being damaged by electromagnetic waves reflected back. The waveguide is located between the magnetron and the resonant cavity, used to transmit microwaves into the resonant cavity. It is usually made of metal and has minimal loss of microwaves. The resonant cavity is made of special materials that can reflect microwaves. The material is heated inside the resonant cavity, and the resonant cavity size decides the resonant mode and resonant frequency. The entire heating process is as follows: when current passes through the cathode of the magnetron, the cathode of the magnetron is heated and emits electrons. As the cathode is surrounded by the anode, an electric field is created, and a magnetic field is also formed due to the principle of electromagnetic induction. The generated electrons move rapidly between the two poles and generate a changing electric field at the anode cavity, thus generating an oscillating electromagnetic field to emit microwaves. The microwaves enter the resonant cavity through a waveguide and interact with the material, thereby achieving the purpose of heating.

At present, there are two different views on the reasons for microwave accelerated chemical reactions [35-40]. One viewpoint holds that although microwave is an internal heating method with fast-speed heating, no hysteresis effect and uniform heating without temperature gradient. In chemical reactions, it is only a heating way just like traditional heating methods. The kinetics of specific reaction remain unchanged regardless of the heating method, as the reactants, catalysts, and products remain constant. They believe that the frequency of 2450 MHz used in chemical reactions by microwaves belongs to non-ionizing radiation, which cannot cause chemical bond breakage when resonating with molecules, nor can it excite molecules to higher rotational or

Materials Research Foundations 189 (2026) 237-267 https://doi.org/10.21741/9781644904039-8

vibrational energy levels. The acceleration of chemical reactions by microwaves is mainly dues to the selective heating of polar substances, namely the heating effect of microwaves. Another viewpoint holds that microwave's effect on the chemical reactions is very complex. The reactant molecules absorb microwave energy, which increases the motion speed of the molecules, resulting in disordered molecular motion and an entropy increase. At the same time, the effect of microwaves on polar molecules forces them to move in an electromagnetic field, resulting in a decrease in entropy. Therefore, the mechanism of microwave action on chemical reactions cannot be described solely by the microwave heating effect. In other words, microwaves also have a non-thermal effect that is not resulting from temperature, which may alter the kinetics of the reaction and reduce its activation energy during the chemical reaction process. Under these two perspectives, there are four mechanisms of process enhancement caused by microwave irradiation: micro hotspot mechanism, local overheating mechanism, molecular radiation mechanism, and non thermal effect mechanism [38-42].

(1) Microscopic Hot Spots Mechanism. It refers to the phenomenon in which strong absorbing materials (such as metal nano-particles and carbon materials) in microwave fields generate instantaneous ultra-high temperatures at the nano-scale due to dielectric loss and plasma resonance effects. When microwave is applied to carbon supported catalysts (such as Pd/C), carbon materials preferentially absorb microwave energy due to extremely high dielectric loss (tan $\delta>0.5$), and form nano-scale high-temperature cores at defect sites through electron relaxation and dipole polarization, with temperatures up to 10-20 times higher than traditional heating. This type of hotspot has a sub microsecond existence time, but can significantly minimize the reaction activation energy. For example, in the methane dry reforming reaction, the surface hotspot of Ni/CNT catalyst increases the CO_2 conversion rate by 8 times. The key difference between this mechanism and local overheating is that the scale of action is at the nano-meter level (vs micrometer level); Dependent on plasma resonance (vs dielectric heating); Higher temperature (> 800 °C vs 200-500 °C). Through in-situ Raman spectroscopy and high-temperature fluorescence probes, it has been confirmed that hotspots can trigger reaction pathways that traditional heating cannot achieve, such as direct activation of carbon-carbon bonds, which have unique advantages in fields such as energy catalysis and high-temperature material synthesis.

(2) Local Overheating Mechanism. It refers to the overheating zones' formation due to the volume heating characteristics of microwave radiation and the uneven distribution of the electromagnetic fields. During microwave heating, the phenomenon of micro-scale temperature gradients in the system is caused by differences in the dielectric properties of substances. For a solution, the temperature gradient depends not only on the electromagnetic field strength and dielectric properties of the solution, but also on the liquid flow turbulence [43,44]. When a microwave is applied to multi-phase systems, high dielectric loss components (such as polar solvents, metal catalysts, or carbon materials) will preferentially absorb microwave energy and rapidly heat up, while low loss media (such as non-polar solvents or reactor walls) will remain relatively low temperature, thus forming a transient high-temperature region around the absorbing material. This non-uniform heating is due to the combined effect of the body heating characteristics of microwaves and thermal conduction hysteresis. Microwave energy directly penetrates the medium and is converted into thermal energy by strong absorbing components, while the surrounding weak absorbing areas cannot balance the temperature difference in a timely manner due to limited thermal diffusion rate. This mechanism exhibits significant performance in Pd/C catalytic hydrogenation reactions, with local surface temperatures of up to 500 °C on the catalyst, promoting

reaction rate bout 10-100 fold, while the overall temperature of the system is only maintained at 80-120 °C. It has been confirmed through infrared thermography and molecular probe technology that the presence of this micro scale high-temperature zone can reduce reaction activation energy, alter reaction pathways, and avoid product thermal decomposition caused by traditional heating. At present, this mechanism has been widely used in fields such as organic synthesis, nano-material preparation, and environmental pollutant degradation. However, its difference from the "hotspot mechanism" lies in its larger scale of action (micrometer vs. nano-meter) and relatively lower temperature (< 600 °C). Accurately regulating local overheating required optimizing microwave parameters (power, frequency) and matching the dielectric properties of the reaction system.

(3) The Molecular Radiation Mechanism of Microwaves. It refers to the process in which polar molecules oscillate synchronously with the electric field through dipole moment orientation polarization under the action of microwave electromagnetic field, converting electromagnetic energy into molecular kinetic energy and ultimately dissipating it in the form of thermal energy. When microwave radiation frequency with a 2.45 GHz (equivalent to a 12.24 cm of wavelength) acts on polar media such as water and ethanol, the molecular dipole will repeatedly rotate in the direction of the alternating electric field at a frequency of 2.45 billion times per second. This forced orientation motion is hindered by inter-molecular forces (such as hydrogen bonds), resulting in frictional heat generation of molecules. Its energy dissipation rate (P) follows the Debye relaxation model: $P=2\pi f\varepsilon_0\varepsilon''E^2$, where ε'' is the dielectric loss factor. Experiments have shown that the relaxation time of water molecules in a microwave field ($\tau \approx 9.3$ ps) is highly matched with a frequency of 2.45 GHz, making it an ideal molecular radiator with an energy conversion efficiency of over 70%. This mechanism is different from ion conduction heating and has three major characteristics: selective heating (only exciting polar molecules), uniform heating of the bulk phase (no electrode effect), and molecular-level energy transfer (directly activating chemical bonds). It is the core principle of microwave chemical synthesis, food processing, and other applications.

(4) Non thermal effects of microwaves. The first three mechanisms all derived from the same model and be owned by the selective heating behavior of microwave radiation in different scales. While a substance rapidly heats up under microwave action, the high-speed and intense movement of polar molecules inside the substance improves the kinetic conditions of chemical reactions, reducing the activation energy required for many chemical reactions and allowing them to proceed at relatively low temperatures. At the same time, the reaction rate is greatly increased, and the long-term collision and friction of polar molecules in the microwave field inevitably generate a large amount of heat. Microwaves not only have a heating effect, but also have a 'non thermal effect'. There is still controversy over the non thermal effects in microwave fields. Supporters believe that the interaction between wave field and materials can alter the dipoles' orientation, which may influence the reaction kinetics and entropy of mass transfer. Under lower power or shorter microwave exposure, microwaves do not cause substances to generate high temperatures. Although the temperature of the substance has not significantly increased, the movement of its internal molecules is still very intense. This intense exercise can cause changes in the chemical bonds in a substance, altering its physical or chemical properties. Therefore, microwaves can be used to treat substances at lower temperatures with the aim of activating them. However, most people believe that the quantum energy of microwaves is too small (only about 1.0×10^{-5} eV at a frequency with 2.45 GHz) to impact the hydrogen or covalent bonds. The root of the debate lies in the difficulty of separating interference and possible non thermal effects from microwave heating effects. Therefore, in the study of microwave chemistry, inappropriate temperature measurements

always mislead researchers, and basic research on the effects of microwave requires precise control and rational design experimental condition.

Regardless of the controversy surrounding the microwave heating mechanism, it is widely believed that microwave heating has the following advantages [32-42,45-50]. (1) selective heating. The dielectric loss and hysteresis loss of different media are different, resulting in different absorption efficiencies of microwaves by different media. Water has a high dielectric loss, making it an excellent medium for microwave absorption, with a very high absorption efficiency for microwaves. The dielectric loss of fats, proteins, and other substances is lower than that of water, and their absorption of microwaves is weaker compared to water. Metal is an excellent conductor, but it usually exhibits weak absorption efficiency in microwave heating, so copper and other metals can be used as the cavity of microwave heating equipment. Materials such as glass and ceramics have extremely low dielectric loss and good wave permeability, and cannot be heated. Therefore, these materials are commonly used as containers for containing the heated medium. By utilizing this feature, targeted heating can be applied to certain media. (2) The reaction rate increases quickly and the heating efficiency is high. Microwave heating can directly generate heat inside the heated medium, unlike traditional convection and conduction heating methods, without waiting for the heat conduction process. At the same time, microwave heating penetrates the medium and causes molecular vibrations throughout the entire volume, reducing the temperature gradient problem that may occur in traditional heating processes. The high-frequency characteristics of microwaves enable them to raise the temperature of the entire medium in a relatively short period, improving heating efficiency. In addition, the energy transfer loss during microwave heating is relatively small, mainly generated through molecular vibration and friction, which reduces energy loss and further improves the heating effect. (3) Low energy consumption, clean and environmentally friendly. Traditional fuel is not required during microwave heating, as microwaves generate heat directly inside objects through electromagnetic waves. Compared to traditional heat sources such as stoves and gas stoves, microwave heating reduces the generation of exhaust gases and residues related to combustion, thereby reducing the complexity of the cleaning process. And in the process of designing the heating chamber of microwave, by designing the chamber door reasonably and equipping some limit switches, it is ensured that the microwave leakage is below the national standard and will not cause radiation damage to the human body. At present, microwave heating has been widely used in fields such as materials science [45,46], metallurgical engineering [47,48], and food-processing [49-50].

5. Microwave-assisted synthesis of ionic liquids

5.1 Research status

In 2001, Varma et al. [51] first synthesized the imidazole based ILs in a household microwave oven. The reaction time is reduced from the traditional 24 hours under a solvent-free condition to 4 minutes, pioneering a new microwave-assisted synthesis method for of ILs. Khadilkar et al. [52] achieved a revolutionary breakthrough by using a professional microwave synthesizer (CEM Discover system). They systematically optimized the synthesis conditions of pyridine ILs (reaction time shortened from traditional 20 hours to 1 hour) and imidazole ILs (reaction only takes a few minutes), and the reaction efficiency was improved by over 95%, when the microwave power was controlled at 300W. Namboodri et al. [53] subsequently developed a new microwave- assisted ion exchange process to prepare [Bmim]BF_4 at 80 °C for 10 minutes, with a yield increased to 91%

(traditional methods account for about 76%). They further synthesized Lewis acidic [Bmim]$AlCl_4$ using microwave method (150W, 5 minutes) [54]. Deetlefs et al. [55] confirmed through comparative experiments (microwave 300W vs traditional oil bath 80 °C) that for C4-C8 alkyl chain pyridine/imidazole ILs, the average reaction time of microwave method was shortened from 24 hours to 2-4 hours (83% reduction), and the yield remained stable at 95-98%. Of particular note, Thanh et al. [56] innovatively synthesized chiral ephedrine based ILs using a solvent-free microwave method (60℃, 30 minutes). The enantiomer excess value (EE) detected by circular dichroism was found to be greater than 99%, an increase of about 12% compared to traditional solvent methods, providing a highly stereo-selective new ionic liquid carrier for an asymmetric catalysis.

The microwave synthesis process of [Bmim]HSO_4 developed by Singh et al. [57] has significant advantages. This process can be synthesized in just 5 minutes at a power of 300W, which is 24 times more efficient than traditional heating methods (2h). When an acidic catalyst is used in the synthesis of coumarin compounds, the reaction yield is as high as 92%, and the catalyst can be recycled 5 times with activity still maintaining over 90%. Bica et al. [58] innovatively synthesized chiral imidazole based ILs derived from camphor using microwave-assisted technology. The ionic liquid obtained by combining (1S) - (+) -10-camphor sulfonic acid anion with chiral imidazole cation exhibits excellent enantio-selectivity in Diels Alder reaction, with a non enantiomeric excess value (EE) exceeding 95%. The threonine p-toluenesulfonic acid salt (ThrTsOH) ionic liquid designed by Li et al. [59] exhibits a unique hydrogen bonding catalytic mechanism. In the esterification reaction, the ionic liquid is activated by dual functions (simultaneously activating carboxyl and hydroxyl groups) to increase the yield to 92%, and can be recovered and utilized through simple phase separation after the reaction. Erdmenger et al. [60] took the lead in developing the first microwave continuous production system, using a 500mL reactor to achieve large-scale preparation of [Bmim]Cl with a production capacity of up to 460 g/h, successfully completing the leap from laboratory to pilot scale. At the same time, Cravotto et al. [61] established a new system for microwave one-step synthesis of PF_6^- and BF_4^- class ILs, significantly reducing the traditional reaction time of 8 hours to 15 minutes and improving the synthesis efficiency. In terms of energy-efficiency optimization, Horikoshi et al. [62] innovatively synthesized [Bmim]BF_4 using a 5.8 GHz high-frequency microwave device, which not only achieved a product yield of 87%, but also achieved a significant effect of reducing energy consumption by 60%. These technological advancements have jointly promoted the development of green synthesis processes for ILs, laying an important foundation for industrial production.

Zheng et al. [63] used microwave-assisted synthesis technology to successfully prepare 1-hydroxyethyl-3-methylimidazolium tetrafluoroborate ([HOEtmim]BF_4). This method significantly shortens the reaction time to 30 minutes through microwave radiation, achieving a high yield of 89.38%. Differential scanning calorimetry characterization shows that the IL has a low melting point of -4.2 °C, indicating its excellent low-temperature fluidity. Yang et al. [64] developed a more simplified microwave-assisted one-step synthesis method for [Bmim]PF_6. By optimizing the reaction molar ratio (1:1.05) and temperature (80 °C), the high performance liquid chromatography (HPLC) purity was 99.2% and the yield was 92.7%, which was about 15% higher than the traditional two-step method. Huang et al. [65] systematically studied the microwave-assisted synthesis and solvent properties of 1-butyl-3-methylimidazolium dicyandiamide ([Bmim]DCA), and obtained a yield of 85.2% using an acetonitrile solvent system under nitrogen protection. They particularly found that the ionic liquid had a high solubility for glucose of 1.2 g/g

(25 °C), which was nearly three times higher than the solubility of traditional dimethyl sulfoxide (DMSO), providing a new solvent platform for biomass conversion. Zhao et al. [66] studied coordination ILs synthesis under microwave conditions and obtained the optimal process conditions. Quaternary ammonium salts and caprolactam were used as raw materials and intermittently irradiated for 90 seconds under microwave (210 W) conditions to obtain coordination ILs containing O-C coordination bonds. Compared with conventional thermal synthesis methods, this method has shorter reaction time, lower energy consumption, and does not require the addition of large amounts of solvents, making it an efficient and convenient new method. Xu et al. [67] synthesized ionic liquid 1-methyl-3-butylimidazole bromide using microwave technology and obtained optimized conditions from an industrial application perspective, with a yield of up to 97.2%; Through industrial simulation, using both intermittent and continuous processes, the yield is relatively high. Industrial simulation shows that the continuous process reduces energy consumption by 38%

Chen et al. [68] used an unmodified household microwave oven to synthesize the intermediate 1,3-di-n-butylimidazolium bromine salts([Dnbim]Br) through quaternization reaction under solvent-free conditions using trimethylsilane imidazole and 1-bromobutane as raw materials, with a yield of 92%; Subsequently, a double decomposition reaction was carried out with potassium hexafluorophosphate to obtain 1,3-di-n-butylimidazolium hexafluorophosphate ([Dnbim]PF_6) within 30 minutes, with a yield of 68%. HPLC confirmed a purity of over 96%. The microwave synthesis process of [Bmim]PF_6 developed by Wu et al. [69] shortened the reaction time to 10 minutes and increased the yield to 82.3%. The TiO_2 photo-catalyst prepared by using this ILs as a reaction medium exhibits excellent performance, with a 27% increase in degradation efficiency of methyl orange in contrast to traditional methods, attributed to the regulatory effect of ionic liquid on TiO_2 lattice defects. Zhang et al. [70] achieved rapid microwave synthesis of 1-ethyl-3-methylimidazole-aluminum trichloride ([Emim]Br/$AlCl_3$), achieving a high yield of 94.2% in just 1000 seconds. This method significantly improved the reaction efficiency by optimizing the microwave power (350W) and material ratio n([Emim]Br): n($AlCl_3$)=1:1.05). Messali et al. [71] compared the traditional heating and microwave heating in the preparation of pyridazine based ILs. They used pyridazine and halogenated hydrocarbons as raw materials to prepare N-alkylpyridazine halides under microwave conditions, and carried out anion exchange with sodium tetrafluoroborate (or potassium hexafluorophosphate) to ultimately prepare fluorinated pyridazine ILs. Isopropyl pyridinium tetrafluoroborate ([Py-C3]BF_4) requires 21 hours of conventional oil bath heating with a yield of 98%. After 30 minutes of microwave intermittent radiation heating, the yield reaches 95%. The conventional 21-hour yield of isopropylpyridine hexafluorophosphate ([Py-C3]PF_6) is 97%, and the microwave 30-minute yield can reach 93%. The conventional oil bath 21-hour yield of tert butylpyridine tetrafluoroborate ([Py-C4]BF_4) and tert-butylpyridine hexafluorophosphate ([Py-C4]PF_6) are 98%, and the microwave intermittent radiation heating 30-minute yield is 94% for both.

Three ILs including 1-butylquinoline bromide was synthesized by Chen et al. [72] using microwave method. The reaction time was shortened to 1/10 of the traditional method, and 1H NMR showed a proton chemical shift of 9.87 ppm for imidazole salt. Zhang et al. [73] compared the conventional and microwave and synthesis of four hydroxyl-functionalized ionic liquid chlorides. It was shown that the reaction time of the mono-hydroxy products [C2(OH)im]Cl and [C2(OH)Py]Cl was reduced from 24 hours to 110 seconds and 390 seconds, respectively (a reduction of 99.8%), and the yield increased by 5-6% (up to 94.2%). The reaction time for the di-hydroxy products [$C3(OH)_2im$]Cl and [$C3(OH)_2Py$]Cl was shortened from 96 hours to 120 seconds and 135 seconds, respectively (99.96% reduction), with a yield of about 90%. Microwave synthesis can adopt an intermittent radiation mode to achieve solvent-free reaction, reducing the energy consumption of synthesis by more than 80%. Huang et al. [74] studied a new synthetic pathway for a benzaldehyde functionalized ionic liquid under microwave conditions. They obtained benzaldehyde functionalized alkyl bromides by alkylating phenol containing aldehyde groups and 1,4-dibromobutane under microwave irradiation and alkaline catalysis conditions. Then, they interacted with N-methylimidazole under microwave irradiation conditions to obtain three benzaldehyde functionalized alkyl-imidazolium bromides with a yield of over 93%. The aldehyde groups on them are very active and can be used as chemical intermediates for other functionalized ILs preparation. Du et al. [75] studied the synthesis pathway of fluoroborate anion imidazolium ionic liquid under microwave radiation conditions, and obtained [Bmim]BF_4 under microwave radiation conditions. The viscosity of this ionic liquid is lower than that of similar ILs without fluorine, and microwave radiation only takes a few tens of seconds to achieve a high yield.

Gong et al. [76] achieved a microwave solvent-free batch synthesis of [Bmim]Cl, with a reaction time of only 30 minutes (traditional methods require 4 hours). The melting point of the product was 78 °C, consistent with the literature value, and the water content was less than 0.5wt%. Messali et al. [77] achieved a record-breaking single batch synthesis of 24 short chain functionalized fluorinated ILs and symmetric double cation fluorinated functionalized ILs. Antibacterial tests showed that [C2mim]Br had a MIC as low as 32 μg/mL. Xu et al. [78] synthesized immobilized ILs intermediates, and TGA showed that their thermal stability reached 280 °C. Zhu et al. [79] successfully synthesized five functionalized organic phosphonate ILs using a two-step microwave irradiation method for the rare-earth extraction for neodymium ions [Nd(III)]. The use of microwave radiation can shorten the traditional heating time from 24 hours to several 10 minutes, without easily producing by-products, and the overall yield is greater than 88%. The relationship between the extraction rate and factors such as solution pH and ionic liquid dosage was explored, and it was found that the extraction effect was optimal under neutral solution and an ionic liquid dosage of 0.8 mmol. Roshan et al. [80] synthesized four amino acids functionalized ILs using a water bath microwave-assisted method. Compared with common imidazole and pyridine ILs, amino acid ILs have better environmental friendliness and bio-compatibility, and are heated faster and more evenly using water bath microwave heating. Under microwave radiation of 100 W, the reaction can be completed in only 2 minutes. In addition, this ILs has large catalytic activity for the cyclo-addition reaction of epoxyethyleneandCO_2, making it a potential CO_2 epoxyethylene cyclo-addition catalyst.

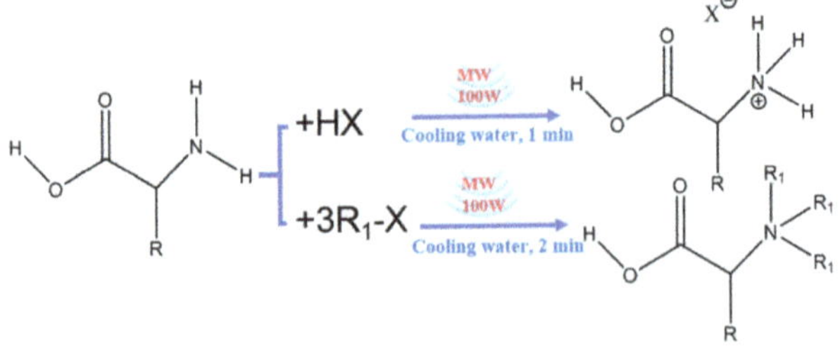

Figure 4 Synthesis Scheme of Amino Acid based ILs with the microwave

Lv et al. [81] used the Box Behnken response surface design method to systematically optimize the microwave synthesis process of tetradecylmethyl imidazolium bromide ([C14mim]Br). By establishing a three factor three-level experimental model with reaction temperature (70-90 °C), reaction time (2-4 hours), and material molar ratio (1:1.05-1:1.15), the optimal conditions were ultimately determined to be 85 °C, 3 hours, and 1:1.1, resulting in a yield of 94.14%. The surface tension test showed that the critical micelle concentration (CMC) of the long-chain ILs was 0.86 mmol/L, which was 28.3% lower than the conventional [C12MIM]Br (1.2 mmol/L), indicating its stronger surface activity and self-assembly ability. Wang et al. [82] developed an alkaline ionic liquid catalytic system based on 1,3-dialkylimidazolium hydroxide ([RNHmim]OH). This system exhibits excellent performance in the synthesis of chalcones. In ethanol solvent, the reaction time is significantly reduced from 6 hours required by traditional base catalysis to 40 minutes, and the conversion rate remains above 99%; Through kinetic studies, the total number of revolutions (TON) was measured to be as high as 5800, which is nearly 15 times higher than the traditional NaOH catalytic system.

Su et al. [83] studied a simple synthesis method of benzimidazole ILs under microwave radiation conditions. The reaction pathway is under microwave radiation conditions, using phase transfer catalyst polyethylene glycol-6000 to transfer the alkyl group of brominated hydrocarbons to benzimidazole, resulting in monoalkylated benzimidazole. Under microwave conditions, it reacts with benzyl bromide to obtain benzyl benzimidazole ionic liquid. This type of ILs can be applied as an additive for flame retardant resin synthesis for adhesive clothing. Under the condition of 0.2 parts addition, the ultimate oxygen index of universal adhesive clothing resin can be increased by 9%. Bhatt et al. [84] reported the preparation of quaternary ammonium ILs with double cations under microwave radiation and conventional heating conditions. They obtained ILs with double cations by reacting organic tertiary amines with dibromoalkanes as raw materials. Conventional heating takes several hours, while under 250 W microwave radiation conditions, it only takes a few tens of minutes. The yields of [2TEA-C]Br_2, [2TEA-C2]Br_2, and [2TEA-C3]Br_2 synthesized by conventional oil bath heating for 6 hours were 81%, 78%, and 76%, respectively, while the yields of [2TEA-C3]Br_2 synthesized by microwave intermittent radiation heating for 30 minutes were 92%, 95%, and 90%, respectively. The yields of conventional synthesis of hexaoctylmethylenediamine dibromide ([TOA-C]Br_2), hexaoctylethylenediamine dibromide ([TOA-C2]Br_2), and hexaoctylpropanediamine dibromide ([TOA-C3]Br_2) after 48 hours was of 85%, 83%, and 80%, respectively, while the yields after 40 minutes of microwave intermittent radiation heating is 96%, 93%, and 91%, respectively. The innovation of this work lies in the use of microwave radiation, which achieves high efficiency and low energy consumption in synthesis of quaternary ammoniumILswith solvent-free methods with double cations.

Li et al. [85] studied the synthesis process of surface active ILs containing silicon under microwave radiation. Various ILs with silicon containing surface active were synthesized using N-alkylimidazole and chloromethyltrimethylsilane under intermittent microwave radiation conditions. Taking chlorinated 1- (trimethylsilyl) -3-octylimidazole ([TMSiC_8im]Cl) as an example, the optimized optimal operating conditions were: 240W, irradiation for 8 minutes, and a molar ratio of chloromethyltrimethylsilane: N-octylimidazole of 1.2:1. The average yield of the four reactions was 89.8%. Conventional heating takes 28 hours and yields 88.1%, indicating that microwave heating greatly improves the reaction rate. Eika et al. [86] synthesized six acidic pyridine ILs and two acidic imidazole ILs under microwave radiation conditions to investigate the most suitable ILs for cellulose saccharification treatment. Taking 1-allylpyridine chloride as an example, heating in an oil bath for 24 hours resulted in a yield of only 48.7%, while microwave output of 150W radiation for 3.5 minutes achieved a yield of 86.2%. The other ILs also achieved high yields within a few minutes. Zhang MM [87] synthesized coordination functionalized ILs using microwave method (320 W/75 °C), reducing the reaction time from the traditional 24-hour to 3-5 hours and achieving a yield of over 90%. This confirms that the viscosity of [N_3mim]Cl significantly increases with the increase of alkyl chains (from 285 cP to 412 cP at 25 °C). Wang B [88] developed a microwave two-step synthesis method for [Bmim]PF_6, and extraction experiments showed that its distribution coefficients for Pb^{2+}/Zn^{2+}/Ni^{2+} reached 184/167/152, which is three times more efficient than traditional heating method.

Wu et al. [89] successfully prepared carboxyl functionalized 1-carboxymethyl-3-methylimidazolium chloride ([Cmmim]Cl) using microwave irradiation (300W, 1h) with a yield of 86%. The reaction efficiency was 23 times higher than that of traditional heating method. Xie et al. [90] developed a microwave synthesis process under nitrogen protection (60 °C, 1 h), which improved the yield of [Bmim]OH intermediate to 94%. Density tests showed that the product density was 1.1713 g/mL, and its strong polarity made it more soluble in polar organic compounds than conventional solvents. Li Y et al. [91] used a stepwise microwave intermittent method (200W pulse mode) to synthesize [Omim]Br with a yield of 96.62%, and then obtained [Omim]PF_6 through anion exchange. The final product purity reached 99.3%, and the surface tension test showed a critical micelle concentration (CMC) of 0.28 mmol/L, indicating its excellent surface activity. Zeng et al. [92] synthesized the [Bmim]PF_6 via microwave method was confirmed to have a purity of 99.3% by HPLC analysis (retention time 4.32 min). This process avoids using organic solvents in traditional methods and provides an environmentally friendly solution for heavy-metal extraction such as cadmium and lead.

Zhang et. al. [93] innovatively synthesized the [Bmim]PF_6 trioctylphosphine oxide (TOPO) composite extraction system using microwave technology. The extraction rates of rare earth elements Nd (III) and Pr (III) by this system at 25 °C reached 97.8% and 94.2%, respectively, with a selectivity coefficient ($\beta_{Nd/Pr}$) of 3.2. Titi et al. [94] designed and synthesized a series of benzyl substituted imidazolium ILs using microwave technology. Among them, compound 4 (1-phenylethyl-3-dodecylimidazolium bromide) exhibits excellent antibacterial activity, with a minimum inhibitory concentration (MIC) as low as 1.95 μg/mL against Staphylococcus aureus. Zuo et al. [95] demonstrated unique advantages in natural product extraction using microwave synthesized double chain ionic liquid 1,3-di(1-butylimidazole)bromide ([Bmbim]Br). Microwave assisted extraction (300 W, 80 °C, 15 mins) was used to extract flavonoid from pine bark, with an extraction rate of up to 41.07 mg/g, which is 62% higher than traditional ethanol extraction. After 5 cycles, the ILs still maintained a recovery rate of 73.14%. The dodecylpyridine bromide ([C12Py]Br) synthesized by the Romero et al. [96] has outstanding performance as a natural-gas hydrate inhibitor. At a concentration of 0.1wt%, the formation temperature of methane hydrate can be reduced by 11.2K, surpassing the commercial inhibitor Inhibitex 101 (ΔT=8.7 K). Molecular dynamics simulations reveal that its inhibitory effect originates from the adsorption of pyridine cations on the surface of crystal nuclei, while long alkyl chains form steric hindrance, jointly hindering the lattice growth of hydrates. Donato et al. [97] synthesized [C6-*x*-pic]TfO series ILs (*x*=2,3,4) using microwave method, and found that [C6-4-pic]TfO as an additive in PEG200 can reduce the friction coefficient of steel silicon friction pairs by 42%. AFM confirmed that it can form an ordered adsorption layer with a thickness of 5-8 nm.Yang et al. [98] developed a heteropolyanionic ILs catalytic system and achieved a three component "one pot" synthesis of

Materials Research Foundations 189 (2026) 237-267 https://doi.org/10.21741/9781644904039-8

dihydroquinazolinone under microwave solvent-free conditions. The reaction time was shortened from the traditional 12 hours to 30 minutes, and the yield remained above 85% after 5 catalyst cycles. Alqurashy et al. [99] synthesized 16 carboxylic acid ester functionalized imidazole ILs using microwave method (600 W), with an inhibition zone of 20 mm against Escherichia coli (better than ampicillin 15 mm), but thermal analysis showed poor thermal stability ($T_{onset} \approx 150$ °C).

Zhao et al. [100] compared the traditional method with the microwave method for synthesizing [Bmim]X (X= BF_4, PF_6, Cl, Br), and confirmed that the microwave method can shorten the reaction time from 24 h to 2 h, increased the yield of [Bmim]PF_6 from 68% to 92%, and the water content was less than 50 ppm. Liu et al. [101] synthesized thiourea based imidazole ILs using microwave-assisted method (800 W) and found that the extraction rate of Zn^{2+} could reach 90.4%. The composite extraction system (90% functional+10% non functional ILs) could further increase the extraction rate by 7%. Wang et al. [102] synthesized [C12mm]BS and [C12mm]NS ILs using microwave method, and measured CMC < 0.06 mmol/L. After compounding with alkali, the interfacial tension between Daqing crude oil and water can be reduced to 0.01 mN/m, with naphthalene sulfonate type showing better performance. Hijji et al. [103] developed a 1-minute microwave synthesis method to prepare ionic liquid of [O2N-Ph-CH2-Py=N(Me)2]Cl. Horikoshi et al. [104] synthesized [Bmim]$FeCl_4$ and [Gmmim]$FeCl_4$ magnetic ILs under microwave irradiation, and obtained magnetic responsive materials (magnetic moment 0.222 A·m²/kg) by composite with cellulose nanofibers, achieving magnetic separation and recovery. Wang H. [105] systematically studied the microwave synthesis of twin imidazole ILs and determined that the optimal conditions for [C12-2-C12mim]Br_2 were 400 W/9 mins (yield 88.49%). It was found that the growth of linking groups could reduce CMC, while the lowest interfacial tension of 2-methylimidazole derivatives was 0.00727mN/m.

Cao et al. [106] developed microwave continuous flow synthesis technology to prepare ultra-pure ILs (such as [Bmim]BF_4) with a water-free rate of over 95%. The production efficiency reached 26.2 g/h and the E factor was as low as 0.8, making it suitable for household microwave equipment. Madgula et al. [107] synthesized [Bmim]Br ionic liquid using microwave method for 3 hours. The benzimidazole derivative prepared by catalysis showed an inhibitory activity of -8.03 kcal/mol against 17-HSD enzyme.The catalyst reused 4 times without deactivation. Dutta et al. [108] synthesized a bifunctional [DDQM]HSO_4 ionic liquid, which was used to synthesize quinazolinone under microwave-assisted conditions in 3.5 minutes. The efficiency was increased by 137 times compared to traditional thermal reactions of 8 hours. Singer et al. [109] improved microwave green synthesis technology to achieve efficient preparation of m-s-m type Gemini surfactants. For the first time, their layered stacking structure (inter-layer spacing of 20-35 Å) was analyzed by X-ray crystallography. Santos et al. [110] synthesized pyrimidine liquid crystal compounds (such as [C12Pyr]Br) using microwave-assisted synthesis, the reaction time was reduced from 48 h to 35 minutes. DSC confirmed that the temperature range for the near crystalline phase transition reached 80-120 °C. Mah et al. [111] compared the traditional (5-day) and microwave (3-hour) synthesis of pyridine functionalized imidazole salts and found that compound 4e had a MIC as low as 3.91 μg/mL against Gram positive bacteria, and compound 3f had an IC50 of 0.71 μM against colon cancer cells, both of which were superior to clinical control drugs.

Thi et al. [112] synthesized various sulfonic acid functionalized ionic liquid (SAIL), which achieved an efficient synthesis of ethyl oleate (conversion rate>90%) using microwave-assisted catalysis. The best catalytic activity for lipophilic SAILs under microwave conditions was founded and could be reused 5 times without deactivation. Titi et al. [113] synthesized fluorinated imidazolium ILs using microwave radiation and found that ILs containing CF_3 groups had the best drug similarity (QED > 0.6) through molecular docking technology. Their ADMET prediction showed low toxicity risk (LD50 > 500 mg/kg). Huang et al. [114] synthesized imidazole ILs such as $[Bmim]PF_6$ using microwave method (1 min/83.48% yield), and confirmed that when used as a cationic polymerization solvent, it can achieve a molecular weight of 3.4×10^4 for isobutyl vinyl ether polymerization, with a reaction efficiency 27% higher than traditional solvents. Shi et al. [115] developed microwave-assisted synthesis of dicycloimidazole ILs (400 W, 9 mins) with a yield of 88.49%. Surface tension testing showed a CMC value of 0.28 mmol/L, demonstrating its excellent surface activity.

Khalili et al. [116] used thermal conditions/microwave radiation to selectively synthesize thiadiazole chromenone compounds through a cup-shaped aromatic ionic liquid catalytic system (Calix-AIL-MN). The catalyst was found to have high recovery efficiency and solvent-free green synthesis characteristics, with a yield of 92%. Fernando et al. [117] developed a microwave-assisted synthesis method to prepare 1-hexylpyridine acetate ionic liquid (1-HP-AC) within 30 minutes. They found that the critical ratio of its interaction with $RuCl_3$ was 1:2, and confirmed through MD simulation that the ionic liquid remained stable at 200 °C. Haddad et al. [118] designed a series of diimidazolium ILs (DILs) using microwave synthesis and anion exchange method. They found that DILs containing Tf_2N^- anions had a thermal stability of 500 °C, and the effect of electron delocalization on physico-chemical properties was revealed through DFT calculations. Yap et al. [119] compared the preparation of deep eutectic solvents (DES) and found that the microwave method reduced energy consumption by 92.8% (only 5 minutes), the DPPH radical scavenging rate exceeded 80%, and the polysaccharide extraction rate was three times higher than traditional methods. Ratnani et al. [120] synthesized methyl imidazole ILs under microwave irradiation, and electro-chemical tests showed that [BzMIm]Br had an inhibition rate of corrosion with 95.49% on low carbon steel in 0.5M H_2SO_4.

5.2 Unresolved Challenges

Although important progress has been made in microwave-assisted synthesis and application of ILs, the application of microwave technology in ILs is limited to laboratory scale and still has a certain gap from industrial application. The main problems as follows.

(1) insufficient research on the reaction mechanism. The reaction mechanism of microwave-assisted synthesis of ILs has not been fully elucidated, especially the mechanism in molecular level such as the interaction between microwave fields and ionic liquid precursors lacks systematic research. Existing literature mostly stays at the level of phenomenon description, with insufficient quantitative analysis of non thermal effects of the microwave. The influence of microwave on the transition state energy barrier of reactions lacks experimental data support, leading to problems such as decreased reaction selectivity and increased by-products during process amplification. For example, in the synthesis of $[C_nmim]Cl$ salts, the stability of intermediates can vary by up to 30% under different power densities, but there is a lack of theoretical models to explain this phenomenon. The quantity structure-property (activity) relationship is unclear, and the relationship between the stable configuration of cations such as $[C3(OH)_2im]^+$ and microwave parameters had not yet been established. The limitations of in-situ characterization techniques make it difficult to monitor molecular conformational changes in microwave fields in real-time using existing methods.

(2) Insufficient precision in process control. The existing microwave synthesis systems commonly suffer from uneven temperature field distribution and difficulty in controlling hotspots. Research has shown that during the synthesis of [Bmim]Cl, the temperature difference between the edge and center of the reactor can reach over 25 °C, resulting in a fluctuation range of 5-8% in product purity between batches. Although an intermittent radiation mode can alleviate local overheating, it will prolong the reaction time (such as the 390 second intermittent radiation required for [C2(OH)Py]Cl synthesis), weakening the efficiency advantage of microwave technology. Unstable power control, such as power fluctuations of ± 15% in household microwave retrofit systems; In addition, due to the lack of online monitoring methods, about 67% of industrial testing projects still rely on offline sampling analysis.

(3) The separation and purification of the product are difficult. Although microwave synthesis shortens the reaction time, the product separation step still takes 60-70% of the total process time. The main problems include incomplete removal of catalysts, and during the purification process of functionalized ILs (such as phosphine oxide modified types), the residual rate of catalysts in functionalized ILs is greater than 15%, which affects the subsequent application performance. Taking [TENC3P(O)Ph(OEt)]Tf_2N as an example, the rare-earth extraction efficiency is reduced by 15-20% due to the presence of impurities. In addition, the by-products generated by solvent-free reactions (such as hydrogen halides) are prone to form azeotropes with the products, making it difficult to separate the by-products and increasing separation energy consumption. Purification energy consumption is high, with some processes accounting for 75% of the total energy consumption.

(4) Poor compatibility between equipment and process. The commonly used improved equipment in ionic liquid synthesis is household microwave ovens. Therefore, there is power instability (fluctuation ± 15%) in household microwave modification systems. The existing equipment has a single electromagnetic field mode, and most devices only support a fixed frequency of 2.45 GHz. The material tolerance of microwave ovens is insufficient, and their tolerance to corrosive anions such as PF_6^- is poor, which can easily cause equipment corrosion and product contamination. It is usually difficult to meet the specific requirements for the different kind of ILs synthesis. In industrial-grade equipment, the mass transfer efficiency of continuous flow reactors is insufficient. For example, the continuous synthesis capacity of [Omim]Cl is limited to 460g/h, which is only 60% of the theoretical value. There is a difference of about 10-15% between laboratory scale yield and industrial production.

Conclusion and Future Directions

From the previous review, we found that in the ILs synthesis, microwave-assisted synthesis has significant advantages compared to traditional heating methods: the reaction efficiency is significantly improved. For example, when synthesizing 1-methyl-3-hydroxyethylimidazole chloride, the reaction time is reduced from 24 h to 110 seconds. The product yield has improved, and the yield of some products is increased by 5-6%. Energy consumption has been significantly reduced, with solvent-free reactions reducing energy consumption by over 80%. Environmentally friendly, reducing the organic solvents' usage and lowering the discharge of three wastes. The purity of the product is improved, and microwave selective heating reduces the occurrence of side reactions. In preparing process of ILs, the use of microwave radiation can significantly improve the reaction rate and synthesize more target ILs in a shorter time. Application of microwave radiation can promote reaction progress, improve yield, reduce solvent usage, and almost all can be regenerated for reuse. The main improvement directions for the existing problems in the future are given as follows.

(1) Multi scale study of synthesis processes mechanism. The establishment of a three-level research system of "molecule mesoscopic device" is an important research direction for revealing the mechanism of microwave enhancement. Using molecular simulations such as density functional theory to calculate the microscopic mechanisms of ion pair conformational changes, interactions, and reactions in microwave fields. Develop or adopt in-situ characterization methods and techniques, such as developing microwave-Raman coupled systems, to monitor key intermediates in real-time and provided a basis for elucidating the microscopic mechanisms of reactions. Adopting multi-scale simulation for process simulation, such as constructing an electromagnetic field reaction dynamics coupling model to further optimize the strategy of power modulation.

(2) Develop new technologies for precise temperature control. Precise temperature control is an important way to control the refinement of chemical reactions and is also a research direction for future efforts. ILs have a strong ability to absorb microwaves, and continuous microwave output may cause local overheating of the ionic liquid, leading to uneven mass transfer and reactions in various parts of the system. Therefore, further research is needed on how to achieve uniform heating of the system and evenly disperse microwaves inside the system. Accurate temperature control can avoid overheating of the system, achieving fast, precise, and efficient synthesis reactions. Precision temperature control technology includes multi-mode resonant cavity design, intelligent temperature control system, pulse radiation optimization, etc. Using a tunable antenna array to achieve uniform distribution of three-dimensional fields (temperature difference controlled within ± 3 °C). Integrated fiber optic temperature sensor (with a resolution of 250 °C and 0.1 °C) and efficient adaptive PID algorithm (response time < 0.5 s) and other intelligent temperature control systems. Develop a millisecond-level fast switching technology to shorten response time to below 80 s.

(3) Research on new processes and technologies for separation and purification. Develop online separation devices, such as using molecular sieve membranes for dehydration (reducing water content to less than 50 ppm). Adopting an electro-chemical impurity removal system (with a metal ion removal efficiency greater than 99%). From the perspective of green chemistry, developing new green purification technologies such as supercritical CO_2 extraction (suitable for Tf_2N^- class hydrophobic ILs), ion-exchange chromatography (purity > 99.5%), etc.Simultaneously realizing the utilization of by-products, such as converting the generated HCl into Cl_2 through membrane electrolysis for reuse, thus achieving closed-loop production.

(4) Optimization and research and development of specialized microwave equipment. Modular reactor design, developing a reaction system with adjustable frequency (2.45-5.8 GHz) and power density gradient control (0-20 W/cm^3). Providing industrial-grade solutions, develop ceramic lined reaction vessels (resistant to HF corrosion), optimize continuous flow system design, and improve mass transfer efficiency. The application of digital twin technology (real-time simulation of reaction processes) and machine learning algorithms (yield prediction accuracy > 95%) to build intelligent control systems is also a focus of research and improvement. In addition, the use of cheap raw material substitution, reduction of synthesis, microwave heat pump coupled energy integration system to reduce energy consumption, development of micro-channel reactors to

miniaturize equipment and significantly reduce reagent usage, formulation of microwave synthesis process specifications for different types of ILs, microwave specific impurity detection methods, and product performance evaluation systems, as well as quality, environmental, and safety standards, are also directions that need attention. Through the above improvements, it is expected that the comprehensive cost of ILs microwave-assisted synthesis can be reduced by 35%, expanding the industrial application to high-end fields. Especially in the field of functionalized ILs, the precise control advantage of microwave technology will promote the industrialization of special ILs such as temperature sensitive and chiral recognition types.

In summary, microwave-assisted synthesis technology provides a revolutionary solution for the large-scale and green production of ILs. In the future, through reactor innovation and interdisciplinary integration, ILs are expected to achieve abroad applications in energy, materials, catalysis, electro-chemistry, extraction and separation fields, etc.

Acknowledgement

This work is partially supported by Yunnan Fundamental Research Projects (NO. 202501AS070127) and supported by the National Natural Science Foundation of China (No.52364042).

References

[1] A.G. Saskia, Ionic liquids in catalytic systems, Chem. Soc. Rev. 26 (1997) 233-238 https://doi.org/10.1039/cs9972600233

[2] T. Zhou, C. Gui, L. Sun, Y. Hu, H. Lyu, Z.Wang, G. Yu,Energy applications of ionic liquids: recent developments and future prospects. Chem. Rev. 123 (2023) 12170-12253. https://doi.org/10.1021/acs.chemrev.3c00391

[3] K.R.J. Seddon, Applications of ionic liquids in industrial chemistry, Chem. Tech. Biotechnol. 68 (1997) 351-356. https://doi.org/10.1002/(SICI)1097-4660(199704)68:4<351::AID-JCTB613>3.0.CO;2-4

[4] A.S. Larsen, J.D. Holbrey, F.S. Tham, C.A. Reed, Designing ionic liquids: imidazolium melts with inert carborane anions, J. Am. Chem. Soc. 122 (2002) 7264-7272. https://doi.org/10.1021/ja0007511

[5] D. Zhao, M. Wu, Y. Kou, E. Min, Ionic liquids: applications in catalysis, Catal. Today 74 (2002) 157-189. https://doi.org/10.1016/S0920-5861(01)00541-7

[6] R.D. Rogers, G.A. Voth, Ionic liquids, Acc. Chem. Res. 40 (2007) 1077-1078. https://doi.org/10.1021/ar700221n

[7] M. Freemantle, Designer solvents-Ionic liquids may boost clean technology development, Chem. Eng. News 76 (1998) 32-37. https://doi.org/10.1021/cen-v076n013.p032

[8] R.X. Li, Green solvents: synthesis and application of ionic liquids, Chemical Industry Engineering Press, Beijing, 2004, pp. 287-289.

[9] Y.Q. Deng, Ionic liquids: properties, preparation and application, China SINO-PEC Press, Beijing, 2006, pp. 20-45.

[10] S.J. Zhang, X.M. Lv, Ionic liquids from fundamental study to industrial application, Science Press, Beijing, 2006, pp. 35-95.

[11] S.J. Zhang, Nano-micro structure of ionic liquid in process intensification, Chemical Industry Engineering Press, Beijing, 2020, pp. 242-245.

[12] M. Rostiomali, P.A. Andrew, S.R. Karl, Electrodeposition of Al-Mg alloys from acidic AlCl3-EMIC-MgCl2 room temperature ionic liquids, J. Electrochem. 21 (2015) 172-180.

[13] H. Yang, J. Chen, W. Wang, H. Cui, D. Zhang, Y. Liu, Extraction kinetics of lanthanum in chloride medium by bifunctional ionic liquid [A336][CA-12] using a constant interfacial cell with laminar flow, Chin. J. Chem. Eng. 22 (2014) 1174-1177. https://doi.org/10.1016/j.cjche.2014.09.010

[14] R. Li, Ionic liquids-green solvents for industrialization, Mod. Chem. Eng. 23 (2003) 17-21.

[15] Y. Wang, X. Li, Y. Tian, Research progress on microwave synthesis of ionic liquids, Chem. New Mater. 44 (2016) 33-35.

[16] M. Zhang, X. Xu, Z. Zhao, Application of microwave assisted ionic liquid method in liquid phase synthesis, Chem. Bull. (2007) 513-520. In Chinese

[17] Y.J. Zhu, W.W. Wang, R.J. Qi, X.L. Hu, Microwave-assisted synthesis of single-crystalline tellurium nanorods and nanowires, Angew. Chem. Int. Ed. 43 (2004) 1410-1414. https://doi.org/10.1002/anie.200353101

[18] Y. Shen, J.K. Sun, Y.X. Yi, B. Wang, F. Xu, R.C. Sun, One-pot synthesis of levulinic acid from cellulose in ionic liquids, Bioresour. Technol. 192 (2015) 812-816. https://doi.org/10.1016/j.biortech.2015.05.080

[19] A.A. Otaibi, A. McCluskey, Ionic liquids, microwave irradiation, and the synthesis of aryl weinreb amides, Monatsh. Chem. 149 (2018) 519-525. https://doi.org/10.1007/s00706-017-2089-3

[20] S. Sugden, H. Wilkins, CLXVII-The parachor and chemical constitution, Part XII, fused metals and salts, J. Chem. Soc. (1929) 1291-1298. https://doi.org/10.1039/JR9290001291

[21] M. Hirao, H. Sugimoto, H. Ohno, Preparation of novel room-temperature molten salts by neutralization of amines, J. Electrochem. Soc. 147 (2000) 4168-4172. https://doi.org/10.1149/1.1394036

[22] P. Bonhote, A.P. Dias, N. Papageorgiou, K. Kalyanasundaram, M. Grätzel, Hydrophobic, highly conductive ambient-temperature molten salts, Inorg. Chem. 35 (1996) 1168-1178. https://doi.org/10.1021/ic951325x

[23] N. Karodia, S. Guise, C. Newlands, J.A. Andersen, Clean catalysis with ionic solvents-phosphonium tosylates for hydroformylation, Chem. Commun. (1998) 2341-2342. https://doi.org/10.1039/a805376f

[24] P. Wasserscheid, W. Keim, Ionic liquids-new solutions for transition metal catalysis, Angew. Chem. Int. Ed. 39 (2000) 3772-3789. https://doi.org/10.1002/1521-3773(20001103)39:21<3772::AID-ANIE3772>3.0.CO;2-5

[25] D.R. MacFarlane, J. Huang, M. Forsyth, Lithium-doped plastic crystal electrolytes exhibiting fast ion conduction for secondary batteries, Nature 402 (1999) 792-795. https://doi.org/10.1038/45514

[26] P. Wang, S.M. Zakeeruddin, M. Grätzel, W. Kantlehner, J. Mezger, E.V. Stoyanov, O. Scherr, Novel room temperature ionic liquids of hexaalkyl substituted guanidinium salts for dye-sensitized solar cells, Appl. Phys. A 79 (2004) 73-77. https://doi.org/10.1007/s00339-003-2505-x

[27] W. Wu, B. Han, H. Gao, Z. Liu, T. Jiang, J. Huang, Desulfurization of flue gas: SO2 absorption by an ionic liquid, Angew. Chem. Int. Ed. 43 (2004) 2415-2417. https://doi.org/10.1002/anie.200353437

[28] J. Huang, A. Riisager, P. Wasserscheid, R. Fehrmann, Reversible physical absorption of SO2 by ionic liquids, Chem. Commun. (2006) 4027-4029. https://doi.org/10.1039/b609714f

[29] Y.F. Zhang, C.J. Li, F.S. Bao, Y.A. Zhang, Classification, synthesis, and application of ionic liquids, Chin. J. Chem. Edu.. 26 (2005) 7-12. In Chinese.

[30] X. Zhao, H.B. Xing, R.L. Li, Q.W. Yang, B.G. Su, Q.L. Ren Gas separation based on ionic liquids[J]. Prog. Chem. 23(2011) 2258-2268 In Chinese.

[31] S.J. Zhang, X.M.Liu, X.Q. Yao, H.F. Dong, X.P. Zhang, Frontiers, progresses and applications of ionic liquids. Sci China Ser B-Chem. 39 (2009) 1134-1144.

[32] C. Liang, Y. Xie, B. Guan, Concise Microwave, Higher Education Press, Beijing, 2006.

[33] J.M. Osepchuk, A history of microwave heating applications, IEEE Trans. Microw. Theory Tech. 32 (2003) 1200-1224. https://doi.org/10.1109/TMTT.1984.1132831

[34] Q. Jin, S. Dai, K. Huang, Microwave Chemistry, Science Press, Beijing, 1999. In Chinese.

[35] G. Chen, J. Peng, J. Chen, Development and Application of Microwave Heating Equipment, Science Press, Beijing, 2018, pp. 8-9. In Chinese.

[36] H. Huang, Research on Modeling of Microwave Heating Process Based on Model Reduction Method, Kunming University of Science and Technology, 2024. In Chinese.

[37] Z. Han, Research on Temperature Uniformity Optimization of Multi-source Microwave Heating Based on Power Frequency Coordination, Kunming University of Science and Technology, 2024. In Chinese.

[38] Z. Zhao, H. Li, X. Gao, Microwave encounters ionic liquid: Synergistic mechanism, synthesis and emerging applications, Chem. Rev. 124 (2023) 2651-2698. https://doi.org/10.1021/acs.chemrev.3c00794

[39] Z. He, Q. Pang, Y. Jin, et al., Microwave Metallurgy Advanced Technology, Metallurgical Industry Press, Beijing, 2017. In Chinese.

[40] J. Dong, P. Lin, H. Deng, Microwave Technology, Machinery Industry Press, Beijing, 2010. In Chinese.

[41] J. Peng, B. Liu, Microwave Calcination Technology and Application, Science Press, Beijing, 2013. In Chinese.

[42] J. Peng, H. Xia, Microwave Metallurgy, Science Press, Beijing, 2017. In Chinese.

[43] L. Ricciardi, W. Verboom, J.P. Lange, J. Huskens, Local overheating explains the rate enhancement of xylose dehydration under microwave heating, ACS Sustain. Chem. Eng. 7 (2019) 14273-14279. https://doi.org/10.1021/acssuschemeng.9b03580

[44] X. Shen, H. Li, Z. Zhao, X. Li, K. Liu, X. Gao, Imaging of liquid temperature distribution during microwave heating via thermochromic metal organic frameworks, Int. J. Heat Mass Transf. 189 (2022) 122667. https://doi.org/10.1016/j.ijheatmasstransfer.2022.122667

[45] M. Dua, Q. Zhang, P. Mertiny, Microwave heating of graphene nanoplatelet polymer composites: Experimental and finite element study, Polym. Compos. 44 (2023) 4924-4936. https://doi.org/10.1002/pc.27460

[46] M. Said, N.A. Salleh, M.F.M. Nazeri, H. Akbulut, S. Kheawhom, A.A. Mohamad, Microwave hybrid heating for lead-free solder: A review, J. Mater. Res. Technol. 26 (2023) 6220-6243. https://doi.org/10.1016/j.jmrt.2023.08.299

[47] S. Chen, S. Cen, R. Liu, C. Tao, S. Guo, G. Chen, Microwave reactor with combined rigid and flexible stirring paddles for improving fluid heating uniformity in soft manganese ore leaching processes, ACS Omega 8 (2023) 42367-42378. https://doi.org/10.1021/acsomega.3c04679

[48] Y. Xiong, K. Wang, D. Qiu, M. Omran, R. Huang, Y. Li, S. Wei, I.U. Khan, D. Zhang, A. Ahmed, Y. Yu, Recent developments on the removal of zinc from electric arc furnace dust by using microwave heating, Environ. Sci. Pollut. Res. 31 (2024) 17481-17493. https://doi.org/10.1007/s11356-024-32346-6

[49] L. Xue, R. Gao, L. Shen, X. Zheng, M. Gao, Dependence of degradation of anthocyanins on non-uniformity of microwave heating in blueberry puree, Food Bioprod. Process. 139 (2023) 129-143. https://doi.org/10.1016/j.fbp.2023.03.009

[50] T. Sofizadeh, J. Khodaei, H. Darvishi, N. Behroozi-Khazaei, M. Koushesh Saba, Process parameters of microwave heating-assisted vacuum evaporation of tomato juice: quality, energy consumption, exergy performance, and kinetic processing, J Microwave Power EE 57 (2023) 203-218. In Chinese. https://doi.org/10.1080/08327823.2023.2235551

[51] R.S. Varma, V.V. Namboodiri, An expeditious solvent-free route to ionic liquids using microwaves, Chem. Commun. (2001) 643-644. https://doi.org/10.1039/b101375k

[52] B.M. Khadilkar, G.L. Rebeiro, Microwave-assisted synthesis of room-temperature ionic liquid precursor, Org. Process Res. Dev. 6 (2002) 826-828. https://doi.org/10.1021/op025551j

[53] V.V. Namboodiri, R.S. Varma, An improved preparation of 1,3-dialkylimidazolium tetrafluoroborate ionic liquids, Tetrahedron Lett. 43 (2002) 5381-5383. https://doi.org/10.1016/S0040-4039(02)01075-4

[54] V.V. Namboodiri, R.S. Varma, Microwave-assisted preparation of dialkylimidazolium tetrachloroaluminates, Chem. Commun. (2002) 342-343. https://doi.org/10.1039/b110565e

[55] M. Deetlefs, K.R. Seddon, Improved preparations of ionic liquids using microwave irradiation, Green Chem. 5 (2003) 181-186. https://doi.org/10.1039/b300071k

[56] G.V. Thanh, B. Pegot, A. Loupy, Solvent-free microwave-assisted preparation of chiral ionic liquids, Eur. J. Org. Chem. (2004) 1112-1116. https://doi.org/10.1002/ejoc.200300601

[57] V. Singh, S. Kaur, V. Sapehiyia, J. Singh, G.L. Kad, Microwave accelerated preparation of [bmim][HSO4] ionic liquid, Catal. Commun. 6 (2005) 57-60. https://doi.org/10.1016/j.catcom.2004.10.011

[58] K. Bica, G. Gmeiner, C. Reichel, B. Lendl, P. Gaertner, Microwave-assisted synthesis of camphor-derived chiral ionic liquids, Synthesis (2007) 1333-1338. https://doi.org/10.1055/s-2007-966018

[59] W. Li, H. Rong, X. Wu, Microwave synthesis of p-toluenesulfonate and its esters with threonine, Acta Phys. -Chim. Sin. 24 (2008) 868-872. In Chinese. https://doi.org/10.3866/PKU.WHXB20080523

[60] T. Erdenger, R.M. Paulus, R. Hoogenboom, U.S. Schubert, Scaling-up the synthesis of 1-butyl-3-methylimidazolium chloride under microwave irradiation, Aust. J. Chem. 61 (2008) 197-203. https://doi.org/10.1071/CH07345

[61] G. Cravotto, E..C Gaudino, L. Boffa, J.M. Lévêque, J. Estager, W. Bonrath, Preparation of second generation ionic liquids by efficient solvent-free alkylation, Molecules 13 (2008) 149-156. https://doi.org/10.3390/molecules13010149

[62] S. Horikoshi, T. Hamamura, M. Kajitani, M. Yoshizawa-Fujita, N. Serpone, Green chemistry with a novel 5.8 GHz microwave apparatus, Org. Process Res. Dev. 12 (2008) 1089-1093. https://doi.org/10.1021/op800135t

[63] J. Zheng, Q.S. Chen,T. Yang, C.Y. Liu, Microwave assisted synthesis of ionic liquid 1-hydroxyethyl-3-methylimidazolium tetrafluoroborate, Chem. New Mater. 38 (2010) 37-39. In Chinese

[64] M. Yang, G. Chen, One step synthesis of 1-butyl-3-methylimidazolium hexafluorophosphate ionic liquid under microwave radiation, Appl. Chem. Eng. 39 (2010) 1177-1179.

[65] M. Huang, E.G. Duan, X.T. Xu, R.S. Wei, Microwave-assisted synthesis of functional ionic liquid [bmim] [dca] and its solubility for some carbohydrates, Chem. Res. Appl. 22 (2010) 1061-1065. In Chinese.

[66] D.S. Zhao, X.L. Bao, J.J. Shan, T. Wu, R.W. Wang, Synthesis of coordinated ionic liquids under microwave irradiation, J. Hebei Univ. Sci. Technol. 31 (2010) 128-131. In Chinese.

[67] M. Xu, Y.F. Zhang, D.X. Zhu, J. Zhang, H.F. Lu, A research on microwave-assisted synthesis of[BMIm]Brand industrialized scheme, J. Northeast Norm. Univ. (Nat. Sci. Ed.) 42 (2010) 81-85. In Chinese.

[68] J. Chen, X. Ma, Y. Dang, Z. Gao, Microwave-assisted synthesis and characterization of 1,3-di(n-butyl)imidazolium hexafluorophosphate ionic liquid, Funct. Mater. 42(S1) (2011) 34-35+46. In Chinese.

[69] Y. Wu, W. Guan, Y. Li, et al., Microwave synthesis and characterization of ionic liquid [Bmim]PF6 and preparation of TiO2 photocatalyst using [Bmim]PF6 as a medium, Chem. New Mater. 39(3) (2011) 82-85.

[70] L. Zhang, X. Yu,D. Zhao, Y. Dong, X. Wang, D. Zhang, Synthesis of [EMIM]Br-AlCl3 ionic liquid, CIESC J.. 62 (2011) 173-178. In Chinese.

[71] M. Messali, S.A. Ahmed, A green microwave-assisted synthesis of new pyridazinium-based ionic liquids as an environmentally friendly alternative, Green Sustain. Chem. 1 (2011) 70-75. https://doi.org/10.4236/gsc.2011.13012

[72] L. Chen, L. Chen, Q. Li, M. Zhang, C. Li, Microwave synthesis of three kinds of nitrogenous ionic liquids, J. Guizhou Norm. Univ. (Nat. Sci. Ed.) 30(2012) 83-85.

[73] W. Zhang, J. Ma, Y. Wang, Y. Hu, X. Song, Study on the microwave-assisted synthesis of hydroxyl functionalized ionic liquids, Chem. Bull. 75 (2012) 357-360.In Chinese.

[74] Q. Huang, B. Zheng, Facile synthesis of benzaldehyde-functionalized ionic liquids and their flexible functional group transformations, Org. Chem. Int. 2012 (2012) 1-5. https://doi.org/10.1155/2012/208128

[75] X.G. Du, J.P. Du, J. Zhang, X.M. Guo, W.W. Lu, Z.Y. Yang, Synthesis of ionic liquids [BMIM]BF4 and [BMIM]PF6 under microwave irradiation by one-pot, Adv. Mater. Res. 496 (2012) 84-87. https://doi.org/10.4028/www.scientific.net/AMR.496.84

[76] G. Gong, W. Li, B. Xiao,Microwave solvent-free synthesis of ionic liquid [BMIM]Cl, Chem. Adhes. 35(4) (2013) 41-42+50.In Chinese.

[77] M. Messali, Z Moussa, A.Y. Alzahrani, M.Y. El-Naggar, A.S. ElDouhaibi, Z.M.A. Judeh Synthesis, characterization and the antimicrobial activity of new eco-friendly ionic liquids, Chemosphere 91(2013) 1627-1634. https://doi.org/10.1016/j.chemosphere.2012.12.062

[78] L. Xu, M. Yang, D. Wang, Q. Peng, N. Liu, S. Cao,Synthesis of a intermediate for the immobilization of imidazolium ionic liquids: 1-methyl-3-(3-triethoxysilylpropyl) imidazolium chloride using microwave irradiation, Chem. New Mater. 42(2014) 132-134. In Chinese.

[79] T. Zhu, M. Guo, S. Wang, J. Jiang, K. Hu, L. Zhao, H. Liang, Microwave-assisted Synthesis of Phosphine-Functionalized Ionic Liquids and Investigation on the Extraction of Rare Earth, Appl. Chem. 31 (2014) 529-535. In Chinese. https://doi.org/10.3724/SP.J.1095.2014.30371

[80] K.R. Roshan, T. Jose, D. Kim, K.A. Cherian, D.W. Park, Microwave-assisted one pot-synthesis of amino acid ionic liquids in water: simple catalysts for styrene carbonate synthesis under atmospheric pressure of CO2, Catal. Sci. Technol. 4 (2014) 963-970. https://doi.org/10.1039/C3CY00769C

[81] C. Lv, Y. Sun, W. Ding, Optimization of microwave radiation synthesis of long-chain imidazole ionic liquids using response surface analysis, Adv. Fine Petrochem. 16(2) (2015) 55-57. In Chinese.

[82] Z. Wang, Microwave-assisted alkaline ionic liquid catalyzed synthesis of chalcone, Synth. Chem. 23 (2015) 202-205.

[83] L. Su, C. Jin, S. Zhang, L. Gu, Microwave-assisted synthesis, characterization and application of 1-benzyl-3-alkylbenzimidazolium ionic liquids, Chemical Industry Times 29(1) (2015) 6-10.In Chinese.

[84] D. Bhatt, K.C. Maheria, J. Parikh, A microwave-assisted one-pot synthesis of novel ammonium based dicationic ionic liquids, RSC Adv. 5(16) (2015) 12139-12143. https://doi.org/10.1039/C4RA16251J

[85] E.W. Qian, H. Tominaga, T.L. Chen, R. Isoe, Synthesis of functional ionic liquids and their application for the direct saccharification of cellulose, J. Chem. Eng. Jpn. 49(2016) 466-474. https://doi.org/10.1252/jcej.15we060

[86] J. Li, Synthesis and properties of silicon-containing surface active ionic liquids, Hebei University of Science and Technology, Shijiazhuang, 2016. In Chinese.

[87] M. Zhang, Synthesis of coordination functional ionic liquids and their application in ATRP Reaction, Hebei University of Science and Technology, Shijiazhuang, 2016. In Chinese.

[88] B. Wang, Preparation of Ionic Liquids and Extraction of Metals from Electroplating Sludge, Nanchang University, Nanchang, 2016.In Chinese.

[89] Z. Wu, Research on the synthesis of polymerizable ionic liquids, Yunnan University, Kunming, 2017.In Chinese.

[90] Y. Xie, X. Li, L. Cao, L. Geng, J. Cheng, Q. Zhao, Study on the microwave-assisted synthesis and mechanism of a hydroxide 1-butyl-3-methyl-imidazole, Shandong Chem. Ind. 46 (2017) 1-4. In Chinese.

[91] Y. Li, Synthesis and application of [Omim]PF6 ionic liquid, Nanjing University of Science and Technology, Nanjing, 2018. In Chinese.

[92] X. Zeng, K. Yang, Microwave synthesis and characterization of [BMIM][PF6] ionic liquids, Jiangxi Chem. Ind. (04) (2019) 89-92. In Chinese.

[93] Z. Zhang, Synthesis of Functionalized Ionic Liquids and Extraction of Neodymium and Praseodymium from Rare Earth Elements, Nanchang University, Nanchang, 2020. In Chinese.

[94] A. Titi, S.M. Almutairi, I. Ali, Novel phenethylimidazolium based ionic liquids: Design, microwave synthesis, in-silico modeling and biological evaluation studies, J. Mol. Liq. 315 (2020) 113761. https://doi.org/10.1016/j.molliq.2020.113778

[95] L. Zuo, X. Ao, Y. Guo, Study on the synthesis of dual-chain ionic liquids and their application in the extraction of flavonoids, J. Chromatogr. A 1628 (2020) 461486. https://doi.org/10.1016/j.chroma.2020.461446

[96] A. Romero-Martínez, E. Hernández-Guerrero, R. Martínez-Palou, Synthesis and evaluation of a pyridinium-based ionic liquid-type surfactant as a new low-dosage methane hydrate inhibitor, Energy Fuels 34(8) (2020) 9243-9251. https://doi.org/10.1021/acs.energyfuels.0c00426

[97] M.T. Donato, F. Caetano, B. Saramago, Picolinium-based hydrophobic ionic liquids as additives to PEG200 to lubricate steel-silicon contacts, ChemistrySelect 5(2020) 5864-5872. https://doi.org/10.1002/slct.202000613

[98] Y. Yang, R. Fu, X. Zeng, Microwave-promoted one-pot three-component synthesis of 2,3-dihydroquinazolin-4(1H)-ones catalyzed by heteropolyanion-based ionic liquids under solvent-free conditions, Tetrahedron 76 (2020) 131284. https://doi.org/10.1016/j.tet.2020.131312

[99] B.A. Alqurashy, An efficient preparation and characterization of 1-(2-ethoxy-2-oxoethyl)imidazolium based ionic liquids derivatives as potential bioactive agents, J. Saudi Chem. Soc. 24 (2020) 631-635. https://doi.org/10.1016/j.jscs.2020.06.005

[100] L. Zhao, Z. Gong, Research on the synthesis and application of imidazole ionic liquids, Tianjin Chem. Ind. 35 (2021) 48-50.

[101] Y. Liu, W. Shi, H. Zhan, Synthesis of thiourea imidazole ionic liquid and its study on the extraction mechanism of Zn2+, Mod. Chem. Eng. 41 (2021) 114-119.

[102] H. Wang, B. Long, Y. Ma, B. Li, R. Niu, Microwave-assisted synthesis and characterization of 1-alkyl-3-methylimidazole-aromatic sulfonate, Chemical Engineer 35 (2021) 6-8+81. In Chinese.

[103] Y. Hijji, R. Rajan, I. Warad, One minute microwave synthesis of [O2N-Ph-CH2-Py=N(Me)2]+[Cl]- ionic liquid: XRD/HSA-interactions, physicochemical, optical, thermal and a DFT/TD-DFT analysis, J. Mol. Liq. 339 (2021) 116794. https://doi.org/10.1016/j.molliq.2021.116737

[104] S. Horikoshi, S. Yamazaki, N. Serpone, Synthesis of recyclable magnetic cellulose nanofibers from ionic liquids for practical applications in separation science, J. Oleo Sci. 70 (2021) 737-743. https://doi.org/10.5650/jos.ess21087

[105] H. Wang, Synthesis and properties of series gemini imidazole ionic liquid surfactants, Northeast Petroleum University, Daqing, 2021. In Chinese.

[106] L. Cao, H.W. Kim, J.K. Park, Rapid continuous-flow water-free synthesis of ultrapure ionic liquids assisted by microwaves, Org. Process Res. Dev. 26(2022) 207-214. https://doi.org/10.1021/acs.oprd.1c00412

[107] K. Madgula, S. Dandu, P. Halady, Microwave synthesized ionic liquids as green catalysts for the synthesis of benzimidazoles: Spectral and computational studies for potential anticancer activity, Inorg. Chem. Commun. 138 (2022) 109273. https://doi.org/10.1016/j.inoche.2022.109218

[108] B. Dutta, N. Dutta, D. Sarma, [DDQM][HSO4]/TBHP as a multifunctional catalyst for the metal free tandem oxidative synthesis of 2-phenylquinazolin-4(3H)-ones, J. Org. Chem. 87 (2022) 10276-10287. https://doi.org/10.1021/acs.joc.2c01625

[109] O.M. Singer, J.W. Campbell, R.D. Singer, Improved green synthesis and crystal structures of symmetrical cationic gemini surfactants, ACS Omega 7 (2022) 34048-34057. https://doi.org/10.1021/acsomega.2c05073

[110] A.F.M. Santos, C. Cruz, L.C. Branco, Synthesis and characterisation of ionic liquid crystals based on substituted pyridinium cations, Liq. Cryst. 49 (2022) 1809-1821. https://doi.org/10.1080/02678292.2022.2070783

[111] W.L. Mah, X.J. Tan, Y.L. Cheow, Microwave-assisted synthesis of bioactive pyridine-functionalized N-alkyl-substituted (benz)imidazolium salts, ChemistrySelect 7 (2022) e202203753. https://doi.org/10.1002/slct.202203864

[112] T.H.N. Thi, J. Koutecká, J. Cermak, Design and synthesis of new sulfonic acid functionalized ionic liquids as catalysts for esterification of fatty acids with bioethanol, Molecules 28 (2023) 5123.3 https://doi.org/10.3390/molecules28135231

[113] A. Titi, S.M. Almutairi, R. Touzani, New fluorinated imidazolium ionic liquids: Ecofriendly synthesis, structural, in silico analysis and identification of antitumor pharmacophore sites, Inorg. Chim. Acta 556 (2023) 121643. https://doi.org/10.1016/j.ica.2023.121619

[114] K. Huang, W. Zhang, T. Liu, et al., Synthesis of imidazole ionic liquids and their application in cationic polymerization, Chem. New Mater. 51(11) (2023) 263-266. In Chinese.

[115] P. Shi, D. Peng, B. Bu, et al., Synthesis and properties of series gemini imidazole ionic liquid surfactants, Shanxi Chem. Ind. 43(6) (2023) 1-3. In Chinese.

[116] A. Khalili, I. Mohammadpoor-Baltork, H. Sabzyan, Selective synthesis of thiadiazolo-2H-chromen-2-ones and chromeno-thiadiazolo-pyrimidines under thermal conditions and microwave irradiation catalyzed by calixarene-based ionic liquid immobilized on magnetic nanoparticles, J. Mol. Liq. 397 (2024) 124015. https://doi.org/10.1016/j.molliq.2024.124118

[117] P.U.A.I. Fernando, G.K. Kosgei, A.N. Kimble, Efficient microwave-assisted synthesis of 1-hexylpyridin-1-ium bromide: Anion exchange to acetate and fundamental tools to probe its interaction with RuCl3, J. Mol. Liq. 397 (2024) 124016. https://doi.org/10.1016/j.molliq.2024.124080

[118] B. Haddad, M. Boumediene, S. Bresson, Synthesis, vibrational and thermal studies of new 3,3′-dibutyl-1,1′-(1,4-phenylenedimethylene)-bis(1H-imidazolium) ionic liquids: An experimental and quantum computational investigation, J. Mol. Struct. 1300 (2024) 137274. https://doi.org/10.1016/j.molstruc.2023.137325

[119] P.G. Yap, Y. Wang, C.Y. Gan, Microwave-assisted preparation of an acidic deep eutectic solvent: Green analysis, characterization and application, Chemistry Select 9 (2024) e202400438. https://doi.org/10.1002/slct.202400836

[120] S. Ratnani, S.K. Sharma, S. Gurjar, Evaluation of inhibition performance of methylimidazolium ionic liquids on the surface of mild steel in sulfuric acid, Surf. Coat. Technol. 485 (2024) 130843. https://doi.org/10.1016/j.surfcoat.2024.130875

Materials Research Foundations 189 (2026) 268-309 https://doi.org/10.21741/9781644904039-9

Chapter 9

Microwave-Assisted Synthesis of Radiopharmaceuticals

Riya Middha[1], Noopur Srivastava[2], and Mridula Guin[2*]

[1]Independent Author, Vaishali Colony, Rudrapur, Uttarakhand, 263153, India

[2]Department of Chemistry and Biochemistry, Sharda University, Greater Noida, India

mridula.guin@sharda.ac.in

Abstract

Radiopharmaceuticals, combining biologically active molecules with radionuclides, are critical tools in diagnostic imaging and targeted therapy. However, the synthesis of these compounds is challenged by the extremely short half-lives of key radionuclides such as Carbon-11 and Fluorine-18, necessitating ultra-rapid, high-yielding, and reproducible synthetic methods. Conventional heating techniques often result in protracted reaction times, low yields, and product degradation. Microwave-assisted synthesis (MAS) has emerged as a powerful alternative, offering rapid volumetric heating through dipolar polarization and ionic conduction. This accelerates reaction kinetics, enhances radiochemical yields and purity, and minimizes thermal decomposition. MAS enables better reproducibility, automation compatibility, and high-throughput workflows—key advantages in radiopharmaceutical production. This chapter explores the mechanistic foundations, technological innovations, regulatory considerations, and clinical applications of MAS in radiochemistry. It demonstrates how MAS is reshaping radiopharmaceutical development, offering a scalable and efficient approach aligned with the demands of modern nuclear medicine.

Keywords

Radiopharmaceuticals, Microwave-Assisted Synthesis, Radiolabeling, Positron Emission Tomography (PET), Single-Photon Emission Computed Tomography (SPECT), Radionuclides, Green Chemistry, [^{18}F]Fluoride, [^{11}C]Carbon, [^{99m}Tc]Technetium, [^{68}Ga]Gallium, Dipolar Polarization

Contents

Materials Research Foundations 189 (2026) 268-309 https://doi.org/10.21741/9781644904039-9

1. Introduction

The landscape of modern medicine is continually being reshaped by advancements in precision diagnostics and targeted therapeutics [1]. This chapter presents a comprehensive introduction to radiopharmaceuticals and examines the scientific and technical challenges associated with their synthesis, particularly those arising from the short half-lives of radionuclides. Emphasis is placed on the evolution of synthetic methodologies and the limitations of conventional approaches [2]. The chapter underscores the emergence of Microwave-Assisted Synthesis (MAS) as a transformative solution, offering significant improvements in reaction speed, efficiency, and reproducibility [3]. By integrating advanced microwave technology, MAS marks a paradigm shift in radiopharmaceutical production, enabling more rapid and reliable development of clinically valuable compounds. The overarching objective is to provide a comprehensive and insightful resource on the transformative potential of microwave technology in the field of radiopharmaceutical chemistry [4,5].

1.1 Defining Radiopharmaceuticals: Diagnostic and Therapeutic Applications

Radiopharmaceuticals are meticulously engineered bioactive molecules that incorporate radioatoms, serving as powerful tools that provide non-invasive, precise information about molecular processes, organ function, and disease states within the human body [6]. Their significance stems from their dual utility in both diagnostic imaging and targeted therapy [7].

Diagnostic radiopharmaceuticals [8], used in PET and SPECT imaging [9], enable early detection, accurate disease staging, and treatment monitoring by targeting specific tissues. Examples include [^{18}F]FDG for metabolic imaging in oncology [10] and [^{99m}Tc]-sestamibi for myocardial perfusion [11]. Their expanding role includes applications in neurology, cardiology, and inflammation.

Therapeutic radiopharmaceuticals [8], or radiotheranosticscombine diagnostic precision with targeted therapy, delivering radioactive payloads to diseased cells while sparing healthy tissue. Agents like [^{177}Lu]Lu-DOTA-TATE [12] and [^{223}Ra]$RaCl_2$ (Xofigo) [13] exemplify this targeted approach, especially in cancer treatment. Innovations in radiolabeled antibodies and peptides are further advancing treatment for solid and neuroendocrine tumors [14].

1.2 Unique Constraints of Radiopharmaceutical Chemistry

Radiopharmaceutical synthesis is uniquely constrained by the short half-lives of radionuclides like carbon-11 (~20 min) and fluorine-18 (~110 min) [2], requiring rapid synthesis, purification, and quality control to preserve radiochemical yield and specific activity. Radionuclide incorporation is ideally performed at the final synthetic step or penultimate step. Additionally, radiation safety mandates specialized facilities, strict protocols, and trained professionals such as Radiation Safety Officers (RSOs) and Authorized Nuclear Pharmacists (ANPs) [15]. The synthesis must retain the compound's biological integrity while managing radioactive risk. Economic challenges also persist due to high development costs, short shelf-lives, and limited market demand, making large-scale production and commercialization particularly difficult [16].

1.3 Historical Overview of Radiopharmaceutical Synthesis Techniques

The origins of radiopharmaceuticals trace back to the 1920s–1930s, with early diagnostic applications using radioisotopes to assess blood flow and thyroid function, notably with radio-iodine [17]. The field advanced significantly after World War II, driven by the availability of artificially produced radionuclides via nuclear reactors [18]. Traditionally, organic and radiopharmaceutical syntheses relied on inefficient conventional heating methods like oil baths and Bunsen burners [5-8], which transferred heat slowly through vessel walls. The 1980s saw the emergence of microwave applications in radiochemistry, initially in gas-phase reactions [19]. A pivotal moment came in 1986 when Gedye, Giguere, and Majetich demonstrated the acceleration of liquid-phase organic reactions using microwave energy [4,6,7,11,12]. Early experiments using domestic microwaves raised safety and reproducibility concerns due to limited control [6,7,11,20,21]. This led to the development of dedicated laboratory microwave reactors in the 1990s [11,17,20], incorporating fiber-optic temperature sensors, magnetic stirrers, pressure monitoring, and automated software [8,11]. These systems established the foundation for precise, reproducible, and high-throughput microwave-assisted radiopharmaceutical synthesis [12,16,22].

1.4 The Need for Technological Innovation in Synthetic Radiochemistry

The inherent limitations of conventional synthesis methods, particularly their slow reaction times and inefficient energy transfer, posed significant barriers to the rapid production of radiopharmaceuticals, especially those involving short-lived isotopes [5-7]. The burgeoning demand for large, diverse libraries of novel compounds in drug discovery and the increasing need for patient-specific therapies further strenthened the urgency for technological innovation in synthetic radiochemistry. Traditional heating often leads to prolonged reaction times, which is highly undesirable for radionuclides that decay rapidly, resulting in reduced yields and specific activities [11,12]. Moreover, the potential for side product formation due to prolonged exposure to suboptimal temperatures in conventional heating necessitated laborious purification steps, further impacting efficiency and cost. The "library validation" and "proof-of-principle"[8] in drug discovery are particularly time-intensive, often spanning weeks, thereby becoming bottlenecks in the overall process [8,9]. Addressing these challenges required a shift towards technologies that could dramatically reduce reaction times, enhance yields, improve reproducibility, and enable high-throughput synthesis in a safe and controlled manner [8,9].

1.5 Emergence and Scope of Microwave-Assisted Synthesis (MAS)

MAS has come up as a revolutionary technology in radiopharmaceutical chemistry [4-7,9,14-16,21-23], fundamentally changing how these complex molecules are produced. The core principle of MAS involves microwave energy undergoing direct coupling with polar molecules such as catalysts, reagents, or solvents, within the reaction mixture [5-7,16]. This contrasts sharply with conventional heating, which relies on indirect heat transfer from an external source through the vessel walls (Fig. 1). The primary mechanisms of microwave heating are dipolar polarization and ionic conduction [5-7,14,16], leading to rapid, volumetric heating [5,8,9,11,14-16].

Microwave-Assisted Synthesis (MAS) offers significant advantages in radiopharmaceutical development, including dramatically reduced reaction times—from hours to seconds—enhanced radiochemical yields [22-26] and improved reproducibility [16-19]. It enables precise control of reaction conditions, supports automation and high-throughput workflows, and aligns with green chemistry principles through solvent-free and energy-efficient processes. MAS is versatile across diverse chemical transformations, including those involving sensitive biomolecules [5,8,9,17,18]. This includes metal-catalyzed processes (e.g., Heck, Suzuki, Stille reactions), reductions, acylations, and various heterocyclic syntheses [5,8,9,11,12]. Advantages are further explained in the subsequent sections of this chapter.

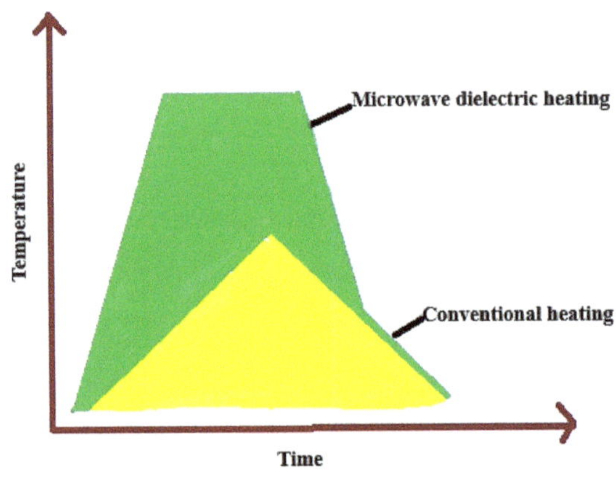

Figure 1. Temperature-Time Profiles: Conventional vs. Microwave Dielectric Heating [Adapted from Tierney, J., & Lidström, P. (Eds.). (2009). Microwave-assisted organic synthesis. John Wiley & Sons.]

Materials Research Foundations 189 (2026) 268-309 https://doi.org/10.21741/9781644904039-9

2. Fundamentals of Microwave-Assisted Chemistry in Radiopharmaceutical Applications

2.1 Physical Basis of Microwave Energy

Microwave radiation is a type of electromagnetic energy, occupying the electromagnetic spectrum region between infrared and radio frequencies [15,16,21]. It is generally defined in terms of frequency, ranging from 300 to 300,000 MHz, which corresponds to 1 cm to 1 meter (m) wavelength. To avoid interference with telecommunication and radar activities, most commercial and domestic microwave ovens work at 2.45 GHz, a specific frequency that corresponds to a 12.25 cm wavelength.

Non-ionizing factor is a crucial distinction of microwave energy with other forms of radiation, such as X-rays and gamma-rays [6,7,15]. This means that microwave energy does not alter the compounds' molecular structure that is being heated; instead, it provides only thermal activation [14-18]. In chemical applications, microwave energy is widely used as a source of heating to drive chemical reactions. Mechanisms of Microwave Heating

The heating effect plays a crucial role in microwave-assisted organic transformations, primarily because of dielectric polarization (also known as dipole rotation) and ionic conduction [5-7,16]. These are the two fundamental mechanisms for transferring energy from microwaves to the substance being heated [5,16].

Dipole Rotation: When molecules with a permanent dipole moment are exposed to an oscillating electric field, as is the case with microwave radiation, they attempt to align themselves with the rapidly changing field [6,7,16]. The rapid oscillation of the electric field (at 2.45×10^9 Hz for common microwave ovens) [15] causes the molecules to continuously reorient themselves, leading to strong agitation and generating heat. This process is a collective property involving aggregates of molecules, not a quantum mechanical phenomenon [16] localized at a single molecular center.

Ionic Conduction: This mechanism involves dissolved charged particles, typically ions, within a sample [5-7,16]. Under the influence of the microwave field, these ions oscillate back and forth, colliding with neighboring molecules or atoms [16]. These collisions create agitation and motion, which in turn generate heat due to frictional losses [16].

Also, the dielectric properties determine the ability of a material to transform electromagnetic energy into thermal energy [8,14,21-22], specifically its dielectric constant and loss tangent (tan δ) [22]. The larger the dielectric constant, the greater the coupling with microwaves, leading to more rapid heating. Solvents such as ethyl acetate, methanol, N,N-dimethylformamide (DMF), water, acetic acid, and acetone all couple effectively with microwaves and heat rapidly. In contrast, solvents with low dielectric constants, like carbon tetrachloride, hexane, and toluene, do not couple effectively and therefore do not heat rapidly under microwave irradiation [6,7,16,20,21]. Compared to conventional heating, which transfers heat slowly from an external source (like an oil bath) through the vessel walls by conduction and convection, [5,8,12,18] microwave heating is fundamentally different and significantly more rapid. Microwaves couple directly with the molecules throughout the entire reaction mixture, leading to a rapid, instantaneous, and volumetric temperature increase. This "bulk heating" means the process is not limited by the thermal conductivity of the vessel [5,8,12,16].

2.2 Thermo-Kinetics and Non-Thermal Effects

Since the initial days of MAOS [11,12,16], the observed accelerations in reaction rates, and sometimes product distribution discrepancies compared to conventionally heated experiments, have led to speculations about the existence of "specific" or "non-thermal" microwave effects [11,12,16,22]. Historically, claims of non-thermal effects arose when the outcome of a microwave synthesis differed from a conventionally heated counterpart at the same apparent temperature. The literature presents conflicting reports on the involvement of such "microwave effects" [11,12].

Some research teams favor a purely heat-derived mechanism, suggesting that increased reaction rates are solely due to rapid and efficient heating. However, others suggest that it may not be possible to generalize this assumption.

Arguments for non-thermal or specific microwave effects include:

Polar Transition States: It has been suggested that transition states in bimolecular reactions, which may be polar, could be susceptible to the direct influence of microwave irradiation [12].

Selective Heating: Under microwave conditions, selective heating of specific components can occur, where only components that couple effectively with microwaves are heated directly [11,12]. Non-absorbing components are then indirectly heated via energy transfer. This can create localized "hot spots" within the sample, characterized by temperature gradients [5,8,12,27].

Enhanced Transport Properties/Diffusion: Ceramic processing suggests that transport properties, such as increased diffusion rates, can be enhanced and occur under microwave conditions [12,21,22]. This is supported by observations of a significant activation energy reduction for diffusion under microwave conditions compared to conventional thermal conditions [12,21].

A specific phenomenon often associated with microwave heating is superheating, where solvents can be heated above their normal boiling points without boiling. This can lead to increased reaction rates due to the exponential dependence of reaction rate on temperature [12,16,18].

2.3 Microwave Reactor Technologies

The field of microwave chemistry has seen a significant evolution from early experiments conducted in domestic kitchen-type microwave ovens to the development of sophisticated, dedicated instruments for chemical synthesis [11,12,16]. Today's commercial microwave reactors for synthesis incorporate features like built-in magnetic stirrers [11], direct temperature control (using fiber-optic probes or IR sensors) [11,15], and software for on-line temperature/pressure control via microwave power regulation [11], (See Table 1).

Materials Research Foundations 189 (2026) 268-309 https://doi.org/10.21741/9781644904039-9

Table 1: Role and characteristics of microwave reactors in radiopharmaceutical synthesis:

Aspect	Description & Significance	Key Examples & Data
Reactor Type	• Microfluidic/micro-reactor chips with integrated microwave heating [8,14]. • Single-mode or continuous-flow microwave cavities tailored for radiochemistry [14].	• PDMS complementary split-ring resonator (CSRR) microfluidic chip: 2.8–6.5 μL volume, 5–8 μL/min flow, 120 °C in seconds at ≤4.4 W. • CEM Voyager CF (continuous-flow) and SF (stop-flow) reactors used since, early 2000s [14].
Volume & Throughput	• Nano- to microliter volumes enable rapid, sequential production. • Scaling via "numbering-up" multiple micro-reactors, not enlarging size, preserving kinetics [8,14].	• Lab-on-chip devices synthesize [18F] FDG and other tracers in concise systems. [3]. • Parallel reactor arrays are used to increase throughput while maintaining efficiency.
Heating Dynamics & Control	• Localized, volumetric microwave heating enables uniform and rapid temperature rise. • Precise control over temperature, pressure, and residence time reduces side-products.	• Uniform heating simulation in PDMS-CSRR device. • Microwave dielectric heating in micro-droplets achieved 30°C rise in ~15 ms.
Reaction Speed & Yield	• Microwave-assisted organic synthesis (MAOS) accelerates reactions from hours/minutes to seconds. • High yields (~quantitative), improved selectivity, and reduced by-products.	• Suzuki cross-coupling [8,12] and fluorinations routinely completed in seconds–minutes under micro-MW conditions [12]. • Heck coupling and Stille coupling also preferred. • $[^{18}F]$T807 and $[^{18}F]$FDG synthesized with high purity.
Scalability	• Continuous-flow reactors replicate micro-conditions at scale [3,8,11,14, 16]. • Switch from stop-flow to continuous regimes for production-scale quantities.	• Voyager CF & SF systems scaled up in collaboration with BASF, including 1 × 60 cm tubes for industrial use [14]. • Flow reactors translating. lab micro-MW to kilogram-scale
Safety & Sustainability	• Small volumes enhance safety (hazard handling, radioprotection). • Green chemistry: Minimal solvent use, reduced energy and waste [3,5,11,12,14,16].	• Micro-reactors mitigate exothermic hazards in bromide-to-aldehyde conversion without cryogenics [3,11,12,14]. • Microfluidic MW aligns with green principles—low waste, solvent-free options

Materials Research Foundations 189 (2026) 268-309 https://doi.org/10.21741/9781644904039-9

Flexibility & Customizability	• Modular designs (e.g., interchangeable flow cells) [8,14,15] • Tunable microwave frequency (2–12 GHz) to match solvent dielectric properties [26].	• CSRR microfluidic reactor offers interchangeable flow cell volumes (2.8–6.48 µL). • Tunable frequency enhances heating efficiency.
Applications in Radiopharmacy	• Ideal for short-lived isotopes (C-11, F-18) [3,8,11,12,16,21] theranostics, and novel radio-nuclides [1, 25]. • Enables rapid synthesis, purification, and formulation [1,3,4,12,16].	• First human imaging with micro-flow synthesized [^{18}F]T807 [3]. • Integration with MAOS for alpha emitters emerging [1,25].

This table brings together state-of-the-art reactor platforms, performance metrics, and practical benefits—offering a clear depiction of how microwave reactors are revolutionizing radiopharmaceutical synthesis.

The microwave batch reactor (MBR) is a laboratory-scale reactor with a 25–200 mL capacity is capable of operating at temperatures up to 260 °C and pressures up to 10 MPa (100 atm) [11]. Beyond batch reactors, continuous-flow and stop-flow [14] processing approaches have emerged for scale-up. Continuous microwave reactors (CMRs) are designed to heat reaction mixtures under controlled conditions, in a safe and rapid manner along with rapid cooling after exiting the irradiation zone.

Microwave reactors are broadly classified into two main designs (Fig. 2, Table 2) :

Table 2 - Comparison of Mono vs. Multi-Reactor Microwave Systems in Radiopharmaceutical Synthesis

Feature	Mono-Reactor Microwave System	Multi-Reactor Microwave System
Definition	A single microwave cavity or waveguide focusing energy on one reaction at a time; includes monomode/single-mode microwave systems [14].	Systems with larger cavities or multiple microwave paths handling several reactions simultaneously; includes multimode or parallel reactorsystems [14].
Microwave Mode	**Monomode** (Single-Mode): • One electromagnetic mode focused directly on the reaction vessel. • Highly homogeneous energy field. • Precise, rapid heating.	**Multimode**: • Multiple, uncontrolled modes; microwaves bounce chaotically, creating less uniform energy distribution. • Heating may vary across vessels.
Heating Uniformity	• High control and uniform heating [14]. • Suitable for sensitive radiochemical reactions.	• Less uniform—often mitigated with turntables or mode stirrers. • Requires careful validation for radiopharmaceutical consistency [8].
Throughput [R8]	• Lower simultaneous throughput. • But can be scaled via robotic automation (e.g., Biotage Initiator 60).	• Higher simultaneous throughput using multi-vessel rotors or deep-well plates (e.g., CEM MARS 6) [16].

Materials Research Foundations 189 (2026) 268-309 https://doi.org/10.21741/9781644904039-9

Reaction Control [R14]	• Precise control over time, temperature, pressure for each individual reaction. • Ideal for method development and short-lived isotopes (e.g., [^{18}F]) [26].	• Simultaneous control is complex. • Some systems allow individual reaction programming across units [22].
Example Systems [Current Trends in the Development of Microwave Reactors for the Synthesis of Nanomaterials in Laboratories and Industries: A Review Sylwia Dąbrowska 1,2,*, Tadeusz Chudoba 1, Jacek Wojnarowicz 1 and WitoldŁojkowski 1 1 Polish Academy of Sciences, Institute of High Pressure Physics, Soko]	• CEM Discover SP-D • Biotage Initiator 60 • Monomode flow systems like CEM Liberty	• Milestone Combi CHEM Anton Paar Synthos 3000 • CEM MARS 6 • Capable of processing multiple samples per cycle
Automation	• Integration with robotic arms or automated sample changers allows quasi-continuous processing in single-vessel mode.	• Full automation of multi-vessel rotors and sample queues possible. • Higher complexity of maintenance and calibration.
Scalability & Flow Processing	• Compatible with continuous-flow and stop-flow systems. • Continuous Microwave Reactors (CMRs) are emerging for scale-up in radio pharma.	• Designed for batch-scale synthesis of multiple agents. • Less adaptable to continuous-flow setups.
Application in Radio pharmaceuticals	• Ideal for short-lived PET tracers requiring precise, rapid heating and small-scale GMP production. • Widely used in academic and translational settings.	• Suitable for high-throughput synthesis of libraries or batch production for clinical centers. • Increasing use in industrial radio pharmacies.
Cost & Infrastructure	• Lower cost, compact footprint, easier shielding. • Ideal for research labs and hospital radio pharmacies.	• Higher initial investment. • Best suited for centralized or industrial radio pharmacy units.

Materials Research Foundations 189 (2026) 268-309 https://doi.org/10.21741/9781644904039-9

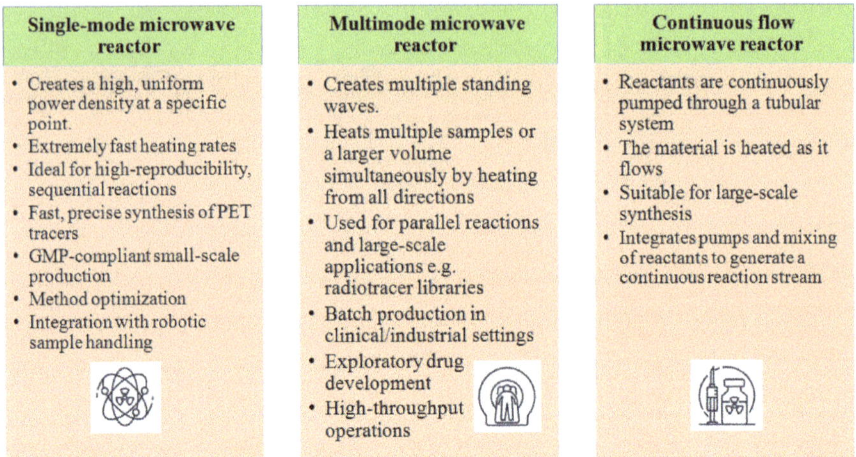

Figure 2 -Characteristics of Microwave Reactors: Single-mode, Multimode, and Continuous Flow

For radiopharmaceutical preparations, specialized microwave reactor systems have been developed, such as those with solid-state microwave devices instead of magnetron sources, and resonant cavities to efficiently heat small solution volumes. Examples include the integration of microwave reactors with automation platforms like Synthia for fully automated radiofluorination.

2.4 Key Features Compared to Conventional Synthesis

Microwave-assisted organic synthesis (MAOS) offers several key advantages over traditional thermal heating methods:

Dramatic Reduction in Reaction Times: This is one of the most significant benefits, often reducing reaction times to even seconds [1]. This speed provides a competitive advantage in areas like drug discovery, allowing for faster exploration of SARs and timely delivery of medicines.

Improved Yields and Purity: As compared to conventional methods, MAOS frequently leads to higher product yields and with greater purity. This is partly attributed to the rapid heating, minimizing side reactions and decomposition that can occur with prolonged heating at suboptimal temperatures.

Environmental Benefits (Green Chemistry): Microwave chemistry aligns with green chemistry principles by enabling solvent-free conditions, minimizing solvent usage, reducing waste, and offering higher energy efficiency than traditional heating. This also leads to lower operating costs [6-7].

Enhanced Reaction Control and Reproducibility: Modern dedicated microwave reactors allow for precise on-line control and monitoring of temperature and pressure profiles [4,8]. This leads to

Materials Research Foundations 189 (2026) 268-309 https://doi.org/10.21741/9781644904039-9

more reproducible reaction conditions and allows for immediate stopping of reactions at specific times for better control over product characteristics, such as nanoparticle size distribution [4,26].

Access to New Chemistry: Microwave heating can enable reactions that were previously difficult, resulted in low yields, or were impossible under conventional conditions. This includes specific applications like selective heating of catalysts or components in multiphase systems [1].

Scalability for Drug Discovery: While full industrial production scale is still evolving, microwave technology supports scalability from milligrams to kilograms range, especially through continuous flow and parallel batch processing, which is crucial for drug discovery and lead optimization efforts.

3. Synthetic Chemistry of Radiopharmaceuticals

Radiopharmaceuticals represent a remarkable fusion of science and innovation, playing a transformative role in medicine. These compounds are meticulously engineered by combining biologically active molecules with radioisotopes, providing non-invasive insights into molecular processes at the cellular level. They are crucial for both diagnostic imaging and targeted therapies.

3.1 Common Radionuclides in Radiopharmaceutical Applications

The radioactive nuclide component is responsible for producing a detectable signal outside the organism for visualization using nuclear medicine methods, making these truly "molecular imaging" modalities [20]. For therapeutic applications, radionuclides directly deliver radiation to targeted lesions [2] (Table 3).

Table 3- Overview of Therapeutic Radionuclides categorized by emitter type (decay modes), with diagnostic, therapeutic, and theranostic applications

Emitter Type	Radionuclide	Key Features/Applications
Alpha (α)	**Radium-223 (Ra-223)**	• First and only FDA-approved α-emitter (Xofigo). • Used for prostate cancer bone metastases; causes irreversible DNA damage with minimal crossfire effect.
	Astatine-211 (At-211)	• Short-range, high-energy α-particles • Under investigation for targeted alpha therapy (TAT).
	Actinium-225 (Ac-225)	• Promising α-emitter; used in targeted alpha therapy • Emits multiple α-particles in decay chain, enhancing cytotoxicity.
	Thorium-227 (Th-227)	• α-emitter in preclinical/clinical development. • Being studied for targeted therapies using antibody-conjugates.
Beta (β)	**Lutetium-177 (Lu-177)**	• Widely used in theranostics (e.g., [^{177}Lu]Lu-DOTA-TATE, [^{177}Lu]Lu-PSMA-617). • Emits both β- and γ-radiation; moderate range and energy; causes reversible DNA damage.
	Yttrium-90 (Y-90)	• High-energy β-emitter. • Used in radio immunotherapy and liver cancer treatments (e.g., SIR-Spheres, Thera Sphere).
	Phosphorus-32 (P-32)	• Early β-emitter.

		• Used in hematologic disorders and palliative cancer therapy.
	Strontium-89 (Sr-89)	• Bone-seeking β-emitter. • Used in bone metastasis pain relief.
	Samarium-153 (Sm-153-EDTMP)	• Targets skeletal metastases. • Emits β- and γ-radiation, allowing imaging and therapy.
	Iodine-131 (I-131)	• Dual β/γ-emitter. • Used in thyroid cancer therapy and neuro blastoma (e.g., [^{131}I]MIBG).
Gamma (γ)&Auger Electron	**Technetium-99m (99mTc)**	• Most widely used diagnostic SPECT radionuclide (~70–80% usage in U.S.). • Ideal 6 h half-life, 0.140 MeV γ-emission; easily sourced from ^{99}Mo/^{99m}Tc generators.
	Iodine-123 (I-123)	• Used in thyroid imaging. • Emits low-energy γ-rays with minimal radiation burden.
	Indium-111 (In-111)	• Used in leukocyte labeling, infection imaging, and neuroendocrine tumors.
	Gallium-67 (Ga-67)	• Employed in lymphoma, infection, and tumor imaging.
	Xenon-133 (Xe-133)	• Inhaled agent for lung ventilation imaging.
	Thallium-201 (Tl-201)	• Myocardial perfusion imaging; analogue of potassium, used in cardiac SPECT.
Theranostic Pairs	**[^{68}Ga]Ga-DOTA-TATE / [^{177}Lu]Lu-DOTA-TATE**	• Shared peptide ligand. • Used in neuroendocrine tumor diagnosis and therapy.
	[^{68}Ga]Ga-PSMA-11 / [^{177}Lu Lu]Lu-PSMA-617	• PSMA-targeting ligands for prostate cancer imaging and therapy.

A growing trend in the field isradiotheranosticpairsdevelopment, where the same targeting ligand is used for both diagnostic and therapeutic purposes, often employing different isotopes of the same element or chemically similar elements. Examples include [^{177}Lu]Lu-DOTA-TATE for therapy paired with [^{68}Ga]Ga-DOTA-TATE for diagnosis, and [^{177}Lu]Lu-PSMA-617 with [^{68}Ga]Ga-PSMA-11.

Radiopharmaceuticals are compounds that combine biologically active molecules with radioisotopes, used for non-invasive insights into molecular processes, early disease detection, and targeted therapies [1]. They play crucial roles in diagnostic imaging techniques such as Positron Emission Tomography (PET) and Single-Photon Emission Computed Tomography (SPECT), as well as in targeted radionuclide therapy [1-2]. Over 100 radioactive substances are used in nuclear medicine [20] (Table 4, Table 5).

Materials Research Foundations 189 (2026) 268-309 https://doi.org/10.21741/9781644904039-9

Table 4 - For Positron Emission Tomography (PET)

Radionuclide	Half-Life	Production	Applications
Fluorine-18 (F-18)	109.8 min	Cyclotron	Metabolic imaging, cancer, neurodegeneration, cardiovascular
Carbon-11 (C-11)	20.4 min	Cyclotron	Brain tumors, metabolic studies, receptor mapping and neurotracer development
Nitrogen-13 (N-13)	10 min	Cyclotron	cardiac perfusion imaging, general PET use
Oxygen-15 (O-15)	122.24 sec	Cyclotron	Myocardial perfusion PET ($H_2{}^{15}O$), cerebral blood flow studies
Gallium-68 (Ga-68)	68 min	Generator	Theranostics, general PET imaging, DOTA-conjugated peptide imaging (e.g. [^{68}Ga]Ga-PSMA-11, [^{68}Ga]Ga-DOTA-TATE).
Rubidium-82 (Rb-82)	75 sec	Generator	Cardiac perfusion, mimics K^+ ions
Copper-64 (Cu-64)	12.7 hr	Cyclotron	Next-gen diagnostics, therapy, theranostic potential
Yttrium-86 (Y-86)	14.7 hr	Cyclotron	Next-gen diagnostics/theranostics, PET surrogate for Y-90-based therapies.
Zirconium-89 (Zr-89)	78.4 hr	Cyclotron	Excellent for immuno-PET and antibody imaging.
Iodine-124 (I-124)	100.2 hr	Cyclotron	Next-gen diagnostics (thyroid, oncology), PET isotope with long half-life (~4.2 days), Suitable for long-term pharmacokinetics of antibodies.

Table 5 - For Single-Photon Emission Computed Tomography (SPECT)

Radionuclide	Half-Life	Gamma Energy	Production Method	Clinical Applications
Technetium-99m (Tc-99m)	6 hours	140 keV	^{99}Mo/^{99m}Tc Generator	Most widely used (~80% of SPECT procedures); ideal energy and half-life for imaging.
Iodine-123 (I-123)	13.2 hours	159 keV	Cyclotron	FDA-approved; widely used for diagnostic medical imaging.
Indium-111 (In-111)	2.8 days	173 keV & 247 keV	Cyclotron	Used in infection imaging and cancer diagnostics.

3.2 Radiolabeling Strategies

Radiolabeling is the process of attaching a radioactive tag to a special compound, transforming it into a powerful tool for nuclear medicine applications. This process requires a careful blend of chemistry, biology, and physics to ensure the radioactive tag is attached without compromising the compound's natural activity [17].

Key considerations when selecting a radionuclide for labeling include its physical decay characteristics (half-life and decay modes), availability, and the ease with which it can be produced, purified, and isolated in a chemical form suitable for subsequent radiochemistry.

The very short half-lives of common PET radionuclides, such as ^{11}C (~20 minutes) and 18F (~2 hours), pose significant challenges for radiolabeling. To maximize radiochemical yields and specific activities, routine radiopharmaceutical production typically requires the incorporation of the isotope in the last or penultimate step of synthesis, ideally within one half-life of the radionuclide, to minimize losses due to radioactive decay. This time constraint means that

Materials Research Foundations 189 (2026) 268-309 https://doi.org/10.21741/9781644904039-9

synthetic routes, purification, and drug formulation methods must be exceptionally rapid and efficient. Unlike traditional chemistry, radiochemistry often prioritizes reduced reaction time over maximizing relative yields.

Specific radiolabeling methods vary depending on the radionuclide:

Fluorine-18 (^{18}F) radiochemistry is predominantly based on nucleophilic substitution (SN_2) reactions, where the ^{18}F- anion replaces a leaving group [2,17]. The process often involves converting cyclotron-produced aqueous [^{18}F] fluoride (which is chemically unreactive) into a more reactive species through azeotropic drying with acetonitrile in the presence of an amino polyether, such as Kryptofix 2.2.2, which acts as a phase transfer catalyst. Newer approaches also employ sulfonium salts and diarylsulfoxides, or hypervalent iodine(III) ylides for regiospecific radiofluorination of non-activated aromatic rings [18].

Carbon-11 (^{11}C) radiochemistry often involves the direct replacement of an atom within a drug. For instance, [^{11}C]WAY-100635 can be prepared at its O-methyl position using [^{11}C]CH_3I in a single step. Strategies like ^{11}C-carbonylation via [^{11}C]CO_2 fixation are utilized for non-conventional radiochemistry, enabling the preparation of compounds like ^{11}C-carbamates and ^{11}C-ureas from in situ generated ^{11}C-isocyanates, and ^{11}C-oxazolidinones through intramolecular cyclization reactions [3,18].

Radiometals, such as gallium-68 (^{68}Ga) and technetium-99m (^{99m}Tc), are typically fixed to the molecular vehicle using a metal-ligand complex, often involving a chelator. This complexation can induce significant changes in the steric and electronic configuration of the radiopharmaceutical, potentially affecting its pharmacokinetics and pharmacodynamics. An example is ^{68}Ga-DOTATOC, used for neuroendocrine tumors. For ^{99m}Tc-MAG3 synthesis, microwave irradiation has been shown to significantly improve radiochemical purity to over 99% within 3 minutes, compared to traditional boiling methods, which can result in up to 45% impurities [18].

Isotopic labeling techniques, such as hydrogen-transfer methodology with solid deuterium or tritium donors, are used for the selective isotopic labeling of pharmacologically relevant compounds [12].

3.3 Molecular Vectors and Targeting Agents

Radiopharmaceuticals are composed of two main parts: a molecular structure, referred to as the vector, vehicle, or ligand, and the radioactive nuclide. The molecular structure is crucial as it determines the radiopharmaceutical's fate within the organism, including its pharmacokinetics and pharmacodynamics [1].

The primary role of these targeting vectors is to achieve accurate delivery of radionuclides to specific targeted cells or lesions, thereby minimizing off-target deposition and enhancing both the efficiency and biosafety of the therapeutic or diagnostic intervention.

These targeting vectors come in various forms:

Small molecules: These can be effectively utilized as targeting agents. However, due to their typically rapid clearance from the body, they require careful design to ensure sufficient specificity and retention at the target site [2].

Materials Research Foundations 189 (2026) 268-309 https://doi.org/10.21741/9781644904039-9

Peptides: Radiolabelled peptides are widely employed in targeted cancer therapy, such as peptide receptor radionuclide therapy for neuroendocrine tumors. Peptide-based drugs are often favored for imaging due to their faster clearance and better tissue penetration compared to larger molecules. An example is the ^{99m}Tc-labelled cyclic peptide, [^{99m}Tc]Tc-3PRGD2, which has shown value in detecting human cancers, including lung cancer, with high accuracy. Somatostatin analogues like ^{99m}Tc-octreotide are used for imaging somatostatin receptor-positive tumors [2].

Antibodies: Radiolabelled antibodies are utilized for targeting solid tumors. For instance, zirconium-89 (^{89}Zr) is a promising PET radionuclide for labeling monoclonal antibodies (mAbs) used in immuno-PET imaging. While antibody-based radiopharmaceuticals offer high specificity, their larger size can present challenges regarding pharmacokinetics (PK) and tumor penetration.

Nanoparticles: These represent a cutting-edge innovation for next-generation radiopharmaceutical formulations. Engineered as sophisticated carriers, nanoparticles enable the precise delivery of both therapeutic and diagnostic payloads. They provide crucial protection for the delicate contents, enhance compatibility, and extend circulation time within the body. By incorporating tailored surface modifications and disease-targeting ligands, these nanoparticles can accurately navigate biological systems, evade clearance mechanisms, and specifically home in on cellular receptors or microenvironments in pathological tissues. This technology also allows for the co-delivery of multiple agents, such as therapeutic agents, contrast agents, or multiple radionuclides, offering remarkable flexibility and synergistic therapeutic potential. Nanoparticle-based delivery systems generally improve drug availability, reduce toxicity, and enhance efficiency, while being customizable for precise targeting. However, challenges remain regarding their potential toxicity, stability, and the scalability of their production.

Research efforts are continuously focused on developing novel targeting ligands to improve uptake in lesions, prolong retention time, and minimize toxicity to healthy organs. High-throughput screening platforms, including phage display, mRNA display, and DNA-encoded compound libraries, are instrumental in identifying ligands with strong binding affinities for these new radiopharmaceuticals.

3.4 Solid-Phase vs Solution-Phase Radiosynthesis

Microwave-assisted solid-phase organic synthesis (MAOS) has gained significant attention due to its wide applicability. It is particularly effective for "dry media" reactions, where reactants are adsorbed onto solid supports like silica, montmorillonite clay, alumina, and zeolites. In these reactions, the organic compounds absorb microwaves while the inorganic support does not, leading to relatively low bulk temperatures despite rapid heating. This approach has demonstrated practical feasibility for solvent-free protocols in various chemical transformations, including protection, deprotection, condensation, oxidation, reduction, rearrangement reactions, and the synthesis of heterocyclic systems.

The advantages of combining solid-phase techniques with microwave heating include the production of cleaner reaction mixtures compared to traditional solution-phase methods. This combination also enables rapid synthetic transformations [12].

However, challenges exist in solid-phase radiochemistry. Issues such as the retention of the product within the resin can lead to lower yields. Additionally, increased precipitation and crystallization of materials involving polymer species are common. To overcome these issues, a range of solvents often needs to be screened to identify the optimal combination for specific reactions and resin types.

While **solution-phase synthesis** is also widely used, most published examples of microwave-assisted organic synthesis performed on this type have been conducted on a small scale, typically with reaction volumes of 1-5 mL and product yields below 1 gram, often utilizing single-mode microwave reactors.

Parallel synthesis, which involves performing multiple reactions simultaneously, is greatly facilitated by microwave heating, allowing for the efficient generation of chemical libraries. Automated sequential library synthesis, often employing robotized microwave reactors, has proven to be a valuable tool in this context [8].

3.5 Microwave-Assisted Synthesis of Radiopharmaceuticals

MAS has evolved as a transformative technique in the field of radiopharmaceutical chemistry, particularly due to its capacity to significantly accelerate chemical transformations, which is essential for radionuclides characterized by short half-lives. The efficiency of MAS in radio labeling is primarily attributed to its rapid and uniform heating mechanism, which facilitates high radiochemical yields (RCY), improved purity, and shortened synthesis time—key parameters in radiopharmaceutical production.

General Advantages and Mechanisms:

- Microwaves couple directly with molecules, leading to rapid and localized superheating through dipole rotation and ionic conduction, which is more efficient than conventional external heating methods [5].
- This rapid thermal transition enhances productivity and shortens production times.
- It's particularly useful for microscale organic chemistry like radiolabeling, where the sample size is comparable to the microwave field's penetration depth [23].
- Microwave heating enables a less aggressive treatment of reagents, which is critical for unstable molecules and for achieving high product purity [23].

- **Specific Applications and Examples**:
  - **Fluorine-18 (F-18) Labeling**:
    - Microwave heating is often used for the nucleophilic substitution reactions required for F-18 radiolabeling, which generate inert C-^{18}F bonds [17].
    - Automated radiofluorination modules integrating microwave reactors have been developed for efficient heating and automation, even for azeotropic drying of [^{18}F]fluoride ions [23].
    - Studies show that microwave heating can significantly improve radiochemical yields for F-18 labeling of substituted benzenes (e.g., from 21% to 47% or 8% to 13%) compared to thermal heating [23].
    - Microwave-assisted synthesis of 16α-[^{18}F]fluoroestradiol ([18F]FES), 4-[^{18}F]fluoro-N-[2-(1-methoxyphenyl)-1-piperazinyl]ethyl-N-2-pyridinylbenzamide ([^{18}F]MPPF), and N-succinimidyl 4-

[^{18}F]fluorobenzoate ([^{18}F]SFB)have shown improved yields and reduced synthesis times [31]. One synthesis of [^{18}F]FES achieved >14 GBq/µmol radioactivity in 25 ± 5 minutes total synthesis time [27].

    - Novel hypervalent iodine(III) ylides as precursors for regio specific radio fluorination of non-activated and sterically-demanding (hetero)aromatic rings using [^{18}F]fluoride have been developed, with one radiopharmaceutical, [^{18}F]FPEB, showing a 10-fold increase in radiochemical yield (20% vs. 2%, uncorrected; relative to [^{18}F]fluoride) and high specific activity [3].

  - **Carbon-11 (C-11) Labeling**:
    - Microwave heating has been used to speed up Stille coupling reactions in the preparation of radiopharmaceuticals containing short-lived positron-emitting isotopes, such as 1-(2'-deoxy-2'-fluoro-β-d-arabinofuranosyl)-[Methyl-11C]thymine ([11C]FMAU), prepared in 3 minutes in modest radiochemical yields [12,18,27].
    - The conventional approach for C-11 radiopharmaceutical production often involves "last or penultimate step" incorporation of the isotope to maximize yields due to its short half-life [3]. Microwave heating supports this by drastically shortening reaction times.

  - **Gallium-68 (Ga-68) Complexation**:
    - Microwave heating has shown potential in speeding up ^{68}Ga complexation with DOTA bifunctional chelators [24].
    - In combination with microwave heating, peptide conjugates of 0.5-1 nmol quantities could be labeled within 10 minuteswith the full ^{68}Ga activity of a generator, resulting in improved specific radioactivity by a factor of ~100 compared to previous techniques [2]. This integrates into a "kit-like technology" for ^{68}Ga-tracer production [2].

  - **Technetium-99m (Tc-99m) Labeling**:

- Microwave treatment has proven very useful in solving specific problems in radiochemistry, such as the production of diagnostic radiopharmaceuticals with Tc-99m [26].
    - For example, in the synthesis of ^{99m}Tc-MAG3, microwave irradiation achieved a radiochemical purity higher than 99% after just 3 minutes of irradiation, whereas traditional procedures resulted in high levels of labeled impurities (45%). Microwave treatment also enabled the successful labeling of rifamicine, which yielded unsatisfactory results with conventional methods.
    - The effectiveness is attributed to direct microwave heating generating "hot spots" that mimic the effects of "hot atoms" produced during radioactive decay, providing the energy needed to reduce pertechnetate ions and facilitate binding.

Materials Research Foundations 189 (2026) 268-309 https://doi.org/10.21741/9781644904039-9

- **Isotopic Labeling (Deuterium, Tritium):**
  - Microwave-assisted dehalogenation with deuterated formates and catalysts provides a rapid route to deuterium-labeled compounds with enhanced isotopic incorporation. This can be significantly faster (40-90 seconds) compared to thermal conditions (2 hours).
  - The use of solid deuterium or tritium donors with hydrogen-transfer methodology is suitable for selective isotopic labeling of pharmacologically relevant compounds under microwave irradiation.

Overall, microwave synthesis has become an essential tool in the pharmaceutical industry for generating new chemical entities, enabling rapid lead generation, optimization, and the production of diverse compound libraries in a high-throughput format (Fig. 3).

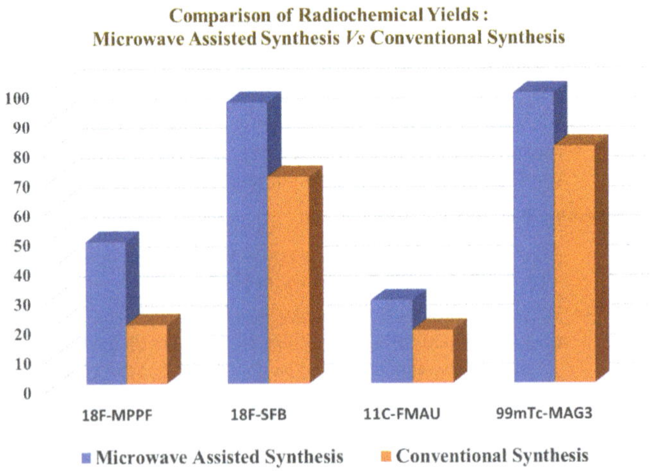

Figure 3:Comparison of Radiochemical Yields: Conventional vs. Microwave-Assisted Synthesis

These gains demonstrate MAS's effectiveness in accelerating reaction kinetics and minimizing decomposition or side reactions.

3.6 Challenges in Multi-Step Radio Synthesis

The synthesis of radiopharmaceuticals incorporating short-lived radionuclides, such as carbon-11 (~20 minutes) and fluorine-18 (~2 hours), is inherently challenging due to the stringent time constraints imposed by their rapid decay. Radiochemical syntheses must be completed swiftly—ideally within one half-life—to ensure high radiochemical yields and specific activities. These constraints hinder the feasibility of complex multi-step syntheses and limit the diversity of available labeled building blocks. Consequently, many pharmaceutical compounds remain inaccessible for labeling with ^{11}C or ^{18}F, due to inadequate radiosynthetic methodologies.

Materials Research Foundations 189 (2026) 268-309 https://doi.org/10.21741/9781644904039-9

To address this, the concept of *Total Radiosynthesis* is gaining prominence [3]. This strategy aims to enable the radiolabeling of virtually any compound and offers several advantages: (i) early-stage incorporation of diverse building blocks, allowing for broader synthetic flexibility; (ii) labeling at distinct molecular positions, essential for studying metabolic pathways in PET imaging; and (iii) the development of robust, rapid, and clinically adaptable methodologies suitable for routine and multi-center application [2].

Radiolysis presents another major limitation, whereby exposure to intense radioactivity leads to degradation of radiolabeled agents. This is often exacerbated by excessive heating and evaporation during synthesis and purification. To mitigate this, milder methods such as microwave heating and solid-phase extraction are preferred. A careful balance is required between high starting activity (to ensure sufficient product yield) and minimal radiolytic degradation [24].

Furthermore, high specific activity is essential for targeting low-density biological sites, as low specific activity—due to carrier isotopes—can compromise imaging efficacy. Finally, PET radiochemistry demands specialized equipment, on-site cyclotron facilities, and trained personnel, while microwave synthesis, though effective, remains cost-intensive and difficult to scale for industrial applications.

4. **Advantages** of Microwave-Assisted Synthesis in Radiopharmaceutical Development

Microwave-assisted synthesis (MAS) has evolved as a transformative technology in chemical synthesis, particularly gaining significant traction in the pharmaceutical and radiopharmaceutical sectors. It offers a compelling alternative to conventional heating methods by leveraging the ability of certain liquids and solids to directly convert electromagnetic radiation into heat. This direct coupling of microwave energy with molecules in the reaction mixture leads to a rapid and uniform temperature increase, which is fundamentally different from traditional external heating that relies on slow and inefficient heat transfer through vessel walls. The underlying mechanisms typically involve dipolar polarization and ionic conduction [25].

The adoption of MAS has revolutionized organic synthesis, allowing small molecules to be built in a fraction of the time as compared to time required by conventional thermal methods. This technology is considered a "green" chemistry tool due to its efficiency, cleanliness, and economic benefits. Its impact is particularly notable in drug discovery, from the generation of discovery libraries to hit-to-lead efforts and lead optimization, primarily because of the increased speed it offers, providing a competitive advantage and effective resource utilization [26].

4.1 Accelaration of Reaction Kinetics and Shortened Synthesis Time

One of the most profound benefits of MAS is the significant reduction in reaction times. Reactions that traditionally require hours or even days can be completed in minutes or seconds using microwave irradiation. Historically, rate enhancements of up to three orders of magnitude were first reported in 1986, establishing the kinetic superiority of MAS. For example, radiolabeling DOTA-conjugated peptides with ^{68}Ga using MAS can be completed in just one minute, compared to the 15–20 minutes needed using conventional heating methods [27].

This acceleration is attributable to several key mechanisms:

Materials Research Foundations 189 (2026) 268-309 https://doi.org/10.21741/9781644904039-9

Rapid and Instantaneous Heating: Microwave energy directly interacts with polar molecules, bypassing the limitations of external heat transfer and enabling "flash heating"—a steep and immediate increase in temperature.

Superheating: Sealed microwave reactors allow reaction temperatures to exceed the solvent's boiling point, significantly enhancing reaction rates.

Efficient Energy Transfer: Unlike conventional heating, where thermal gradients originate at the vessel wall, MAS delivers energy volumetrically, ensuring uniform heating.

Enhanced Molecular Mobility: In solid-state reactions, MAS reduces the diffusion barrier, thereby enhancing reaction kinetics and lowering the activation energy.

Examples of accelerated processes include:

- Esterification of sterically hindered 2,4,6-trimethylbenzoic acid, reduced from days to 1 hour.
- Fluorescein synthesis in 35 minutes versus 10 hours by conventional methods.
- Solid-phase peptide synthesis steps reduced from 30 minutes to 2–6 minutes each.
- Synthesis of 1,6-anhydrosugars under solvent-free conditions completed in 7 minutes.
- Efficient synthesis of [^{18}F]SFB and ^{68}Ga-labeled peptides in under one minute.

4.2 Enhanced Radiochemical Yields and Specific Activity

In radiopharmaceutical chemistry, the short half-lives of isotopes such as fluorine-18 (^{18}F), gallium-68 (^{68}Ga), and carbon-11 (^{11}C) necessitate rapid and high-yielding synthetic routes. MAS consistently achieves superior radiochemical yields (RCY) and higher specific activity due to minimized decomposition and reduced formation of by-products [25].

Minimization of Side Reactions: The fast ramp-up and rapid cooling (a "square" temperature profile) reduce the thermal exposure window, thus limiting side reactions and degradation.

Enhanced Purity: This leads to simplified downstream purification. For instance, EMS protocols for α-keto amides yield cleaner product profiles than conventional methods.

Improved Specific Radioactivity: In ^{68}Ga radiochemistry, MAS allows quantitative incorporation in a single step with superior reproducibility, enhancing specific activity up to 100-fold in certain instances.

Notably, microwave-assisted synthesis of ^{99m}Tc-MAG3 achieved >99% radiochemical purity in 3 minutes, compared to conventional methods that produced 45% impurity-laden product.

4.3 Reproducibility and Precise Control Over Reaction Parameters

Modern microwave reactors are equipped with feedback-regulated temperature and pressure sensors, facilitating high reproducibility, which is essential for clinical-grade radiopharmaceuticals [28-29].

Uniform Heating: MAS provides consistent temperature profiles across the entire reaction mixture, minimizing temperature gradients that can lead to inconsistencies in reaction outcome.

Real-Time Monitoring: Systems equipped with Raman or IR spectroscopy allow in-situ reaction tracking.

Materials Research Foundations 189 (2026) 268-309 https://doi.org/10.21741/9781644904039-9

Reactor Design: Sealed vessels in commercial systems (e.g., Biotage Initiator, CEM Discover) offer robust control over reaction variables, ensuring safe and replicable outcomes.

This level of control represents a significant evolution from early attempts using household microwaves, which were unsafe and unreliable due to lack of modulation and temperature feedback.

4.4 Compatibility with Automation and High-Throughput Systems

MAS technology integrates seamlessly with automated synthesis platforms and high-throughput workflows, both essential in radiopharmaceutical and drug discovery pipelines.

Automated Microwave Synthesizers: Robotic platforms, such as the Synthia and Biotage Initiator EXP, automate multi-step radiolabeling protocols, enhancing throughput.

Parallel Synthesis: Multi-well plates and rotor-based systems allow parallel reactions under uniform conditions, ideal for combinatorial chemistry.

Continuous-Flow Microwave Reactors: These enable real-time, scalable synthesis for PET tracers and are being explored for routine clinical supply.

Microfluidic Integration: Lab-on-a-chip systems with microwave modules are under development for miniaturized, rapid, and reproducible PET tracer synthesis.

4.5 Suitability for Combinatorial and Diversity-Oriented Chemistry

The speed and reproducibility of MAS make it particularly advantageous for generating compound libraries [27].

Rapid Lead Generation: MAS allows the synthesis of compound libraries in hours, e.g., a 48-member dihydropyrimidine (DHPM) library synthesized in 12 hours.

Reaction Scouting: Multiple reaction parameters can be tested quickly, expediting optimization cycles.

Solid-Phase Synthesis: MAS overcomes slow kinetics in bead-based synthesis (e.g., peptides, β-peptides).

Multi-component Reactions (MCRs): Biginelli and Ugi reactions benefit from MAS due to reduced times and improved yields, enabling rapid assembly of heterocyclic structures.

4.6 Safety and Operational Benefits

Modern microwave systems mitigate early safety concerns through robust engineering [27].

Pressure and Temperature Regulation: Dedicated systems prevent overpressure and allow reactions in volatile solvents under controlled conditions.

Miniaturization: Microscale reactions reduce reagent consumption and minimize risk associated with radioactive or hazardous materials.

Cleaner Chemistry: Reduced by-products translate to easier work-up, often involving only filtration or crystallization.

4.7 Contributions to Green Chemistry

MAS aligns with sustainable chemistry principles, offering both environmental and economic advantages.

Microwave-assisted synthesis is a significant contributor to **green chemistry**, promoting more environmentally friendly chemical processes.

Reduced Solvent Usage: A major benefit is the ability to perform reactions under solvent-free conditions, or with significantly reduced solvent volumes. This minimizes emissions, reduces the need for redistillation, and lowers hazardous waste.

Use of Benign Reagents and Media: MAS encourages the use of milder, less toxic, or readily recyclable reagents and media, aligning with green chemistry principles. This includes ionic liquids [20], which offer advantages like high boiling points, low vapor pressure, thermal stability, and efficient microwave interaction, while also facilitating catalyst recycling [30].

Energy Efficiency: Microwave heating is more energy efficient than conventional methods due to direct molecular heating. For example, the synthesis of biodiesel using microwaves required significantly less energy and achieved high conversion in minutes compared to hours conventionally.

Recyclable Supports: The use of recyclable solid supports instead of polluting mineral acids and oxidants further contributes to environmental benefits.

Waste Minimization: By increasing yields and reducing side products, MAS leads to less overall waste generated.

Microwave-Assisted Synthesis represents a transformative advancement and offers multifaceted advantages in radiopharmaceutical chemistry. The advantages of MAS are particularly critical in radiochemistry, where time constraints due to radionuclide decay, coupled with the need for high purity and yield, demand the most efficient synthetic strategies available. As microwave technology continues to evolve, it is poised to play an increasingly central role in the scalable, sustainable, and safe production of next-generation diagnostic and therapeutic radiopharmaceuticals.

5.1 Case Studies and Applications of MAS in Radiopharmaceutical Production

Radiopharmaceuticals are specialized medicinal compounds combining radioisotopes with pharmaceutical agents to enable precise diagnostic imaging and targeted therapy. They play a vital role in nuclear medicine by enabling precise diagnostic imaging, such as Positron Emission Tomography (PET) and Single-Photon Emission Computed Tomography (SPECT), and by delivering radioactive payloads directly to diseased tissues, thus minimizing harm to healthy cells. The development of radiopharmaceuticals requires efficient radiolabeling, where radioisotopes are attached to biologically active compounds without altering their function. However, producing these agents—especially those with short-lived isotopes like fluorine-18 or carbon-11—is challenging due to the need for rapid synthesis, purification, and quality control within narrow timeframes. Microwave-Assisted Synthesis (MAS) is a game-changing technology in this domain. By utilizing microwave dielectric heating, MAS enables rapid, uniform, and volumetric heating, drastically reducing reaction times and improving yields and reproducibility compared to conventional methods. This ensures that high-quality radiopharmaceuticals can be produced

Materials Research Foundations 189 (2026) 268-309 https://doi.org/10.21741/9781644904039-9

swiftly and efficiently, maximizing their clinical utility before radioactive decay limits their effectiveness.

5.1 [^{18}F] Fluoride Radio labelling

[^{18}F]Fluoride radiolabeling is a cornerstone of positron emission tomography (PET) imaging, typically involving nucleophilic substitution reactions to form stable carbon-fluorine bonds. The radionuclide fluorine-18 (^{18}F) has a half-life of approximately 109.77 minutes, which, while relatively long for a PET radionuclide still necessitates rapid and efficient synthetic routes [31]. Microwave technology has been successfully applied to enhance these radiolabeling reactions, significantly speeding up nucleophilic radiofluorinations and radioiodinations and increasing end-of-synthesis yields.

Key Applications and Advantages of MAS in [^{18}F] Radiolabeling:

Reduced Reaction Times and Increased Yields: Microwave heating allows chemical reactions to proceed over much shorter reaction times and in higher yields than conventional thermal conditions. For instance, the microwave-accelerated complexation reaction with ^{68}Ga, another important radionuclide, can be completed within 1 minute, compared to 5 minutes with conventional heating.

Automated Synthesis: For routine radiopharmaceutical production, especially of widely used agents like [^{18}F]fluorodeoxyglucose ([^{18}F]FDG), fully automated synthesis modules are utilized [32]. The production of [^{18}F]FDG, an analogue of glucose used for imaging cellular metabolism, primarily relies on cassette-based modules employing catalyst-based nucleophilic substitution reactions. These automated systems are crucial due to the short half-life and stringent regulatory requirements for drug release.

Novel Radiolabeling Strategies: Recent advancements include the single-step direct radiosynthesis of a powerful sigma-1 receptor radioligand,[^{18}F]FBFP, from an iodoniumylide precursor, demonstrating significant brain uptake and appropriate tissue kinetics in preclinical studies [33].

Regiospecific Radiofluorination: A significant breakthrough in radiochemistry has been the discovery of a methodology utilizing spirocyclic iodonium ylide precursors for regiospecific radiofluorination of non-activated and sterically-demanding (hetero)aromatic rings using [^{18}F]fluoride. This approach allows for the design of radiotracers without the need for electron-withdrawing arenes or prosthetic groups [2]. A radiopharmaceutical, [^{18}F]FPEB, synthesized by this method, demonstrated a 10-fold increase in radiochemical yield (20% vs. 2%, uncorrected) and high specific activity (>18 Ci/μmol), making it suitable for human use [34].

Ideal Criteria for [^{18}F] Radiolabeling: The ideal criteria for radiofluorination of non-activated arenes with [^{18}F] fluoride include a one-step labeling process, stable precursors, metal-free reactions, no fluorinated counterions, regioselectivity, compatibility with a wide variety of functional groups, easy automation, and ready translatability to routine clinical use. The iodonium ylide strategy effectively meets these criteria.

Addressing Supply Shortages: During shortages of ^{99}Mo (the parent radionuclide for ^{99m}Tc), [^{18}F]NaF (sodium fluoride) has been used as a PET alternative for bone imaging.

Materials Research Foundations 189 (2026) 268-309 https://doi.org/10.21741/9781644904039-9

Chemical Modifications: Advances in chemical modification strategies for radiopharmaceuticals include ^{18}F-trifluoromethylation of unmodified peptides at tryptophan and tyrosine residues, and the formation of [^{18}F]aryl-CF_3 from Aryl-$CF_2SO_2CF_3$ or ArylCF_2Br, both achieving high radiochemical yields under relatively mild conditions.

5.2 [^{11}C] Radiochemistry

Carbon-11 (^{11}C) [24] is another vital positron-emitting radionuclide used in PET, characterized by an even shorter half-life of 20 minutes. This extremely short decay time presents formidable challenges for radiopharmaceutical synthesis, demanding exceptionally rapid and efficient chemical transformations, purification, and formulation methods to minimize radioactive decay losses and maximize the amount of active product for clinical use.

Impact of MAS and Total Radiosynthesis in [^{11}C] Radiochemistry:

Concise and Rapid Syntheses: MAS, combined with the concept of "total radiosynthesis," aims to simplify and accelerate multi-step reactions involving short-lived radionuclides. This approach transforms radiopharmaceutical synthesis by allowing the exploration of complex multi-step reactions that would otherwise be impractical within the short half-life window.

Automated and Streamlined Processes: For routine clinical impact, [^{11}C] radiochemical methods must be easily automated and validated. Examples include the efficient and simplified automated production of [^{11}C] Ritalin and [^{18}F]-AV1451 (T807). Similarly, [^{11}C]methionine production has benefited from simple, fast, and automated online approaches, sometimes without the need for preparative HPLC.

Multi-Step Reaction Efficiency: The total radiosynthesis vision emphasizes achieving multiple reaction steps quickly. It has been demonstrated that over 5-step reactions can be achieved using ^{11}C within 40 minutes, leading to the discovery of new [^{11}C]-labeled building blocks and radiotracers.

Specific [^{11}C] Labeling Applications:

- **[^{11}C]CO_2 Fixation**: A key area of research involves exploring [^{11}C]CO_2 fixation for library radiosynthesis and for labeling non-activated arenes. This technique is pivotal for introducing a^{11}C label into complex molecules efficiently.
- **[^{11}C]WAY-100635**: This compound was initially prepared by one-step O-methylation using [^{11}C]CH_3I. Further advancements aimed to simplify its synthesis for routine use and address issues like brain-penetrating metabolites.
- **[^{11}C]Lapatinib**: This radiotracer, important for human PET imaging studies, was synthesized in four steps using a [^{11}C]fluorobenzylazide building block.

Copper-Mediated Radiocyanation: Novel methods for radiocyanation of aryl halides have been developed to introduce [^{11}C]CN into peptides and unprotected amino acids, achieving high radiochemical yields under relatively mild conditions.

5.3 Chilator-Based Radio-metal Labeling

Metallo radionuclides [24,34] are increasingly vital in radiopharmaceutical development, requiring the use of bifunctional chelating groups (BFCAs) to stably bind the metallic radioisotope

Materials Research Foundations 189 (2026) 268-309 https://doi.org/10.21741/9781644904039-9

to a biologically active molecule. These chelators, such as DOTA, DTPA, and NOTA, are designed to provide high thermodynamic and kinetic stability with the desired metalloradionuclide and can be derivatized for covalent bonding to targeting vectors like peptides and antibodies. The short half-lives of many relevant radionuclides, like gallium-68 (^{68}Ga), cobalt-57 (^{57}Co), indium-111 (^{111}In), and copper-64 (^{64}Cu), pose significant challenges for their incorporation into radiopharmaceuticals due to the need for rapid and efficient labeling.

Enhanced Labeling with MAS for Metallo radionuclides:

- **Gallium-68 (^{68}Ga)**:
  - Generator-produced ^{68}Ga (T1/2 = 68 min) is of significant clinical interest for PET. However, its generator eluate often contains large volumes and metal traces that interfere with fast, reliable, and quantitative labeling [34].
  - Microwave heating has been shown to dramatically shorten the reaction time and improve reproducibility for ^{68}Ga complexation reactions, even with macrocyclic chelators. For example, quantitative ^{68}Ga incorporation with DOTATOC (DOTA-D-Phe1-Tyr3-octreotide), a test tracer, was achieved within 1 minute with microwave heating, compared to 5 minutes with conventional heating.
  - A simple anion-exchange technique can be integrated to purify and concentrate ^{68}Ga from the generator eluate in one step, meeting the requirements for direct, highly efficient labeling of modern bioconjugates under microwave heating conditions. This technique can reduce the eluate volume by a factor of 30, significantly improving the specific radioactivity of the final ^{68}Ga-labeled peptide conjugate by a factor of ~100.
  - This integrated approach often eliminates the need for further purification of the labeled peptide conjugate, allowing the overall preparation process to be completed in 10-12 minutes.
  - FDA approvals of [^{68}Ga]Ga-DOTA-TATE and [^{68}Ga]Ga-PSMA-11 underscore the increasing clinical implementation of ^{68}Ga-based theranostics in oncology [35].
- **Copper-64 (^{64}Cu)**:
  - ^{64}Cu has a longer half-life (12.8 h), high resolution, and detection rate in PET/CT imaging, making it suitable for labeling antibodies with high binding affinity and for facilitating transportation to distant hospitals [24].
  - MAS has been used in the efficient production of high specific activity ^{64}Cu using a biomedical cyclotron.
  - Examples of ^{64}Cu-labeled therapeutic agents include ^{64}Cu-TETA-Tyr_3-octreotate and ^{64}Cu-ATSM (diacetyl-bis(N4-methylthiosemicarbazone), which have been studied for radiotherapy.
- **Other Radionuclides**: The increasing availability and clinical translation of "nonstandard" PET radionuclides such as ^{89}Zr and ^{124}I, alongside ^{64}Cu and ^{68}Ga, highlights a growing trend in nuclear medicine.
- **Chemical Modification Strategies**: To improve binding affinity, retention time, and accumulation of radiopharmaceuticals in lesions while reducing toxicity to normal organs, chemical modification strategies are crucial. These include the use of covalent warhead incorporation, unnatural amino acid replacement, and cyclic peptides. For instance, a GPC3-targeting radiopharmaceutical, RAYZ-8009, labeled with ^{68}Ga, ^{177}Lu, and ^{225}Ac, showed high uptake in hepatic carcinoma and significant anti-tumor effects.

Materials Research Foundations 189 (2026) 268-309 https://doi.org/10.21741/9781644904039-9

5.4 Peptide and Antibody Radio labeling

Radiolabeling of peptides and antibodies [34] has revolutionized targeted cancer therapy and other medical domains by enabling precise delivery of diagnostic and therapeutic radionuclides to specific molecular targets. This approach leverages the high specificity of monoclonal antibodies and peptides for overexpressed receptors in diseased tissues, minimizing collateral damage to healthy organs. Microwave-assisted synthesis plays a vital role in accelerating these complex biomolecule conjugations.

Case Studies and Advantages of MAS in Peptide and Antibody Radiolabeling:

- **Radiolabelled Antibodies for Solid Tumors**:
  - **^{177}Lu-PSMA-617**: A notable example is its use in metastatic castration-resistant prostate cancer (mCRPC). The Phase III VISION trial demonstrated a substantial improvement in overall survival and radiographic progression-free survival when ^{177}Lu-PSMA-617 was combined with standard care[34].
  - **^{177}Lu-DOTATATE**: This radiolabeled somatostatin analogue is effectively used to treat neuroendocrine tumors (NETs), targeting somatostatin receptors overexpressed on NET cells to deliver radiation precisely to tumor sites [36].
  - **Peptide Receptor Radionuclide Therapy (PRRT)**: PRRT can act synergistically as neo-adjuvant therapy before surgery or alongside other targeted agents to enhance therapeutic outcomes in neuroendocrine tumors [37].
  - **Radiohybrid (rh) PSMA-targeting ligands**: A novel series, including [^{177}Lu]Lu-rhPSMA-7.3, has shown significantly increased tumor uptake compared to [^{177}Lu]Lu-PSMA I&T, although concerns about higher accumulation in healthy organs like the kidney and bone marrow persist [38].
- **Nanobody and IgG-Binding Strategies**: To enhance tumor uptake and prolong tumor residence time, strategies involve the addition of albumin-binding and immunoglobulin-binding domains to targeting radioligands and HER2 nanobodies. This prolongs their blood circulation time and improves therapeutic efficacy.
- **MAS in Solid-Phase Peptide Synthesis (SPPS)**:
  - Microwave irradiation significantly enhances deprotection, coupling, and cleavage reactions in SPPS.
  - It addresses inherent difficulties in traditional SPPS, such as intermolecular aggregation, β-sheet formation, and steric hindrance, by promoting rapid heating and thermal disruption of peptide aggregation.
  - This allows for the production of peptides, such as the acyl carrier peptide, ACP (65–74), in high yields (>95%) [39].
  - Microwave flash heating is particularly beneficial for reactions involving polymer-supported reagents, ensuring rapid heating and cooling in sealed vessels while maintaining the integrity of the polymer matrix.
- **Automated Synthesis of Labeled Peptides**: The automated synthesis of [^{18}F]Ga-rhPSMA-7/-7.3 demonstrates the integration of microwave technology for routine production of complex radiopharmaceuticals. The use of bioconjugation chemistry is fundamental to creating these targeted agents [40].

5.5 Heterocyclic Scaffold Synthesis

Heterocyclic compounds are ubiquitous in natural products and target drug compounds, making their efficient synthesis crucial in drug discovery. Microwave heating has emerged as a powerful tool in heterocyclic chemistry, offering distinct advantages over conventional thermal methods, including faster reaction times, higher yields, and simplified purification processes.

Advantages and Examples of MAS in Heterocyclic Synthesis:

- **Accelerated Reactions**: Microwave-promoted synthesis allows for the rapid generation of diverse heterocyclic systems. Many reactions that are difficult or impossible to perform traditionally can be achieved with microwave irradiation.
- **High-Yielding Protocols**: Microwave heating can generally achieve higher yields and faster reaction times compared to conventional thermal conditions for polyheterocyclic structures, many of which are inspired by natural marine or terrestrial alkaloids with pharmaceutical potential.
- **Convenient Scale-Up and Purification**:
  - The synthesis of various benzothiazoles from N-arylimino-1,2,3-dithiazoles demonstrated that focused microwave methodologies were more productive and provided convenient methods for scale-up, often yielding products in high purity. Comparisons between solvent-free and solvent-present conditions showed that while no general rule applies, microwave heating consistently offered benefits.
  - The preparation of pyrazole and isoxazole libraries on cellulose beads was achieved through a two-step, open-vessel microwave application, leading to excellent yields and purities (>95% yield, >98% purity) [41].
  - Thioamide synthesis via the Kindler reaction under sealed vessel microwave conditions offered an ideal route, producing products in very high purity (96%) with convenient isolation by simple precipitation in water [42].
  - A one-pot microwave-assisted cyclisation procedure was successfully used to prepare 5-substituted-2-amino-oxadiazole compounds, often yielding excellent purity after simple filtration. This method can also extend to the preparation of N-protected sulphonamides [43].

5.6 Solid-Phase MAS for Radiotracer Libraries

Solid-phase organic synthesis (SPOS) [62] offers significant advantages in chemical synthesis, particularly in simplifying product isolation and purification. The integration of microwave-assisted synthesis (MAS) with solid-phase techniques has further amplified these benefits, making it an invaluable tool for accelerating the production of chemical libraries, including those for radiotracers.

Enhanced Productivity and Efficiency with Solid-Phase MAS [44]:

- **Accelerated Reactions on Solid Supports**: Microwave heating significantly accelerates reactions on solid supports such as alumina, silica, montmorillonite clay, and zeolites, even under solvent-free conditions, providing pure products with quantitative yields and high selectivity. In these "dry media" reactions, organic compounds adsorbed on inorganic oxides absorb microwaves, while the support itself does not, leading to relatively low bulk temperatures.
- **Improved Solid-Phase Peptide Synthesis (SPPS)**: As discussed previously, microwave-assisted SPPS enhances deprotection, coupling, and cleavage reactions, overcoming issues

like intermolecular aggregation, β-sheet formation, and steric hindrance. This is particularly effective using microwave flash heating in sealed vessels, which rapidly heats and cools the sample, improving the integrity of the polymer matrix itself [45].

- **Library Generation and High-Throughput Synthesis**:
  - Microwave-assisted solid-phase synthesis (MASS) has been successfully applied to the synthesis of 2,6,9-trisubstituted purines, where key aromatic nucleophilic substitution steps were performed rapidly by microwave heating [46].
  - Solid-supported amide synthesis using the safety catch principle benefited significantly from microwave assistance, reducing scavenging periods from 4-6 hours at ambient temperature to less than 10 minutes under directed microwave heating.
  - The solid-supported Ugi-4CC (four-component condensation) [47] was effectively performed under microwave conditions, affording products in high purities.
  - Automated sequential library synthesis, utilizing robotized microwave reactors, has become a powerful tool for generating small- and medium-sized compound libraries efficiently.
  - Parallel microwave library synthesis, using plate-based systems (e.g., 96-deep-well microtitreplates) [26] within multimode microwave reactors, enables high-throughput production on a 0.5–4 ml scale. This allows for the preparation of large compound libraries, although continuous instrument operation might be required for very large sets.
- **Significant Time-Savings**: One of the main benefits of microwave synthesis is the dramatic reduction in reaction times, often from days or hours to minutes or seconds [48]. This allows for the optimization of reaction conditions in a few hours instead of days or weeks. This increased speed has a profound impact on the "proof-of-principle" and "library validation" phases of drug discovery, where experimental designs and synthetic routes are evaluated and optimized [48].
- **Scale-Up Capabilities**: While microwave technology for organic synthesis at a production scale level (>1000 kg) is still a developing area, advances in continuous-flow and stop-flow microwave reactors are addressing the need for larger-scale MAS techniques, enabling continuous production of materials for lead development and eventual large-scale production [22]. Specialized microwave devices designed for laboratories and industries, capable of controlling temperature and pressure, are replacing domestic microwave ovens for reliable and reproducible chemical synthesis. Modern microwave apparatus can operate at variable temperatures and power settings, facilitating rapid heating and enhancing productivity. The integration of microwave heating with microfluidic technology is also being explored for high-throughput synthesis in microreactors, although challenges in applying and controlling heating in these systems remain.

5.7 Technical and Methodological Limitations of MAS

Despite its many advantages, microwave-assisted synthesis (MAS) faces several limitations that hinder its broader adoption in large-scale radiopharmaceutical production [26]. One key challenge is scalability—most microwave systems are optimized for small to medium batch sizes, and very few examples exist of consistent production at industrial scales (>1000 kg) [27]. Temperature control and measurement [43] remain problematic, especially in multimode systems, where non-uniform electromagnetic fields can cause gradients and reduce reproducibility. Additionally,

Materials Research Foundations 189 (2026) 268-309 https://doi.org/10.21741/9781644904039-9

accurate temperature monitoring [49] is difficult; metal thermocouples can spark, while fiber-optic probes are delicate and may be misled by absorbing materials.

Safety concerns are heightened under microwave conditions due to the combination of high temperatures, pressures, and flammable solvents, with risks such as arcing. Material limitations [43,49] also persist, as commonly used vessels like PTFE have restricted thermal and pressure tolerances. Certain solvents or reagents may not absorb microwaves efficiently, complicating uniform heating.

Moreover, post-reaction processing, including cooling and purification, can become a bottleneck, limiting the benefits of rapid synthesis. Finally, equipment cost remains a barrier; although prices have dropped over time, microwave reactors are still significantly more expensive than conventional heating systems, which may discourage adoption in resource-constrained settings.

In conclusion, microwave-assisted synthesis is not merely an alternative heating method but a fundamental shift in radiopharmaceutical production. Its ability to accelerate reactions, increase yields, and integrate seamlessly with automated and solid-phase techniques addresses critical challenges in radiochemistry, particularly for short-lived isotopes. This technology is becoming an indispensable tool in drug discovery, speeding up the development of novel diagnostic and therapeutic radiopharmaceuticals and contributing to the advancement of personalized medicine.

6. Regulatory, GMP, and Legal Framework for Microwave-Based Radio synthesis

Microwave-Assisted Synthesis (MAS) has revolutionized radiopharmaceutical chemistry, offering fast, reproducible, and high-yield methods crucial for the production of time-sensitive radioactive drugs. However, its integration into clinical and commercial practice demands strict adherence to Good Manufacturing Practice (GMP), regulatory guidelines, and legal frameworks that ensure patient safety and therapeutic efficacy [50].

6.1 GMP Requirements in Radiopharmacy

Radiopharmaceuticals are considered drugs and must meet comprehensive regulatory standards similar to conventional pharmaceuticals. This includes GMP compliance in areas such as raw material qualification, facility validation, staff training, and product quality assurance. The cost and complexity of radiopharmaceutical development are significantly heightened by these stringent requirements. Additionally, because of the short half-lives of radionuclides, synthesis and administration must be completed within tight time frames, limiting opportunities for full sterility and endotoxin testing prior to patient use.

To address these constraints, organizations like the European Association of Nuclear Medicine (EANM) and the Society of Nuclear Medicine and Molecular Imaging (SNMMI) have developed specialized guidelines tailored for small-scale and rapid radiopharmaceutical preparation. In India, the Atomic Energy Regulatory Board (AERB) enforces strict rules governing radioactive materials, including mandatory facility authorization and personnel certification.

Personnel competency is central to ensuring safety and regulatory compliance. Credentialing mechanisms such as Authorized Nuclear Pharmacist (ANP), Certified Health Physicist (CHP), and Radiation Safety Officer (RSO) ensure a skilled workforce capable of managing the unique risks involved in radioactive drug manufacturing.

6.2 Validation and Qualification of MAS Protocols

MAS enable rapid and efficient radiolabeling under tightly controlled and reproducible conditions. To ensure safety and consistency, MAS protocols must meet both national and international regulatory requirements. Modern MAS systems, often housed in shielded hot cells, are fully automated and include advanced safety mechanisms such as magnetic stirring, infrared or fiber-optic temperature sensors, pressure monitoring, and emergency shut-off systems.

MAS offers substantial advantages over conventional heating in radiolabeling reactions. For example, MAS can achieve >99% radiochemical incorporation of ^{68}Ga-labeled peptide conjugates with increased specific activity. Furthermore, integrating purification and preconcentration steps enables quality control via HPLC to be completed in under 10 minutes—a crucial feature given ^{68}Ga's short half-life (67.7 minutes) [51,52].

Because full sterility testing is often impractical within the time available, parametric release based on robust, validated processes is increasingly accepted. This approach ensures clinical safety using data from well-characterized production protocols.

6.3 Quality Control and Batch Documentation

Quality control (QC) in radiopharmaceuticals [53] is adapted to account for radioactive decay, while still ensuring product safety and efficacy. Essential QC tests include:

Radiochemical Purity – confirms that radioactivity is in the desired chemical form.

Radionuclidic Purity – verifies the correct radionuclide.

Chemical Purity – measures residual chemical contaminants.

Specific Activity – radioactivity per unit mass.

Radioactive Concentration/Total Activity – important for dosage calculations.

Sterility and Endotoxin Testing – though often performed post-administration, justified by validated aseptic processes.

Under GMP, detailed batch records are mandatory for reproducibility and traceability. These records document everything from raw materials and synthesis parameters to QC results. Automation enhances reliability, while rigorous staff training in QC protocols remains essential. Regulatory pharmacopoeias like the European Pharmacopoeia and Indian Pharmacopoeia outline critical standards and definitions, including half-life, generator systems, isotopic carriers, and precursor specifications.

6.4 Regulatory Perspectives (FDA, EMA, IAEA)

Radiopharmaceuticals, being parenteral drugs containing radioactive components, fall under stringent international regulations covering formulation, labelling, packaging, and clinical use.

In India, the AERB oversees all aspects of radiopharmaceutical usage, focusing on laboratory certification, personnel training, and radiation safety [54].

In the EU, the European Pharmacopoeia offers monographs specific to PET radiopharmaceuticals, supplemented by the EANM's cGRPP guidelines tailored for small-scale production under GMP [55]. In the USA, the FDA has approved several advanced radiotherapeutics like [^{177}Lu]Lu-

Materials Research Foundations 189 (2026) 268-309 — https://doi.org/10.21741/9781644904039-9

DOTA-TATE, [^{177}Lu]Lu-PSMA-617, and Radium-223, establishing the foundation for theranostics in oncology [56].

The IAEA supports global harmonization through infrastructure development, training, and guidance for regulatory compliance [27].

However, regulatory inconsistencies, especially within Europe, where member states may interpret EU directives differently, continue to hinder standardized cross-border production and distribution.

6.5 Integration into Clinical and Commercial Settings

Microwave-Assisted Organic Synthesis (MAOS) is now widely adopted beyond research labs into industrial settings due to its speed, yield, and scalability. Sophisticated microwave reactors, featuring integrated magnetic stirrers, IR or fiber-optic temperature monitoring, and programmable control software, allow reactions to be completed in seconds or minutes.

Modern MAS systems support production across a wide range—from milligram-scale lab experiments to kilogram-scale manufacturing [57]. Flow-through and stop-flow microwave reactors allow both continuous and batch processes. For instance, Monash University's microwave batch reactor can handle 25–200 mL at up to 260°C and 10 MPa, demonstrating the feasibility of MAS for preclinical and clinical-scale production [58].

6.6 Legal Barriers and Limitations

Despite the promise of MAS in radiopharmaceuticals, several legal and practical obstacles [58-59] impede its widespread use:

High Development Costs: Like other pharmaceuticals, radiopharmaceuticals require full clinical trials (Phases I–III), which are cost-prohibitive given the typically small patient populations [60].

Short Shelf Life and Generics: These compounds cannot be stockpiled and face swift generic competition, reducing commercial incentives.

Regulatory Inconsistencies: Divergent interpretations of guidelines across regions create complexity for multinational collaboration.

Safety Risks in MAS: Operating at high temperatures and pressures introduces risks such as microwave absorption by metallic particles and arcing. Thus, robust safety systems—including shielding, leak detection, and operator training—are mandatory.

In conclusion, Microwave-Assisted Synthesis represents a breakthrough in radiopharmaceutical manufacturing. It aligns well with the sector's need for rapid, high-yield, and reproducible processes. However, to fully realize MAS's potential, it is essential to harmonize global regulatory frameworks, validate synthesis protocols, and overcome economic and legal hurdles. Only through these combined efforts can MAS become a standardized approach in both clinical and commercial radiopharmacy.

7. Future Trends and Innovations in Radiopharmaceutical Synthesis

The field of radiopharmaceuticals is rapidly evolving, driven by significant advancements in chemistry, physics, and medicine. Future trends and innovations are poised to revolutionize how these agents are synthesized and utilized, promising enhanced precision, efficacy, and sustainability in medical applications.

The landscape of radiopharmaceutical synthesis is undergoing a transformative shift, propelled by innovations in chemistry, engineering, and data science. As demand for targeted diagnostics and therapeutics grows, emerging technologies such as microfluidics, artificial intelligence, theranostics, and novel radionuclides—alongside microwave-assisted synthesis (MAS)—are reshaping the field with an emphasis on precision, speed, and sustainability.

7.1 Synergy with Microfluidics, Flow Chemistry, and Microwave-Assisted Synthesis

To address the challenge of synthesizing short-lived radionuclides like carbon-11 and fluorine-18, the integration of microfluidic and continuous-flow chemistry systems with MAS has emerged as a powerful solution. Microfluidic platforms—or "lab-on-a-chip" systems [22]—enable miniaturized, high-throughput synthesis under tightly controlled conditions. These technologies significantly accelerate reaction rates and improve yields by facilitating localized, uniform heating through microwave irradiation. Automated microwave synthesizers operating in single-mode cavities can rapidly perform multiple sequential reactions, aided by robotic manipulation of sealed vessels. Continuous-flow microwave reactors, such as the Voyager system, exemplify scalable synthesis for clinical and industrial needs, offering enhanced reproducibility and safety over early domestic-microwave setups [61].

MAS also enables complex transformations, such as Suzuki cross-coupling, within capillary-based micro-reactors [62]. These innovations allow for parallel and combinatorial synthesis with minimal reaction times, supporting the production of structurally diverse radiopharmaceuticals. Notably, microfluidic systems have already translated into clinical use—for instance, the first human imaging study using [^{18}F]T807 synthesized via flow chemistry—demonstrating their real-world applicability [63].

7.2 AI-Powered Radio Synthesis and Diagnostic Precision

Artificial Intelligence (AI) is revolutionizing both the development and clinical utility of radiopharmaceuticals. Machine learning algorithms excel in analyzing vast imaging datasets, detecting subtle anomalies, and improving diagnostic accuracy beyond human capabilities. AI-driven image analysis enhances lesion detection, disease staging, and monitoring of treatment response [64].

Moreover, AI supports personalized therapy planning by integrating radiotracer biodistribution, dosimetry data, and patient-specific parameters to optimize therapeutic efficacy. In research, AI accelerates the identification and development of novel radiotracers by modelling molecular interactions and predicting in vivo behavior. The convergence of AI with radiopharmaceutical science is poised to transform patient care, making personalized molecular medicine a clinical standard (Fig. 4).

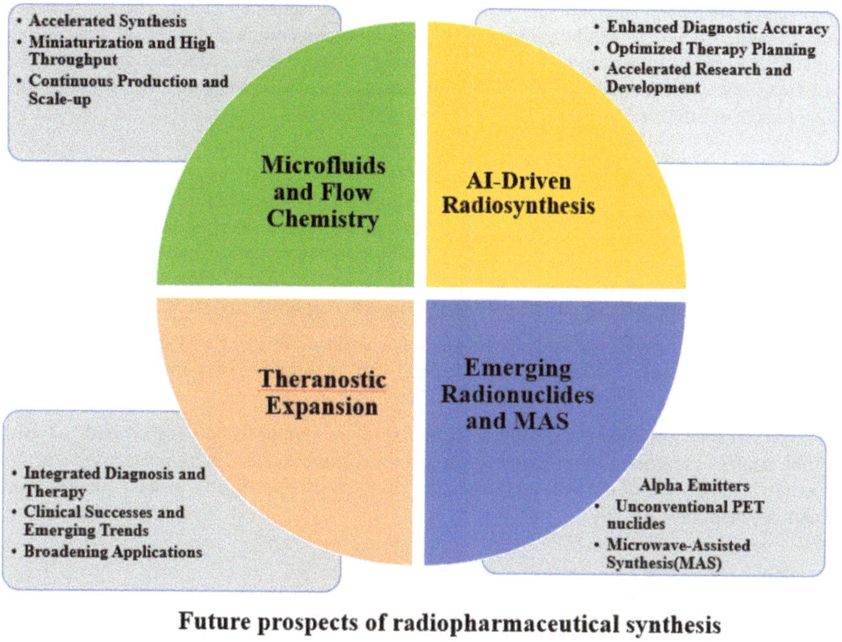

Figure 4 – Future prospects of Radiopharmaceutical synthesis

7.3 Theranostic Expansion

Theranostics—combining diagnostic imaging with therapeutic functionality in a single agent—is redefining treatment strategies in nuclear medicine. Radiopharmaceuticals such as [^{68}Ga]Ga-PSMA-11 and [^{177}Lu]Lu-PSMA-617 allow for precise disease localization followed by targeted therapy using the same ligand scaffold [65]. This seamless integration reduces variability in pharmacokinetics and enables real-time therapy monitoring.

Clinically, agents like [^{177}Lu]Lu-DOTA-TATE have demonstrated exceptional outcomes in oncology. The theranostic approach is expanding beyond cancer to include neurodegenerative, cardiovascular, and inflammatory diseases, aligning with the goals of precision medicine.

7.4 Novel Radionuclides and the Expanding Role of MAS

Emerging radionuclides, especially alpha emitters such as actinium-225 and thorium-227, offer superior therapeutic profiles due to their high linear energy transfer and localized tissue damage [66]. However, their adoption is limited by production, cost, and radiochemical challenges. Additionally, nonconventional PET nuclides like ^{64}Cu, ^{89}Zr, and ^{124}I open new diagnostic possibilities but require specialized handling due to their complex decay schemes and radioprotection needs [67].

Materials Research Foundations 189 (2026) 268-309 https://doi.org/10.21741/9781644904039-9

MAS plays an increasingly critical role in handling these sensitive radionuclides. Its benefits include rapid, high-yield synthesis, precise control over temperature and pressure, compatibility with diverse organic transformations, and potential for solid-phase and multi-component reactions. MAS systems, when coupled with automation and continuous flow, are well-suited for both small-scale radiotracer development and industrial-scale production.

7.5 Sustainability Goals in Radio Pharmacy

Sustainability is gaining prominence in radiopharmaceutical manufacturing. MAS aligns with green chemistry principles by enabling solvent-free reactions and reducing hazardous waste. The miniaturization of synthesis platforms, supported by relocatable microwave reactors, minimizes chemical stockpiles and enhances safety [68]. By targeting disease sites with high precision, modern radiopharmaceuticals reduce off-target effects and optimize resource usage.

Compliance with international safety standards, such as those from the SNMMI and EANM, along with rigorous training, patient-specific dosing, and real-time monitoring, are integral for maintaining quality in radiopharmacy practice [69].

The convergence of these innovations—from advanced synthesis techniques and AI-driven optimization to the expansion of theranostics and the exploration of new radionuclides—is set to redefine the landscape of medical progress, offering more precise, effective, and patient-centered healthcare solutions.

8. Perspective and Conclusion

The evolution of radiopharmaceutical synthesis is marked by a transformative blend of scientific disciplines, poised to significantly reshape medical practices and patient care. This concluding perspective reflects on the impactful strides made, the pivotal role of enabling technologies, and the critical areas necessitating future focus.

Radiopharmaceuticals lie at the confluence of nuclear physics, chemistry, and medicine, playing a pivotal role in advancing molecular imaging and targeted therapy. Recent innovations have significantly enhanced the precision and efficacy of these agents. Notably, theranostics—combining diagnostic and therapeutic functions within a single radiopharmaceutical—has revolutionized personalized medicine. FDA approvals of agents like [^{177}Lu]Lu-DOTA-TATE and [^{177}Lu]Lu-PSMA-617 underscore this progress, enabling patient-specific treatment informed by real-time imaging [70]. Nanoparticle-based formulations further enhance this paradigm by serving as advanced carriers with improved biodistribution, extended circulation, and the potential for multifunctional payload delivery [35].

Artificial Intelligence (AI) integration marks a transformative advance, optimizing image interpretation, therapy planning, and the discovery of novel radiotracers through high-throughput data analysis [62,71]. These technological strides collectively drive a healthcare future that is more precise, effective, and patient-centric.

Microwave-Assisted Synthesis (MAS) has become a powerful translational tool in radiopharmaceutical development. By reducing reaction times from hours to seconds, MAS is especially advantageous for synthesizing compounds labeled with short-lived radionuclides such as carbon-11 and fluorine-18. It consistently delivers higher radiochemical yields and purity, with greener chemistry through solvent-free protocols. MAS also supports diverse chemical transformations, including multi-component and solid-phase reactions. The integration of automated systems and continuous-flow microreactors facilitates scale-up for industrial applications. The clinical preparation of [^{18}F]T807 using microfluidic flow MAS exemplifies its translational potential [72].

Despite progress, several challenges remain. These include the limited supply and production capacity for promising alpha-emitting radionuclides (e.g., astatine-211, actinium-225), and the need for robust infrastructure and redundancy planning to ensure consistent availability. Dosimetry and radiobiology must be further refined to improve individualized radiation delivery while minimizing normal tissue exposure [73].

Off-target effects—such as uptake in healthy organs causing toxicity—remain a major limitation. Designing ligands with higher specificity and favorable pharmacokinetics is a priority. Furthermore, imaging resolution must advance toward sub-cellular sensitivity through device innovation. Clinical translation remains hindered by regulatory and licensing complexities, calling for stronger collaboration among stakeholders. Specific radionuclide challenges include poor tumor retention (e.g., with FAPIs), unwanted hepatic uptake ([^{99m}Tc]Tc-MIP-1404), and high healthy organ doses ([^{177}Lu]Lu-rhPSMA-7.3), as well as the logistical and cost burdens of antibody-based tracers.

9. Final Remarks

The future of radiopharmaceuticals promises a continuous story of scientific innovation and transformative medical progress. The field is characterized by a dynamic and innovative nature, constantly exploring new avenues for cancer treatment and disease diagnosis. This progress is critically dependent on multidisciplinary collaboration, bringing together radiopharmacists, nuclear medicine specialists, chemists, bioengineers, nuclear physicists, and device engineers. As research and clinical trials continue to expand, these advancements are expected to redefine the landscape of medical care, offering more precise, effective, and patient-centered healthcare solutions. The harmonious integration of molecular precision, advanced imaging technologies, and collaborative insights will continue to push the boundaries of possibility, ultimately leading to improved patient outcomes and a redefinition of disease management.

References

[1] S. Dhoundiyal, S. Srivastava, S. Kumar, G. Singh, S. Ashique, R. Pal, N. Mishra, F. Taghizadeh-Hesary, Radiopharmaceuticals: Navigating the frontier of precision medicine and therapeutic innovation, Eur. J. Med. Res. 29 (2024) 26. https://doi.org/10.1186/s40001-023-01627-0

[2] S.H. Liang, N. Vasdev, Total radiosynthesis: Thinking outside 'the box', Aust. J. Chem. 68 (2015) 1319-1328. https://doi.org/10.1071/CH15406

[3] B.L. Hayes, Recent advances in microwave-assisted synthesis, Aldrichimica Acta 37 (2004) 66-77.

[4] J.P. Holland, Chemical kinetics of radiolabelling reactions, Chem.-Eur. J. 24 (2018) 16472-16483. https://doi.org/10.1002/chem.201803261

[5] J. Wang, R.M. van Dam, High-efficiency production of radiopharmaceuticals via droplet radiochemistry: A review of recent progress, Mol. Imaging 19 (2020) 1536012120973099. https://doi.org/10.1177/1536012120973099

[6] G.J. Wasserburg, Isotopic adventures-geological, planetological, and cosmic, Annu. Rev. Earth Planet. Sci. 31 (2003) 1-74. https://doi.org/10.1146/annurev.earth.31.100901.141409

[7] J.W. Bai, S.Q. Qiu, G.J. Zhang, Molecular and functional imaging in cancer-targeted therapy: Current applications and future directions, Signal Transduct. Target. Ther. 8 (2023) 89. https://doi.org/10.1038/s41392-023-01366-y

[8] C.W. Lindsley, C.E. Müller, S. Bongarzone, Diagnostic and therapeutic radiopharmaceuticals, ACS Pharmacol. Transl. Sci. 5 (2022) 835-837. https://doi.org/10.1021/acsptsci.2c00173

[9] G. Crişan, N.S. Moldovean-Cioroianu, D.G. Timaru, G. Andrieş, C. Căinap, V. Chiş, Radiopharmaceuticals for PET and SPECT imaging: A literature review over the last decade, Int. J. Mol. Sci. 23 (2022) 5023. https://doi.org/10.3390/ijms23095023

[10] D. Delbeke, W.H. Martin, Metabolic imaging with FDG: A primer, Cancer J. 10 (2004) 201-213. https://doi.org/10.1097/00130404-200407000-00001

[11] J.W. Hicks, Fluorous-Phase Purification of ^{99m}Tc Radiopharmaceuticals, Mastars Thesis, McMaster University, (2010).

[12] C. Schedeit, N. Gözlügöl, R.S. Alsheikh, M. Shelan, R. Seifert, F. Caobelli, U. Borner, et al., Radioactivity levels in the saliva of patients undergoing targeted radioligand therapy with [177Lu]Lu-PSMA-I&T and [177Lu]Lu-DOTA-TOC, Clin. Oral Investig. 29 (2025) 241. https://doi.org/10.1007/s00784-025-06300-w

[13] E. Deshayes, M. Roumiguie, C. Thibault, P. Beuzeboc, F. Cachin, C. Hennequin, D. Huglo, F. Rozet, D. Kassab-Chahmi, X. Rebillard, N. Houédé, Radium-223 dichloride for prostate cancer treatment, Drug Des. Devel. Ther. 11 (2017) 2643-2651. https://doi.org/10.2147/DDDT.S122417

[14] P. Kręcisz, K. Czarnecka, L. Królicki, E. Mikiciuk-Olasik, P. Szymański, Radiolabeled peptides and antibodies in medicine, Bioconjug. Chem. 32 (2020) 25-42. https://doi.org/10.1021/acs.bioconjchem.0c00617

[15] D. Le, Regulatory oversight of radiopharmaceuticals, in: F.C.L. Wong (Ed), Radiopharmaceuticals in the Management of Leptomeningeal Metastasis, Springer, Cham, 2022, pp. 199-213. https://doi.org/10.1007/978-3-031-14291-8_11

[16] F.F. Knapp, A. Dash, Radiopharmaceuticals for Therapy, Springer, New Delhi, India, 2016. https://doi.org/10.1007/978-81-322-2607-9

[17] P. Brugarolas, J. Comstock, D.W. Dick, T. Ellmer, J.W. Engle, S.E. Lapi, S.H. Liang, E.E. Parent, N.V.K. Pillarsetty, S. Selivanova, X. Sun, A. Vavere, P.J.H. Scott, Fifty years of radiopharmaceuticals, J. Nucl. Med. Technol. 48 (Suppl. 1) (2020) 34S-39S.

[18] N.R. Kar, Production and applications of radiopharmaceuticals: A review, Int. J. Pharm. Investig. 9 (2019) 78-85. https://doi.org/10.5530/ijpi.2019.2.8

[19] S. Stone-Elander, N. Elander, The use of microwaves for the automated production of radiopharmaceuticals, in: Chemists' Views of Imaging Centers, Springer US, Boston, MA, 1995, pp. 445-454. https://doi.org/10.1007/978-1-4757-9670-4_48

[20] M. Gaba, N. Dhingra, Microwave chemistry: General features and applications, Ind. J. Pharm. Educ. Res. 45 (2011) 175-183.

[21] S.H.O.M.O.K.H. Alsharef, M.A.S.H.A.E.L. Alanazi, F.A.T.I.M.A.H. Alharthi, D.A.N.A. Qandil, M.O.N.A. Qushawy, Review about radiopharmaceuticals: Preparation, radioactivity, and applications, Int. J. App. Pharm.12 (2020) 8-15. https://doi.org/10.22159/ijap.2020v12i3.37150

[22] T.N. Glasnov, C.O. Kappe, Microwave-assisted synthesis under continuous-flow conditions, Macromol. Rapid Commun. 28 (2007) 395-410. https://doi.org/10.1002/marc.200600665

[23] M. Willert-Porada, Advances in Microwave and Radio Frequency Processing, Springer-Verlag Berlin Heidelberg, 2006. https://doi.org/10.1007/3-540-32944-7

[24] W. Wadsak, M. Mitterhauser, Basics and principles of radiopharmaceuticals for PET/CT, Eur. J. Radiol. 73 (2010) 461-469. https://doi.org/10.1016/j.ejrad.2009.12.022

[25] H.K. Solanki, V.D. Prajapati, G.K. Jani, Microwave technology-a potential tool in pharmaceutical science, Int. J. Pharmtech Res. 2 (2010) 1754-1761.

[26] C.O. Kappe, D. Dallinger, The impact of microwave synthesis on drug discovery, Nat. Rev. Drug Discov. 5 (2006) 51-63. https://doi.org/10.1038/nrd1926

[27] B.A. Roberts, C.R. Strauss, Toward rapid, "green", predictable microwave-assisted synthesis, Acc. Chem. Res. 38 (2005) 653-661. https://doi.org/10.1021/ar040278m

[28] S. Ravichandran, E. Karthikeyan, Microwave synthesis-a potential tool for green chemistry, Int. J. ChemTech Res. 3 (2011) 466-470.

[29] H. Kimura, Y. Yagi, N. Ohneda, H. Odajima, M. Ono, H. Saji, Development of resonant-type microwave reactor and its application to PET radiochemistry, (2014) 680-686. https://doi.org/10.1002/jlcr.3232

[30] O. Eskola, Fluorine and 18F-fluorine in radiopharmaceutical preparation, (2013). Ph.D. Thesis, Department of Chemistry and Turku PET Centre, University of Turku, Turku, Finland, ISBN 978-951-29-5318-9 (PRINT)

[31] H.J.L.M. Timmers, C.C. Chen, J.A. Carrasquillo, M. Whatley, A. Ling, G. Eisenhofer, K.S. King, et al., Staging and functional characterization of pheochromocytoma and paraganglioma by 18F-fluorodeoxyglucose (18F-FDG) positron emission tomography, J. Natl. Cancer Inst. 104 (2012) 700-708. https://doi.org/10.1093/jnci/djs188

[32] T. Wang, Y. Zhang, X. Zhang, L. Chen, M. Zheng, J. Zhang, P. Brust, W. Deuther-Conrad, Y. Huang, H. Jia, Synthesis and characterization of the two enantiomers of a chiral sigma-1 receptor radioligand: (S)-(+)- and (R)-(-)-[18F] FBFP, Chin. Chem. Lett. 33 (2022) 3543-3548. https://doi.org/10.1016/j.cclet.2022.03.099

[33] B.H. Rotstein, L. Wang, R.Y. Liu, J. Patteson, E.E. Kwan, N. Vasdev, S.H. Liang, Mechanistic studies and radiofluorination of structurally diverse pharmaceuticals with

spirocyclic iodonium (III) ylides, Chem. Sci. 7 (2016) 4407-4417. https://doi.org/10.1039/C6SC00197A

[34] S. Zhang, X. Wang, X. Gao, X. Chen, L. Li, G. Li, C. Liu, Y. Miao, R. Wang, K. Hu, Radiopharmaceuticals and their applications in medicine, Signal Transduct. Target. Ther. 10 (2025) 1. https://doi.org/10.1038/s41392-024-02041-6

[35] A.R. Jalilian, J. Osso Jr., The current status and future of theranostic Copper-64 radiopharmaceuticals, Iran. J. Nucl. Med. 25 (2017) 1-8.

[36] E. Merola, C.M. Grana, Peptide receptor radionuclide therapy (PRRT): Innovations and improvements, Cancers 15 (2023) 2975. https://doi.org/10.3390/cancers15112975

[37] A. Wurzer, F. De Rose, S. Fischer, M. Schwaiger, W. Weber, S. Nekolla, H.-J. Wester, M. Eiber, C. D'Alessandria, Preclinical comparison of [177Lu] Lu-rhPSMA-10.1 and [177Lu] Lu-rhPSMA-10.2 for endoradiotherapy of prostate cancer: Biodistribution and dosimetry studies, EJNMMI Radiopharm. Chem. 9 (2024) 18. https://doi.org/10.1186/s41181-024-00246-2

[38] M.M. Da'san, O.A. Jaradat, F. Albericio, Advances in solid-phase peptide synthesis in aqueous media (ASPPS), Green Chem. 24 (2022) 6360-6372. https://doi.org/10.1039/D2GC02319A

[39] A. Wurzer, D. Di Carlo, M. Herz, A. Richter, S. Robu, R. Schirrmacher, A. Mascarin, et al., Automated synthesis of [18F] Ga-rhPSMA-7/-7.3: Results, quality control and experience from more than 200 routine productions, EJNMMI Radiopharm. Chem. 6 (2021) 1-15. https://doi.org/10.1186/s41181-021-00120-5

[40] M. Erdélyi, Solid-phase methods for the microwave-assisted synthesis of heterocycles, Microwave-Assisted Synthesis of Heterocycles (2006) 79-128. https://doi.org/10.1007/7081_001

[41] O.I. Zbruyev, N. Stiasni, C.O. Kappe, Preparation of thioamide building blocks via microwave-promoted three-component Kindler reactions, J. Comb. Chem. 5 (2003) 145-148. https://doi.org/10.1021/cc0200538

[42] I.R. Baxendale, A.-L. Lee, S.V. Ley, Integrating microwave-assisted synthesis and solid-supported reagents, Blackwell Publishing, Oxford, UK, 2005. https://doi.org/10.1002/9781444305548.ch6

[43] J. Tierney, P. Lidström (Eds.), Microwave Assisted Organic Synthesis, John Wiley & Sons, 2009.

[44] H.M. Yu, S.T. Chen, K.T. Wang, Enhanced coupling efficiency in solid-phase peptide synthesis by microwave irradiation, J. Org. Chem. 57 (1992) 4781. https://doi.org/10.1021/jo00044a001

[45] R.E. Austin, J.F. Okonya, D.R.S. Bond, F. Al-Obeidi, Microwave-assisted solid-phase synthesis (MASS) of 2,6,9-trisubstituted purines, Tetrahedron Lett. 43 (2002) 6169-6171. https://doi.org/10.1016/S0040-4039(02)01271-6

[46] Y. Jiang, Y. Yuan, The tribromolanthanoids (LnBr3) catalysed reactions of benzyl alkyl ethers and carboxylic acids promoted by microwave irradiation, Synth. Commun. 24 (1994) 1045. https://doi.org/10.1080/00397919408020781

[47] S.A. Galema, Microwave chemistry, Chem. Soc. Rev. 26 (1997) 233. https://doi.org/10.1039/cs9972600233

[48] J.M. Kremsner, A. Stadler, C.O. Kappe, The scale-up of microwave-assisted organic synthesis, in: Microwave Methods in Organic Synthesis, 2006, pp. 233-278. https://doi.org/10.1007/128_048

[49] M. Poli, M. Quaglierini, A. Zega, S. Pardini, M. Telleschi, G. Iervasi, L. Guiducci, Risk management in Good Manufacturing Practice (GMP) radiopharmaceutical preparations, Appl. Sci. 14 (2024) 1584. https://doi.org/10.3390/app14041584

[50] N. Loktionova, Development and Evaluation of a 44Ti-44Sc Radionuclide Generator and Labeling of Biomolecules with 44Sc and 68Ga for PET Imaging, PhD Thesis, University of Mainz, 2010.

[51] L. Wang, H.-T. Kuo, C.-C. Chen, D. Chapple, N. Colpo, P. Ng, W.S. Lau, S. Jozi, F. Bénard, K.-S. Lin, Synthesis and evaluation of the first ^{68}Ga-labeled C-terminal hydroxamate-derived gastrin-releasing peptide receptor-targeted tracers for cancer imaging with positron emission tomography, Molecules, 29 (2024) 3102. https://doi.org/10.3390/molecules29133102

[52] S. Molavipordanjani, S.J. Hosseinimehr, Fundamental concepts of radiopharmaceuticals quality controls, Pharm. Biomed. Res. 4 (2018) 1-8. https://doi.org/10.18502/pbr.v4i3.538

[53] Atomic Energy Regulatory Board, AERB Regulation 2, no. 03, 2012.

[54] N. Gillings, O. Hjelstuen, J. Ballinger, M. Behe, C. Decristoforo, P. Elsinga, V. Ferrari, et al., Guideline on current good radiopharmacy practice (cGRPP) for the small-scale preparation of radiopharmaceuticals, EJNMMI Radiopharm. Chem. 6 (2021) 1-22. https://doi.org/10.1186/s41181-021-00123-2

[55] R. Chakravarty, S. Chakraborty, A review of advances in the last decade on targeted cancer therapy using 177Lu: Focusing on 177Lu produced by the direct neutron activation route, Am. J. Nucl. Med. Mol. Imaging 11 (2021) 443.

[56] M.O. Adelodun, E.C. Anyanwu, Global standards in radiation safety: A comparative analysis of healthcare regulations, Int. J. Eng. Inv. 13 (2024) 216-227.

[57] K.D. Raner, C.R. Strauss, R.W. Trainor, J.S. Thorn, A new microwave reactor for batchwise organic synthesis, J. Org. Chem. 60 (1995) 2456-2460. https://doi.org/10.1021/jo00113a028

[58] S.E. Lapi, P.J.H. Scott, A.M. Scott, A.D. Windhorst, B.M. Zeglis, M. Abdel-Wahab, R.P. Baum, et al., Recent advances and impending challenges for the radiopharmaceutical sciences in oncology, Lancet Oncol. 25 (2024) e236-e249. https://doi.org/10.1016/S1470-2045(24)00030-5

[59] L. Greene, S. Srinivas, S. Park, N. Hatami, T. Nobashi, L. Baratto, A. Toriihara, S.S. Gambhir, Role of imaging in early-phase trials, in: Novel Designs of Early Phase Trials for

Cancer Therapeutics, Academic Press, 2018, pp. 129-149. https://doi.org/10.1016/B978-0-12-812512-0.00010-5

[60] P.S. Dittrich, A. Manz, Lab-on-a-chip: Microfluidics in drug discovery, Nat. Rev. Drug Discov. 5 (2006) 210-218. https://doi.org/10.1038/nrd1985

[61] E. Comer, M.G. Organ, A microreactor for microwave-assisted capillary (continuous flow) organic synthesis, J. Am. Chem. Soc. 127 (2005) 8160-8167. https://doi.org/10.1021/ja0512069

[62] S.H. Liang, D.L. Yokell, M.D. Normandin, P.A. Rice, R.N. Jackson, T.M. Shoup, T.J. Brady, G. El Fakhri, T.L. Collier, N. Vasdev, First human use of a radiopharmaceutical prepared by continuous-flow microfluidic radiofluorination: Proof of concept with the tau imaging agent [18F] T807, Mol. Imaging 13 (2014) 1-5. https://doi.org/10.2310/7290.2014.00025

[63] O.V. Unanah, S. Akoji, Integrating AI and radioactive pharmacy nuclear imaging to tackle healthcare and mental health disparities in underserved US populations: Business models, ethical considerations, and policy implications, Int. J. Res. Pub. Rev. 6 (2025) 2756-2769. https://doi.org/10.55248/gengpi.6.0325.1199

[64] C. Meyer, V. Prasad, A. Stuparu, P. Kletting, G. Glatting, J. Miksch, C. Solbach, et al., Comparison of PSMA-TO-1 and PSMA-617 labeled with gallium-68, lutetium-177 and actinium-225, EJNMMI Res. 12 (2022) 65. https://doi.org/10.1186/s13550-022-00935-6

[65] U.B. Hagemann, K. Wickstroem, S. Hammer, R.M. Bjerke, S. Zitzmann-Kolbe, O.B. Ryan, J. Karlsson, et al., Advances in precision oncology: Targeted thorium-227 conjugates as a new modality in targeted alpha therapy, Cancer Biother. Radiopharm. 35 (2020) 497-510. https://doi.org/10.1089/cbr.2020.3568

[66] G. Hao, A.N. Singh, W. Liu, X. Sun, PET with non-standard nuclides, Curr. Top. Med. Chem. 10 (2010) 1096-1112. https://doi.org/10.2174/156802610791384289

[67] P. Priecel, J.A. Lopez-Sanchez, Advantages and limitations of microwave reactors: From chemical synthesis to the catalytic valorization of biobased chemicals, ACS Sustain. Chem. Eng. 7(2018) 3-21. https://doi.org/10.1021/acssuschemeng.8b03286

[68] K. Herrmann, L. Giovanella, A. Santos, J. Gear, P. OzgenKiratli, J. Kurth, A.M. Denis-Bacelar, et al., Joint EANM, SNMMI and IAEA enabling guide: How to set up a theranostics centre, Eur. J. Nucl. Med. Mol. Imaging 49 (2022) 2300-2309. https://doi.org/10.1007/s00259-022-05785-x

[69] E. Yazdani, P. Geramifar, N. Karamzade-Ziarati, M. Sadeghi, P. Amini, A. Rahmim, Radiomics and Artificial Intelligence in Radiotheranostics: A Review of Applications for Radioligands Targeting Somatostatin Receptors and Prostate-Specific Membrane Antigens, Diagnostics 14 (2024) 181. https://doi.org/10.3390/diagnostics14020181

[70] S. Anwarkhan, J. Koilpillai, D. Narayanasamy, Utilizing multifaceted approaches to target drug delivery in the brain: From nanoparticles to biological therapies, Cureus 16 (2024) e68419. https://doi.org/10.7759/cureus.68419

[71] G.D. Bowden, Y. Lu, J. Jones, R.M. van Dam, G. Pascali, P.J.H. Scott, High-throughput techniques in radiochemistry, in: Automated Technologies for the Development and

Production of Radiopharmaceuticals, Springer, Cham, 2025, pp. 459-499. https://doi.org/10.1007/978-3-031-84632-8_18

[72] D.S. Wilbur, The radiopharmaceutical chemistry of alpha-emitting radionuclides, in: Radiopharmaceutical Chemistry, Springer, Cham, 2019, pp. 409-424. https://doi.org/10.1007/978-3-319-98947-1_23

[73] M. Desrosiers, L. DeWerd, J. Deye, P. Lindsay, M.K. Murphy, M. Mitch, F. Macchiarini, S. Stojadinovic, H. Stone, The importance of dosimetry standardization in radiobiology, J. Res. Natl. Inst. Stand. Technol. 118 (2013) 403-417. https://doi.org/10.6028/jres.118.021

Materials Research Foundations 189 (2026) 310-347 | https://doi.org/10.21741/9781644904039-10

Chapter 10

Metal-Ferrite Synthesis via Microwave for Targeted Drug Delivery

Tooba Irfan[1], Iqra Urooj[1], Zia Ur Rehman[2], Muhammad Arshad Raza[1*], Aamna Majeed[1], Adnan Ashraf[3], Gulzar Muhammad[1*]

[1]Department of Chemistry, Institute of Chemical Sciences, Government College University Lahore, 54000 Lahore, Pakistan

[2]Department of Chemistry, College of Physical Sciences & Engineering, University of Wyoming, Laramie

[3]Department of Chemistry, University of Lahore, Lahore, Pakistan

Arshad.raza@gcu.edu.pk, mgulzar@gcu.edu.pk

Abstract

Ferrites are homogeneous, ceramic-based magnetic materials primarily comprising iron oxide, known for achieving full magnetization along a preferred axis when subjected to an applied magnetic field. Microwave-assisted synthesis provides a rapid and efficient route for producing controlled-shape and size ferrites by aligning dipoles through microwave excitation. The materials exhibit high electrical resistivity, low dielectric loss, moderate permittivity, biocompatibility, and cost-effectiveness. The unique magnetic properties make ferrites promising candidates for biomedical applications, particularly in targeted drug delivery, where they serve as drug carriers, contrast agents, imaging tracers, and reagents for immune assays. This chapter explores the synthesis and biomedical potential of ferrites.

Keywords

Ferrite, Magnetic Field, Microwave Synthesis, Targeted Drug Delivery, Electric Field, Drug Delivery System

List of Abbreviations

MRI	Magnetic Resonance Imaging
MPI	Magnetic Particle Imaging
Dex	Dexamethasone

Contents

Materials Research Foundations 189 (2026) 310-347 https://doi.org/10.21741/9781644904039-10

1. Introduction

Ferrite materials date back centuries to when stones were discovered. The ferrite stones occurred mainly in the "Magnesia region of Asia". Therefore, the stones were given the name *magnetite* (Fe_3O_4). In ancient times, early navigators used magnetite as "Lodestones" to find the direction of north [1]. Naturally, Lodestone is an iron oxide that is used extensively by geologists and engineers. About 2000 years ago, its properties fascinated the Greeks. Long ago, the Chinese used a compass made of lodestone, also known as ferrite, to find direction [2]. In 1600, William Gilbert was the first to publish his study on "*De Magnete*" magnetism [3]. Hans Christian Oersted (1819) described that an electric current passing through a wire affects a magnetic compass [4]. The field of electromagnetism has been developed due to the contributions of Faraday, Hertz, Maxwell, and many other scientists. The word "ferrite" originates from the Latin "ferrum", denoting "iron". Ferrites are homogeneous ceramic-based materials synthesized from various oxides, mainly iron oxide [5].

In recent years, ferrite materials have been exploited due to their moderate magnetic properties in various practical applications [6]. Recently, research on iron oxide-based magnetically assembled ferrimagnetic materials, particularly hematite (Fe_2O_3) and magnetite (Fe_3O_4), has been conducted due to their unique magnetic properties, focusing on their synthesis and various applications. Y. Kato and T. Takei, in 1930, reported the synthesis of the first ferrite at the Tokyo Institute of Technology [7]. *Three different crystal systems* of ferrites can be obtained via varying synthesis methods. The flexibility of preparing various solid solution compositions enables the tailoring of desired properties for a wide range of applications. Ferromagnetic metals cannot replace ferrites. This makes ferrites a suitable candidate for different applications. They are also comparable to many metals due to economic reasons. Ferrites are dielectric, demonstrating that although they are non-conductive electrically, they can be magnetized [2]. The ferrites have magnetic properties but are insulators, which gives them an advantage compared to transition metals (Ni, Fe, and others). Transition metals have magnetic properties but are not insulators. Ferrites have high resistivity, shape versatility, a wide range of frequencies (10 kHz to 50 MHz), huge material selection, small package, time, temperature stability [8], an inexpensive nature, are adaptable, stable over various conditions, and are less toxic towards living creatures [9]. Compared to other magnetic materials, ferrites exhibit negligible eddy current resistance losses over a wide frequency range (10 kHz to 50 MHz) with enhanced electrical resistance [8]. Ferrite synthesis is a highly complex process. The researchers have spent over 50 years understanding the mechanisms involved in synthesis. However, further research is still needed to determine the most efficient method. Ferrite can be synthesized using a variety of processes, including co-precipitation, ion-exchange, standard ceramic, salt melt, sol-gel, citrate, combustion, hydrothermal, glass crystallization, spray drying, self-propagating high-temperature synthesis, microemulsion, and industrial manufacturing methods [10-14].

The quantitative (temporally and spatially) controlled drug release in the human body is known as a *drug delivery system*. The conventional system was primarily focused on optimizing drugs (in terms of organ compatibility and drug kinetics). The delivery of anti-cancer drugs in high concentration was difficult, as most drugs were removed from the body through the blood system before reaching their target. This resulted in complications in normal tissues and a decreased therapeutic response to the drug. The use of magnetism for the targeted delivery of the drug is one method that addresses this issue by injecting a ferromagnetic drug into a blood vessel, utilizing magnetic force to navigate the drug to the diseased body part [15,16]. The role of magnetite particles in targeted drug delivery was confirmed using permanent magnets [17]. Ferrites are utilized in various applications, including drug delivery, sensors, catalysis, pharmaceuticals, and wastewater treatment [18]. Among the most remarkable applications is their role in targeted drug delivery. Site-specific delivery of drugs treats the infected body organ directly without affecting other organs of the human body. Ferrite materials are a boon in this area, as they carry an appropriate drug concentration and release it to the targeted organ [19]. The study reported the deeper accumulation of ferromagnetic particles on tissues using superconducting magnets. They display fewer side effects compared to conventional techniques. Proper positioning of magnets is essential for controlled and precise drug delivery [20]. Microwave-assist]ed synthesis is a clean

Materials Research Foundations 189 (2026) 310-347 https://doi.org/10.21741/9781644904039-10

route to synthesize high-efficiency volumetric modes. Ferrites are synthesized at a temperature (900-1000 °C) much lower than that of conventional heating (1200-1250 °C). Low temperature enhances the electromagnetic performance of ferrites [21]. Literature reports that microwave-assisted synthesis was used to synthesize ferric oxide nanoparticles (average particle size: 10 nm) in a very short time (25 min) at a low temperature (900-1000 °C). Ferrite oxide nanoparticles have been depicted as having prospective potential in medicinal chemistry across various drug delivery phases. This chapter discusses the general properties, synthesis of metal ferrite via the microwave-assisted method, and its applications, particularly in the field of targeted drug delivery.

1.1 General Properties of Ferrites

The ferrite properties largely depend upon "*how they are synthesized*" and "*what their chemical composition is*". Adding a minor impurity resulted in variations of properties, making ferrites extremely valuable in various applications. The following are some common properties of ferrites:

- The closely packed structure, resulting from oxygen ions, gave ferrites a significantly high dielectric constant ranging from thousands (at low frequencies) to 10-20 (at microwave frequencies) [22].
- The loss of oxygen ions at high temperatures makes determining melting points impossible. However, melting points can be measured using an oxyhydrogen flame [23].
- They have closely packed oxygen ions in spatial formation, forming a strong ionic bond. As a result, they are highly brittle and hard [24].
- They do not conduct electricity. However, under the influence of an applied electric field, it acts as a semiconductor [25].
- Their color is generally black or grey. This is due to the same energies of orbitals (mainly of 3d and 4s states). Furthermore, the ferrite crystal structure is characterized by light absorption within the visible region of electromagnetic radiation, giving it a characteristic dark black or grey color [26].
- Ferrites are famous for their magnetic properties. When the magnetic field is applied, they can become fully magnetized along the "*preferred axis*", also known as "magnetic anisotropy" [22]. They also have square or fragile hysteresis loops known as *permeability and hysteresis* [27].

2. Classification of Ferrite Materials

Ferrites are also known as "*ferrite ceramic materials*"; their structure mainly comprises iron oxide. With the change in the oxygen anion arrangement around the metal cation, the crystal structure changes too. Ferrites can be classified based on their structure or magnetic behavior. Based on their magnetic behavior, ferrites are classified into two types: hard and soft. While based on structure, they can be classified as spinel, garnet, hexagonal, and orthorhombic ferrites.

2.1 Classification of Ferrites Based on Their Magnetic Behavior

Ferrites are divided into hard and soft ferrites based on their resistance to demagnetization and magnetic coercivity, respectively [28]. Hard ferrites, also known as "permanent magnets," can retain their magnetism once they are magnetized. Soft ferrites, on the other hand, cannot maintain

Materials Research Foundations 189 (2026) 310-347 https://doi.org/10.21741/9781644904039-10

their magnetism after they are magnetized. Soft ferrites include ferrites of cobalt, zinc, manganese, nickel, and magnesium [29].

2.1.1 Soft Ferrites

Soft ferrite is an electromagnetically active ceramic material. According to the crystallographic aspect, they possessed a homogenous cubic spinel structure. Soft ferrite is composed of divalent metallic oxides and iron oxide. The term *soft ferrite* does not imply the ferrite's physical properties. The term is based on its magnetic nature. They are grey or black in appearance. They are known for their extremely brittle and complex nature. Soft ferrites do get magnetized, but they are unable to retain their magnetization. Japan and the Netherlands had been studying the properties of soft ferrite since the 1930s. In 1945, "*J.L. Snoek of Phillips Research Laboratories, Netherlands*" successfully synthesized soft ferrites, which would be used commercially. The notable examples of soft ferrites are "Ni-Zn($NiZnFe_2O_4$)" and "Mn-Zn ferrites ($MnZnFe_2O_4$)" [30]. In the spinel crystalline structure, the interaction of metallic ions occupying specific positions relative to oxygen ions gives rise to the characteristic magnetic properties of soft ferrites [31].

The magnetic domain theory states that interactions among the microscopic magnetizing areas in a material produce magnetic domains. Without any magnetizing force, the position of the magnetic domains is random. When the magnetizing force is zero, the net contribution to the flux is also zero. However, local domains are completely magnetized. In the presence of a magnetic field, the magnetic domains arrange themselves towards the magnetic field. As a result, the contribution of net reflux is enormous. Therefore, soft ferrites are semiconductors as their ability to conduct the flow of electrons through any material is between a conductor and a semiconductor [32]. Soft ferrites have inherently high resistivity, which enables fewer eddy current reductions over a frequency range. Additionally, they demonstrated significant magnetic permeability and stability. Soft ferrites find applications in the transformer and inductor cores as they are stable at high frequencies, compared to the range of magnetic materials [8].

2.1.2 Hard Ferrites

"Permanent magnetic ferrites" or "*hard ferrites"* have permanent magnetism. They are composed of iron, barium, or strontium oxides. They have high magnetic flux and magnetic permeability in a magnetically saturated state. This property enables the magnets to store a stronger magnetic field than iron [33]. In hard ferrites, when the applied magnetizing field is no longer available, a strong magnetization remains. Residual magnetization remains unaffected by the application of a demagnetizing field [34]. They are widely used in radio applications as they have an optimum magnetic field (0.35 T) and magnetic strength (30-160 kA/m) at a magnetic field of 400-2000 Oe. They have a hexagonal structure and are classified as M-type ($RFe_{12}O_{19}$), X-type ($RMeFe_{28}$), Y-type ($R_2Me_2Fe_{12}O_{22}$), W-type (RMe_2Fe_{16}), and Z-type ($R_3Me_2Fe_{12}O_{41}$) ferrites, where R is Barium, strontium, or lead, and Me could be divalent cations (Fe^{2+}, Ni^{2+}, Mn^{2+}, etc.) [33].

Hard ferrites constitute a significant portion of ceramic-based materials. The hard ferrites have a color range from dark grey to black and are generally brittle and hard. An example of naturally occurring hard ferrite is magnetite (weak hard ferrite) [34]. Hard ferrites play a significant role in the permanent magnet market, primarily due to their wide availability, higher chemical stability, permanent magnetism, and lower cost per unit of energy [35]. The important practical examples of hard ferrites are strontium ($SrO.6Fe_2O_3$) and barium ($BaO.6Fe_2O_3$) ferrite [36]. Compared to

Materials Research Foundations 189 (2026) 310-347   https://doi.org/10.21741/9781644904039-10

metallic magnetic materials, they offer a large coercive force. This property enables the synthesis of fragile magnets. Compared to the soft ferrites, hard magnetic ferrites are structurally sensitive and easily influenced by impurities [37].

2.2. Classification of Ferrites Based on Structure

Scientists usually classify ferrites based on their crystal structure. Based on structure, ferrites have four types: spinel, hexa, garnet, and orthoferrite.

2.2.1 Spinel ferrite

Spinel ferrites are "*a prototype of naturally occurring ferrite*", i.e., $FeO.Fe_2O_3$. Spinel structure is based on mineral spinel ($MgAl_2O_4$ or $MgO.Al_2O_3$), which, upon crystallization, displayed a cubic system [38]. Bragg [39] and Nishikawa [40] were the pioneers of determining the spinel crystal structure. Comparable to mineral spinel, the general formula of magnetic spinel is "$MeFe_2O_4$ or $MeO.Fe_2O_3$" (Divalent metal ion usually with iron oxides). The spinel lattice's unit cell has cubic symmetry, with eight $MeFe_2O_4$ molecules [41]. The small metal ion (0.7-0.8 Å) inhabits the spaces between the large-sized oxygen ion (1.4 Å) in the face-centered cubic lattice structure. The cubic close-packed structure has two types of interstitial sites (octahedral and tetrahedral), which are surrounded by oxygen ions (six and four, respectively). In the cubic unit cell of metallic spinel, 64 tetrahedral and 32 octahedral sites are present. Only 8 tetrahedral and 16 octahedral sites are filled with metal ions. In the mineral spinel, Mg^{2+} and Al^{3+} are present at tetrahedral and octahedral sites, respectively. $MeO.Fe_2O_3$ has a similar structure to that of the divalent metal ion present at the tetrahedral site and Fe^{3+} at the octahedral site, respectively.

In zinc ferrite, zinc ions occupy the tetrahedral sites and are non-magnetic (have no anti-ferromagnetic orientation). In contrast, Fe^{3+} occupies the octahedral sites. The Fe^{3+} at the octahedral position has weak interactions; thus, zinc ferrites are non-ferrimagnetic. Cadmium ferrites are also non-magnetic (paramagnetic). Barth and Posnjak have studied several cases of spinel ferrites and reported a preference for trivalent ions at tetrahedral sites. They are also the first to occupy the positions compared to other ions. However, many ferrites with inverse spinel structure are seen, in which divalent ions are present at the octahedral site. In contrast, trivalent ions are equally present on both the tetrahedral and octahedral sites. Some critical commercial ferrites (Fe, Ni, and Co) confirm the inverse spinel structure. The ferrites have two different types of divalent ions known as mixed ferrites, for example, $(NiZn)O.Fe_2O_3$ is a mixed ferrite comprising a "solid solution of $NiO.Fe_2O_3$ and $ZnO.Fe_2O_3$". Commercial ferrites are usually present in a mixed ferrite state.

2.2.2 Garnet Ferrite

Garnet ferrites incorporate a large trivalent ion (usually yttrium) with a significant magnetic moment [42]. Garnet ferrite structure is based on the "*silicate mineral garnet*". Like mineral garnet, magnetic garnet crystallizes into a dodecahedral (12-sided) structure. Their general formula is $Me_3Fe_5O_{12}$. Metal ions present at all sites of the crystal structure are trivalent, unlike other classes [43]. Bertaut and Forrat were the first to report the structure of garnet ferrite crystal and the dimensions of a unit cell of metallic garnet [44]. The garnet structure is highly stable as the cations occupy all the sites [45].

2.2.3 Hexagonal Ferrite

Hexagonal ferrites have *magnetoplumbite structure* [46] with a general formula of "$MeFe_{12}O_{19}$", where the metal is generally Ba, Pb, or Sr. The structure of magnetoplumbite exhibits hexagonal symmetry, characterized by major and minor axes, designated as the c-axis and a-axis, respectively [47]. In permanent magnet material, the preferred direction is followed, usually by ferrimagnetic compounds. Went [48], Fahlenbranch, and Heister [49] initially synthesized anti-ferrimagnetic compounds. Braun was the first to establish the complex structure of the compounds [50]. Commonly known hexagonal ferrites are $PbFe_{12}O_{19}$, $BaFe_{12}O_{19}$, and $SrFe_{12}O_{19}$. Further research on the ternary system "BaO-FeO-Fe_2O_3" has led to the discovery of many new hexagonal ferrites [51]. A range of compounds having a hexagonal structure is synthesized by mixing spinel ferrites with magnetoplumbite in various ratios. Hence, the spinel ferrite layers are sandwiched between the layers of magnetoplumbite in hexagonal ferrites [46].

2.2.4 Ortho Ferrite

Forestier and Guillot-Guillain first studied orthoferrites, also known as perovskites. Later, Pauthenet and Blum disclosed that perovskites have *parasitic ferromagnetic behavior* [52]. After successfully synthesizing a perovskite crystal in the laboratory, information about its magnetic properties was obtained [53]. Their general formula is "$RFeO_3$"(R is yttrium or a rare earth ion) [54]. They are sometimes considered cubic ferrites, but the structure is "slightly distorted perovskites". The perovskite structure has an arrangement of "RMO_3 oxides"[55]. The structure determination of $GdFeO_3$ marked a milestone in understanding the structure of this material [56].

Coppens and Eibschutz refined the crystal structure of $GdFeO_3$ and $YFeO_3$ [57]. Bozorth et al. determined the magnetic properties of single crystals [58]. Koehler and Wollan established the magnetic structure of the materials using neutron diffraction [59]. This structure is present in many complicated materials with unique molecular arrangements along a steady cubic symmetry. The resultant structure is not cubic but rather orthorhombic. The non-parallel organization of anti-ferromagnetically coupled ions resulted in relatively weak ferromagnetism in perovskite materials. The rare-earth iron perovskites are generally referred to as *orthoferrite* [60]. The involvement is essential as they have an original bubble memory structure [61].

3. Synthesis of Metal Ferrites

The synthesis plays a vital role in producing high-quality ferrite-based materials. The solution-based chemical synthesis yielded robust and simplified routes to the synthesis. The methods are skillful but not specific, thus allowing the synthesis of various nanomaterials. Various synthetic methods are employed in ferrite preparation; each synthesis method has its pros and cons. Additionally, there is a difference in the properties of the same magnetic nanoparticles synthesized using different methods. For example, Table 1 summarizes the differences in the properties of $ZnFe_3O_4$ nanoparticles synthesized by various techniques, including sol-gel, hydrothermal, precipitation, solvothermal, and microwave-assisted synthesis.

Table 1. A comparison of differences in magnetic and morphological properties of zinc ferrite nanoparticles synthesized by sol-gel, hydrothermal, precipitation, solvothermal, and microwave-assisted methods

Synthesis method	**Crystalline size [nm]**	**Particle size [nm]**	**RM [emu/g]**	**H [Oe]**	**M [emu/g]**	**Ref.**
Sol-gel	22	-	3.4	352.1	12.9	[62]
Hydrothermal	26	-	2	100	10	[63]
Precipitation	19.95	51.8	-	-	-	[64]
Solvothermal	11	10	0	0	50.4	[65]
Microwave	37.6	-	0.01	7.5	2.6	[66]

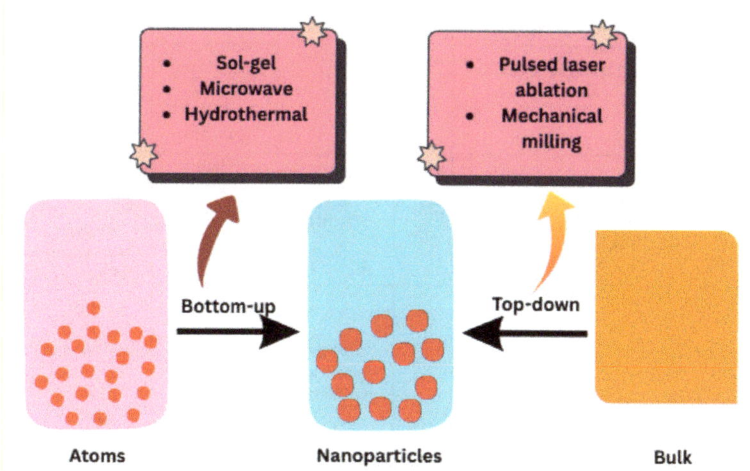

Figure 1. General illustration of bottom-up and top-down synthesis techniques.

Synthetic methods are categorized into two main types: bottom-up and top-down. When chemical mixing of ions forms a particle, the process is referred to as "bottom-up." Many synthetic techniques follow the "bottom-up" approach in synthesis, including co-precipitation, hydrothermal synthesis, thermal decomposition, sol-gel synthesis, sonochemical synthesis, solvothermal synthesis, microwave synthesis, polyol synthesis, microemulsion synthesis, flame spray pyrolysis, and vapor deposition. On the contrary, the method is called "top-down" when materials are broken down into smaller particles. Pulsed laser ablation and mechanical milling techniques fall under the "top-down" synthesis category. A general representation of how top-down and bottom-up occur is illustrated in Fig. 1. Table 2 compares the processing conditions, morphology, and size of different ferrite synthesis methods. This chapter discusses various bottom-up synthesis methods.

Materials Research Foundations 189 (2026) 310-347 https://doi.org/10.21741/9781644904039-10

Table 2. Comparison of processing conditions, size range, morphology, advantages, and disadvantages of a few ferrite synthesis techniques

Methods	Processing conditions	Size [nm]	Morphology	Advantages	Disadvantages	Ref.
Sol-gel	Ambient conditions, water (solvent), and surface capping agents are needed	20-200	Spherical	Desirable shape and length of particle (controlled morphology, nanoparticle hybrids can be produced easily	Purity is hard to achieve as the sol-gel matrix occupies their surfaces	[67]
Hydrothermal	Water-ethanol (solvent), 220 °C, may take hours to days, and needs surface capping agents	520	Spherical	Reasonable shape control with narrow particle size distribution, highly crystalline particles	The yield is not high, has an irregular spherical shape, a large size distribution, and a long reaction time	[68]
Precipitation	Water (solvent), 20-90 °C, a few minutes to complete the reaction, surface capping agents are added during the reaction	15-200	Spherical	High production rate, narrow size distribution of particles	Polydispersity, no control on size and shape, irregular morphologies of particles	[69]
Solvothermal	Organic solvents, 400 °C	20-30	Cuboidal, spherical, and fibrous	Control over final product properties, control the growth of the crystal, and improved purity	High temperature and pressure are needed, complex reaction conditions	[70]
Microwave	350-450 °C, high pH (12)	20-40	Spherical, clustered	Easy synthesis, fast reaction, pure product, selective heating, and high temperature are achieved within minutes	Low yield, expensive apparatus	[71]

3.1 Sol-gel Synthesis

Sol-gel synthesis utilizes metal alkoxide solutions that are hydrolyzed, followed by reaction of condensation polymerization to synthesize a gel at room temperature. Gel is converted into the final crystalline state using heat treatments, which remove volatile by-products within a range of

Materials Research Foundations 189 (2026) 310-347 https://doi.org/10.21741/9781644904039-10

temperatures (25-200 °C) [67]. The stoichiometric amount (1:1 M) of citric acid and metal nitrate was dissolved in distilled water. Ammonia solution is used to adjust the pH to 7. The solution is then heated until a gel is formed. Upon ignition, combustion begins, resulting in the release of the loose citrate precursor powder.

The sol-gel method has the benefits of an economical process, short processing time at temperatures < 100 °C, no fancy instruments needed, and reasonable stoichiometric control. The sol-gel method is generally a preferred route to control purity, shape, composition, size distribution, and microstructure of ferrite materials by adjusting the parameters (stirring rate, annealing temperature, and sol concentration). The procedure's advantages and simplicity make this method very attractive. Common ferrites synthesized by this method are $CuFe_2O_4$, $NiFe_2O_4$, $CdFe_2O_4$, and $ZnFe_2O_4$ [67].

3.2 Hydrothermal Synthesis

Hydrothermal synthesis is a heterogeneous reaction. During the process, materials are dissolved and recrystallized under high pressure, temperature, and aqueous solvents. The hydrothermal method is typically employed when ferrites of various shapes and sizes need to be synthesized on a large scale for production. This method offers several benefits, including the synthesis of highly pure, homogeneous, metastable compounds with crystal symmetry, a limited particle size distribution, a smooth and rapid reaction, a short residence time, simple equipment, and low energy consumption. This method is employed for synthesizing the crystals with low solubility or polymorphic modifications [68].

3.3 Precipitation Synthesis

Precipitant is the chemical responsible for converting a liquid solution into a solid. Precipitant is used for the synthesis of uniform-sized nanoparticles. The method uses a mixture of aqueous solutions of transition metals, mainly divalent and trivalent, respectively (1:2 mole). The soluble salts of ferric (III) are used as the source of trivalent metal ions. The reaction medium is alkaline. The pH of the reaction is carefully controlled for the high-quality ferrite-based material synthesis. Sodium hydroxide or ammonium solution is used to adjust the pH. Under inert conditions, the solution is stirred vigorously, either with or without heat. The obtained powder is then subjected to different temperatures to obtain pure-phase ferrites [69].

3.4 Solvothermal Synthesis

Solvothermal synthesis is employed to prepare ferrites with enhanced chemical and physical properties, which are utilized in biomedical and industrial applications. This method utilizes both aqueous and non-aqueous solvents to synthesize ferrite materials with the desired shape, crystalline phase, and size distribution. Physical characteristics are changed by changing the experimental parameters (reaction time, solvent, precursors, surfactant, and temperature) [70].

3.5 Microwave-Assisted Synthesis

Microwave-assisted synthesis utilizes microwave electromagnetic radiation to produce excitation, which helps align the dipoles of materials in the direction of an external field. This technique is employed in conjunction with a well-established synthesis strategy. This process can produce metal ferrites of desired shape and size [71]. Fig. 2 illustrates the schematic diagram of Fe_2O_3

Materials Research Foundations 189 (2026) 310-347 https://doi.org/10.21741/9781644904039-10

synthesis via the green synthesis method (microwave-assisted). The external electric field aligns the molecules in a specific orientation, producing internal heat. This heat provides the required energy and helps reduce reaction time. There is a fair reduction in reaction time using microwave-assisted synthesis compared to other processes. The method helps influence the ferrite nanoparticle's magnetic properties by jiggling in the experiment [72]. The green synthesis of Fe_2O_3 involved the following reactions (Eq. 1-5) [73]:

$$NaOH \rightarrow Na^+ + OH^- \quad (1)$$

$$FeSO_4 \rightarrow Fe^{2+} + SO_4{}^{2-} \quad (2)$$

$$Fe^{2+} + 2OH^- \rightarrow FeOOH + H^+ \quad (3)$$

$$FeSO_4 + 2NaOH \rightarrow FeOOH + H^+ + 2Na^+ + SO_4{}^{2-} \quad (4)$$

$$2FeOOH \rightarrow Fe_2O_3 + H_2O \quad (5)$$

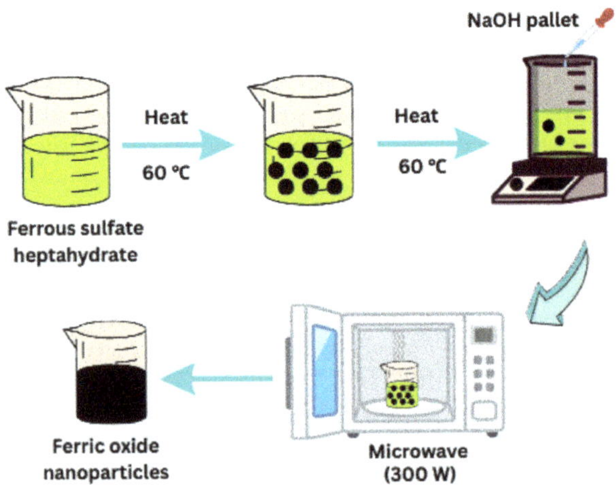

Figure 2. Schematic illustration of Ferric oxide nanoparticles synthesized via the microwave-assisted method.

3.5.1 Principle of Microwave-Assisted Synthesis

Not all materials are susceptible to microwave heating. Different materials have different responses to microwave heating. Based on the reaction towards microwave, the materials are classified into three categories;

- Materials that are transparent to microwaves.
- Materials that reflect microwave.
- Material that absorbs microwave.

Materials Research Foundations 189 (2026) 310-347 https://doi.org/10.21741/9781644904039-10

Microwave-absorbing materials have pivotal significance in microwave chemistry. There are three different heating mechanisms involved in microwave synthesis techniques: conduction, dipolar polarization, and interfacial polarization [71]. In the case of ferrite synthesis using microwaves, dipolar polarization and conduction mechanisms are generally employed (Fig. 3).

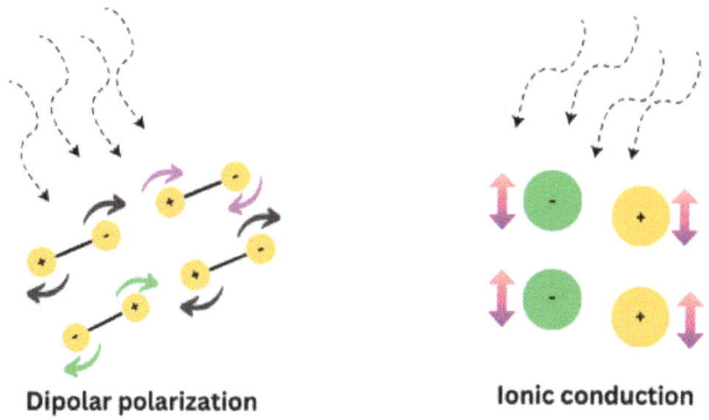

Figure 3. Microwave synthesis based on dipolar polarization and ionic conduction.

3.5.1.1 Dipolar Polarization

When microwaves are directed onto a substance, it only produces heat if it has a dipole moment. In the presence of an electric field, a dipole redirects itself in the direction of the electric field, and during this process, molecular friction occurs, resulting in heat loss. The frequency range (0.3-30 GHz) of microwave radiation is sufficient to oscillate the polar molecules and produce inter-particle connections, making it an ideal choice for heating the polar solution. Dipolar polarization can produce heat by interacting with polar solute molecules (ammonia and formic acid, etc.) or by polar solvent molecules (water, ethanol, methanol, etc.). In dipolar polarization, the oscillating field frequency should be high enough to generate inter-particle interaction. However, when the range of frequency is too high, inadequate inter-particle interactions occur, as the polar molecule motion is retarded by intermolecular interactions before the electric field can act. On the contrary, in the low-frequency range, polar molecules have sufficient time to align in the direction of the electric field [74,75].

3.5.1.2 Conduction Mechanism

The conduction mechanism generates heat by resisting the electric current. An electromagnetic field generates an oscillation of ions or electrons in a conductor, which results in an electric current. The current encounters internal resistance, resulting in the heating of the conductor. Under the influence of an electric field, all the ions will move across the solution. This process generates energy. As solution polarity increases, its ability to absorb microwaves increases at higher temperatures. In the case of an electrical conductor, microwave radiation falling on the material generates the charge carriers (ions, electrons, etc.), which move through the material, causing polarization. Induced currents face electrical resistance, resulting in the heating of the sample. This method is unsuitable for good conductors, as they reflect most of the energy irradiated [74].

3.5.2 Advantages of Microwave-Assisted Synthesis

The morphology, chemical composition, and size of ferrites play a significant role in their application usage. These characteristics rely on the synthesis techniques. Several techniques have been employed to synthesize ferrites, including sol-gel, hydrothermal, solvothermal, and precipitation methods. The methods have several limitations, including the need for careful pH adjustment of the reaction, difficulty in controlling particle size, and limitations in achieving a uniform size distribution [76,77]. The primary issue with these processes are the potential risk of contamination from reaction by-products and the requirement for post-treatment, such as thermal treatment of the products [78]. Additionally, conventional techniques require significantly longer to achieve the target temperature, rendering ferrites inefficient in energy transfer [79].

The microwave-assisted technique can significantly reduce energy costs, solvent usage, and reaction time [80], as it provides built-in heating in a highly selective and homogeneous manner, compared to classic methods. A study reported the synthesis of ferric oxide nanoparticles (average particle size: 10 nm) in a very short time (25 min) using microwave heating. This study proved that microwave heating needed less processing time to produce comparable results [81]. As it utilizes less solvent and energy, it efficiently synthesizes complex compounds. It impacted three major drug delivery phases in medicinal chemistry: "*lead generation, lead optimization, and hit-to-lead efforts*". Microwaves are efficiently used in dedicated rotors and microtiter plate systems. A multimode microwave apparatus can achieve hundreds of reactions within one microwave experiment [82]. The main advantages offered by microwave-assisted synthesis are:

- The ferrite synthesis reactions using microwave are more "*reproducible*" because of better control over parameters and a uniform heating system. The reaction temperature can be regulated easily.
- Microwave-assisted synthesis reactions are "*cleaner and eco-friendly*". The solvent consumption is very low as the microwave directly heats the compounds. Solvent-less synthesis (reagents are adsorbed on the mineral surface) has been getting much attention recently because of its eco-friendly protocol in synthesis technology. It also eliminates the need for purifying end-products.
- Microwave-assisted reactions are approximately 1000 times "*faster*" than conventional methods. They also easily achieve higher temperatures, resulting in reaction completion within minutes. *Less by-product* formation is observed, resulting in a higher yield of the product.
- "*Selective and uniform heating*" is achieved as the reaction only has solvent and solute particles. *Selective heating* depends on the ability of different materials to respond differently to microwaves [83].

4. Applications of Metal Ferrites

Ferrites are synthesized by the reaction of iron oxide with other metal oxides. They have been important to mankind for centuries [84]. Ferrite's practicability reached its maximum potential in

Materials Research Foundations 189 (2026) 310-347 https://doi.org/10.21741/9781644904039-10

the 1930s after the discovery of its structural, electrical, and magnetic properties. Thereafter, ferrites became the focus of the scientific community. Ferrites are insulating materials with moderate permittivity, enhanced electrical resistivity, lower eddy current, decreased dielectric losses, reasonable permeability, and adequate saturation magnetization. All properties are challenging to find within a single material, making ferrite-based materials highly useful in various applications across diverse fields. Ferrite materials can be easily synthesized at a low cost with high efficiency, which gives them a significant advantage over other materials [85,86].

Ferrites are semiconductors with a magnetic nature. This gave ferrites remarkable technological preference. Microwave technology has applications ranging from fields such as microwave and computer technology, antenna rods, memory chips, permanent magnets, activators, and high-density magnetic recording media, among others [87-89]. The appearance of quantum confinement and surface effects is due to size dependence, and a large surface-to-volume ratio has led the spotlight of the whole scientific community to ferrite materials [90,91]. Ferrites in the nano-crystalline form have applications in various fields, including magnetic resonance imaging (MRI), catalytic reduction, photocatalysis, and humidity sensing. It is also used in magnetic drug delivery, hyperthermia, magnetic fluids, gas sensors, and for detecting alcohol vapors, among other applications [92-94].

Polycrystalline ferrite is essential in various technological applications and is hard to replace with substituted materials. To modify the characteristics of ferrite, controlling the processing conditions of polycrystalline ferrite synthesis is crucial for achieving standards in developmental and research-related applications [95]. Moreover, ferrites are famous for their dielectric nature, making ferrites suitable as "*electromagnetic waveguides*" as the electromagnetic waves penetrate ferrites without producing electrical conduction. The dielectric nature of ferrite materials provides them with advantages over conventional transition metals (Ni and Fe) in various applications. Ferrites are also known for high porosity, which is further enhanced by novel synthesis techniques (soft templating, nano-casting, surfactant-assisted chemotherapy, etc.). Due to their properties, studies on ferrites have been ongoing for years, and these properties are being fine-tuned, which is expected to provide a brighter future for ferrite materials [96].

Ferrite powder is employed as a coating for electrical applications (radars, magnetic recording tapes, and aircraft) due to its spontaneous magnetic moment occurrence below the Curie temperature [95]. It has diverse applications in our daily lives, ranging from its role as an adsorbent in contaminant removal to enzyme mimics, magnetic site-specific drug delivery, and photocatalysis. This chapter will discuss the application of ferrite in drug delivery.

4.1 Drug Delivery Application of Metal Ferrite

Ferrites are magnetic nanoparticles with remarkable site-specific drug-delivering properties, utilizing magnetic probes [97]. Nanomaterials incorporated into polymers are often employed in drug delivery, as the unavailability of constant observation, by facilitating imaging, results in poor contrast capability. Ferrites address this issue by providing high-contrast MRIs. The iron oxide nanoparticles, or "*superparamagnetic iron oxide nanoparticles,*" confirmed super-paramagnetic behavior at the nanoscale. Hence, they are instrumental in biomedical applications [98,99]. One approach involved loading magnetic nanoparticles under a controlled external field into the bloodstream for site-specific drug delivery. The strong dipole-dipole magnetic attraction among

Materials Research Foundations 189 (2026) 310-347 https://doi.org/10.21741/9781644904039-10

particles of ferrites produces large aggregates. During the drug delivery process, surfactants are employed to accelerate the stability and biocompatibility of ferrites [100].

Colloidal magnetic nanoparticles are highly useful in localized drug delivery. Ferrites can hold pharmaceutical drugs within their volume or on their surface, allowing for targeted drug release to the specified organ or tissue. Any change in the size, shape, surface chemistry, and charge of a ferrite colloid alters ferrite chemistry, which helps improve blood circulation and drug bioavailability in the body [101]. The capacity of particle internalization strongly depends on the particle size of ferrite [102]. Particles larger than 0.2 μm get segregated by mechanical filtering. As a result, the drug has a shorter circulation time as it is removed by phagocytic system cells; however, particles with a smaller size ($\leq$ 10 nm) dissolve easily into the bloodstream. Therefore, the effective delivery of the drug is achieved through particles (10-100 nm), as small particles easily enter the reticuloendothelial system and capillaries [103]. The particle range (10-100 nm) can be easily synthesized via microwave technology. Particle's large surface area to volume ratio enables them to adsorb swiftly into the plasma proteins. The hydrophobic surface of the reticuloendothelial system helps the uptake of ferrites [104].

Ferrites are highly biocompatible, have a wide range of particle sizes, and are suitable for targeted drug delivery applications. The magnetic behavior makes them superior targeted drug delivery agents and excellent for use in MRI with external magnetic fields. The particles produce heat, which targets the tumor region through fluid hyperthermia, causing the release of the drug and inhibiting tumor cells via thermal apoptosis [105]. Ferrites are getting popular not only in targeted drug loading and delivery but also in other biomedical fields, including theranostic aids and drug delivery vehicles [106-110]. They facilitate monitoring via MRI and healing through the controlled release of specific drugs in cancer. However, if not controlled, the agglomeration could be a significant disadvantage [111,112]. The amalgamation of ferrites with polymers prevents the formation of aggregates in nanostructures, which traps the ferrites within magnetic micelles [113]. Polymer addition in ferrites offers additional advantages, including the ability of secondary drug functionalization and protection against the body's immune system. This increases the drug's circulation time [114,115].

The addition of a polymer to ferrite enables it to perform both non-covalent and covalent drug delivery processes. This allows ferrites access to several drug release mechanisms, such as external stimulus control. Ferrites are highly sensitive to particle size; a slight change in the particle size of γ-Fe_2O_3 (30 nm) renders them super-paramagnetic below the threshold value [116,117]. The dexamethasone (Dex) drug regimen is used in the treatment of coronavirus disease. Dex (5% w/w) was functionalized using nanocomposites (30% $ZnFe_2O_4$/Halloysite and $NiFe_2O_4$/Halloysite). The $ZnFe_2O_4$/halloysite/Dex/polyethylene glycol demonstrated high drug release efficiency (17.5%) at pH 5.6 compared to release efficiency (< 5%) at neutral pH using the Korsmeyer-Peppas model. The compatibility of the nanocomposite was assessed on HEK_{293} cells. The Dex-loaded nanocomposite exhibits improved cell viability compared to Dex alone. This nanocomposite reduced the drug's toxic effect and has significant potential for pulmonary drug delivery in targeted therapeutics for lung infections [118]. In conclusion, the magnetite nanoparticles are functionalized in different groups. They are coated with other polymers to encapsulate various drugs, enabling pH- and redox-responsive delivery of drugs in living organisms, which results in higher cell viability compared to the drug delivered alone.

4.1.1 Mechanism of Targeted Drug Delivery

Targeted drug delivery significantly improves therapeutic response and delivery [119]. The drug delivery process involves the synthesis of a formulation designed to introduce pharmaceutical agents into targeted cells, thereby achieving therapeutic effects. This route involved two essential steps: active drug transportation to the targeted area and control over the rate of release of the medicinal substance [120,121]. The involvement of magnetic systems in drug delivery involves loading drugs onto magnetic nanoparticles, which then travel to the tumor cells. The magnetic nanoparticles accumulate at the targeted tumor site with the assistance of a magnetic field until the process is complete. After the role, they are detected, seized, and removed from the body by the immune system's action [122]. Fig. 4 illustrates a schematic representation of ferrite synthesis, loading, and delivering the drug onto ferrite nanoparticles for targeted drug delivery to specific areas.

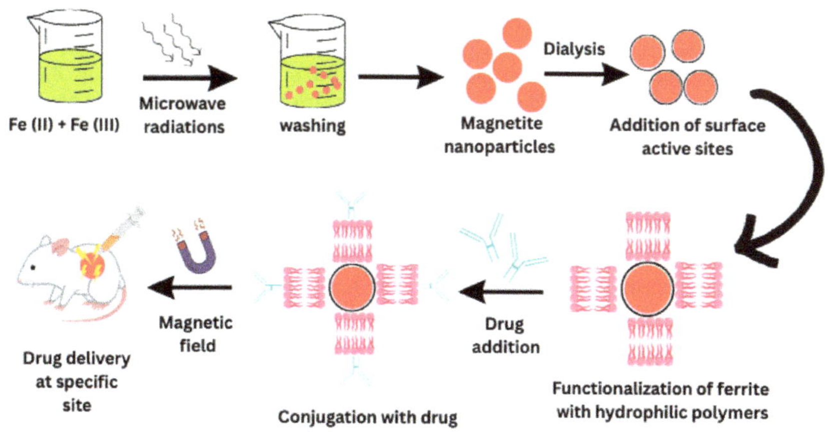

Figure 4. A schematic representation of how ferrite nanoparticles are involved in the drug delivery mechanism.

For successfully delivering drugs into the body, magnetic nanoparticles must possess stable paramagnetism, biocompatibility, and monodispersity [123]. The size of magnetic nanoparticles must be controllable (within a range of a few nm) [124]. A single magnetic domain is required in superparamagnetic materials. Magnetism is retained as long as an external magnetic field is present [116,125]. Superparamagnetism facilitates the quick detection and capture of tumors in human blood [126]. The magnetic core of the nanoparticle must be stable against air oxidation and prevent the stopping of aggregate formation. Hence, ensuring permanent biocompatibility. The organic shells (polymer or inorganic compound) are coated on nanoparticles to enhance the applications in the medicinal field [116].

Materials Research Foundations 189 (2026) 310-347 https://doi.org/10.21741/9781644904039-10

4.1.2 Ferrite Role in Targeted Drug Delivery in Cancer Therapy

According to the WHO, cancer is the leading cause of death (with more than 10 million deaths in a year) all over the world [127]. Hence, it posed a serious threat to human health. Cancer can occur in any body organ or tissue of a living organism. It resulted from "*uncontrolled cell growth along rapid cell division in the body*" [128]. Currently, the most widely used and effective treatment is surgery in combination with chemotherapy and radiotherapy. However, the research on implementing magnetic, metallic, and semiconductor nanoparticles is revolutionizing cancer medicine [129]. Cancer nanotechnology not only limits tumor growth but also removes malignant tissues resulting from metastasis in the body. After proper functionalization, the tiny particles become suitable biomaterials that can deliver the drug to cells and tissues within the cancer microenvironment. The small size of the particle is responsible for enhanced binding affinity with chemotherapeutic drugs, antibodies, and ligands, which promotes the safe release of controlled and targeted drugs [130].

Among the various types of nanoparticles, magnetic nanoparticles are getting considerable attention in biomedical applications, such as MRI [131] and targeted drug delivery [130]. The small particle size (below 20-30 nm) of nanoparticles resulted in shortened domain walls; thus, the particle acts as a single magnetic domain. Magnetic energy overcomes thermal energy, producing superparamagnetism in the magnetic nanoparticles. Several magnetic particles of varying compositions are evaluated for biomedical applications. The slight change in the nanoparticle's structure, composition, morphology, size, and surface chemistry can improve its magnetic properties and impact its behavior in *in vivo* applications [116]. Nano-ferrites are smart nanoparticles with various biomedical applications, including cell signaling, hyperthermia, nanorobots, drug delivery, anti-cancer function, biosensors, anti-fungal, and anti-bacterial properties, brain stimulation, wound healing, and MRI. Nano-ferrites are best known for their anti-cancer features towards different cancer cells, primarily in breast cancer. They are employed extensively in both drug release and drug delivery. A successful approach in cancer treatment is magnetic hyperthermia, in which nano-ferrites are employed under the influence of an external magnetic field. They are used as MRI contrast agents [132]. The role of various metal ferrite nanoparticles in carrying different anti-cancerous drugs and delivering them to targeted cells in the body has been explored (Table 3).

Recently, transition metal ferrite nanoparticles (MFe_2O_4, where M can be any divalent cation) have been extensively explored in advanced biomedical applications. The role of ferrite metal in chemotherapy has been extensively studied [152]. Fine ferromagnetic particles incorporated into lymphocytes or erythrocytes could be transported to targeted sites via an external magnetic field [153]. Ferrite nanoparticles should be prepared under proper conditions to pass through capillaries and deliver the drug to selected sites [154]. The experiments on animals concluded that ferrite nanoparticles can provide medications to the vascular system via magnetic field [155]. Rats were used as experimental animals to determine the capability of magnetic microspheres, demonstrating that the total brain concentration of drugs can be improved by properly conditioning the synthesis of magnetic spheres [156]. The performance of magnetically controlled ferrites in site-specific drug delivery is considered a high-performance material that could treat affected human organs with greater efficiency in less time [113].

Materials Research Foundations 189 (2026) 310-347 https://doi.org/10.21741/9781644904039-10

Table 2. A summary of the role of metal ferrite as a drug delivery agent in cancer therapy

Ferrite composition	Model drug	Drug loading and release profile	Ref.
Fe_3O_4 nanoparticle	Methotrexate	6 and 25% drug release efficiency at 7.4 and 5.4 pH, respectively	[133]
Fe_3O_4 nanoparticle	Cisplatin	Drug release efficiency [96%]	[134]
Poly (*N*-isopropyl acrylamide) metal ferrite	Doxorubicin	Drug release efficiency [20.9-76.5%]	[135]
Cobalt ferrite nanoparticle	Doxorubicin	Drug release efficiency [70-75%]	[136]
Ytterbium ferrite	Camptothecin	Drug adsorption efficiency [≥ 90%], 50 and 60% drug release efficiency at 7.4 and 6.8 pH, respectively	[137]
Cobalt ferrite nanoclusters	Doxorubicin	Drug loading capacity [0.43 mg/mg]	[138]
Mesoporous cobalt ferrite nanoparticles	Etoposide	Drug release efficiency [64%]	[139]
Zinc ferrite/mesoporous zinc oxide/nitrogen-doped graphene quantum dots	Curcumin	Drug loading efficiency [76.45%], drug release [64.71%]	[140]
Cobalt ferrite nanoparticles	Doxorubicin	Drug release [60%]	[141]
Cobalt ferrite nanoparticles	Doxorubicin	Drug release [52%]	[142]
Manganese ferrite nanoparticle	Doxorubicin	Loading capacity [97%]	[143]
Manganese ferrite nanoparticle	Methotrexate	Drug loading [48%] and release efficiency [25%]	[133]
Calcium ferrite nanoparticle	Curcumin	Loading capacity [98.3%]	[144]
Calcium ferrite nanoparticle	Epirubicin	Drug loading capacity [97.5%]	[145]
Zinc ferrite nanoparticle	Imatinib	Loading capacity [94%]	[142]
Calcium ferrite nanoparticles	Cinnamaldehyde	Loading capacity [86.5%]	[140]
Cobalt ferrite nanoparticles	Doxorubicin	Loading capacity [78%]	[146]
Cobalt ferrite nanoparticles	Doxorubicin	Loading capacity [68.8%]	[147]
Zinc ferrite	Cinnamaldehyde	Loading capacity [76.4%]	[148]
Nickel ferrite nanohybrid	Zidovudine	Loading capacity [63.2%]	[149]
Calcium ferrite nanoparticles	Hesperidin	Loading capacity [62.9%]	[150]
Nickel ferrite/polyethylene glycol/lipid polymer nanoparticle hybrid	Doxorubicin	Loading capacity [64.3%]	[151]

$MnFe_2O_4$ exhibits enhanced heating efficiency, making it a suitable candidate for the magnetic hyperthermia-mediated cancer therapy. They behaved as novel drug carriers with highly controlled targeted release (80-90%) of anti-cancer drugs [157]. The chitosan-pegylated coated manganese ferrite nanoparticle encapsulated (encapsulation efficiency 92.8%) the methotrexate to improve its pH-sensitive drug delivery. Methotrexate is a well-known anti-cancer drug. The *in vitro* drug release revealed a difference in drug release (6 and 25%, respectively) at different pH levels (7.4 and 5.4, respectively) after 72 hours. This specific system can be used as a novel carrier for targeting the tumor tissues in cancer [133]. Another study reported the use of $MnFe_2O_4$@Au nanoparticles for docking anti-cancer drugs via a coupling molecule, targeting cancerous tissues. A high binding efficiency (> 90%) of $MnFe_2O_4$@Au nanoparticles with drugs was observed, with maximum release occurring at low pH [158,159].]

Superparamagnetic nanoparticles make very stable colloids due to the negligible remanent magnetization. In the first trial of magnetically controlled drug delivery, advanced cancer (14 cases) was treated using magnetic nanoparticles (100 nm) loaded with epirubicin. The treatment focuses on utilizing high-energy permanent magnets (neodymium block) that direct the drug using magnetism. Depending on the patient's tumor stage, adjustments were made. A minimum distance of 0.5 cm was kept constant throughout the treatment between the magnet and the tumor. Treatments were effective in treating skin cancer, but hematological studies revealed that drug concentration was insufficient to cure the tumor completely [160].

Cobalt ferrite nanoparticles are synthesized (260 °C for 1.0 h) by the reaction shown in Eq. 6;

$$\text{Cobalt acetylacetone} + \text{iron acetylacetone} + \text{Triton X 100} \rightarrow \text{Triton X 100 capped Cobalt ferrite nanoparticles} \quad (6)$$

Cobalt ferrite is known for its high coercivity and magneto-crystalline anisotropy at room temperature, making it a valuable asset for cancer therapy [159,160]. The cytotoxicity of cobalt ferrite nanoparticles (10 nm) at the concentration (5-40 mg/mL) is targeted at the "triple-negative breast cancer cell line" using microwaves (2.45 GHz). With the specific absorption rate (68.28 W/g), cobalt ferrite nanoparticles, if appropriately targeted, can kill the tumor cells in a shorter time [161]. Sabahat et al. evaluated the cytotoxic potential of cobalt nanoparticles in the MCF-7 cell line. Bare cobalt nanoparticles exhibit cell viability (60%) and cytotoxicity (40%), whereas doxorubicin shows viability (20%) and toxicity (80%) [162]. In another study, $CoFe_2O_4$ nanoparticles were incorporated into meso-2,3-dimercaptosuccinic acid. The functionalized nanoparticles were conjugated with doxorubicin. The intracellular acidic environment enables $CoFe_2O_4$ to release doxorubicin, which assembles at localized sites in large amounts. Due to the combined use of chemo and thermal therapy, an enhanced cell cytotoxicity produced a synergistic effect [163] (Fig. 5).

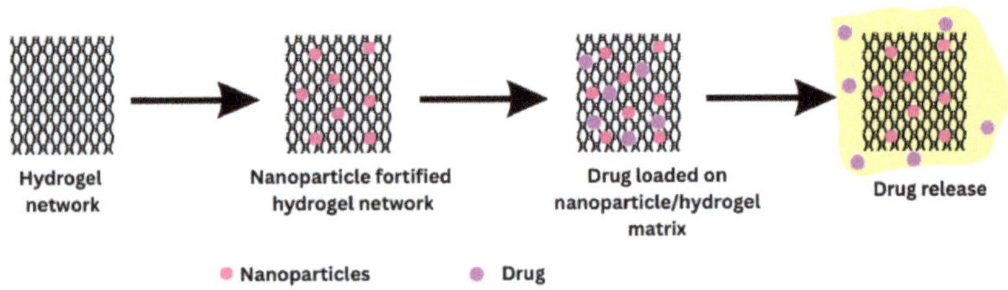

Figure 5. Illustration showing functionalization of a hydrogel network with nanoparticles loaded with a drug, which helps control drug release.

Another study reported the promising nature of ferrite nanoparticles as a potential nano-carrier, as they effectively deliver the drug to the targeted site [164]. The monodisperse cobalt ferrite nanoparticles have a diameter of 280 nm with a uniform and porous structure, an average pore size of 7.21 nm, a pore volume of 0.13 cm^2/g, a surface area of 58.03 m^2/g, and are highly suitable for etoposide loading. The drug targeting ability is due to the high magnetization (42.2 emu/g) of cobalt ferrite nanoparticles. The drug release rate was fast under the influence of the microwave. After the six repeating cycles of microwave irradiation, the drug release efficiency was about 64% [139]. Polymers are typically added to ferrites to facilitate subsequent chemical modification of the drug and protect the ferrites from immune system recognition. Polymer addition also increases the drug circulation time. Poly (*N*-isopropyl acrylamide) metal ferrite nanocomposites have recently been synthesized to deliver doxorubicin. The amine functionalization was employed to

Materials Research Foundations 189 (2026) 310-347 https://doi.org/10.21741/9781644904039-10

attach proteins/ligands to ferrite nanoparticles without compromising their biological activity. Polymer ferrite nanocomposites exhibited enhanced biocompatibility (81-93% cell viability) and drug delivery activity (20.9-76.5% drug release efficiency) in breast cancer cells. The drug release activity was pH-dependent, providing an additional tool for controlled drug release [135]. Hydrophilic polymers such as "guar gum, Arabic gum, poly-methacrylic acid" coated cobalt ferrite nanoparticles, serve as a nanocarrier, and exhibit magnetic properties. Doxorubicin was released efficiently (70-75%) under the influence of the applied magnetic field. The coated cobalt ferrite nanoparticles were cytocompatible during the *in vitro* studies on the specific cells of "Chinese hamster ovary and Huh-7". Cobalt ferrite nanoparticles coated with hydrophilic polymers have proven to be efficient magnetic nanocarriers in targeted drug delivery mechanisms [136].

Poly(dl-lactic-co-glycolide)-coated magnetic ytterbium ferrite nanoparticles were employed as a nano-carrier agent for loading camptothecin. The enhanced inhibition efficiency of the drug and sustained drug release demonstrated the suitability of "magnetic nanocarriers" for drug transport [137]. Zinc ferrite nanoparticles (22.36 ± 2.21 nm) effectively carried imatinib by efficiently loading it onto magnetic liposome nanocomposites, delivering it to specified cells or tissues [143]. Veronica et al. synthesized a multifunctional complex (μ-RNA/polyethyleneimine/superparamagnetic nanoparticles) using angiogenesis manipulation. The conjugation of polyethyleneimine with superparamagnetic iron oxide nanoparticles typically enhances toxicity. However, in the present case, it caused a decrease in toxicity with an increase in high transfection efficiency. The rise in iron-loaded nanoparticles reduced the cytotoxicity of polyethyleneimine. This eventually reduced the polyethyleneimine needed for μ-RNA [165,166].

The multi-functional magnetite nanoparticles play a crucial role as a "drug carrier system" for targeted drug delivery to treat different types of cancer, including breast, liver, colon, cervical, prostate, renal, osteosarcoma, skin, fibrosarcoma, and gastric cancer. The multifunctional magnetite nanoparticles are incorporated with specific polymers to enhance the cytotoxicity (Fig. 6). Concluding, magnetite, cobalt, calcium, nickel, zinc, manganese, and ytterbium nanoparticles were used extensively to carry the anti-cancer drugs (methotrexate, doxorubicin, cinnamaldehyde, imatinib, epirubicin, curcumin, hesperidin, zidovudine, cisplatin, etoposide, camptothecin, and so on) to targeted cells with the release efficiency reaching up to 98.3%. However, limitations still need to be resolved to convert this possibility into a tangible reality for humans.

4.1.3 Magnetic Resonance Imaging Contrast Agents

MRI is a high-resolution, non-invasive imaging technique used for visualizing anatomical structures. Unlike X-ray systems, it does not use radioactive materials. MRI is often considered a safe technique; however, it also has some inherent risks associated with the magnetic resonance environment, which should be addressed and avoided [167]. The MRI technique usually has signals with "*low intensity and sensitivity*", which causes difficulty in correct diagnosis. The contrast agents are generally employed to enhance the image [168]. Conventionally, Gadolinium chelates are used in MRI due to the low toxicity, high image augmentation, and small systemic half-life [169,170]. Recently, several studies have explored the potential applications of ferrite nanoparticles as "*contrast media*". Magnetic nanoparticles of ferrites, such as "manganese, cobalt-zinc, cobalt, and chitosan-coated nickel-based ferrites", are widely used as "MRI contrasting agents". Co-Zn nanoparticles (3-34 nm) functionalized with silica are investigated for potential as an MRI contrast agent in human breast cancer magnetic imaging [171]. Pegylated, coated manganese-zinc nanoparticles (14 nm) were evaluated against L929 and 4T1 cells, showing lower

Materials Research Foundations 189 (2026) 310-347 https://doi.org/10.21741/9781644904039-10

cytotoxicity levels than the commercially used Sinerem contrast agent towards both cell types [172]. Chitosan-carboxymethyl-cellulose-albumin-nickel ferrite nanoparticles also effectively act as contrast agents against the MCF-7 cell line [173]. Magnetic nanoparticles have established themselves as effective contrast media in MRI due to their small particle size, low cytotoxicity, and cancer-specific properties.

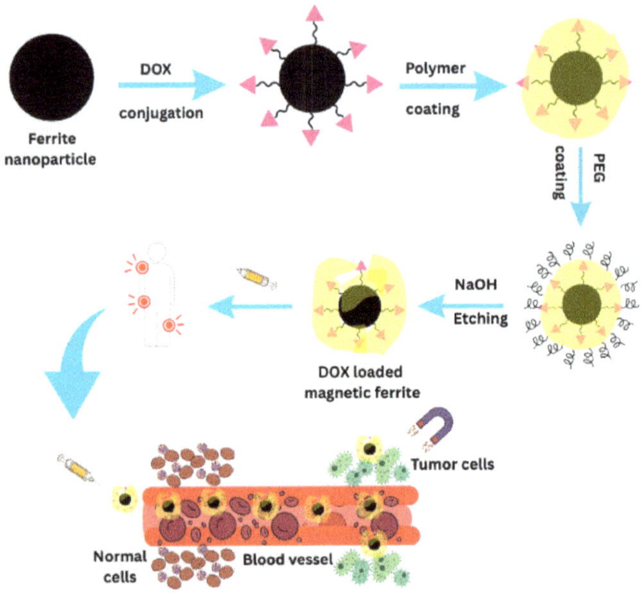

Figure 6. Illustration showing the schematic scheme involved in synthesizing doxorubicin-loaded magnetite nanoparticle and the release at targeted cells under the influence of a magnetic field.

4.2 Ferrites as Magnetic Particle Imaging Tracer

Magnetic particle imaging (MPI) was proposed in 2001 [174]. Superparamagnetic iron oxide nanoparticles generate signals using a dynamically shifting magnetic field-free zone, which serves as the detection point in MPI [175]. Gleich and colleagues, in 2005, pointed to the possibility of quantitative-based imaging having greater sensitivity (100 mol Fe/L) and spatial resolution (1 mm). This signal can be used to trace the spatial positions and concentrations [176]. MPI was evaluated using ferrite nanoparticles containing MgO and TiO_2. The nanoparticles demonstrated high cell viability and induced hyperthermia in MDA-MB-231 cells at a concentration of 150 μg/mL. The nanoparticles behaved as "hyperthermia mediators" at high concentrations in magnetic fluid against the MDA-MB-231 cell lines [177]. The nanoparticles have shown enhanced and uniform delivery for efficacious magnetic hyperthermia in cancer treatment [178].

Materials Research Foundations 189 (2026) 310-347 https://doi.org/10.21741/9781644904039-10

4.3 Ferrites in Immune Assays

Before homologous transplantation in the treatment of adolescent neuroblastoma, magnetized beads facilitate the separation of neoplastic cells from bone marrow aspirates [179]. To minimize the proportion of tumor cells in aspirates, researchers add nanoparticles to antibodies, enabling them to detect the "antigen base" of tumor cells. Modern beads of magnetic nanoparticles and the blends offer higher performance than conventional magnetic beads. *Ferrite, especially cobalt ferrite, has replaced iron oxide as a magnetic material due to its high saturation and responsiveness to applied fields* [180]. In the 1970s, magnetic separation of biological materials was first performed. This phenomenon is known as "*magnetophoretic*" [181]. This is used in separating specified cells from the body field, primarily for cancer diagnosis [182]. Research has demonstrated the utility of antibody-conjugated ferrite nanoparticles in identifying and isolating specific breast cancer cells, including "CF-7 and SK-BR-3", which are used to distinguish between different cancer cell subtypes in blood [183]. The major limitation of using this technique lies in its cost, low processing speed, limited sample throughput, and loss of cell viability [176].

5. Limitations of Ferrites in Site-Specific Drug Delivery

In the 1990s, experiments on-site directed drug delivery were performed on humans. Still, the technique has not been extensively developed in the medical industry. The magnetic properties cause the particles to aggregate, affecting the drug circulation time. The functional modification of the surface has been employed to resolve this concern, resulting in an influence on the ferrite's biocompatibility. Even complete elimination of ferrite nanocomposites after drug delivery is also a problem, as the remaining particles could enter the vascular system and cause blood flow blockage. A strategy needs to be developed that can dissolve the remaining drug carriers by directing them to a suitable organ system (digestive system). The cytotoxicity of the drug carrier is also concerning, as it varies from individual to individual. A careful test should be done before employing them for drug delivery. Moreover, there are other challenges such as drug circulation time before release, the controlled magnetic field value which does not disrupt any other body organ functioning, uniformity of drug carrying particle, carrier stability in human blood, particle size control in drug carrier, precise directing of drug on to organ, and the extent of drug availability to targeted organ after each cycle must be addressed. Research should be conducted to achieve the desired results in site-specific drug delivery using ferrites [19].

6. Future Perspectives of Ferrites in Targeted Drug Delivery

Over the years, magnetite nanoparticle systems have undergone significant development for targeted drug delivery in research laboratories. As advancements continue in the medical field, expertise is needed to obtain the desired morphology and size of nanoparticles. Several approaches have been developed to protect nanoparticles from agglomeration and enhance their bioavailability. The extensive research led to the development of surface functionalization techniques using diverse ligands to enhance the controlled and targeted drug delivery with minimum toxicity of active materials. This led to the lowered allowed concentration of drug doses without affecting therapeutic efficiency. In the isolated systems, the targeted drug delivery systems are highly responsive to stimuli (external magnetic field, pH, and redox).

Moreover, the functionality of nanocarriers is primarily assessed in the laboratory, focusing on the *in vitro* based drug release under various conditions and cytotoxicity studies of targeted cancerous cells. The knowledge of the interaction and transformation of nanocarriers with the biological system has also not been expanded much. Additionally, the impact of the "protein corona" on reducing on-site delivery to tumor cells remains unclear. The reticuloendothelial system is also responsible for reducing the effectiveness of nanocarriers, working by eliminating them from the system. Implementing the results obtained from a single-component isolated model onto the behavior of nanocarriers in living organisms is challenging. Another significant barrier is the high cost.

The development of analytical methods plays a crucial role in expanding this field, enabling the tracking of complex multicomponent interactions within biological systems. Even clinical trials are required to enhance efficiency, better understanding, and monitoring the release of active substances from nanocarrier materials within the selective on-site drug delivery environment after cellular uptake occurs. Finally, a suitable method for the controlled release of magnetite nanoparticles must be developed under the influence of an external magnetic field. This addresses the limitation of active material effectiveness in subcutaneous tissues [184].

Conclusion

The science of metal ferrites is a rapidly emerging domain of study due to its widespread applications, including magnetic, catalytic, and adsorptive properties. Typically, traditional methods are commonly used to synthesize ferrites on a vast scale; however, ferrites are often in the micrometer regime. Chemical methods were used to synthesize nanosized ferrites with the desired stoichiometry. A chemical synthesis method, such as microwave, is convenient for adhering to the principles of green chemistry. It is an efficient method that takes less time and is solvent-free. Thus, microwave synthesis has advantages over conventional methods. The benefits of this technique are being exploited in medicinal chemistry for drug delivery, contrast agents, immune assays, MPI tracers, and in related fields such as materials science and nanotechnology.

The role of ferrite material in targeted drug delivery has been extensively explored in cancer. Utilizing magnetite-based nanoparticles as a drug carrier is necessary for treating several diseases. The scientific community has faced serious challenges in developing magnetite nanoparticle-based targeted drug delivery systems. It has significant potential for practical implementation and could enhance the treatment efficiency of several diseases. Concluding, the concept has been proven to be feasible in pre-clinical and phase-clinical trials on animals. However, it may take some time to obtain approval for human utilization and make it available in the market. The scale-up of magnetic field-based equipment with optimized parameters tailored for targeted drug delivery will be challenging. Equipment design complexity will also be based on the type of tissue being targeted. The rising awareness of the potential nanotoxicity will also become an obstacle in obtaining approval from health authorities.

This chapter covers the basics of metal ferrites, including their properties, classification, synthesis, and applications in drug delivery. Metal ferrite nanocomposites are considered superior nano components for targeted drug delivery. The ferrite nanocomposites are used in MRI as a contrast agent. Recently, researchers have explored the role of contrasting agents and tracers in MRI and MPI, respectively. This chapter presented an overview of the ferrite chemistry. Different synthesis techniques develop metal ferries with desired sizes, shapes, particle morphology, abilities, and properties are viewed. This chapter discusses the benefits of the microwave-assisted technique over other synthesis methods. In drug delivery, ferrite applications are primarily addressed in pioneering work in the field, providing readers with a comprehensive reference. Different techniques for nanostructure modification formed by ferrites to enhance the shortcomings in drug on-site delivery are discussed. The scientists are focusing on each parameter whose fine-tuning can improve the application of materials in the desired field.

References

[1] A.A. Mills, The lodestone: History, physics, and formation, Ann. Sci. 61 (2004) 273-319. https://doi.org/10.1080/00033790310001642812

[2] J.B. Carlson, Lodestone compass: Chinese or olmec primacy? Multidisciplinary analysis of an olmec hematite artifact from San Lorenzo, Veracruz, Mexico, Science 189 (1975) 753-760. https://doi.org/10.1126/science.189.4205.753

[3] S. Malin, D. Barraclough, Gilbert's de Magnete: An early study of magnetism and electricity, Eos, Trans. Am. Geophys. Union 81 (2000) 233-234. https://doi.org/10.1029/00EO00163

[4] S. Stringari, R.R. Wilson, Romagnosi and the discovery of electromagnetism, Rend. Lincei 11 (2000) 115-136. https://doi.org/10.1007/BF02904378

[5] M. Kaur, N. Kaur, Ferrites: Synthesis and applications for environmental remediation, in: V.K. Sharma, R. Doong, H. Kim, R. S. Varma, D. D. Dionysiou (Eds.), Ferrites and Ferrates: Chemistry and Applications in Sustainable Energy and Environmental Remediation, Am. Chem. Soc., Wahington, DC, 2016, pp. 113-136. https://doi.org/10.1021/bk-2016-1238.ch004

[6] P. Kumar, S. Sharma, M. Knobel, M. Singh, Effect of La3+ doping on the electric, dielectric and magnetic properties of cobalt ferrite processed by co-precipitation technique, J. Alloys Compd. 508 (2010) 115-118. https://doi.org/10.1016/j.jallcom.2010.08.007

[7] R. Yao, S. Liao, C. Dai, Y. Yang, F. Zheng, Dual functions of novel glass-ceramic floor tile design and preparation, Ceram. Int. 40 (2014) 8667-8675. https://doi.org/10.1016/j.ceramint.2014.01.084

[8] R. Srivastava, B.C. Yadav, Ferrite materials: Introduction, synthesis techniques, and applications as sensors, Int. J. Green Nanotechnol. 4 (2012) 141-154. https://doi.org/10.1080/19430892.2012.676918

[9] M. Mayakkannan, V. Siva, A. Murugan, A. Shameem, S. Thangarasu, S.A. Bahadur, Microwave-assisted synthesis of ternary transition metal ferrite: Structural, morphological, optical, magnetic and electrochemical properties, Phys. E Low Dimens. Syst. Nanostruct. 147 (2023) 115573. https://doi.org/10.1016/j.physe.2022.115573

[10] L. Castelliz, K. Kim, P. Boucher, Preparation, stability range and high frequency permeability of some ferroxplana compounds, J. Can. Ceram. Soc. 38 (1969) 57.

[11] E. Neckenburger, H. Severin, J. Vogel, G. Winkler, Ferrite hexagonaler kristallstrustur mit hoher Grenzfrequenz, Z. Angew. Phys. 18 (1964) 65.

[12] M. Chandel, V.P. Singh, R. Jasrotia, K. Singha, R. Kumar, A review on structural, electrical and magnetic properties of Y-type hexaferrites synthesized by different techniques for antenna applications and microwave absorbing characteristic materials, AIMS Mater. Sci. 7 (2020) 244-268. https://doi.org/10.3934/matersci.2020.3.244

[13] S.I. Kuznetsova, E.P. Naiden, T.N. Stepanova, Topotactic reaction kinetics in the formation of the hexagonal ferrite Ba3Co2Fe24O41, Inorg. Mater. 24 (1988) 856-859.

[14] J. Drobek, W.C. Bigelow, R.G. Wells, Electron microscopic studies of growth structures in hexagonal ferrites, J. Am. Ceram. Soc. 44 (1961) 262-264. https://doi.org/10.1111/j.1151-2916.1961.tb15375.x

[15] C. Alexiou, W. Arnold, R.J. Klein, F.G. Parak, P. Hulin, C. Bergemann, W. Erhardt, S. Wagenpfeil, A.S. Lubbe, Locoregional cancer treatment with magnetic drug targeting, Cancer Res. 60 (2000) 6641-6648.

[16] C. Alexiou, R.J. Schmid, R. Jurgons, M. Kremer, G. Wanner, C. Bergemann, E. Huenges, T. Nawroth, W. Arnold, F.G. Parak, Targeting cancer cells: Magnetic nanoparticles as drug carriers, Eur. Biophys. J. 35 (2006) 446-450. https://doi.org/10.1007/s00249-006-0042-1

[17] T. Terada, S. Fukui, F. Mishima, Y. Akiyama, Y. Izumi, S. Nishijima, Development of magnetic drug delivery system using HTS bulk magnet, Phys. C 468 (2008) 2133-2136. https://doi.org/10.1016/j.physc.2008.05.145

[18] V.K. Sharma, L. Chen, R. Zboril, Review on high valent Fe VI (ferrate): A sustainable green oxidant in organic chemistry and transformation of pharmaceuticals, ACS Sustain. Chem. Eng. 4 (2016) 18-34. https://doi.org/10.1021/acssuschemeng.5b01202

[19] M.S. Dahiya, V.K. Tomer, S. Duhan, Metal-ferrite nanocomposites for targeted drug delivery, in: A. Inamuddin, A. Mohammad (Eds.), Applications of Nanocomposite Materials in Drug Delivery, Woodhead Publishing, Cambridge, 2018, pp. 737-760. https://doi.org/10.1016/B978-0-12-813741-3.00032-7

[20] Y. Hirota, Y. Akiyama, Y. Izumi, S. Nishijima, Fundamental study for development magnetic drug delivery system, Phys. C 469 (2009) 1853-1856. https://doi.org/10.1016/j.physc.2009.05.248

[21] B. Liu, Y. Wang, G. Han, L. Zhang, Y. Huang, Facile microwave-assisted synthesis of magnetic ferrite: rapid interfacial reaction underlying intensifying mechanism, J. Clean. Prod. 361 (2022) 132181. https://doi.org/10.1016/j.jclepro.2022.132181

[22] C. Prakash, J. Baijal, Dielectric behaviour of tetravalent titanium-substituted Ni-Zn ferrites, J. Less-Common Met. 107 (1985) 51-57. https://doi.org/10.1016/0022-5088(85)90240-1

[23] K. Lark-Horovitz, V.A. Johnson, Solid state physics: preparation, structure, mechanical and thermal properties, in: F. Seitz, D. Turnbull (Eds.), Solid State Physics, Elsevier, 1959.

[24] M. Niederberger, N. Pinna, J. Polleux, M. Antonietti, A general soft-chemistry route to perovskites and related materials: synthesis of BaTiO3, BaZrO3, and LiNbO3 nanoparticles, Angew. Chem. Int. Ed. 43 (2004) 2270-2273. https://doi.org/10.1002/anie.200353300

[25] M. Khairy, M. Gouda, Electrical and optical properties of nickel ferrite/polyaniline nanocomposite, J. Adv. Res. 6 (2015) 555-562. https://doi.org/10.1016/j.jare.2014.01.009

[26] R.J. Fernandes, C.A. Magalhães, C.O. Amorim, V.S. Amaral, B.G. Almeida, E.M. Castanheira, P.J. Coutinho, Magnetic nanoparticles of zinc/calcium ferrite decorated with silver for photodegradation of dyes, Materials 12 (2019) 3582. https://doi.org/10.3390/ma12213582

[27] M.M. Haque, M. Huq, M. Hakim, Thermal hysteresis of permeability and transport properties of Mn substituted Mg-Cu-Zn ferrites, J. Phys. D: Appl. Phys. 41 (2008) 055007. https://doi.org/10.1088/0022-3727/41/5/055007

[28] D. Roy, C. Shivakumara, P.A. Kumar, Observation of the exchange spring behavior in hard-soft-ferrite nanocomposite, J. Magn. Magn. Mater. 321 (2009) L11-L14. https://doi.org/10.1016/j.jmmm.2008.09.017

[29] S. Hajarpour, A.H. Raouf, K. Gheisari, Structural evolution and magnetic properties of nanocrystalline magnesium-zinc soft ferrites synthesized by glycine-nitrate combustion process, J. Magn. Magn. Mater. 363 (2014) 21-25. https://doi.org/10.1016/j.jmmm.2014.03.027

[30] A. Suarez, J. Victoria, A. Alcarria, J. Torres, P.A. Martinez, J. Martos, J. Soret, R. Garcia-Olcina, S. Muetsch, Characterization of different cable ferrite materials to reduce the electromagnetic noise in the 2-150 kHz frequency range, Mater. 11 (2018) 174. https://doi.org/10.3390/ma11020174

[31] M.G. Naseri, M. Halimah, A. Dehzangi, A. Kamalianfar, E.B. Saion, B.Y. Majlis, A comprehensive overview on the structure and comparison of magnetic properties of nanocrystalline synthesized by a thermal treatment method, J. Phys. Chem. Solids 75 (2014) 315-327. https://doi.org/10.1016/j.jpcs.2013.11.004

[32] D.S. Mathew, R.S. Juang, An overview of the structure and magnetism of spinel ferrite nanoparticles and their synthesis in microemulsions, Chem. Eng. J. 129 (2007) 51-65. https://doi.org/10.1016/j.cej.2006.11.001

[33] F. Kools, P. Van Der Valk, Ferrites, in: R.E. Kirk, D.F. Othmer (Eds.), Kirk-Othmer Encyclopedia of Chemical Technology, J. Wiley & Sons, New York, 10 (2000) 200-215. https://doi.org/10.1002/0471238961.0605181811151512.a01

[34] D. Hoydick, E. Palmiere, W. Soffa, Microstructural development in MnAl-base permanent magnet materials: New perspectives, J. Appl. Phys. 81 (1997) 5624-5626. https://doi.org/10.1063/1.364619

[35] K. Buschow, F. de Boer, Permanent Magnets, J. Magn. Magn. Mater. (2003) 105-129. https://doi.org/10.1007/0-306-48408-0_12

[36] J.S. Reed, R.M. Fulrath, Characterization and sintering behavior of Ba and Sr ferrites, J. Am. Ceram. Soc. 56 (1973) 207-211. https://doi.org/10.1111/j.1151-2916.1973.tb12458.x

[37] M. Sugimoto, The past, present, and future of ferrites, J. Am. Ceram. Soc. 82 (1999) 269-280. https://doi.org/10.1111/j.1551-2916.1999.tb20058.x

[38] A.H. De Aza, P. Pena, S. De Aza, Ternary system Al2O3-MgO-CaO: I, primary phase field of crystallization of spinel in the subsystem MgAl2O4-CaAl4O7-CaO-MgO, J. Am. Ceram. Soc. 82 (1999) 2193-2203. https://doi.org/10.1111/j.1151-2916.1999.tb02062.x

[39] W. Bragg, The structure of magnetite and the spinels, Nature 95 (1915) 561-561. https://doi.org/10.1038/095561a0

[40] R.J. Hill, J.R. Craig, G. Gibbs, Systematics of the spinel structure type, Phys. Chem. Miner. 4 (1979) 317-339. https://doi.org/10.1007/BF00307535

[41] I. Szczygieł, K. Winiarska, Synthesis and characterization of manganese-zinc ferrite obtained by thermal decomposition from organic precursors, J. Therm. Anal. Calorim. 115 (2014) 471-477. https://doi.org/10.1007/s10973-013-3281-2

[42] R. Moretti, G. Ottonello, An appraisal of endmember energy and mixing properties of rare earth garnets, Geochim. Cosmochim. Acta 62 (1998) 1147-1173. https://doi.org/10.1016/S0016-7037(98)00033-7

[43] M. Nur-E-Alam, M. Vasiliev, V. Belotelov, K. Alameh, Properties of ferrite garnet (Bi, Lu, Y)3(Fe, Ga)5O12 thin film materials prepared by RF magnetron sputtering, Nanomater. 8 (2018) 355. https://doi.org/10.3390/nano8050355

[44] J.R. Cunningham Jr, E.E. Anderson, Samarium substitutions in yttrium iron garnet, J. Appl. Phys. 31 (1960) S45-S46. https://doi.org/10.1063/1.1984598

[45] Z. Song, D. Zhou, Q. Liu, Tolerance factor and phase stability of the garnet structure, Acta Crystallogr. Sect. C Struct. Sci. Cryst. Eng. Mater. 75 (2019) 1353-1358. https://doi.org/10.1107/S2053229619011975

[46] D.A. Vinnik, V.E. Zhivulin, E.A. Trofimov, A.Y. Starikov, D.A. Zherebtsov, O.V. Zaitseva, S.A. Gudkova, S.V. Taskaev, D.S. Klygach, M.G. Vakhitov, Extremely polysubstituted magnetic material based on magnetoplumbite with a hexagonal structure: synthesis, structure, properties, prospects, Nanomater. 9 (2019) 559. https://doi.org/10.3390/nano9040559

[47] C. Delacotte, G. Whitehead, M. Pitcher, C. Robertson, P. Sharp, M. Dyer, J. Alaria, J. Claridge, G. Darling, D. Allan, Structure determination and crystal chemistry of large repeat mixed-layer hexaferrites, IUCrJ 5 (2018) 681-698. https://doi.org/10.1107/S2052252518011351

[48] H. Hibst, Hexagonal ferrites from melts and aqueous solutions, magnetic recording materials, Angew. Chem. Int. Ed. 21 (1982) 270-282. https://doi.org/10.1002/anie.198202701

[49] H. Kojima, Fundamental properties of hexagonal ferrites with magnetoplumbite structure, Handb. Magn. Mater. 3 (1982) 305-391. https://doi.org/10.1016/S1567-2719(82)03008-X

[50] D.A. Bardwell, C.S. Adjiman, Y.A. Arnautova, E. Bartashevich, S.X. Boerrigter, D.E. Braun, A.J. Cruz-Cabeza, G.M. Day, R.G. Della Valle, G.R. Desiraju, Towards crystal structure prediction of complex organic compounds-a report on the fifth blind test, Acta Crystallogr. Sect. B Struct. Sci. 67 (2011) 535-551. https://doi.org/10.1107/S0108768111042868

[51] I. Stefan, R. Chiriac, V. Nicoara, N. Cioatera, C. Nicolicescu, M. Trotea, J. Ghercioiu, PM functional materials: research regarding structural transformations of M-type barium ferrite pirosynthesis, in: Proc. Eur. Congr. Exhib. Powder Metall., European Powder Metallurgy Association, 2011, p. 1.

[52] Y.K. Jeong, J.H. Lee, S.J. Ahn, H.M. Jang, Temperature-induced magnetization reversal and ultra-fast magnetic switch at low field in SmFeO3, Solid State Commun. 152 (2012) 1112-1115. https://doi.org/10.1016/j.ssc.2012.04.010

[53] Y. Dang, D. Ju, L. Wang, X. Tao, Recent progress in the synthesis of hybrid halide perovskite single crystals, CrystEngComm 18 (2016) 4476-4484. https://doi.org/10.1039/C6CE00655H

[54] K. Bamzai, M. Bhat, Electrical and magnetic properties of some rare earth orthoferrites (RFeO3 where R = Y, Ho, Er) systems, Integr. Ferroelectr. 158 (2014) 108-122. https://doi.org/10.1080/10584587.2014.957591

[55] E.A.R. Assirey, Perovskite synthesis, properties and their related biochemical and industrial application, Saudi Pharm. J. 27 (2019) 817-829. https://doi.org/10.1016/j.jsps.2019.05.003

[56] S. Geller, Crystal structure of gadolinium orthoferrite, GdFeO3, J. Chem. Phys. 24 (1956) 1236-1239. https://doi.org/10.1063/1.1742746

[57] P.T. Coppens, M. Eibschütz, Determination of the crystal structure of yttrium orthoferrite and refinement of gadolinium orthoferrite, Acta Crystallogr. 19 (1965) 524-531. https://doi.org/10.1107/S0365110X65003833

[58] L. Tarasov, F. Bitter, Precise magnetic torque measurements on single crystals of iron, Phys. Rev. 52 (1937) 353. https://doi.org/10.1103/PhysRev.52.353

[59] W. Koehler, E. Wollan, Neutron-diffraction study of the magnetic properties of perovskite-like compounds LaBO3, J. Phys. Chem. Solids 2 (1957) 100-106. https://doi.org/10.1016/0022-3697(57)90095-1

[60] E. Haye, F. Capon, S. Barrat, P. Boulet, E. Andre, C. Carteret, S. Bruyere, Properties of rare-earth orthoferrites perovskite driven by steric hindrance, J. Alloys Compd. 657 (2016) 631-638. https://doi.org/10.1016/j.jallcom.2015.10.135

[61] T.C. Chen, H. Chang, Magnetic bubble memory and logic, in: M.C. Youvits (Ed.), Advances in computers, Elsevier, New York, 1978, pp. 223-282. https://doi.org/10.1016/S0065-2458(08)60393-9

[62] B.V. Prasad, K. Ramesh, A. Srinivas, Structural and magnetic studies of nano-crystalline ferrites MFe2O4 (M= Zn, Ni, Cu, and Co) synthesized via citrate gel autocombustion method, J. Supercond. Nov. Magn. 30 (2017) 3523-3535. https://doi.org/10.1007/s10948-017-4153-y

[63] M.A. Golsefidi, M. Abrodi, Z. Abbasi, A. Dashtbozorg, M.E. Rostami, M. Ebadi, Hydrothermal method for synthesizing ZnFe2O4 nanoparticles, photo-degradation of Rhodamine B by ZnFe2O4 and thermal stable PS-based nanocomposite, J. Mater. Sci.: Mater. Electron. 27 (2016) 8654-8660. https://doi.org/10.1007/s10854-016-4886-6

[64] I. Nurhasanah, Photodegradation of rhodamine B by using ZnFe2O4 nanoparticles synthesized through precipitation method, IOP Conf. Ser.: Mater. Sci. Eng. 202 (2017) 012044. https://doi.org/10.1088/1757-899X/202/1/012044

[65] A. Manohar, C. Krishnamoorthi, K.C.B. Naidu, C. Pavithra, Dielectric, magnetic hyperthermia, and photocatalytic properties of ZnFe2O4 nanoparticles synthesized by solvothermal reflux method, Appl. Phys. A. 125 (2019) 1-10. https://doi.org/10.1007/s00339-019-2760-0

[66] E. Hema, A. Manikandan, M. Gayathri, M. Durka, S.A. Antony, B. Venkatraman, The role of Mn2+-doping on structural, morphological, optical, magnetic and catalytic properties of spinel ZnFe2O4 nanoparticles, J. Nanosci. Nanotechnol. 16 (2016) 5929-5943. https://doi.org/10.1166/jnn.2016.11037

[67] A. Hussain, T. Abbas, S.B. Niazi, Preparation of N1+xMnxFe2O4 ferrites by sol-gel method and study of their cation distribution, Ceram. Int. 39 (2013) 1221-1225. https://doi.org/10.1016/j.ceramint.2012.07.049

[68] L. Yao, Y. Xi, G. Xi, Y. Feng, Synthesis of cobalt ferrite with enhanced magnetostriction properties by the sol− gel− hydrothermal route using spent Li-ion battery, J. Alloys Compd. 680 (2016) 73-79. https://doi.org/10.1016/j.jallcom.2016.04.092

[69] I. Gul, W. Ahmed, A. Maqsood, Electrical and magnetic characterization of nanocrystalline Ni-Zn ferrite synthesis by co-precipitation route, J. Magn. Magn. Mater. 320 (2008) 270-275. https://doi.org/10.1016/j.jmmm.2007.05.032

[70] A. Kalam, A.G. Al-Sehemi, M. Assiri, G. Du, T. Ahmad, I. Ahmad, M. Pannipara, Modified solvothermal synthesis of cobalt ferrite (CoFe2O4) magnetic nanoparticles photocatalysts for degradation of methylene blue with H2O2/visible light, Results Phys. 8 (2018) 1046-1053. https://doi.org/10.1016/j.rinp.2018.01.045

[71] L. Hu, A. Percheron, D. Chaumont, C.H. Brachais, Microwave-assisted one-step hydrothermal synthesis of pure iron oxide nanoparticles: Magnetite, maghemite and hematite, J. Sol-Gel Sci. Technol. 60 (2011) 198-205. https://doi.org/10.1007/s10971-011-2579-4

[72] F. Jiang, C.M. Wang, Y. Fu, R. Liu, Synthesis of iron oxide nanocubes via microwave-assisted solvolthermal method, J. Alloys Compd. 503 (2010) L31-L33. https://doi.org/10.1016/j.jallcom.2010.05.020

[73] P. Yadav, S. Manori, P. Chamoli, K.K. Raina, R.K. Shukla, Microwave assisted green synthesis of γ-Fe2O3 nanoparticles and their application for photodegradation of ternary dye mixture, J. Mater. Sci.: Mater. Electron. 34 (2023) 1065. https://doi.org/10.1007/s10854-023-10463-1

[74] B. Wathey, J. Tierney, P. Lidström, J. Westman, The impact of microwave-assisted organic chemistry on drug discovery, Drug Discov. Today 7 (2002) 373-380. https://doi.org/10.1016/S1359-6446(02)02178-5

[75] P. Lidström, J. Tierney, B. Wathey, J. Westman, Microwave assisted organic synthesis-a review, Tetrahedron 57 (2001) 9225-9283. https://doi.org/10.1016/S0040-4020(01)00906-1

[76] P. Berger, N.B. Adelman, K.J. Beckman, D.J. Campbell, A.B. Ellis, G.C. Lisensky, Preparation and properties of an aqueous ferrofluid, J. Chem. Educ. 76 (1999) 943. https://doi.org/10.1021/ed076p943

[77] T. Belin, N. Guigue-Millot, T. Caillot, D. Aymes, J. Niepce, Influence of grain size, oxygen stoichiometry, and synthesis conditions on the γ-Fe2O3 vacancies ordering and lattice parameters, J. Solid State Chem. 163 (2002) 459-465. https://doi.org/10.1006/jssc.2001.9426

[78] A.S. Teja, P.Y. Koh, Synthesis, properties, and applications of magnetic iron oxide nanoparticles, Prog. Cryst. Growth Charact. Mater. 55 (2009) 22-45. https://doi.org/10.1016/j.pcrysgrow.2008.08.003

[79] P. Lidstrom, J. Westman, A. Lewis, Enhancement of combinatorial chemistry by microwave-assisted organic synthesis, Comb. Chem. High. Throughput. Screen. 5 (2002) 441-458. https://doi.org/10.2174/1386207023330147

[80] D. Stuerga, P. Gaillard, Microwave heating as a new way to induce localized enhancements of reaction rate. Non-isothermal and heterogeneous kinetics, Tetrahedron 52 (1996) 5505-5510. https://doi.org/10.1016/0040-4020(96)00241-4

[81] V. Sreeja, P. Joy, Microwave-hydrothermal synthesis of γ-Fe2O3 nanoparticles and their magnetic properties, Mater. Res. Bull. 42 (2007) 1570-1576. https://doi.org/10.1016/j.materresbull.2006.11.014

[82] O. Algul, A. Kaessler, Y. Apcin, A. Yilmaz, J. Jose, Comparative studies on conventional and microwave synthesis of some benzimidazole, benzothiazole and indole derivatives and testing on inhibition of hyaluronidase, Molecules 13 (2008) 736-748. https://doi.org/10.3390/molecules13040736

[83] M. Gaba, N. Dhingra, Microwave chemistry: general features and applications, Indian J. Pharm. Educ. Res. 45 (2011) 175-183.

[84] J. Smit, H. Wijn, Ferrites, Philips Technical Library, Eindhoven, The Netherlands. 1959, pp. 157-158.

[85] A. Gadkari, T. Shinde, P. Vasambekar, Structural analysis of Y3+-doped Mg-Cd ferrites prepared by oxalate co-precipitation method, Mater. Chem. Phys. 114 (2009) 505-510. https://doi.org/10.1016/j.matchemphys.2008.11.011

[86] S. Jacobo, S. Duhalde, H. Bertorello, Rare earth influence on the structural and magnetic properties of NiZn ferrites, J. Magn. Magn. Mater. 272 (2004) 2253-2254. https://doi.org/10.1016/j.jmmm.2003.12.564

[87] A. Verma, T. Goel, R. Mendiratta, R. Gupta, High-resistivity nickel-zinc ferrites by the citrate precursor method, J. Magn. Magn. Mater. 192 (1999) 271-276. https://doi.org/10.1016/S0304-8853(98)00592-7

[88] T. Nakamura, T. Miyamoto, Y. Yamada, Complex permeability spectra of polycrystalline Li-Zn ferrite and application to EM-wave absorber, J. Magn. Magn. Mater. 256 (2003) 340-347. https://doi.org/10.1016/S0304-8853(02)00698-4

[89] H. Waqas, A. Qureshi, Influence of pH on nanosized Mn-Zn ferrite synthesized by sol-gel auto combustion process, J. Therm. Anal. Calorim. 98 (2009) 355-360. https://doi.org/10.1007/s10973-009-0289-8

[90] N. Deraz, Glycine-assisted fabrication of nanocrystalline cobalt ferrite system, J. Anal. Appl. Pyrolysis 88 (2010) 103-109. https://doi.org/10.1016/j.jaap.2010.03.002

[91] N. Rezlescu, E. Rezlescu, F. Tudorache, P. Popa, MgCu nanocrystalline ceramic with La3+ and Y3+ ionic substitutions used as humidity sensor, J. Optoelectron. Adv. Mater. 6 (2004) 695-698.

[92] X. Liu, B. Cheng, J. Hu, H. Qin, M. Jiang, Semiconducting gas sensor for ethanol based on LaMgxFe1−xO3 nanocrystals, Sens. Actuators B Chem. 129 (2008) 53-58. https://doi.org/10.1016/j.snb.2007.07.102

[93] K. Raj, B. Moskowitz, R. Casciari, Advances in ferrofluid technology, J. Magn. Magn. Mater. 149 (1995) 174-180. https://doi.org/10.1016/0304-8853(95)00365-7

[94] E. Melagiriyappa, H. Jayanna, Structural and magnetic susceptibility studies of samarium substituted magnesium-zinc ferrites, J. Alloys Compd. 482 (2009) 147-150. https://doi.org/10.1016/j.jallcom.2009.03.112

[95] N.A. Spaldin, M. Fiebig, The renaissance of magnetoelectric multiferroics, Science 309 (2005) 391-392. https://doi.org/10.1126/science.1113357

[96] A. Goldman, Modern Ferrite Technology, Springer, New York, 2006.

[97] A. Zarrin, S. Sadighian, K. Rostamizadeh, O. Firuzi, M. Hamidi, S. Mohammadi-Samani, R. Miri, Design, preparation, and in vitro characterization of a trimodally-targeted nanomagnetic onco-theranostic system for cancer diagnosis and therapy, Int. J. Pharm. 500 (2016) 62-76. https://doi.org/10.1016/j.ijpharm.2015.12.051

[98] N. Singh, G.J. Jenkins, R. Asadi, S.H. Doak, Potential toxicity of superparamagnetic iron oxide nanoparticles (SPION), Nano Rev. 1 (2010) 5358. https://doi.org/10.3402/nano.v1i0.5358

[99] L. Alpsoy, A. Baykal, M. Amir, Z. Ülker, M. Nawaz, SPION@APTES@FA-PEG@ Usnic acid bionanodrug for cancer therapy, J. Supercond. Nov. Magn. 31 (2018) 1395-1401. https://doi.org/10.1007/s10948-017-4333-9

[100] S. Kumari, R.P. Singh, Glycolic acid functionalized chitosan-Au-Fe3O4 hybrid nanoparticle based nanohybrid scaffold for drug delivery, Int. J. Biol. Macromol. 54 (2013) 244-249. https://doi.org/10.1016/j.ijbiomac.2012.12.001

[101] C. Chouly, D. Pouliquen, I. Lucet, J. Jeune, P. Jallet, Development of superparamagnetic nanoparticles for MRI: effect of particle size, charge and surface nature on biodistribution, J. Microencapsul. 13 (1996) 245-255. https://doi.org/10.3109/02652049609026013

[102] J. Chatterjee, Y. Haik, C.J. Chen, Size dependent magnetic properties of iron oxide nanoparticles, J. Magn. Magn. Mater. 257 (2003) 113-118. https://doi.org/10.1016/S0304-8853(02)01066-1

[103] S.E. Pratsinis, S. Vemury, Particle formation in gases: a review, Powder Technol. 88 (1996) 267-273. https://doi.org/10.1016/S0032-5910(96)03130-0

[104] G. Storm, S.O. Belliot, T. Daemen, D.D. Lasic, Surface modification of nanoparticles to oppose uptake by the mononuclear phagocyte system, Adv. Drug Deliv. Rev. 17 (1995) 31-48. https://doi.org/10.1016/0169-409X(95)00039-A

[105] K. Ulbrich, K. Hola, V. Subr, A. Bakandritsos, J. Tucek, R. Zboril, Targeted drug delivery with polymers and magnetic nanoparticles: covalent and noncovalent approaches, release control, and clinical studies, Chem. Rev. 116 (2016) 5338-5431. https://doi.org/10.1021/acs.chemrev.5b00589

[106] W. Chen, N. Xu, L. Xu, L. Wang, Z. Li, W. Ma, Y. Zhu, C. Xu, N.A. Kotov, Multifunctional magnetoplasmonic nanoparticle assemblies for cancer therapy and diagnostics (theranostics), Macromol. Rapid Commun. 31 (2010) 228-236. https://doi.org/10.1002/marc.200900793

[107] N. Ahmed, H. Fessi, A. Elaissari, Theranostic applications of nanoparticles in cancer, Drug Discov. Today 17 (2012) 928-934. https://doi.org/10.1016/j.drudis.2012.03.010

[108] J. Xie, Y. Zhang, C. Yan, L. Song, S. Wen, F. Zang, G. Chen, Q. Ding, C. Yan, N. Gu, High-performance PEGylated Mn-Zn ferrite nanocrystals as a passive-targeted agent for magnetically induced cancer theranostics, Biomater. 35 (2014) 9126-9136. https://doi.org/10.1016/j.biomaterials.2014.07.019

[109] Y. Cohen, S.Y. Shoushan, Magnetic nanoparticles-based diagnostics and theranostics, Curr. Opin. Biotechnol. 24 (2013) 672-681. https://doi.org/10.1016/j.copbio.2013.01.006

[110] A. Zarepour, A. Zarrabi, A. Khosravi, SPIONS as nano-theranostics agents, in: R. Zhang, H. Olin (Eds.), Magnetic Nanoparticles in Biomedical Applications, Springer, Singapore, 2017, pp. 1-44. https://doi.org/10.1007/978-981-10-3563-0_1

[111] T. Neuberger, B. Schöpf, H. Hofmann, M. Hofmann, B. Von Rechenberg, Superparamagnetic nanoparticles for biomedical applications: possibilities and limitations of a new drug delivery system, J. Magn. Magn. Mater. 293 (2005) 483-496. https://doi.org/10.1016/j.jmmm.2005.01.064

[112] V.J. Venditto, F.C. Szoka, Cancer nanomedicine: so many papers and so few drugs, Adv. Drug Deliv. Rev. 65 (2013) 80-88. https://doi.org/10.1016/j.addr.2012.09.038

[113] M. Liong, J. Lu, M. Kovochich, T. Xia, S.G. Ruehm, A.E. Nel, F. Tamanoi, J.I. Zink, Multifunctional inorganic nanoparticles for imaging, targeting, and drug delivery, ACS Nano. 2 (2008) 889-896. https://doi.org/10.1021/nn800072t

[114] C.C. Berry, A.S. Curtis, Functionalisation of magnetic nanoparticles for applications in biomedicine, J. Phys. D: Appl. Phys. 36 (2003) R198. https://doi.org/10.1088/0022-3727/36/13/203

[115] C.C. Berry, Progress in functionalization of magnetic nanoparticles for applications in biomedicine, J. Phys. D: Appl. Phys. 42 (2009) 224003. https://doi.org/10.1088/0022-3727/42/22/224003

[116] A.K. Gupta, M. Gupta, Synthesis and surface engineering of iron oxide nanoparticles for biomedical applications, Biomater. 26 (2005) 3995-4021. https://doi.org/10.1016/j.biomaterials.2004.10.012

[117] J. Estelrich, E. Escribano, J. Queralt, M.A. Busquets, Iron oxide nanoparticles for magnetically-guided and magnetically-responsive drug delivery, Int. J. Mol. Sci. 16 (2015) 8070-8101. https://doi.org/10.3390/ijms16048070

[118] B.R. Jermy, V. Ravinayagam, D. Almohazey, W. Alamoudi, H. Dafalla, S. Akhtar, G. Tanimu, PEGylated green halloysite/spinel ferrite nanocomposites for pH sensitive delivery of dexamethasone: a potential pulmonary drug delivery treatment option for COVID-19, Appl. Clay Sci. 216 (2022) 106333. https://doi.org/10.1016/j.clay.2021.106333

[119] R. Puri, V. Arora, Magnetic nano-ferrites as targeted drug delivery system for the treatment of cancer, AIP Conf. Proc., AIP Publishing, Melville, New York, 2023, 040012. https://doi.org/10.1063/5.0140628

[120] S.C. McBain, H.H. Yiu, J. Dobson, Magnetic nanoparticles for gene and drug delivery, Int. J. Nanomed. 3 (2008) 169-180. https://doi.org/10.2147/IJN.S1608

[121] Q.A. Pankhurst, J. Connolly, S.K. Jones, J. Dobson, Applications of magnetic nanoparticles in biomedicine, J. Phys. D: Appl. Phys. 36 (2003) R167. https://doi.org/10.1088/0022-3727/36/13/201

[122] N. Sulaiman, M. Ghazali, B.Y. Majlis, J. Yunas, M. Razali, Superparamagnetic calcium ferrite nanoparticles synthesized using a simple sol-gel method for targeted drug delivery, Biomed. Mater. Eng., 26 (1_suppl) (2015) S103-S110. https://doi.org/10.3233/BME-151295

[123] M. Arruebo, R. Fernández-Pacheco, M.R. Ibarra, J. Santamaría, Magnetic nanoparticles for drug delivery, Nano today 2 (2007) 22-32. https://doi.org/10.1016/S1748-0132(07)70084-1

[124] F. Dumestre, B. Chaudret, C. Amiens, P. Renaud, P. Fejes, Superlattices of iron nanocubes synthesized from Fe [N(SiMe3)2]2, Science 303 (2004) 821-823. https://doi.org/10.1126/science.1092641

[125] S.R. Mudshinge, A.B. Deore, S. Patil, C.M. Bhalgat, Nanoparticles: Emerging carriers for drug delivery, Saudi Pharm. J. 19 (2011) 129-141. https://doi.org/10.1016/j.jsps.2011.04.001

[126] Q.A. Pankhurst, J. Connolly, S.K. Jones, J. Dobson, Applications of magnetic nanoparticles in biomedicine, J. Phys. D: Appl. Phys. 36 (2003) R167. https://doi.org/10.1088/0022-3727/36/13/201

[127] J. Ferlay, M. Colombet, I. Soerjomataram, C. Mathers, D.M. Parkin, M. Piñeros, A. Znaor, F. Bray, Estimating the global cancer incidence and mortality in 2018: GLOBOCAN sources and methods, Int. J. Cancer 144 (2019) 1941-1953. https://doi.org/10.1002/ijc.31937

[128] G.M. Cooper, K. Adams, The Cell: A Molecular Approach, Oxford Univ. Press, New York, 2022. https://doi.org/10.1093/hesc/9780197583746.001.0001

[129] V.K. Chaturvedi, A. Singh, V.K. Singh, M.P. Singh, Cancer nanotechnology: a new revolution for cancer diagnosis and therapy, Curr. Drug Metab. 20 (2019) 416-429. https://doi.org/10.2174/1389200219666180918111528

[130] Y. Yao, Y. Zhou, L. Liu, Y. Xu, Q. Chen, Y. Wang, S. Wu, Y. Deng, J. Zhang, A. Shao, Nanoparticle-based drug delivery in cancer therapy and its role in overcoming drug resistance, Front. Mol. Biosci. 7 (2020) 193. https://doi.org/10.3389/fmolb.2020.00193

[131] M. Yang, L. Gao, K. Liu, C. Luo, Y. Wang, L. Yu, H. Peng, W. Zhang, Characterization of Fe3O4/SiO2/Gd2O(CO3)2 core/shell/shell nanoparticles as T1 and T2 dual mode MRI contrast agent, Talanta 131 (2015) 661-665. https://doi.org/10.1016/j.talanta.2014.08.042

[132] A.I. Ghoneim, Applications of nano-ferrites in medicine, in: M. Khan (Ed.), Applications of Ferrites, BoD, London, 2024.

[133] Z. Karimi, S. Abbasi, H. Shokrollahi, G. Yousefi, M. Fahham, L. Karimi, O. Firuzi, Pegylated and amphiphilic chitosan coated manganese ferrite nanoparticles for pH-sensitive delivery of methotrexate: synthesis and characterization, Mater. Sci. Eng. C 71 (2017) 504-511. https://doi.org/10.1016/j.msec.2016.10.008

[134] S. Siavashy, M. Soltani, F. Ghorbani-Bidkorbeh, N. Fallah, G. Farnam, S.A. Mortazavi, F.H. Shirazi, M.H.H. Tehrani, M.H. Hamedi, Microfluidic platform for synthesis and optimization of chitosan-coated magnetic nanoparticles in cisplatin delivery, Carbohydr. Polym. 265 (2021) 118027. https://doi.org/10.1016/j.carbpol.2021.118027

[135] G. Bisht, M. Zaidi, S. Rayamajhi, Supercritical carbon dioxide-assisted synthesis of stimuli-responsive magnetic poly (N-isopropylacrylamide)-ferrite biocompatible nanocomposites for targeted and controlled drug delivery, Int. J. Polym. Mater. 66 (2017) 708-716. https://doi.org/10.1080/00914037.2016.1263949

[136] M.W. Mushtaq, F. Kanwal, A. Batool, T. Jamil, M. Zia-ul-Haq, B. Ijaz, Q. Huang, Z. Ullah, Polymer-coated CoFe2O4 nanoassemblies as biocompatible magnetic nanocarriers for anticancer drug delivery, J. Mater. Sci. 52 (2017) 9282-9293. https://doi.org/10.1007/s10853-017-1141-3

[137] K. Kaliyamoorthy, A.S. Pillai, A. Alexander, A. Arivarasu, I.V. Enoch, S. Ramasamy, β-Cyclodextrin-folate functionalized poly (lactic-co-glycolide)-superparamagnetic ytterbium ferrite hybrid nanocarrier for targeted delivery of camptothecin, Mater. Sci. Eng. C 122 (2021) 111796. https://doi.org/10.1016/j.msec.2020.111796

[138] Z. Shi, Y. Zeng, X. Chen, F. Zhou, L. Zheng, G. Wang, J. Gao, Y. Ma, L. Zheng, B. Fu, Mesoporous superparamagnetic cobalt ferrite nanoclusters: synthesis, characterization and application in drug delivery, J. Magn. Magn. Mater. 498 (2020) 166222. https://doi.org/10.1016/j.jmmm.2019.166222

[139] P. Chen, B. Cui, X. Cui, W. Zhao, Y. Bu, Y. Wang, A microwave-triggered controllable drug delivery system based on hollow-mesoporous cobalt ferrite magnetic nanoparticles, J. Alloys Compd. 699 (2017) 526-533. https://doi.org/10.1016/j.jallcom.2016.12.304

[140] M. Hafezi, M. Rostami, A. Hosseini, M. Rahimi-Nasrabadi, M. Fasihi-Ramandi, A. Badiei, F. Ahmadi, Cur-loaded ZnFe2O4@mZnO@N-GQDs biocompatible nano-carriers for smart and controlled targeted drug delivery with pH-triggered and ultrasound irradiation, J. Mol. Liq. 322 (2021) 114875. https://doi.org/10.1016/j.molliq.2020.114875

[141] H. Zhang, J. Wang, Y. Zeng, G. Wang, S. Han, Z. Yang, B. Li, X. Wang, J. Gao, L. Zheng, Leucine-coated cobalt ferrite nanoparticles: Synthesis, characterization and potential biomedical applications for drug delivery, Phys. Lett. A 384 (2020) 126600. https://doi.org/10.1016/j.physleta.2020.126600

[142] M. Ghanbari, F. Davar, A.E. Shalan, Effect of rosemary extract on the microstructure, phase evolution, and magnetic behavior of cobalt ferrite nanoparticles and its application on anti-cancer drug delivery, Ceram. Int. 47 (2021) 9409-9417. https://doi.org/10.1016/j.ceramint.2020.12.073

[143] M. Amiri, T. Gholami, O. Amiri, A. Pardakhti, M. Ahmadi, A. Akbari, A. Amanatfard, M. Salavati-Niasari, The magnetic inorganic-organic nanocomposite based on ZnFe2O4-Imatinib-liposome for biomedical applications, in vivo and in vitro study, J. Alloys Compd. 849 (2020) 156604. https://doi.org/10.1016/j.jallcom.2020.156604

[144] S. Kamaraj, U.M. Palanisamy, M.S.B.K. Mohamed, A. Gangasalam, G.A. Maria, R. Kandasamy, Curcumin drug delivery by vanillin-chitosan coated with calcium ferrite hybrid nanoparticles as carrier, Eur. J. Pharm. Sci. 116 (2018) 48-60. https://doi.org/10.1016/j.ejps.2018.01.023

[145] N. Shahabadi, M. Razlansari, A. Khorshidi, H. Zhaleh, Investigation of controlled release properties and anticancer effect of folic acid conjugated magnetic core-shell nanoparticles as a

dual responsive drug delivery system on A-549 and A-431 cancer cell lines, Res. Chem. Intermed. 46 (2020) 4257-4278. https://doi.org/10.1007/s11164-020-04205-9

[146] M. Rostami, M. Aghajanzadeh, M. Zamani, H.K. Manjili, H. Danafar, Sono-chemical synthesis and characterization of Fe3O4@mTiO2-GO nanocarriers for dual-targeted colon drug delivery, Res. Chem. Intermed. 44 (2018) 1889-1904. https://doi.org/10.1007/s11164-017-3204-0

[147] A. Seyfoori, S.S. Ebrahimi, S. Omidian, S.M. Naghib, Multifunctional magnetic ZnFe2O4-hydroxyapatite nanocomposite particles for local anti-cancer drug delivery and bacterial infection inhibition: an in vitro study, J. Taiwan Inst. Chem. Eng. 96 (2019) 503-508. https://doi.org/10.1016/j.jtice.2018.10.018

[148] R. Mannu, V. Karthikeyan, N. Velu, C. Arumugam, V.A. Roy, A.I. Gopalan, G. Saianand, P. Sonar, K.P. Lee, W.J. Kim, Polyethylene glycol coated magnetic nanoparticles: Hybrid nanofluid formulation, properties and drug delivery prospects, Nanomater. 11 (2021) 440. https://doi.org/10.3390/nano11020440

[149] V. Sawant, S. Bamane, R. Shejwal, S. Patil, Comparison of drug delivery potentials of surface functionalized cobalt and zinc ferrite nanohybrids for curcumin in to MCF-7 breast cancer cells, J. Magn. Magn. Mater. 417 (2016) 222-229. https://doi.org/10.1016/j.jmmm.2016.05.061

[150] Z. Karimi Ghezeli, M. Hekmati, H. Veisi, Synthesis of Imatinib-loaded chitosan-modified magnetic nanoparticles as an anti-cancer agent for pH responsive targeted drug delivery, Appl. Organomet. Chem. 33 (2019) e4833. https://doi.org/10.1002/aoc.4833

[151] K. Joshy, R. Augustine, A. Mayeen, S.M. Alex, A. Hasan, S. Thomas, H. Chi, NiFe2O4/poly (ethylene glycol)/lipid-polymer hybrid nanoparticles for anti-cancer drug delivery, New J. Chem. 44 (2020) 18162-18172. https://doi.org/10.1039/D0NJ01163K

[152] S.K. Sharma, Y. Javed, Magnetic nanoheterostructures: diagnostic, imaging and treatment, Springer Nature, Cham, 2020. https://doi.org/10.1007/978-3-030-39923-8

[153] U. Zimmermann, J. Vienken, G. Pilwat, Development of drug carrier systems: electrical field induced effects in cell membranes, J. Electroanal. Chem. 116 (1980) 553-574. https://doi.org/10.1016/S0022-0728(80)80278-6

[154] J.A. Freeman, J.C. Geer, Intestinal fat and iron transport, goblet cell mucus secretion, and cellular changes in protein deficiency observed with the electron microscope, Exp. Mol. Pathol. 4 (1965) 1004-1023. https://doi.org/10.1007/BF02233375

[155] P.H. Meyers, C.M. Nice JR, G.R. Meckstroth, H.C. Becker, P.J. Moser, M. Goldstein, Pathologic studies following magnetic control of metallic iron particles in the lymphatic and vascular system of dogs as a contrast and isotopic agent, Am. J. Roentgenol. Radium Ther. Nucl. Med. 96 (1966) 913-921. https://doi.org/10.2214/ajr.96.4.913

[156] J.M. Gallo, P. Varkonyi, E.E. Hassan, D.R. Groothius, Targeting anticancer drugs to the brain: II. Physiological pharmacokinetic model of oxantrazole following intraarterial administration to rat glioma-2 (RG-2) bearing rats, J. Pharmacokinet. Biopharm. 21 (1993) 575-592. https://doi.org/10.1007/BF01059115

[157] C.R. Kalaiselvan, S.S. Laha, S.B. Somvanshi, T.A. Tabish, N.D. Thorat, N.K. Sahu, Manganese ferrite (MnFe2O4) nanostructures for cancer theranostics, Coord. Chem. Rev. 473 (2022) 214809. https://doi.org/10.1016/j.ccr.2022.214809

[158] M. Ravichandran, S. Velumani, J.T. Ramirez, A. Vera, L. Leija, Biofunctionalized MnFe2O4@Au core-shell nanoparticles for pH-responsive drug delivery and hyperthermal agent for cancer therapy, Artif Cells Nanomed. Biotechnol. 46 (2018) S993-S1003. https://doi.org/10.1080/21691401.2018.1523182

[159] V.L. Calero-DdelC, C. Rinaldi, Synthesis and magnetic characterization of cobalt-substituted ferrite (CoxFe3− xO4) nanoparticles, J. Magn. Magn. Mater. 314 (2007) 60-67. https://doi.org/10.1016/j.jmmm.2006.12.030

[160] Y. Cedeño-Mattei, O. Perales-Perez, M.S. Tomar, F. Roman, P.M. Voyles, W.G. Stratton, Tuning of magnetic properties in cobalt ferrite nanocrystals, J. Appl. Phys. 103 (2008) 07E512. https://doi.org/10.1063/1.2838215

[161] M.A. Medina, G. Oza, A. Ángeles-Pascual, M. González M, R. Antaño-López, A. Vera, L. Leija, E. Reguera, L. Arriaga, J.M. Hernández Hernández, Synthesis, characterization and magnetic hyperthermia of monodispersed cobalt ferrite nanoparticles for cancer therapeutics, Molecules 25 (2020) 4428. https://doi.org/10.3390/molecules25194428

[162] S. Fiaz, M.N. Ahmed, I. ul Haq, S.W.A. Shah, M. Waseem, Green synthesis of cobalt ferrite and Mn doped cobalt ferrite nanoparticles: Anticancer, antidiabetic and antibacterial studies, J. Trace Elem. Med. Biol. 80 (2023) 127292. https://doi.org/10.1016/j.jtemb.2023.127292

[163] Y. Oh, M.S. Moorthy, P. Manivasagan, S. Bharathiraja, J. Oh, Magnetic hyperthermia and pH-responsive effective drug delivery to the sub-cellular level of human breast cancer cells by modified CoFe2O4 nanoparticles, Biochimie 133 (2017) 7-19. https://doi.org/10.1016/j.biochi.2016.11.012

[164] O. Veiseh, J.W. Gunn, M. Zhang, Design and fabrication of magnetic nanoparticles for targeted drug delivery and imaging, Adv. Drug Deliv. Rev. 62 (2010) 284-304. https://doi.org/10.1016/j.addr.2009.11.002

[165] M.K. Riley, W. Vermerris, Recent advances in nanomaterials for gene delivery-a review, Nanomater. 7 (2017) 94. https://doi.org/10.3390/nano7050094

[166] N. Voronina, H. Lemcke, F. Wiekhorst, J.-P. Kuehn, C. Rimmbach, G. Steinhoff, R. David, Non-viral magnetic engineering of endothelial cells with microRNA and plasmid-DNA-An optimized targeting approach, Nanomedicine 12 (2016) 2353-2364. https://doi.org/10.1016/j.nano.2016.06.015

[167] J.M. Arshad, W. Raza, N. Amin, K. Nadeem, M.I. Arshad, M.A. Khan, Synthesis and characterization of cobalt ferrites as MRI contrast agent, Mater. Today Proc. 47 (2021) S50-S54. https://doi.org/10.1016/j.matpr.2020.04.746

[168] B.R. Smith, S.S. Gambhir, Nanomaterials for in vivo imaging, Chem. Rev. 117 (2017) 901-986. https://doi.org/10.1021/acs.chemrev.6b00073

[169] S. Shen, W. Ding, S. Ahmed, R. Hu, A. Opalacz, S. Roth, Z. You, G.R. Wotjkiewicz, G. Lim, L. Chen, Ultrasmall superparamagnetic iron oxide imaging identifies tissue and nerve

inflammation in pain conditions, Pain Med. 19 (2018) 686-692. https://doi.org/10.1093/pm/pnx267

[170] J. Wahsner, E.M. Gale, A. Rodríguez-Rodríguez, P. Caravan, Chemistry of MRI contrast agents: current challenges and new frontiers, Chem. Rev. 119 (2018) 957-1057. https://doi.org/10.1021/acs.chemrev.8b00363

[171] D. Shigeoka, T. Yamazaki, T. Ishikawa, K. Miike, K. Fujiwara, T. Ide, A. Oshima, T. Hashimoto, D. Aihara, K. Kanda, Functionalization and magnetic relaxation of ferrite nanoparticles for theranostics, IEEE Trans. Magn. 54 (2018) 1-7. https://doi.org/10.1109/TMAG.2018.2845132

[172] T. Sobhani, D. Shahbazi-Gahrouei, M. Rostami, M. Zahraei, A. Farzadniya, Assessment of manganese-zinc ferrite nanoparticles as a novel magnetic resonance imaging contrast agent for the detection of 4T1 breast cancer cells, J. Med. Signals Sens. 9 (2019) 245-251. https://doi.org/10.4103/jmss.JMSS_59_18

[173] S. Bano, T. Zafar, S. Akhtar, S.A. Buzdar, M.M. Waraich, M. Afzal, Biopolymers coated superparamagnetic nickel ferrites: enhanced biocompatibility and MR imaging probe for breast cancer, J. Magn. Magn. Mater. 417 (2016) 284-290. https://doi.org/10.1016/j.jmmm.2016.05.063

[174] J.W. Bulte, Superparamagnetic iron oxides as MPI tracers: A primer and review of early applications, Adv. Drug Deliv. Rev. 138 (2019) 293-301. https://doi.org/10.1016/j.addr.2018.12.007

[175] Y. Du, P.T. Lai, C.H. Leung, P.W. Pong, Design of superparamagnetic nanoparticles for magnetic particle imaging (MPI), Int. J. Mol. Sci. 14 (2013) 18682-18710. https://doi.org/10.3390/ijms140918682

[176] D.D. Stueber, J. Villanova, I. Aponte, Z. Xiao, V.L. Colvin, Magnetic nanoparticles in biology and medicine: past, present, and future trends, Pharmaceutics 13 (2021) 943. https://doi.org/10.3390/pharmaceutics13070943

[177] F. Vurro, M. Gerosa, A. Busato, M. Muccilli, E. Milan, J. Gaudet, P. Goodwill, J. Mansfield, E. Forlin, A. Negri, Doped ferrite nanoparticles exhibiting self-regulating temperature as magnetic fluid hyperthermia antitumoral agents, with diagnostic capability in magnetic resonance imaging and magnetic particle imaging, Cancers 14 (2022) 5150. https://doi.org/10.3390/cancers14205150

[178] Y. Du, X. Liu, Q. Liang, X.-J. Liang, J. Tian, Optimization and design of magnetic ferrite nanoparticles with uniform tumor distribution for highly sensitive MRI/MPI performance and improved magnetic hyperthermia therapy, Nano Lett. 19 (2019) 3618-3626. https://doi.org/10.1021/acs.nanolett.9b00630

[179] J. Kemshead, L. Heath, F. Gibson, F. Katz, F. Richmond, J. Treleaven, J. Ugelstad, Magnetic microspheres and monoclonal antibodies for the depletion of neuroblastoma cells from bone marrow: experiences, improvements and observations, Br. J. Cancer 54 (1986) 771-778. https://doi.org/10.1038/bjc.1986.239

[180] A. Singh, S.K. Sahoo, Magnetic nanoparticles: a novel platform for cancer theranostics, Drug Discov. Today. 19 (2014) 474-481. https://doi.org/10.1016/j.drudis.2013.10.005

[181] D. Cohn, M. Zarek, A. Elyashiv, M.A. Sbitan, V. Sharma, R.V. Ramanujan, Remotely triggered morphing behavior of additively manufactured thermoset polymer-magnetic nanoparticle composite structures, Smart Mater. Struct. 30 (2021) 045022. https://doi.org/10.1088/1361-665X/abeaeb

[182] C.W. Shields Iv, C.D. Reyes, G.P. López, Microfluidic cell sorting: a review of the advances in the separation of cells from debulking to rare cell isolation, Lab Chip 15 (2015) 1230-1249. https://doi.org/10.1039/C4LC01246A

[183] K.J. Kim, H.Y. Cho, W.J. Lee, J.W. Choi, Subtyping of magnetically isolated breast cancer cells using magnetic force microscopy, Biotechnol. J. 13 (2018) 1700625. https://doi.org/10.1002/biot.201700625

[184] J. Kurczewska, B. Dobosz, Recent progress and challenges regarding magnetite-based nanoparticles for targeted drug delivery, Appl. Sci. 14 (2024) 1132. https://doi.org/10.3390/app14031132

Materials Research Foundations 189 (2026) 348-383 https://doi.org/10.21741/9781644904039-11

Chapter 11

Microwave-Assisted Transition Metal-Catalyzed Reactions

Aron Manick Joel J[1], Sharmila Nokku[1] and Gopal Chandru Senadi[1]*

[1]Green and Sustainable Synthesis Laboratory, Department of Chemistry, College of Engineering and Technology, SRM Institute of Science and Technology, Chengalpattu, 603 203, India

chandrug@srmist.edu.in

Abstract

Microwave radiation, which lies between infrared and radio waves in the electromagnetic spectrum, has been widely used in chemical synthesis to accelerate reactions and minimize the formation of by-products. Transition metals have greatly influenced modern chemistry due to their unique ability to exist in multiple oxidation states and to promote a wide range of catalytic transformations. They play a key role in diverse processes, from industrial-scale reactions such as the Haber-Bosch synthesis of ammonia to modern cross-coupling reactions that enable the efficient construction of carbon-carbon and carbon-heteroatom bonds which are essential steps in the synthesis of pharmaceuticals, agrochemicals, and advanced materials. In recent years, the move toward environmentally friendly and sustainable methods in organic synthesis has gained significant momentum. However, conventional transition metal-catalyzed reactions often require long reaction times and show limited efficiency, making them less sustainable. The use of microwave irradiation in such reactions offers a powerful solution, reducing reaction time, increasing yield, and improving selectivity by minimizing side reactions. This chapter explores the synergy between microwave irradiation and transition metal catalysis and its growing importance in developing efficient and sustainable synthetic methodologies.

Keywords

Microwave-Assisted Synthesis, Transition Metal Catalysis, Cross-Coupling Reactions, Sustainable Chemistry, Reaction Efficiency

List of abbreviations:

GHz – Giga Hertz
$Cu(OAc)_2$ – copper(II) acetate
t-AmOH – *tert*-amyl alcohol
$[RuCl_2(p\text{-cymene})]_2$ – dichloro(*p*-cymene)ruthenium(II) dimer
AcOH – acetic acid
$AgSbF_6$ – hexafluoroantimonate
AgOAc – silver acetate
DCE – dichloroethane
$RhCl_3.3H_2O$ – rhodium(III) chloride trihydrate
$NaBH_4$ – sodium borohydride

Materials Research Foundations 189 (2026) 348-383   https://doi.org/10.21741/9781644904039-11

Rh – rhodium
SEM – scanning electron microscopy
TEM – transmission electron microscopy
KF – potassium fluoride
PEG-400 – polyethylene glycol 400
SS-Rh – solid supported rhodium nanoparticles
$[Cp^*RhCl_2]_2$ – pentamethylcyclopentadienylrhodium(III) chloride dimer
$Cu(OAc)_2 \cdot H_2O$ – copper(II) acetate monohydrate
R^1-$(B(OH)_2$ – boronic acid
Pd – palladium
POPd – $[(t\text{-}Bu)_2P(OH)]_2PdCl_2$
K_2CO_3 – potassium carbonate
Pd EnCat 30 – polyurea microencapsulated palladium catalyst
THF – tetrahydrofuran
CsF – cesium fluoride
$Pd(PPh_3)_4$ – tetrakis(triphenylphosphine)palladium(0)
$NaHCO_3$ – sodium bicarbonate
DME – 1,2 dimethoxyethane
MW – microwave
LDH – layered double hydroxide
$[NBu_4]Br$ or TBAB – tetrabutylammonium bromide
XPS – X-ray photoelectron spectroscopy
TGA-MS – thermogravimetric analysis with mass spectrometry
Pd/C – palladium on carbon
Pd/SiO_2 – palladium on silicon dioxide
$Pd(L\text{-proline})_2$ – bis(L-prolinato)palladium(II)
NaOAc – sodium acetate
Nano-Pd@RAGC – rice husk derived activated carbon supported palladium nanaoparticles
EtOH – ethyl alcohol
$Pd(PPh_3)_2Cl_2$ - bis(triphenylphosphine)palladium(II) dichloride
LiCl – lithium chloride
CuI – copper(I) Iodide
DMF – N,N-dimethylformamide
Sn_2Bu_6 – bis(tributyltin)
H_2O – water
Cu – copper
NEt_3 – triethylamine
SPhos – 2-dicyclohexylphosphino-2',6'-dimethoxybiphenyl
CTAB – hexadecyltrimethylammonium bromide
$PdCl_2(PCy_3)_2$ – dichlorobis(triscyclohexylphosphine)palladium(II)
DMA – N,N-Dimethylacetamide
L or L_n – ligand
Zn – zinc
$Pd_2(dba)_3$ – tris(dibenzylideneacetone)dipalladium(0)
$Bu_3P.HBF_4$ – tri-*tert*-butylphosphonium tetrafluoroborate
$ZnCl_2$ – zinc chloride

Materials Research Foundations 189 (2026) 348-383 https://doi.org/10.21741/9781644904039-11

TMEDA – N,N,N',N'-Tetramethylethylenediamine
$Pd(OAc)_2$ – palladium(II) acetate
DPPB – 1,4-bis(diphenylphosphino)butane
$Mo(CO)_6$ – molybdenum hexacarbonyl
$FeCl_3$ – ferric chloride
$Fe_2(acac)_3$ – tris(acetylacetonato)iron(III)
DMEDA – N,N'-dimethylethanediamine
Cs_2CO_3 – cesium carbonate
PhI=NR – iminoiodinane
$FeCl_2$ – iron(II) chloride
Ru – ruthenium
Py – pyridine
MS – molecular sieves
KO^tBu – potassium *tert*-butoxide
NMP – N-Methyl-2-pyrrolidone
$Cp^*Co(CO)I_2$ – carbonyl(pentamethylcyclopentadienyl)cobalt(III) diiodide
DFT – density functional theory
ADC – acceptorless dehydrogenative coupling
KOH – potassium hydroxide
$CoCl_2$ – cobalt(II) chloride
Ni – nickel
C – carbon
NH_4OAc – ammonium acetate
$NiCl_2$ – nickel(II) chloride
NH_3 – ammonia
Cu_2O – copper(I) oxide
CO_2 – carbon dioxide
DMSO – dimethyl sulfoxide
S_N2 – substitution nucleophilic bimolecular
$ZnBr_2$ – zinc bromide
t-$BuNH_2$ – *tert*-butylamine

Contents

Materials Research Foundations 189 (2026) 348-383 https://doi.org/10.21741/9781644904039-11

1. Introduction

Microwave is an electromagnetic radiation that comes between infrared and radio waves in the electromagnetic spectrum. It has a frequency in the range of 0.3 to 300 Giga Hertz (GHz). When a molecule is irradiated with microwaves, the electric field component of the microwave interacts with the molecule through two different mechanisms, including dipolar polarization and conduction mechanism, which results in heating. The heat produced in the reaction through microwave irradiation depends primarily on the characteristics of dielectric heating, which is a phenomenon of using microwave energy to produce heating in the reaction. This microwave-assisted heating depends on how well the reactant or solvent absorbs the incident microwave and efficiently converts it into heat energy, and those molecules with this behavior have better efficiency and high yields. In dipolar polarization, when a molecule with a dipole is exposed to an oscillating electric field of a microwave, the dipole starts to align itself with the oscillating field. As a result, the molecules try to move back and forth and cause friction between the molecules, which releases heat energy. While in ionic conduction, the ions already present in the medium move rapidly when they are subjected to an electric field of a microwave. Consequently, it causes collisions between ions and generates heat energy [1]. In synthetic organic chemistry, transition metals play a pivotal role in many reactions in which coupling reactions including carbon-carbon coupling reactions like Suzuki, Heck, Sonogashira, Negishi, and Stille couplings and carbon-heteroatom coupling reactions such as Buchwald-Hartwig amination, Ullmann, and Chan-Lam reactions have created groundbreaking pathways for discovering new drugs and developing new methodologies for many marketed drugs [2]. Mostly, coupling reactions require moderate to high temperatures to achieve high yields [3]. Conventional mode of heating in coupling reactions typically shows poor efficiency and low yields because, in conventional heating, the heat is transferred from the reaction vessel to the reaction medium, which consumes more time, causes many side reactions, and decreases the efficiency of reactions. In microwave heating, molecules specifically absorb microwave energy and convert it into heat energy, so heat is transferred from the reaction medium to the reaction vessel, which decreases reaction time usually from hours to minutes in most cases, and increases efficiency [4]. In this chapter, we have discussed the microwave-assisted noble transition metal and non-noble transition metal-catalyzed reactions.

Materials Research Foundations 189 (2026) 348-383 https://doi.org/10.21741/9781644904039-11

2. Microwave-assisted transition metal-catalyzed reactions

Transition metal-catalyzed reactions, particularly coupling reactions, play a vital role in constructing complex molecular frameworks and bioactive compounds that are often challenging to synthesize using conventional methods. Although transition metal catalysts significantly facilitate such transformations, these reactions typically require stringent conditions, prolonged reaction times, and careful control to achieve good efficiency. The introduction of microwave-assisted transition metal-catalyzed systems has addressed many of these limitations. Microwave irradiation enhances reaction efficiency by providing rapid and uniform heating, thereby reducing reaction times, minimizing side reactions, and improving overall yields.

2.1 Microwave-assisted noble transition metal-catalyzed reaction

This section discusses microwave-assisted reactions catalyzed by noble transition metals such as ruthenium, rhodium, and palladium. These metals are well known for their remarkable catalytic activity, stability, and ability to access multiple oxidation states, which enable a wide range of organic transformations. Under microwave-assisted conditions, catalytic processes involving these metals often exhibit enhanced reaction rates, improved selectivity, and higher yields compared to conventional thermal methods.

2.1.1 Ruthenium metal-catalyzed reactions

In 2018, Eycken et al. reported a method for synthesizing isoquinolones and alkenylated benzamides from N-benzoyl α-amino ester derivatives [5]. They first investigated the intermolecular cyclization of *N*-benzoyl glycine ester with diphenylacetylene using a Ru(II) catalyst under various oxidants, solvents, and temperatures. Optimization of the reaction parameters led to a high yield of up to 92% when copper(II) acetate [$Cu(OAc)_2$] was used as the oxidant and tert-amyl alcohol (*t*-AmOH) as the solvent at 120 °C under microwave irradiation (Scheme 1a).

The proposed mechanism for the microwave-assisted ruthenium-catalyzed reaction is shown in Scheme 1b. The reaction begins when the ruthenium complex [$RuCl_2$(*p*-cymene)]$_2$ is activated through ligand exchange with $Cu(OAc)_2$, forming the active catalyst. Next, the amide group **1** loses a proton, allowing the ruthenium atom to coordinate with the nitrogen of the amide. This is followed by C–H bond activation, during which acetic acid is released, leading to the formation of a five-membered ring (**A**) containing ruthenium. The alkyne **2** then adds to this ring to form a seven-membered intermediate (**B**), which finally undergoes reductive elimination to produce the desired product **3**.

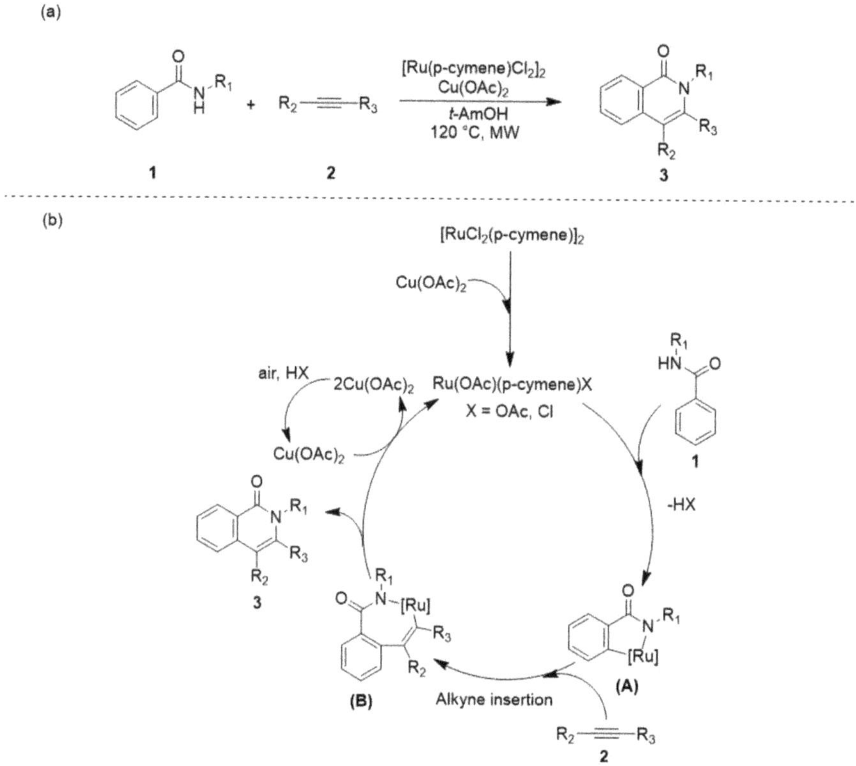

Scheme 1. *(a) Ruthenium-catalyzed C-H activation of α-amino acid esters. (b) Proposed mechanistic pathway.*

In 2022, Song et al. reported the *ortho*-alkenylation of α-amino ester-derived phosphinamides **4** with alkynes **5** under microwave irradiation [6]. In this study, they successfully synthesized a series of new phosphinamide derivatives **6** with high efficiency and excellent *E*-stereoselectivity. The optimized reaction conditions involved the use of $[Ru(p\text{-cymene})Cl_2]_2$ as the catalyst, along with silver hexafluoroantimonate ($AgSbF_6$), silver acetate (AgOAc), and acetic acid (AcOH) in dichloroethane (DCE) as the solvent at 150 °C, affording the alkenylated products in good yields within just 5 minutes of microwave irradiation (Scheme 2a). Furthermore, various α-monosubstituted, α,α-disubstituted, and α-amino acid ester-derived phosphinamides were found to give the corresponding alkenylated products in good yields. In addition, a [4 + 2] annulation reaction was also observed, particularly for glycine methyl ester-substituted phosphinamides.

The proposed mechanism is illustrated in Scheme 2b. The reaction begins with the ligand exchange of $[Ru(p\text{-cymene})Cl_2]_2$ with AgOAc in the presence of $AgSbF_6$, generating the active catalytic species $Ru(L)(OAc)_2$. Subsequently, the phosphinamide substrate **4** coordinates to $Ru(L)(OAc)_2$, followed by *ortho*-C–H bond cleavage to form the ruthenacycle intermediate **(A)**. In the next step, coordination and 1,2-migratory insertion of acetylene compound **5** into the Ru–C bond of intermediate **(A)** occurs to form intermediate **(B)**. Finally, protonation of intermediate **(B)**. by acetic acid (HOAc) affords the alkenylated product **6** along with the regeneration of the $Ru(L)(OAc)_2$ catalyst.

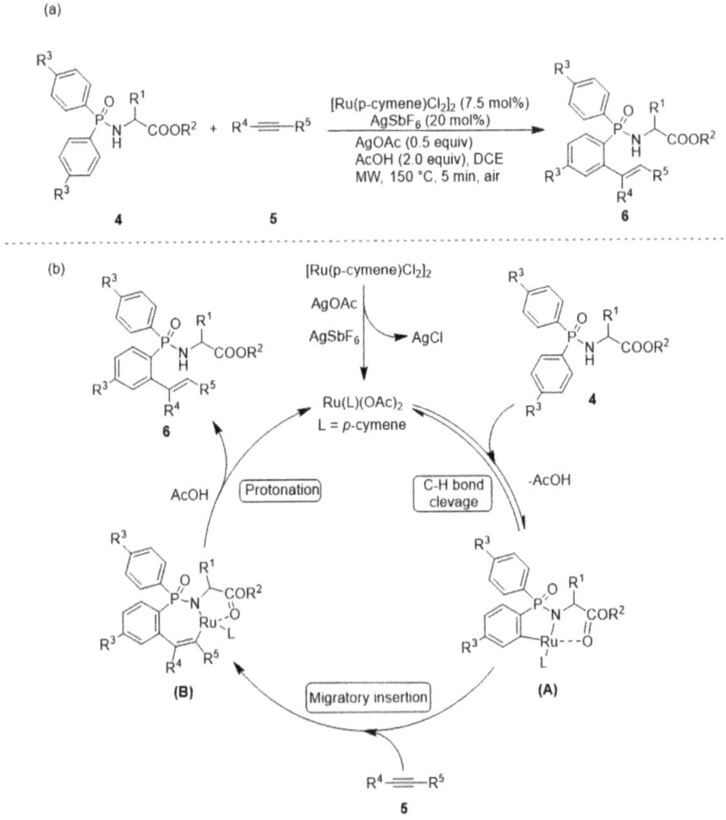

Scheme 2. *(a) Ruthenium metal-catalyzed C-H functionalization of phosphinamides. (b) Proposed mechanism of alkenylation.*

2.1.2 Rhodium metal-catalyzed reaction

In 2012, Das and co-workers reported a Suzuki–Miyaura cross-coupling reaction between aryl halide **7** and boronic acid **8** to form coupled product **9** under microwave irradiation using solid-supported rhodium nanoparticles as a heterogeneous catalyst [7]. The catalyst was prepared by the *in-situ* reduction of rhodium(III) chloride trihydrate ($RhCl_3 \cdot 3H_2O$) on amberlite resin with sodium borohydride ($NaBH_4$), resulting in the formation of spherical rhodium (Rh(0)) particles. The formation and morphology of these particles were confirmed by scanning electron microscopy (SEM) and transmission electron microscopy (TEM) analyses. The optimized reaction conditions involved 3 mol% of Rh nanoparticles as the catalyst, potassium fluoride (KF) as the base, and polyethylene glycol 400 (PEG-400) as the solvent. Under microwave heating at 110 °C for 15 minutes, the reaction afforded up to 97% yield (Scheme 3). Because the catalyst was heterogeneous, it could be easily separated by filtration after the reaction and reused up to twelve times without any loss of activity.

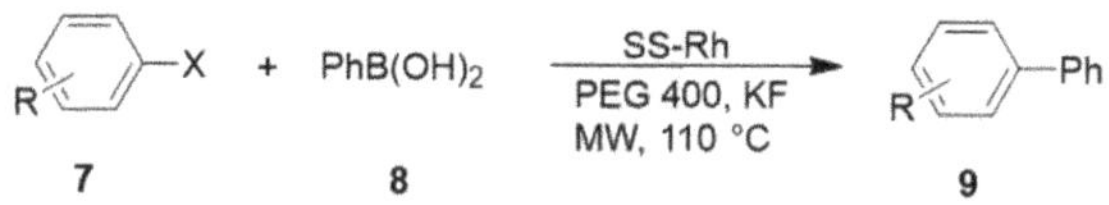

Scheme 3. *Solid-supported rhodium-catalyzed Suzuki cross-coupling reaction of aryl halides with phenylboronic acid.*

In 2014, Jun et al. developed a microwave-assisted Rh(III)-catalyzed N-annulation protocol for the efficient synthesis of isoquinoline and pyridine derivatives **12** via a C–H activation strategy [8]. In this study, aryl and α,β-unsaturated ketones **10** were reacted with alkynes **11** in the presence of ammonium acetate or ammonia under microwave irradiation using pentamethylcyclopentadienylrhodium(III) chloride dimer $[Cp^*RhCl_2]_2$ as the catalyst and copper(II) acetate monohydrate ($Cu(OAc)_2.H_2O$) as the oxidant (Scheme 4a). This method enabled a convenient one-pot synthesis of isoquinolines and pyridines in good yields.

The proposed mechanism for this Rh(III)-catalyzed, microwave-assisted N-annulation reaction is illustrated in Scheme 4b. The reaction begins with the condensation of an α,β-unsaturated ketone **10** with ammonia, generated *in situ* from ammonium acetate, to form a Schiff base imine intermediate **(A)**. The imine then undergoes N–H activation by the Rh(III) catalyst to afford the rhodium–imine complex **(B)**. Subsequently, β-C–H bond activation occurs, followed by coordination of the alkyne **11** to the rhodium center, forming the rhodacycle intermediate **(C)**. This intermediate undergoes alkyne carbometallation to generate a seven-membered rhodacycle **(D)**, which then undergoes reductive elimination to yield the annulated product **12** while reducing Rh(III) to Rh(I). Finally, the Rh(I) species is reoxidized to Rh(III) by copper acetate, completing the catalytic cycle.

Materials Research Foundations 189 (2026) 348-383 https://doi.org/10.21741/9781644904039-11

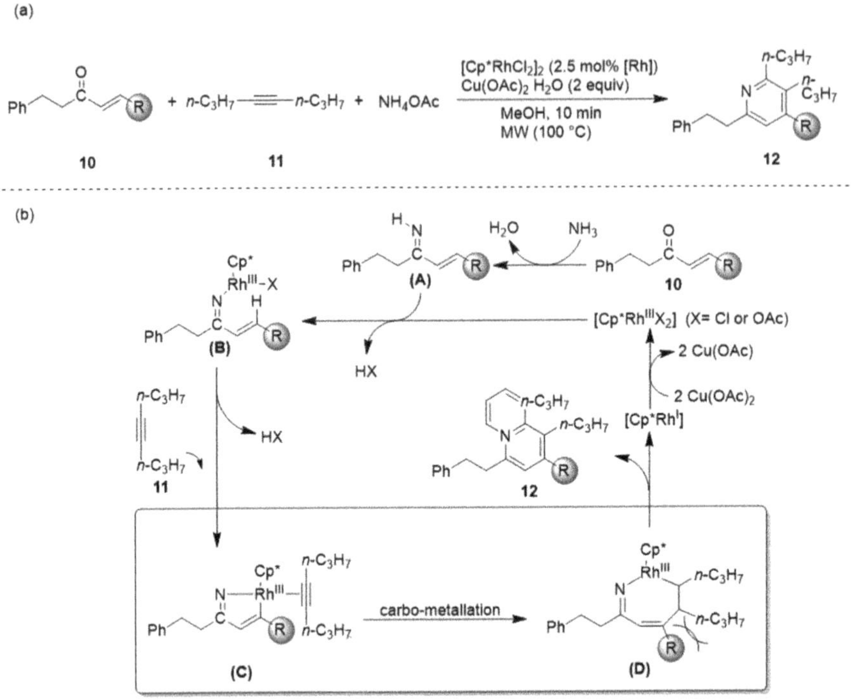

Scheme 4. (a) N-annulation reaction of aryl ketone with alkyne. (b) Proposed mechanism of N-annulation reaction.

2.1.3 Palladium metal-catalyzed reactions

Palladium-catalyzed reactions have become indispensable tools in modern synthetic chemistry, enabling the efficient construction of carbon–carbon and carbon–heteroatom bonds. Among these, coupling reactions such as Suzuki, Heck, Stille, Sonogashira, Negishi, and Hiyama have emerged as fundamental methodologies that serve as key building blocks in organic synthesis. These reactions play a crucial role in the preparation of complex molecules, including pharmaceuticals, natural products, and agrochemicals [9].

2.1.3.1 Suzuki cross-coupling reaction

The Suzuki reaction is one of the most versatile synthetic methods for the preparation of unsymmetrical biaryl compounds [10]. In this cross-coupling reaction, boronic acids or boronic esters react with aryl, vinyl, or alkyl halides in the presence of a palladium catalyst. Compared to other palladium-catalyzed coupling reactions, the Suzuki reaction exhibits excellent tolerance toward a wide range of functional groups.

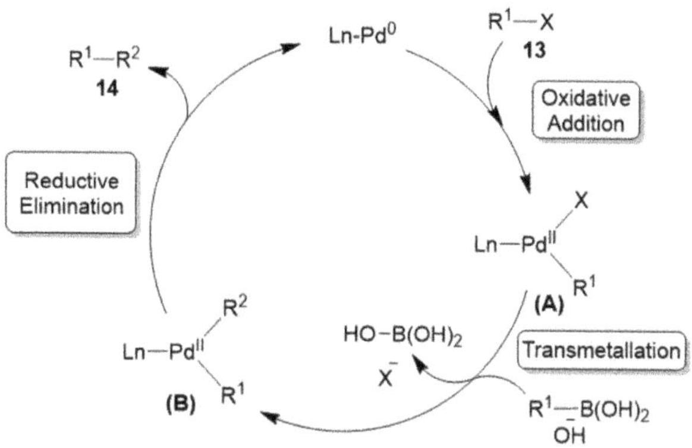

Scheme 5. *General mechanism for Suzuki cross-coupling reaction.*

The general mechanism of the Suzuki coupling reaction is shown in Scheme 5 [11]. It proceeds through three main steps catalyzed by palladium. In the first step, oxidative addition, the palladium(0) complex reacts with an alkyl or aryl halide **13** to form a Pd(II) intermediate **(A)**. In the second step, transmetallation, the organic group from the boronic acid (R^1–$B(OH)_2$) is transferred to the palladium center in the presence of a base, giving a new Pd(II) complex **(B)**. In the final step, reductive elimination, the two organic groups combine to form the coupled product **14**, and the palladium(0) catalyst is regenerated to complete the cycle.

In 2004, Turner and co-workers reported a fast and efficient method for preparing β-aryl and β-alkylarylidene malonates using a microwave-assisted Suzuki coupling reaction [12]. In this method, β-chloroalkylidene or β-arylidene malonates **15** were coupled with arylboronic acids **16** to form coupled product **17** by using 1 mol% of the palladium catalyst $[(t\text{-Bu})_2P(OH)]_2PdCl_2$ (POPd) and potassium carbonate (K_2CO_3) as the base under microwave heating at 100 °C (Scheme 6). The reaction was completed within 30 minutes and provided good to excellent yields.

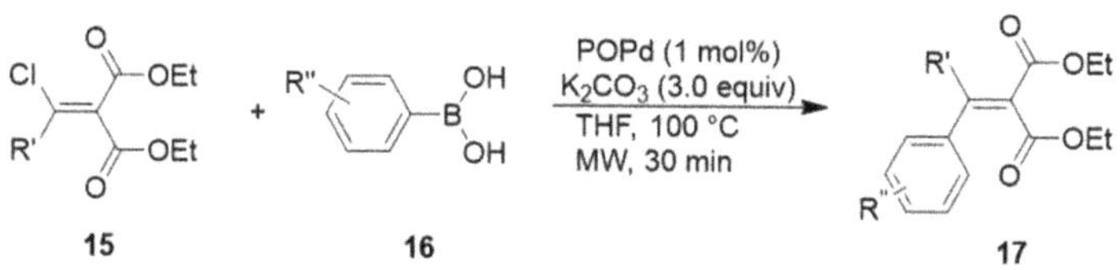

Scheme 6. *Palladium-catalyzed synthesis of β-chloroalkylidene malonates.*

In 2007, Amin et al. reported an efficient microwave-assisted Suzuki cross-coupling method for the synthesis of polycyclic aromatic hydrocarbon **20** by the coupling of aryl halide **18** with boronic acid **19** and their carcinogenic metabolites [13]. In this method, a polyurea microencapsulated palladium catalyst (Pd EnCat 30) was used along with a base and solvent, giving reaction yields of over 98% within just 20 minutes under microwave irradiation (Scheme 7). This approach was significantly faster than conventional heating methods, which typically required up to 24 hours.

Materials Research Foundations 189 (2026) 348-383 https://doi.org/10.21741/9781644904039-11

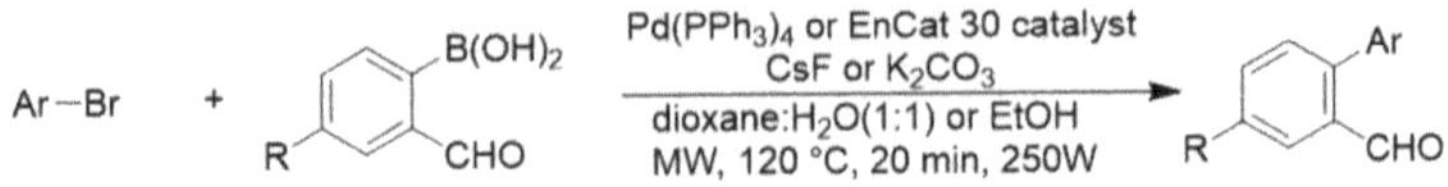

18 **19** **20**

Scheme 7. *Microwave-assisted palladium-catalyzed synthesis of polycyclic aromatic hydrocarbons.*

In 2015, Qu et al. developed a microwave-assisted Suzuki cross-coupling method for the synthesis of 4-methyl-6-arylsalicylates **23** [14]. These compounds are key intermediates in the design of herbicides, particularly acetohydroxyacid synthase (AHAS) inhibitors. In this method, aryl sulfonates **21** were coupled with aryl boronic acids **22** under optimized microwave conditions using tetrakis(triphenylphosphine)palladium(0) [$Pd(PPh_3)_4$] as the catalyst (Scheme 8). The reaction provided high yields in a short time. This approach allowed the efficient synthesis of various substrates that serve as valuable building blocks for developing bioactive compounds effective against herbicide-resistant weeds.

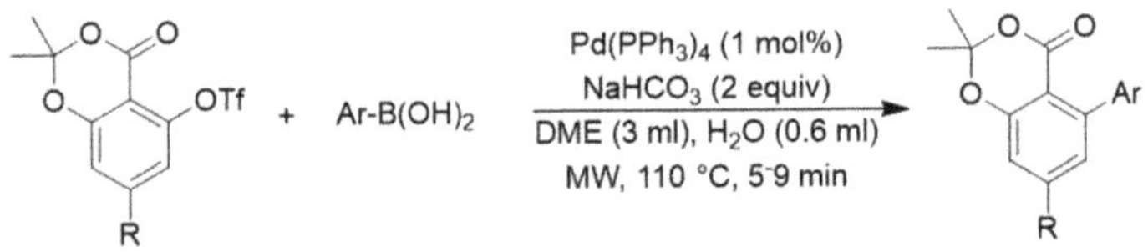

21 **22** **23**

Scheme 8. *Palladium-catalyzed synthesis of 6-arylsalicylates.*

2.1.3.2 Heck cross-coupling reaction

In the early 1970s, Heck and Mizoroki first reported the Heck coupling reactions [15]. It is a palladium-catalyzed coupling of an olefin to an aryl halide substrate in the presence of a base [16]. A general mechanism for the Heck coupling reaction is illustrated in Scheme 9 [17]. The mechanism is initiated from the active Pd(0) catalyst **(A)** obtained from the preactive Pd catalyst. Then **(A)** undergoes oxidative addition with an aryl halide **24** to form a Pd(II) complex **(B)**. The intermediate **(B)** then coordinates to the **25** alkene to form a π-complex **(C)**, which undergoes migratory insertion of the alkene into the Pd–C bond, forming a new carbon-carbon bond between the aryl group and alkene, resulting in an alkyl-palladium intermediate **(D)**. Thereafter, a β-hydride elimination occurs to generate a palladium–alkene π-complex **(E)**. Finally, a reductive elimination occurs to give cross-coupling between the alkene and aryl group to get **26**, and the reduced Pd(0) catalyst is again regenerated to finish the catalytic cycle.

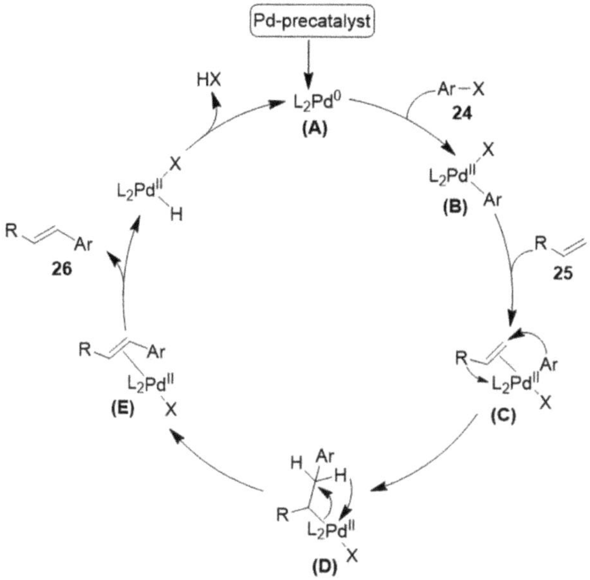

Scheme 9. *Mechanism of Heck cross-coupling reaction.*

In 2002, Choudary et al. reported a layered double hydroxide (LDH-Pd(0)) nanoparticle catalyzed Heck coupling of chloroarenes **27** with olefins **28** in the presence of ionic liquid tetrabutylammonium bromide ([NBu_4]Br) to produce coupled product **29** which is given in Scheme 10 [18]. In this method, the coupled products achieved high yields in the range of 76-98% in 0.5-40 hours under microwave irradiation at 130 °C. This heterogeneous catalyst is confirmed by X-ray photoelectron spectroscopy (XPS) and thermogravimetric analysis with mass spectrometry (TGA-MS) studies. After completion of the reaction, the catalyst can be reused for at least five cycles with no appreciable loss of activity. Compared to typical palladium on carbon (Pd/C) and palladium on silicon dioxide (Pd/SiO_2) catalysts, the LDH-Pd(0) exhibited better activity.

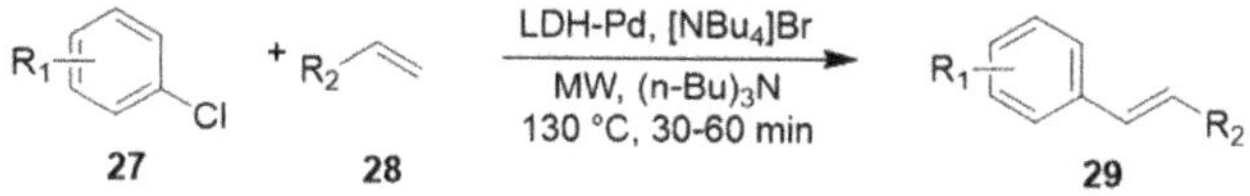

Scheme 10. *Heck Olefination of Chloroarenes with Olefins.*

In 2011, Singh et al. described a phosphine-free Heck reaction that uses bis(L-prolinato)palladium(II) (Pd(L-proline)$_2$) as a catalyst in water under microwave irradiation (Scheme 11) [19]. This reaction couples aryl and benzyl halides **30** with alkenes **31** using TBAB as an additive and sodium acetate (NaOAc) as a base to form a coupled product **32**. The optimized

Materials Research Foundations 189 (2026) 348-383 https://doi.org/10.21741/9781644904039-11

reaction conditions with 1 mol% catalyst, 100 mol% TBAB, 10 mol% NaOAc under 200–300 W microwave irradiation at the temperature range between 80–140 °C, giving 76-98 % yields at a reaction time of 10–50 minutes.

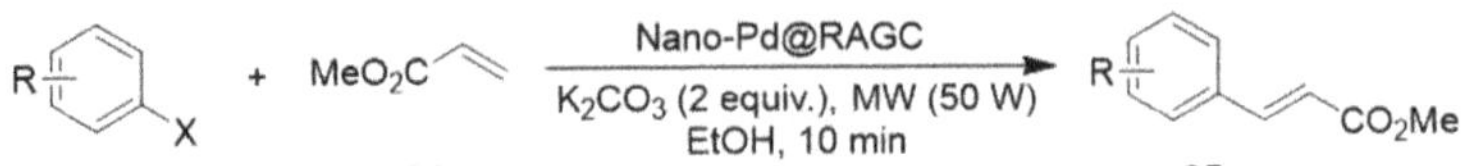

33 34 35

Scheme 11. *Heck coupling between aryl/benzyl halides and various alkenes.*

In 2019, Banerjee et al. reported a Heck cross-coupling reaction of aryl halide **33** and alkene **34** to get coupled product **35** in the presence of a heterogeneous palladium catalyst, which is synthesized by loading palladium nanoparticles onto activated graphitic carbon made from rice husks. (Scheme 12) [20]. The catalyst worked well in ligand-free cross-coupling reactions under microwave irradiation in green solvents like ethanol and water, and it shows high stability and can be recycled up to 10 times with minimal loss of activity. The optimized conditions resulted in satisfactory yields for various substrates. This method provides an environmentally friendly strategy for employing Heck cross-coupling reactions.

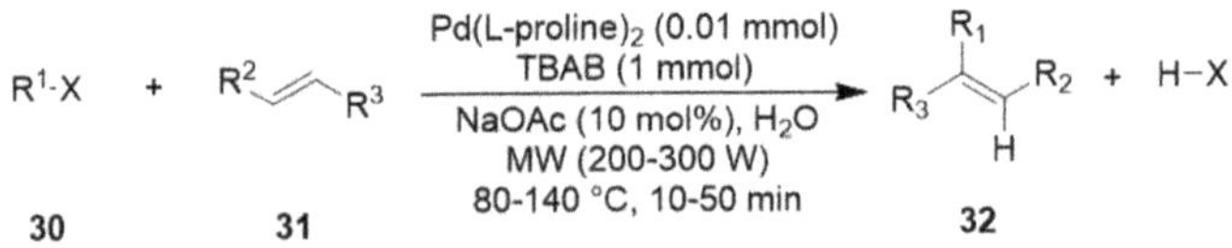

30 31 32

Scheme 12. *Heck coupling of aryl halides and alkenes.*

2.1.3.3 Stille coupling reaction

The Stille coupling reaction is a palladium-catalyzed reaction created by John K. Stille in the late 1970s. It occurs between organostannanes and organic electrophiles, usually aryl or vinyl halides. This reaction leads to the formation of carbon–carbon bonds [21]. The general mechanism for the Stille coupling reaction is depicted in Scheme 13 [22]. In this mechanism, the first step is the oxidative addition, where Pd(0) is inserted between the alkyl or aryl halide **36** and oxidized to Pd(II) to form an intermediate **(A)**. Following this step, transmetallation occurs, in which the alkyl group is transferred from the organotin substrate to the palladium intermediate **(B)**. Finally, reductive elimination leads to occurrence of C-C coupling to give coupled product **37**.

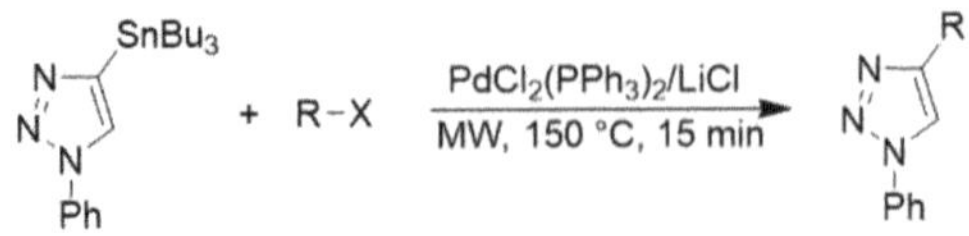

38 39 40

Scheme 13. *Mechanism of Stille coupling reaction.*

In 2009, Cebrain et al. reported the first microwave-assisted Stille cross-coupling reaction in which mono- and disubstituted polyheteroaryl triazoles **40** have been synthesized from different triazole derivatives **38** and aryl halides **39** under solvent-free conditions (Scheme 14) [23]. They used different catalytic systems, such as bis(triphenylphosphine)palladium(II) dichloride ($PdCl_2(PPh_3)_2$) with either lithium chloride (LiCl) or copper(I) iodide (CuI) for selective substitutions. Compared to conventional heating, this method produces triazoles in adequate yields. Additionally, this method reported a generation of symmetrical and unsymmetrical 3,5-disubstituted triazoles, triazole-fused biheteroaryl systems, and 1,2,3-triazole motifs.

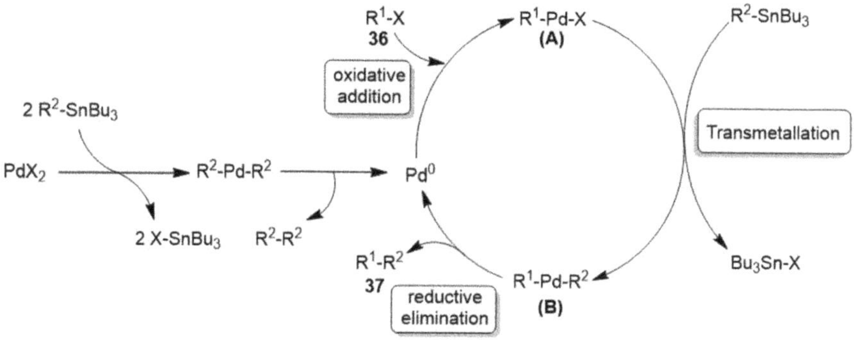

Scheme 14. *Stille coupling reaction of tributyltin triazole and halo derivatives.*

In 2011, Rafiee et al. used a dimeric ortho-palladated complex of tri-benzylamine as a robust and air-stable palladium catalyst to carry out the Stille cross-coupling reaction between various aryl iodides, bromides, and chlorides **41** with phenyl tributyltin **42** to form coupled product **43** under microwave irradiation [24]. The reaction conditions were optimized using K_2CO_3 as the base, affording excellent yields of biaryl products (68–97%) within significantly shorter reaction times of 2–10 minutes (Scheme 15).

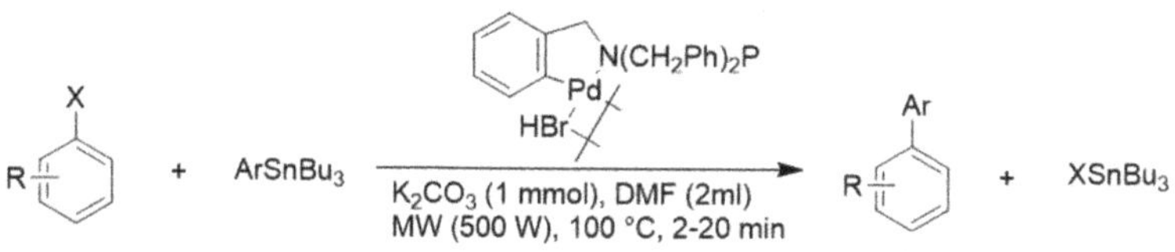

Scheme 15. *Stille reaction of aryl halides with phenyl tributyltins under microwave irradiation.*

In 2014, Duan et al. reported a microwave-assisted Stille cross-coupling reaction of aryl groups using a one-pot, two-step synthesis strategy [25]. In the first step, an aryl halide **44** undergoes stannylation to form an organostannane intermediate **45**, which subsequently reacts with another aryl halide to yield the biaryl coupling product **46** (Scheme 16). The reaction conditions were optimized using water as the solvent under microwave irradiation, affording good yields (74–

Materials Research Foundations 189 (2026) 348-383 https://doi.org/10.21741/9781644904039-11

92%). This method was successfully applied to the synthesis of a Boscalid intermediate and exhibited high regioselectivity in the diarylation reaction.

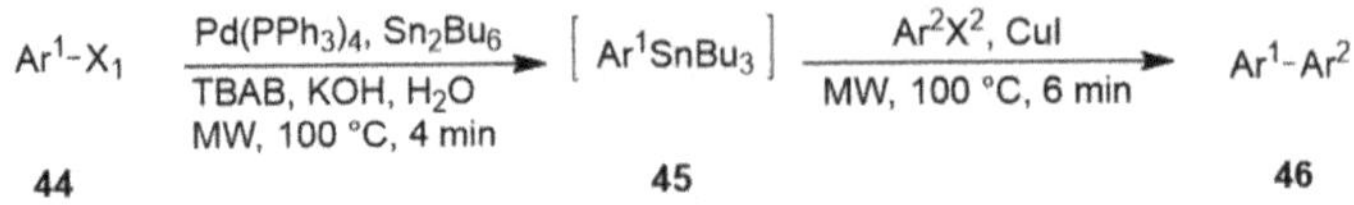

Scheme 16. *One-pot stannylation and Stille cross-coupling reaction.*

2.1.3.4 Sonogashira coupling reactions

The Sonogashira coupling reaction is a palladium-catalyzed cross-coupling method that enables the formation of carbon–carbon bonds between terminal alkynes and aryl or vinyl halides [26]. The general mechanism of the Sonogashira reaction is illustrated in Scheme 17 [27]. This process involves a synergistic catalytic cycle between palladium and copper. It begins with the oxidative addition of an aryl or vinyl halide **47** to a Pd(0) species, forming a Pd(II) complex **(A)** with the halide as the leaving group. The terminal alkyne **48** is then activated by copper(I) to form a copper acetylide complex **(B)** through deprotonation in the presence of a base. In the subsequent transmetallation step, the acetylide group **(B)** is transferred from copper to the palladium complex to produce organopalladium intermediate **(C)** which then undergoes trans to cis isomerization to form cis-intermediate **(D)** to facilitate reductive elimination. Finally, reductive elimination from the intermediate **(D)** yields the desired aryl- or vinyl-substituted alkyne product **49**, regenerating the active Pd(0) catalyst. The copper co-catalyst enhances reactivity by facilitating the formation of the acetylide intermediate, while the palladium cycle drives the overall transformation. The combined action of both metals allows the Sonogashira coupling to proceed under mild conditions with high functional group tolerance.

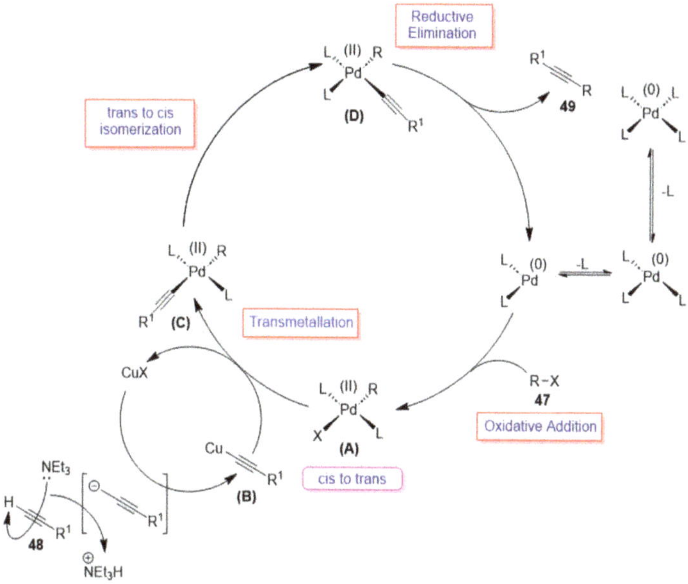

Scheme 17. *General mechanism of Sonogashira coupling reaction.*

In 2009, Balme et al. reported an efficient synthetic method for the preparation of aryl-substituted furo[3,2-c]pyridin-6(5H)-ones **52** through a microwave-assisted Sonogashira coupling of 3-aryl-5-iodo-4-methoxy-2-pyridones **50** with terminal alkynes **51** in a one-pot, sequential Pd/Cu-catalyzed cross-coupling process (Scheme 18) [28]. Under conventional heating, the reactions required long durations to reach completion and demanded high catalyst loadings. However, the use of microwave irradiation significantly shortened the reaction time from about five days to just 1–2.5 hours and enabled the use of much lower catalyst and alkyne loadings without compromising the yield.

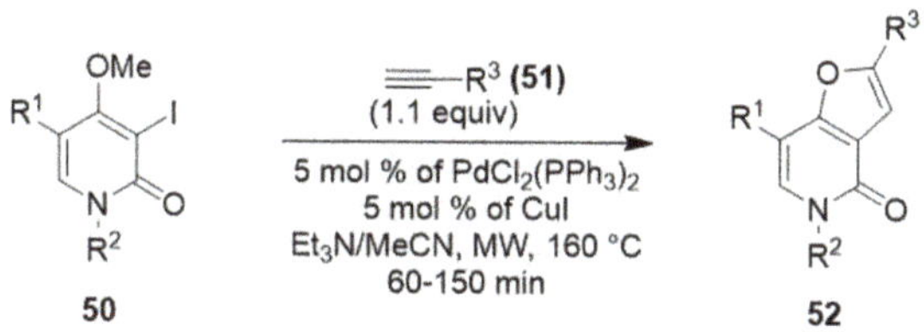

Scheme 18. *Sonogashira coupling reactions between pyridines and terminal alkynes.*

In 2013, Nájera and co-workers developed a highly efficient and environmentally friendly method for performing Sonogashira–Hagihara cross-coupling reactions using aryl imidazolylsulfonates **53** as alternative phenol-derived electrophiles and terminal alkyne **54** to form coupled product **55**

[29]. The reaction proceeds under copper-free conditions in neat water and under microwave irradiation, employing only 1 mol% of a stable oxime–palladacycle catalyst, 2 mol% of the SPhos ligand, triethylamine (NEt_3) as the base, and hexadecyltrimethylammonium bromide (CTAB) as a surfactant (Scheme 19). This protocol enables the efficient coupling of various aryl, alkyl, and silyl acetylenes to afford disubstituted alkynes in high yields within 30 minutes.

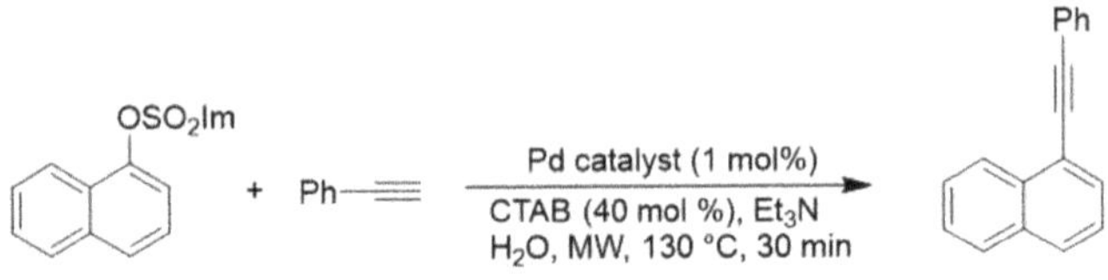

Scheme 19. *Microwave-assisted copper-free Sonogashira coupling of aryl imidazoylsulfonates in water.*

In 2014, Bodke et al. reported a rapid and efficient protocol for the Sonogashira cross-coupling of 4-methyl-7-nonafluorobutylsulfonyloxy coumarins **56** with various terminal alkynes **57** under microwave irradiation, notably in the absence of copper, amine, and ligand additives to form coupled product **58** [30]. In this study, the nonaflate electrophiles demonstrated superior stability and reactivity compared to the corresponding triflates, effectively avoiding issues such as hydrolysis and detriflation. A key factor contributing to the success of this method is the use of $TBAF{\cdot}3H_2O$, which serves as both a mild base and a solvating agent, thereby minimizing side reactions and enhancing product yields (Scheme 20).

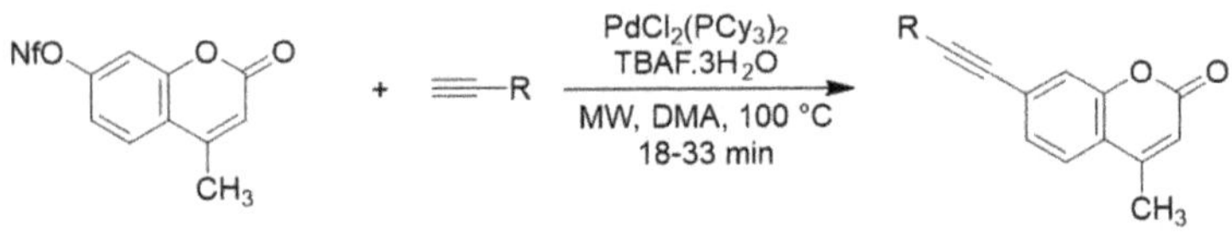

Scheme 20. *Sonogashira coupling of 4-methyl-7-trifluoromethylsulfonyloxy coumarin with phenyl acetylene.*

2.1.3.5 Negishi coupling reactions

Negishi was the first to demonstrate the formation of unsymmetrical biaryl compounds through nickel- or palladium-catalyzed cross-coupling reactions between aryl zinc reagents and aryl bromides or iodides [31-32]. The general mechanism of the Negishi coupling reaction is given in Scheme 21 [33].

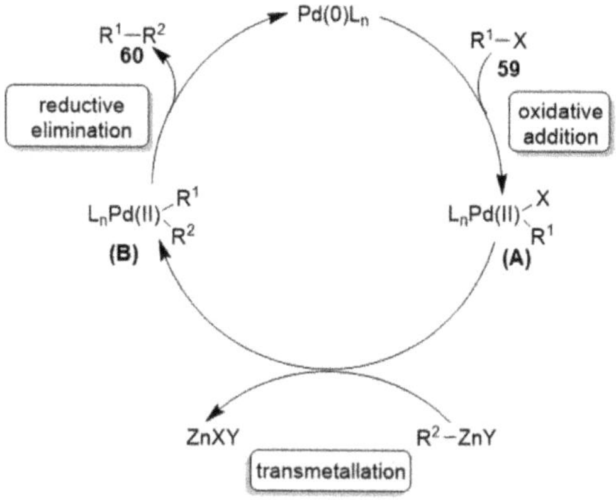

Scheme 21. General mechanism of Negishi coupling reaction.

The Negishi coupling reaction proceeds through a catalytic cycle that begins with an oxidative addition of alkyl halide **59** with palladium catalyst, forming a palladium intermediate **(A)**. This is followed by transmetallation, in which the alkyl or aryl group from the organozinc reagent is transferred to the palladium center with the concurrent exchange of a halide to form palladium (II) intermediate **(B)**. Finally, reductive elimination occurs, producing the cross-coupled product **60** and regenerating the active palladium catalyst.

In 2004, Kappe et al. reported microwave-assisted, palladium-catalyzed Negishi cross-coupling reactions between various aryl chlorides **61** and organozinc reagents **62** to form coupled product **63** [34]. Using an air-stable catalyst system comprising *tris*(dibenzylideneacetone)dipalladium(0) and *tri-tert*-butylphosphonium tetrafluoroborate ($Pd_2(dba)_3/Bu_3P \cdot HBF_4$), the reactions afforded high yields across a broad substrate scope. Under optimized conditions (typically 175 °C for 10 minutes), the process delivered biaryl products in excellent yields of 85–90% (Scheme 22).

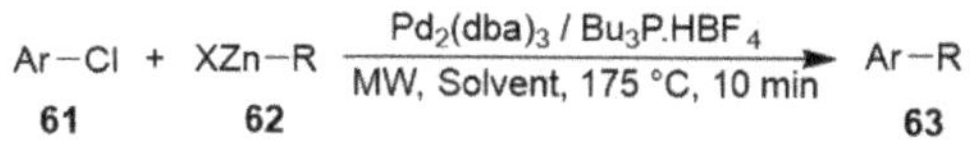

Scheme 22. Microwave-assisted Pd-catalyzed Negishi cross-coupling reactions.

In 2005, Suna et al. reported a microwave-assisted method for the transmetallation of phenyl magnesium bromide **65** to phenyl zinc chloride **64** followed by a Negishi cross-coupling reaction to form coupled product **66** [35]. Under microwave heating at 80 °C for 15 minutes, phenyl magnesium bromide reacts with $ZnCl_2$ in the presence of N,N,N',N'-Tetramethylethylenediamine (TMEDA) to form the phenyl zinc intermediate. This intermediate then undergoes a Negishi cross-coupling with 4-bromobenzaldehyde in a single vessel at room temperature for 5 minutes using $Pd(PPh_3)_2Cl_2$ as the catalyst (Scheme 23). The method also allows the selective formation of 4′-methoxy-biphenyl-4-carbaldehyde, demonstrating efficient transmetallation to the corresponding arylzinc species.

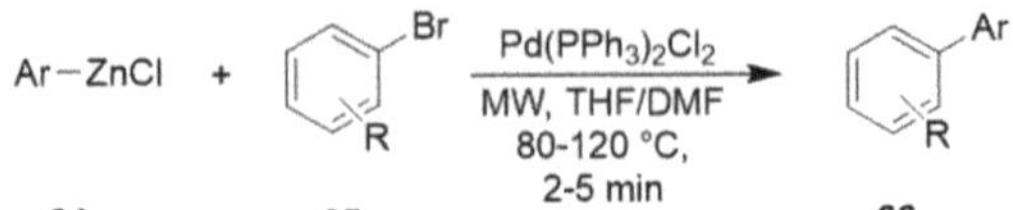

64 65 66

Scheme 23. *Synthesis of biaryls via a three-step reaction through the Negishi coupling reaction*

In 2013, Larhed et al. reported a microwave-assisted, palladium-catalyzed carbonylative Negishi cross-coupling method for a variety of aryl iodides **67** with benzyl zinc bromide **68** in a one-pot reaction using molybdenum hexacarbonyl ($Mo(CO)_6$) as a solid CO source to get coupled product **69** [36]. The reactions were carried out under microwave irradiation at 90 °C for 1 hour, employing a $Pd(OAc)_2$/1,4-bis(diphenylphosphino)butane (DPPB) catalyst/ligand system in tetrahydrofuran (THF) (Scheme 24). A range of aryl iodides was efficiently converted to diarylated ethanones in moderate to high isolated yields. The method also tolerated substituted benzyl zinc bromides, such as 4-cyanobenzyl zinc bromide, which afforded the corresponding product in 66% yield.

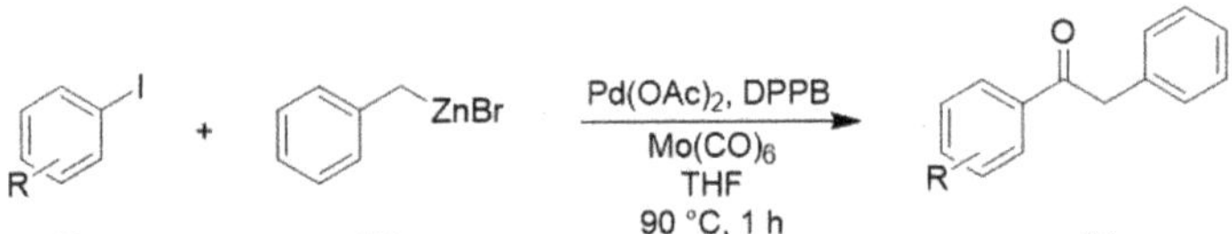

67 68 69

Scheme 24. *Microwave-assisted Palladium-catalyzed carbonylative Negishi cross-coupling of aryl iodides and organozinc compounds.*

2.2 Microwave-assisted non-noble transition metal-catalyzed reactions

In this section, we discussed several chemical reactions catalyzed by non-noble transition metals, including iron, cobalt, nickel, copper, and zinc, under microwave irradiation. These metals are abundant, inexpensive, and widely available, making them attractive alternatives to noble metals. Their unique catalytic properties contribute significantly to the development of green and sustainable synthetic methodologies.

2.2.1 Iron metal-catalyzed reaction

In 2009, Liu et al. developed a greener, faster, and more efficient method for synthesizing quinazolinone derivatives **72** via microwave-assisted, iron-catalyzed intramolecular cyclization of substituted 2-halobenzoic acids **70** with amidines **71** [37]. The reactions were carried out in benign solvents such as water or N,N-dimethylformamide (DMF), under either ligand-free or ligand-assisted conditions, using inexpensive and readily available iron salts like ferric chloride ($FeCl_3$) and tris(acetylacetonato)iron(III) ($Fe_2(acac)_3$) (Scheme 25). The authors reported four optimized protocols (Methods A–D) that demonstrated excellent reactivity and broad functional group tolerance. Methods A and B gave good yields while methods C and D produced good to excellent yields. Microwave irradiation significantly accelerated the reactions, providing good yields of quinazolinone products within minutes. Notably, this study was the first to report iron-catalyzed C–N bond formation in aqueous media for the synthesis of N-heterocycles.

Materials Research Foundations 189 (2026) 348-383 https://doi.org/10.21741/9781644904039-11

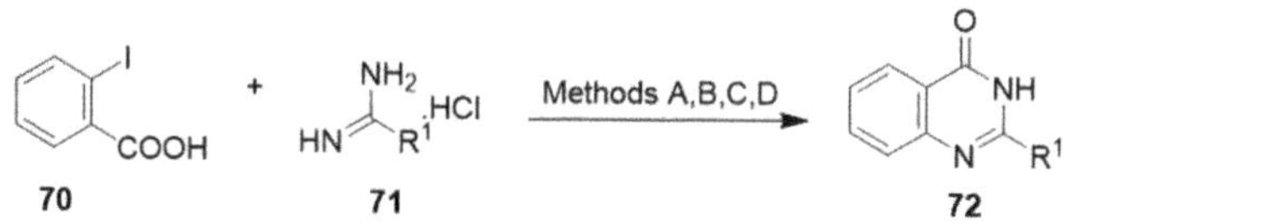

Reaction conditions: 2-iodobenzoic acid (1.0 mmol), amidine hydrochloride (1.5 mmol), catalyst (0.1 mmol),ligand (0.2 mmol), base (2 mmol), solvent (6 ml) with microwave assisted under a nitrogen atmosphere. Method A: $FeCl_3$, Cs_2CO_3, H_2O. Method B: $FeCl_3$, L-proline, Cs_2CO_3, H_2O. Method C: $FeCl_3$, L-proline, Cs_2CO_3, DMF. Method D: $Fe_2(acac)_3$, DMEDA, Cs_2CO_3, DMF.

Scheme 25. *Synthesis of quinazolinone derivative by microwave-assisted iron-catalyzed cyclization reaction.*

In 2011, Chan et al. reported a simple and efficient iron(II)-catalyzed protocol for the direct amidation of aldehydes **73** using iminoiodinanes (PhI=NR) **74** as the nitrogen source to get amide product **75** [38]. The active catalyst is generated *in situ* from inexpensive iron(II) chloride ($FeCl_2$) and pyridine, operating under mild, room-temperature conditions with microwave irradiation (Scheme 26). This method affords acyl sulfonamides in good yields and exhibits excellent chemoselectivity, even in the presence of other reactive functional groups. Mechanistic studies suggest that the reaction proceeds via insertion of an iron–nitrene species into the formyl C–H bond through a hydrogen atom abstraction–radical rebound pathway. Control experiments confirmed that trace metal impurities such as copper or palladium do not contribute to catalysis, establishing that iron alone serves as the active catalyst. Overall, this methodology provides a cost-effective and scalable alternative to traditional Ru- and Cu-based systems for amide bond formation.

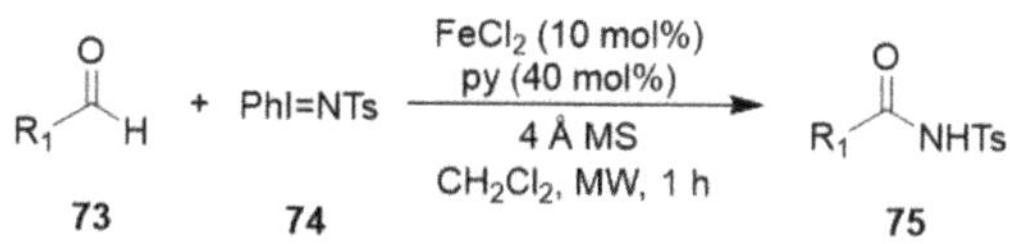

Scheme 26. *Microwave-assisted iron-catalyzed amidation of aldehydes.*

In 2019, Yan and co-workers developed a highly efficient, microwave-assisted iron-catalyzed strategy for the synthesis of 7-azaindole derivatives **78** via the cyclization of *o*-haloaromatic amines **76** with terminal alkynes **77** [39]. Under optimized conditions, $Fe(acac)_3$ served as the catalyst, CuI as the co-catalyst, and potassium *tert*-butoxide (KO^{t}Bu) as the base in *N*-methyl-2-pyrrolidone (NMP) solvent. The reaction, performed under microwave irradiation at 130 °C for 60 minutes, furnished the desired 7-azaindoles **78** in good to excellent yields (Scheme 27). This method offers a rapid, scalable, and efficient route for the synthesis of bioactive 7-azaindole frameworks.

The proposed mechanism for the microwave-assisted, iron-catalyzed synthesis of 7-azaindole derivatives **78** begins with the oxidative addition of the iron catalyst, $Fe(acac)_3$, to the *o*-haloaromatic amine substrate **76**, generating an organoiron intermediate **(A).** This is followed by a transmetalation step mediated by the copper(I) co-catalyst, during which the terminal alkyne **77** is transferred to the iron center to form an iron–alkyne complex **(B)**. Subsequent reductive elimination furnishes the key C–C bond, yielding a Sonogashira-type coupling intermediate **(C)**. Finally, this intermediate undergoes intramolecular cyclization under microwave irradiation to produce the desired 7-azaindole product **78**, with KO^{t}Bu serving as the base and NMP as the reaction medium.

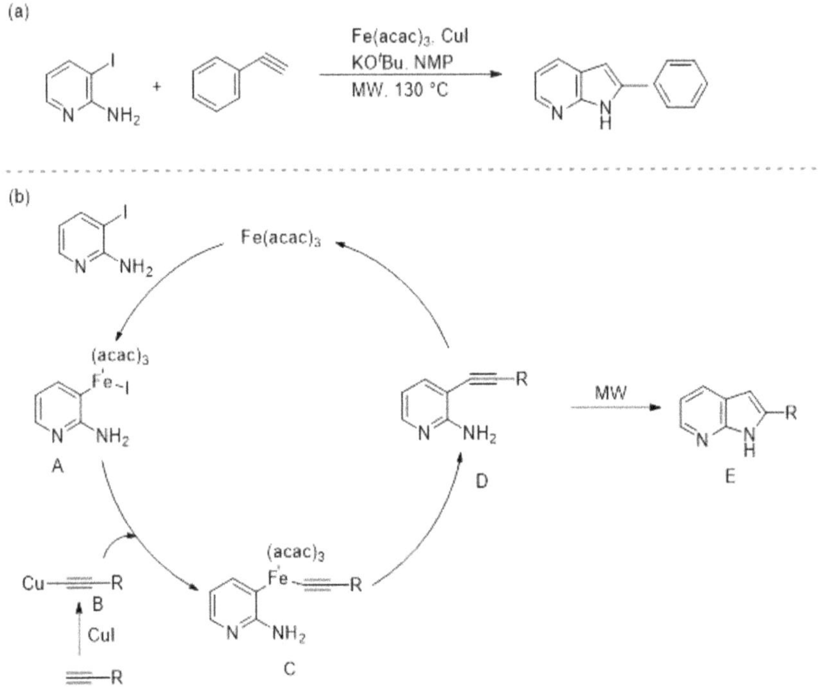

Scheme 27. (a) Synthesis of 7-azaindoles through cyclization of o-haloaromatic amines with terminal alkynes. (b) Proposed mechanism for the synthesis of 7-azaindole derivatives.

2.2.2 Cobalt metal-catalyzed reaction

In 2025, Lete et al. reported an efficient microwave-assisted cobalt (III)-catalyzed C–H amino carbonylation method for synthesizing thiophene carboxamides **81** using thiophene derivatives **79** and isocyanates **80** (Scheme 28(a)) [40]. In this reaction, they used carbonyl(pentamethylcyclopentadienyl)cobalt(III) diiodide (Cp*Co(III)I_2) complexes along with pyridine or pyrimidine directing groups to achieve high site-selectivity under microwave irradiation to accelerate reaction time compared to conventional heating, which consumes more time. A broad range of aromatic isocyanates was found to be suitable for the transformation, while aliphatic variants showed lesser reactivity.

Materials Research Foundations 189 (2026) 348-383 https://doi.org/10.21741/9781644904039-11

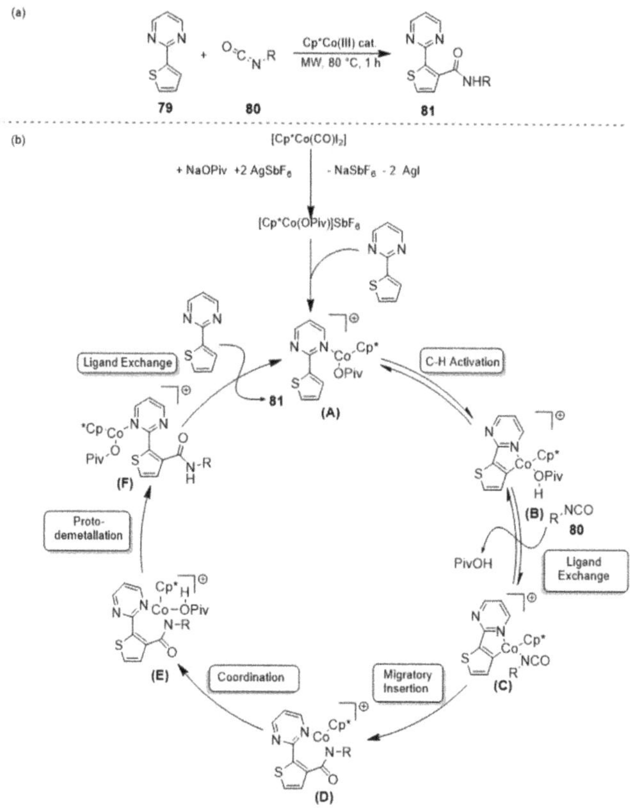

Scheme 28. *(a) Microwave-assisted cobalt-catalyzed amino carbonylation of thiophene derivatives with arylisocyanates. (b) Proposed mechanism.*

The mechanism of the cobalt(III)-catalyzed C–H amino carbonylation is illustrated in Scheme 28(b), which involves several key steps. First, the active catalytic species $[Cp^*Co(OPiv)]^+$ is formed through ligand exchange between the cobalt precursor and sodium pivalate in the presence of a silver salt. This catalyst then binds to the thiophene substrate **79** to generate **(A)**, leading to a reversible C–H activation that forms a cobaltacycle intermediate **(B)**. Next, the isocyanate molecule **80** subsequently coordinates to the cobalt center to form the intermediate **(C)**. Next, the rate-limiting step occurs through migratory insertion into the carbon–nitrogen (C=N) bond of the isocyanate. This results in the formation of a cobalt-bound amide intermediate **(D)**. Pivalic acid then coordinates to the metal center to form intermediate **(E)**, and it facilitates a rapid and irreversible protodemetallation step to form intermediate **(F)**, and that releases the amino carbonylated product **81** after coordination of another molecule of thiophene and regenerates the active catalyst. Density functional theory (DFT) studies confirm the feasibility and energy profile of each step, highlighting the role of microwave irradiation in enhancing the reaction efficiency.

In the same year, Thakur and co-workers developed a microwave-assisted, cobalt-catalyzed acceptorless dehydrogenative coupling (ADC) reaction between various benzyl alcohols **83** and 2-aminobenzamide **82** under optimized solvent-free conditions [41]. The reaction proceeded efficiently, affording quinazolinone derivatives in good to excellent yields (Scheme 29a). The proposed mechanism for this cobalt-catalyzed ADC process (Scheme 29b) begins with the deprotonation of benzyl alcohol **83** to form deprotonated species which coordinates with cobalt(II) chloride ($CoCl_2$). The resulting complex undergoes chloride elimination and hydrogen abstraction, generating benzaldehyde **(A)** in situ along with molecular hydrogen as a byproduct. The condensation of benzaldehyde **(A)** with 2-aminobenzamide **82** then forms an aminal intermediate **(B)**, which coordinates to the cobalt center to produce complex **(C)**. A subsequent dehydrogenation step convert **(C)** into the desired quinazolinone product **84**, accompanied by the liberation of an additional equivalent of hydrogen gas.

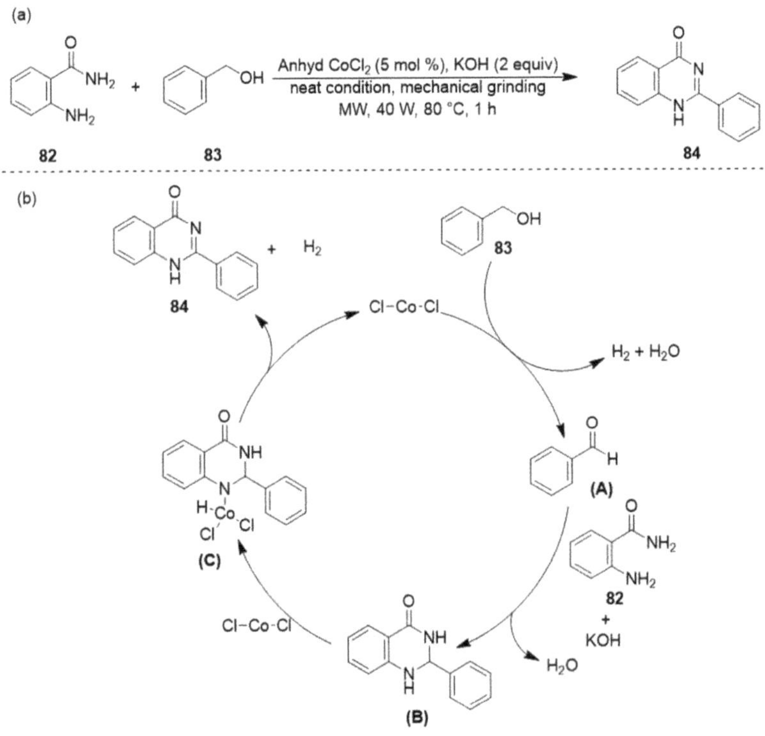

Scheme 29. *(a) Synthesis of quinazolinones via acceptorless dehydrogenative coupling under microwave radiation. (b) Plausible mechanism.*

2.2.3 Nickel metal-catalyzed reaction

In 2006, Lipshutz reported microwave-assisted reactions by utilizing nickel-in-charcoal (Ni/C) as a heterogeneous catalyst to efficiently perform coupling reactions between aryl halides **86** and alkenyl zirconocenes **85** to produce coupled product **87**, which is shown in Scheme 30 [42]. Along with this, various cross-coupling reactions, including Negishi, Suzuki, amination, and vinyl alane–benzylic chloride couplings were also reported. Under microwave irradiation, these reactions proceed rapidly within 10 to 40 minutes at temperatures between 150–200 °C, in contrast to the several hours required under conventional heating.

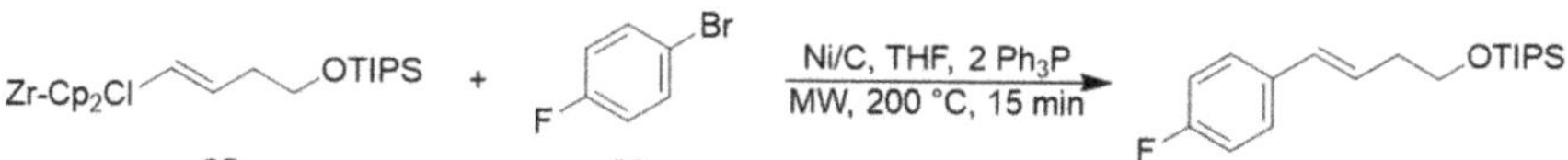

Scheme 30. *Nickel-catalyzed cross-coupling of vinyl zirconocene with aryl halides.*

In 2016, Bhagat et al. reported a highly efficient, one-pot methodology for the synthesis of 2,4,5-trisubstituted imidazoles **90** under microwave irradiation, utilizing a Schiff's base complex of nickel (Ni-C) as a catalyst [43]. In this reaction, a condensation reaction occurs between benzil **88**, aromatic aldehydes **89**, and ammonium acetate in ethanol as the solvent. Microwave heating decreased reaction durations (10–30 min) and enhanced yields (up to 95%) compared to traditional heating. The Ni-C catalyst exhibited remarkable recyclability, maintaining substantial catalytic activity across three cycles. The method provides a quick, environmentally sustainable, and high-yield approach for the synthesis of physiologically relevant imidazole derivatives.

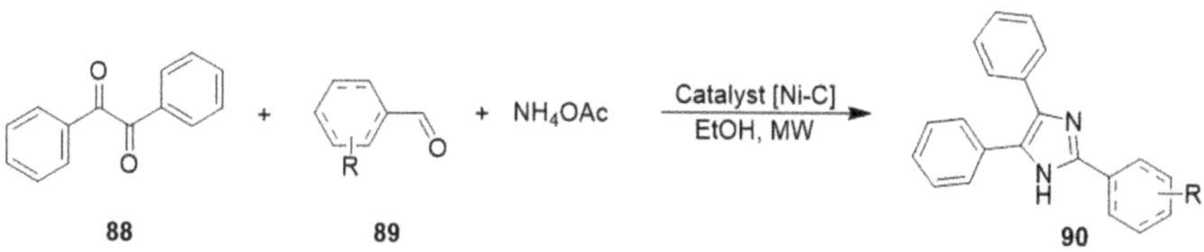

Scheme 31. *Synthesis of 2,4,5-triphenyl-1H-imidazole by microwave-assisted nickel-catalyzed reaction.*

In 2018, Ke et al. presented a highly efficient and environmentally benign technique for the synthesis of benzimidazoles **93** using microwave-assisted nickel catalysis [44]. Herein, they performed a single-step reaction of substituted aniline **91** with benzaldehyde **92** with ammonia as an inexpensive and non-toxic nitrogen source and nickel(II) chloride ($NiCl_2$) as the catalyst, 8-hydroxyquinoline as the ligand, and cesium carbonate (Cs_2CO_3) as the base, under microwave irradiation (130 W, 100 °C, 13 min) (Scheme 32(a)). This process employs ammonia as a cost-effective, non-toxic nitrogen source, yielding 2-substituted benzimidazoles in good to exceptional yields (up to 95%) without byproducts.

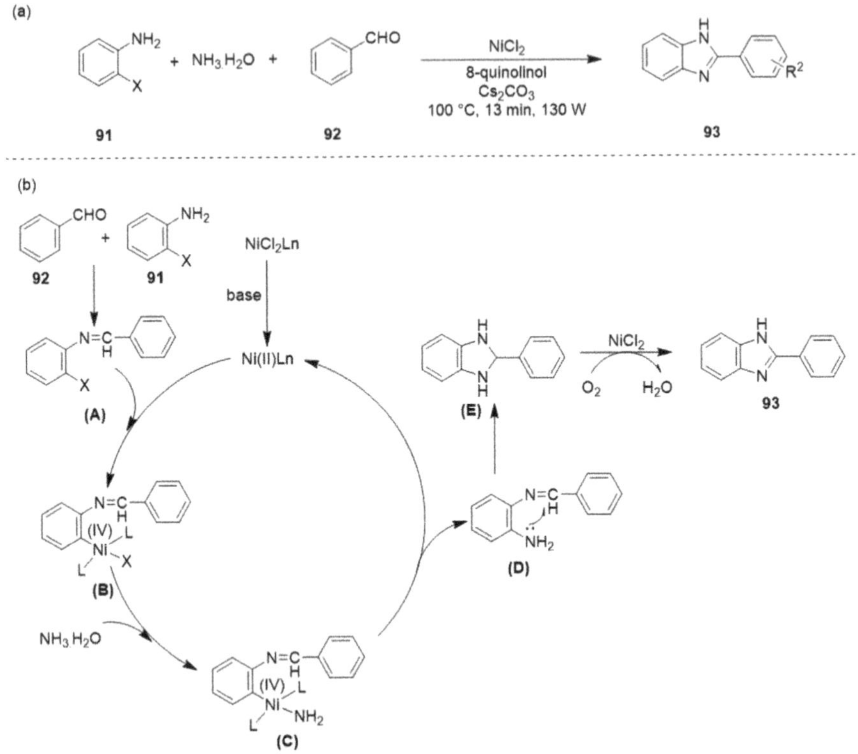

Scheme 32. *(a) Microwave-assisted Nickel-catalyzed synthesis of benzimidazole derivatives. (b) Proposed reaction mechanism.*

The mechanism for the nickel-catalyzed synthesis of benzimidazoles is displayed in Scheme 32(b). The mechanism is initiated with the condensation of 2-haloaniline **91** and an aldehyde **92** to generate an *N*-phenylformimine intermediate **(A)**, which is coordinated to the activated nickel catalyst to form intermediate **(B)**, and then ammonia reacts with this intermediate to form **(C)**. This leads to a intermediate **(D)** which subsequently undergoes intramolecular cyclization to produce a dihydrobenzimidazole intermediate **(E)**. In the final step, oxidation of this intermediate by atmospheric oxygen, which $NiCl_2$ also catalyzes, gives the benzimidazole product **93**.

2.2.4 Copper metal-catalyzed reaction

In 2007, Baqi and Muller reported a highly efficient, microwave-assisted Ullmann coupling catalyzed by copper(0) [45]. In this study, anilinoanthraquinone derivatives **96** were synthesized by reacting bromaminic acid **94** with various aniline derivatives **95** in a phosphate buffer (pH 6–7) under microwave irradiation at 80–120 °C, affording the desired products in just 2–20 minutes

Materials Research Foundations 189 (2026) 348-383 https://doi.org/10.21741/9781644904039-11

(Scheme 33). Compared to conventional Ullmann coupling methods, this microwave-assisted approach offers remarkably higher efficiency and significantly reduced reaction times.

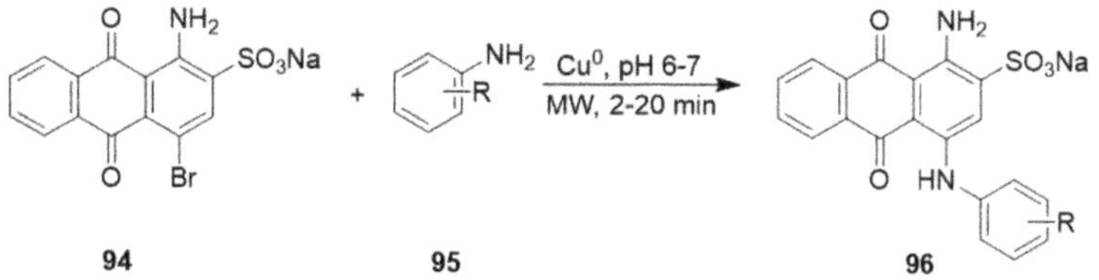

Scheme 33. *Synthesis of anilinoanthraquinone derivatives through microwave irradiation.*

In 2008, Goossen et al. reported a microwave-assisted, copper-catalyzed protodecarboxylation of aromatic carboxylic acids **97**, offering a superior alternative to conventional thermal methods [46]. The reaction employs an inexpensive catalytic system generated *in situ* from copper(I) oxide (Cu_2O) and 1,10-phenanthroline, affording the corresponding arenes **98** in up to 88% yield within 5–15 minutes under microwave irradiation (Scheme 34a). Under optimized conditions, the reaction conducted at 190 °C in a 3:1 mixture of NMP and quinoline efficiently decarboxylated a broad range of substrates. This scalable and environmentally benign protocol avoids the use of stoichiometric heavy metals, minimizes thermal stress on sensitive substrates, and offers a practical and sustainable approach with potential utility in drug discovery and synthetic chemistry.

The mechanism for the copper-catalyzed protodecarboxylation of aromatic carboxylic acids is illustrated in Scheme 34(b). The reaction proceeds via direct insertion of a copper(I) species into the aryl-carboxylate bond to form intermediate **(A)**, without the formation of a π-complex intermediate. The active catalytic species, generated *in situ* from copper(I) oxide and 1,10-phenanthroline, coordinates with the substrate to promote C–COO^- bond cleavage, yielding an aryl-copper intermediate **(B)**. Subsequent protonation of this intermediate affords the desired arene product **98**, with concurrent regeneration of the copper catalyst.

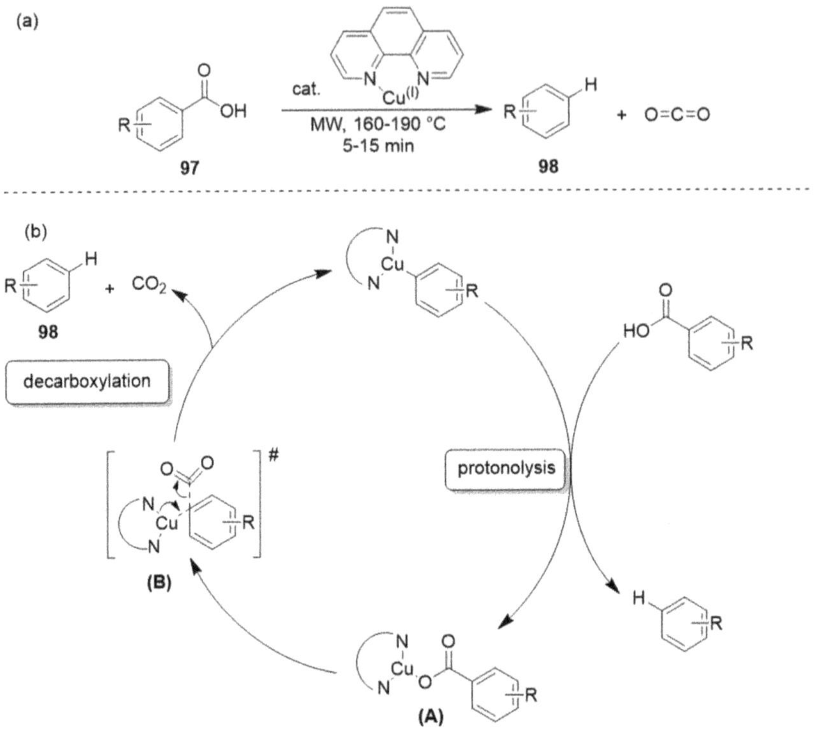

Scheme 34. *(a) Microwave-assisted Cu-catalyzed proto decarboxylation of aromatic carboxylic acids. (b) Proposed mechanism for Cu-catalyzed proto decarboxylation of aromatic carboxylic acids.*

In 2016, Jacob et al. reported an efficient copper-catalyzed intramolecular Ullmann cross-coupling strategy for the synthesis of fused 1,2,3-triazolo[1,3,6]triazonines **101**, a class of nine-membered nitrogen-containing heterocycles [47]. In this method, 5-iodo-1,2,3-triazoyl benzaldehydes **99** were reacted with various substituted diaminobenzenes **100** using copper(I) iodide (CuI) as the catalyst in dimethyl sulfoxide (DMSO) under microwave irradiation (Scheme 35a). Optimal reaction conditions used were 5 mol% CuI, 100 °C in 1 hour provided the final products in good yields.

Materials Research Foundations 189 (2026) 348-383 https://doi.org/10.21741/9781644904039-11

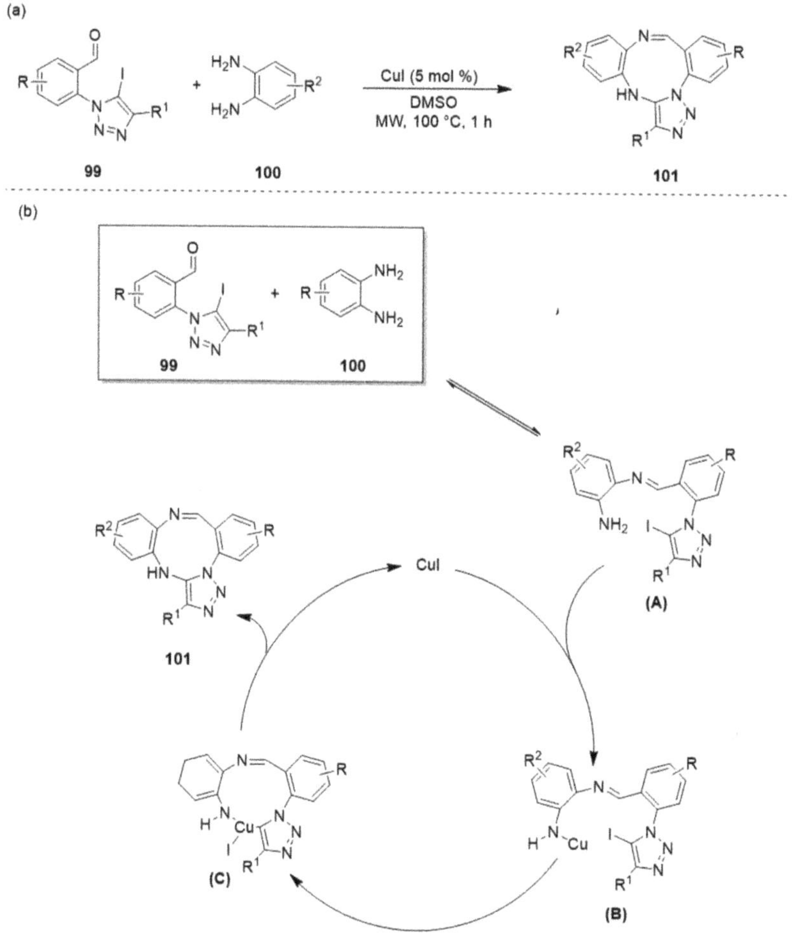

Scheme 35. *(a) Synthesis of fused 1,2,3-triazolo-1,3,6-trazonines through copper-catalyzed Ullmann cross-coupling under microwave irradiation. (b) Proposed mechanism.*

The proposed mechanism (Scheme 35b) begins with the condensation of 5-iodo-1,2,3-triazoyl benzaldehyde **99** with a substituted diaminobenzene **100** to form an imine intermediate **(A)**. This intermediate coordinates with CuI, which activates the adjacent C–I bond in the triazole ring to form intermediate **(B)**. Subsequent copper insertion into this bond generates a cyclic organocopper species **(C)**, which then undergoes reductive elimination to furnish the nine-membered triazolo-triazonine product **101**, concurrently regenerating the active Cu(I) catalyst.

Materials Research Foundations 189 (2026) 348-383 https://doi.org/10.21741/9781644904039-11

2.2.5 Zinc metal-catalyzed reaction

In 2008, Dembiński and co-workers reported a microwave-assisted, $ZnCl_2$-catalyzed method for the synthesis of substituted pyrroles from homopropargyl azides **102** [48]. They developed a ligand-free 5-endo-dig cyclization in which two distinct pyrrole derivatives **103** and **104** were formed (Scheme 36a). $ZnCl_2$ was employed as the catalyst and dichloromethane as the solvent at elevated temperature. At lower catalyst loading (5–10 mol%), the predominant product was pyrrole **103**, whereas increasing the catalyst amount (20–40 mol%) favored the formation of the partially reduced derivative **104**. This suggests that higher $ZnCl_2$ concentrations not only facilitate the 5-endo-dig cyclization of the homopropargyl azide but also promote a subsequent reduction process, possibly involving zinc-mediated hydrogen transfer. Under optimized conditions (40 mol% $ZnCl_2$, 105–130 °C, microwave heating), compound **104** was obtained in 67% yield, while the lower catalyst concentration (5 mol%) afforded **103** in 64% yield. The mechanism of the reaction is illustrated in Scheme 36(b). The reaction goes through a 5-endo-dig cyclization. It involves N coordination of the $ZnCl_2$ to the alkyne moiety of the homopropargyl azide **102** to give complex **(A)**. From this point, there is an intramolecular nucleophilic attack by the azide, which leads to closure and forms vinyl zinc intermediate **(B)**. Next, the nitrogen molecule is eliminated in the gaseous form to form intermediate **(C)** which drives the reaction to form the final product.

Materials Research Foundations 189 (2026) 348-383 https://doi.org/10.21741/9781644904039-11

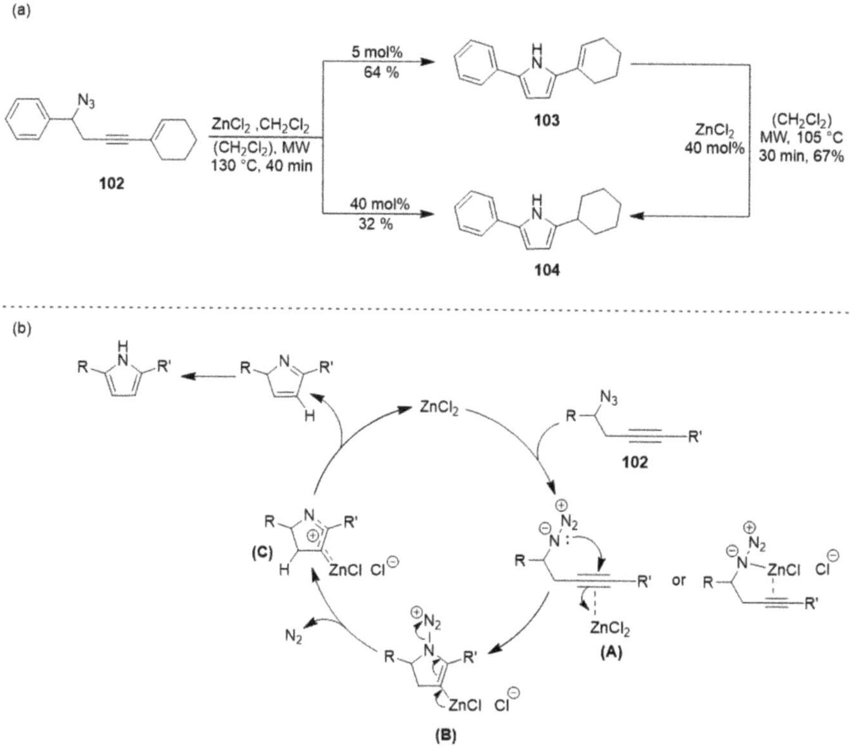

Scheme 36. *(a) Reaction of cyclohexenyl-substituted azide with $ZnCl_2$. (b) Proposed mechanism.*

In 2018, Sieburth et al. reported a novel one-pot cascade reaction for the synthesis of chromeno[4,3-b]pyrroles **112** [49]. Interestingly, this transformation was discovered serendipitously during attempts to prepare benzo[*b*,*g*][1,5]oxazonine building blocks. The process follows a build/couple/pair strategy and begins with the formation of the ether-linked aldehyde precursor **107** via a nucleophilic substitution (S_N2) reaction between 2-hydroxybenzaldehyde **105** and trans-4-bromobutenoate **106**. This aldehyde **107** subsequently undergoes a Ugi four-component condensation with an amine **108**, a carboxylic acid **109**, and an isocyanide **110** to afford the multifunctional Ugi adduct **111**. Initial attempts to convert this intermediate into Diels–Alder-type products **113** under various catalytic conditions were unsuccessful. However, when zinc bromide ($ZnBr_2$) was employed under microwave irradiation in 1,2-dichloroethane (DCE) at 170 °C, an unexpected five-step cascade reaction occurred. The sequence involves initial O-acylation in the presence of $ZnBr_2$ to generate an azomethine ylide, followed by an intramolecular [3 + 2] cycloaddition, ultimately yielding the tricyclic chromeno[4,3-*b*]pyrrole 8a in 65% yield (Scheme 37a).

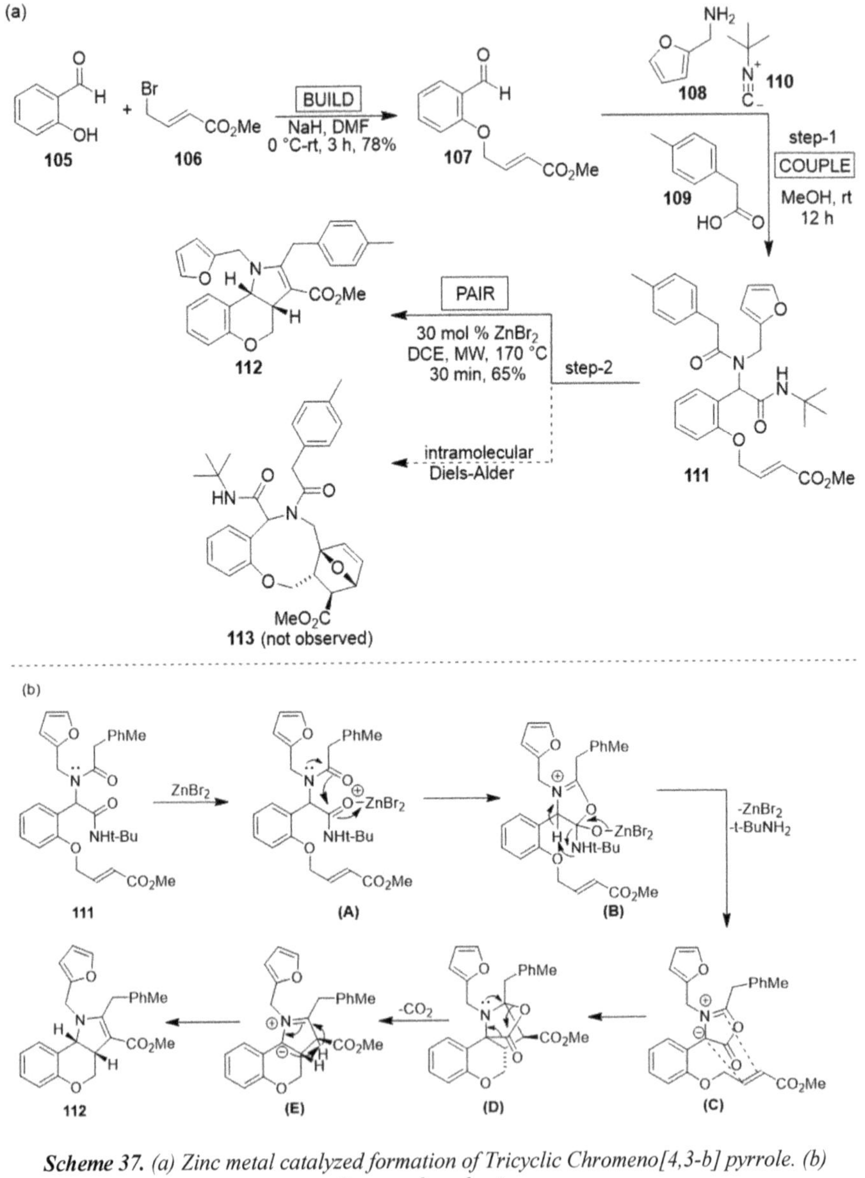

Scheme 37. (a) Zinc metal catalyzed formation of Tricyclic Chromeno[4,3-b] pyrrole. (b) Proposed mechanism.

The proposed mechanism is illustrated in Scheme 37(b). The reaction is initiated from the Ugi adduct, which contains an *o*-hydroxybenzyl moiety, a carboxamide group, and an alkene or alkyne unit. Coordination of the Lewis acid $ZnBr_2$ to the Ugi adduct generates intermediate **(A)**. The coordinated Lewis acid then promotes *O*-acylation of the tertiary amide, forming the tetrahedral intermediate **(B)**. Subsequent rearrangement of **(B)** leads to the elimination of *tert*-butylamine (t-$BuNH_2$) from the amide moiety. The liberated amine acts as a base, abstracting a benzylic proton to yield the azomethine ylide intermediate **(C)**. This reactive species undergoes an intramolecular [3 + 2] dipolar cycloaddition with the alkene or alkyne functionality, affording the oxazolidinone intermediate **(D)**. Decarboxylation of **(D)** then produces a second azomethine ylide intermediate **(E)**, which subsequently undergoes a 1,4-proton shift to furnish the final chromeno[4,3-*b*]pyrrole product **112**.

Conclusion

Microwave-assisted transition metal-catalyzed reactions have emerged as a powerful tool in modern sustainable and green synthetic chemistry. These methods offer significant advantages, including reduced reaction times, higher yields, and cleaner transformations. However, the primary limitation remains their scalability, which poses a challenge for adaptation to industrial-scale synthesis. Future research should therefore prioritize the development of scalable microwave-assisted protocols, potentially through the integration of continuous-flow technology, hybrid heating systems, or microwave-reactor engineering.

At present, microwave-assisted methodologies are extensively employed in cross-coupling reactions to construct essential carbon–carbon and carbon–heteroatom bonds that serve as key frameworks in pharmaceutical synthesis. Moving forward, efforts should be directed toward expanding their scope to enable the efficient synthesis and late-stage modification of complex drug molecules, with an emphasis on enhancing efficacy, selectivity, and atom economy while minimizing by-product formation. Such advancements would further strengthen the role of microwave-assisted catalysis as a cornerstone in green and sustainable drug discovery.

References

[1] C.O. Kappe, Controlled microwave heating in modern organic synthesis, Angew. Chem. Int. Ed. 43 (2004) 6250-6284. https://doi.org/10.1002/anie.200400655

[2] J. Magano, J.R. Dunetz, Large-scale applications of transition metal-catalyzed couplings for the synthesis of pharmaceuticals, Chem. Rev. 111 (2011) 2177-2250. https://doi.org/10.1021/cr100346g

[3] A. Suzuki, Cross-coupling reactions of organoboranes: an easy way to construct c-c bonds (nobel lecture), Angew. Chem. Int. Ed. 50 (2011) 6722-6737. https://doi.org/10.1002/anie.201101379

[4] P. Lidstrom, J. Tierney, B. Wathey, J. Westman, Microwave-assisted organic synthesis - a review, Tetrahedron. 57 (2001) 9225-9283. https://doi.org/10.1016/S0040-4020(01)00906-1

[5] N. Sharma, V. Bahadur, U.K. Sharma, D. Saha, Z. Li, Y. Kumar, J. Colaers, B.K. Singh, E.V.V Eycken, Microwave-assisted ruthenium-catalyzed ortho-C−H functionalization of *N*-

benzoyl α-amino ester derivatives, Adv. Synth. Catal. 360 (2018) 3083-3089. https://doi.org/10.1002/adsc.201800458

[6] X.H. Li, J.F. Gong, M.P. Song, Microwave-assisted ruthenium- and rhodium- catalyzed couplings of α-amino acid ester-derived phosphinamides with alkynes, Chem. Asian. J. 17 (2022) e202101158. https://doi.org/10.1002/asia.202101158

[7] N. R. Guha, C. B. Reddy, N. Aggarwal, D. Sharma, A. K. Shil, Bandna, P. Das, Solid-Supported Rhodium(0) Nano-/Microparticles: An efficient ligand-free heterogeneous catalyst for microwave-assisted Suzuki-Miyaura cross-coupling reaction, Adv. Synth. Catal. 354 (2012) 2911-2915. https://doi.org/10.1002/adsc.201200418

[8] H. Lee, Y.K. Sim, J.W. Park, C.H. Jun, Microwave-assisted, rhodium(iii)-catalyzed n-annulation reactions of aryl and α,β-unsaturated ketones with alkynes, Chem. Eur. J. 20 (2014) 323-333. https://doi.org/10.1002/chem.201302699

[9] K.C. Nicolaou, P.G. Bulger, D. Sarlah, Palladium-catalyzed cross-coupling reactions in total synthesis, Angew. Chem. Int. Ed. 44 (2005) 4442-4489. https://doi.org/10.1002/anie.200500368

[10] I.R. Baxendale, C.M. Griffiths-Jones, S.V. Ley, G.K. Tranmer, Microwave-assisted Suzuki coupling reactions with an encapsulated palladium catalyst for batch and continuous-flow transformations, Chem. Eur. J. 12 (2006) 4407-4416. https://doi.org/10.1002/chem.200501400

[11] M.C.D'Alterio, E. Casals-Cruanas, N.V. Tzouras, G. Talarico, S.P. Nolan, A. Poater, Mechanistic aspects of the palladium-catalyzed Suzuki-Miyaura cross-coupling reaction, Chem. Eur. J. 27 (2021) 13481-13493. https://doi.org/10.1002/chem.202101880

[12] R.R. Poondra, P.M. Fischer, N.J. Turner, Efficient palladium-catalyzed cross-coupling of β-chloroalkylidene/arylidene malonates using microwave chemistry, J. Org. Chem. 69 (2004) 6920-6922. https://doi.org/10.1021/jo049132r

[13] A.K. Sharma, K. Gowdahalli, J. Krzeminski, S. Amin, Microwave-assisted Suzuki cross-coupling reaction, a key step in the synthesis of polycyclic aromatic hydrocarbons and their metabolites, J. Org. Chem. 72 (2007) 8987-8989. https://doi.org/10.1021/jo701665j

[14] R.Y. Qu, Y.C. Liu, Q.Y. Wu, Q. Chen, G.F. Yang, An efficient method for syntheses of functionalized 6-bulkysubstituted salicylates under microwave irradiation, Tetrahedron 71 (2015) 8123-8130. https://doi.org/10.1016/j.tet.2015.08.040

[15] R.F. Heck, J.P. Nolley, Palladium-catalyzed vinylic hydrogen substitution reactions with aryl, benzyl, and styryl halides, J. Org. Chem. 37 (1972) 2320-2322. https://doi.org/10.1021/jo00979a024

[16] I.P. Beletskaya, A. V. Cheprakov, The Heck reaction as a sharpening stone of palladium catalysis, Chem. Rev. 100 (2000) 3009-3066. https://doi.org/10.1021/cr9903048

[17] K.S.M. Salih, Y. Baqi, Microwave-assisted palladium-catalyzed cross-coupling reactions: generation of carbon-carbon bond, Catalysts. 10 (2020) 4. https://doi.org/10.3390/catal10010004

[18] B.M. Choudary, S. Madhi, N.S. Chowdari, M.L. Kantam, B. Sreedhar, Layered double hydroxide supported nanopalladium catalyst for Heck-, Suzuki-, Sonogashira-, and Stille-type coupling reactions of chloroarenes, J. Am. Chem. Soc. 124 (2002) 14127-14136. https://doi.org/10.1021/ja026975w

[19] B.K. Allam, K.N. Singh, An efficient phosphine-free Heck reaction in water using Pd(l - proline)2 as the catalyst under microwave irradiation, Synthesis, (2011) 1125-1131. https://doi.org/10.1055/s-0030-1258452

[20] A.R. Patel, A. Asatkar, G. Patel, S. Banerjee, Synthesis of rice husk derived activated mesoporous carbon immobilized palladium hybrid nano-catalyst for ligand-free Mizoroki-Heck/Suzuki/Sonogashira cross-coupling reactions, ChemistrySelect. 4 (2019) 5577-5584. https://doi.org/10.1002/slct.201900384

[21] J.K. Stille, The palladium-catalyzed cross-coupling reactions of organotin reagents with organic electrophiles, Angew. Chem. Int. Ed. Engl. 25 (1986) 508-524. https://doi.org/10.1002/anie.198605081

[22] P. Espinet, A.M. Echavarren, The mechanisms of the Stille reaction, Angew. Chem. Int. Ed. 43 (2004) 4704-4734. https://doi.org/10.1002/anie.200300638

[23] C. Cebrián, A. de Cózar, P. Prieto, A. Díaz-Ortiz, A. de la Hoz, J.R. Carrillo, A.M. Rodriguez, F. Montilla, Microwave-assisted Stille reactions as a powerful tool for building polyheteroaryl systems bearing a (1h)-1,2,4-triazole moiety, Synlett. (2010) 55-60. https://doi.org/10.1055/s-0029-1218533

[24] A.R. Hajipour, K. Karami, F. Rafiee, An efficient Stille cross-coupling reaction catalyzed by ortho-palladated complex of tribenzylamine under microwave irradiation, Appl. Organometal. Chem. 26 (2012) 27-31. https://doi.org/10.1002/aoc.1860

[25] X. Tan, Z.J. Zhou, J.X. Zhang, X.H. Duan, Efficient one-pot cross-coupling of two aryl halides by stannylation/Stille reaction in water under microwave irradiation, Eur. J. Org. Chem. 2014 (2014) 5153-5157. https://doi.org/10.1002/ejoc.201402404

[26] K. Sonogashira, Y. Tohda, N. Hagihara, A convenient synthesis of acetylenes: catalytic substitutions of acetylenic hydrogen with bromoalkenes, iodoarenes and bromopyridines, Tetrahedron Lett. 16 (1975) 4467-4470. https://doi.org/10.1016/S0040-4039(00)91094-3

[27] M. Karak, L.C.A. Barbosa, G.C. Hargaden, Recent mechanistic developments and next generation catalysts for the Sonogashira coupling reaction, RSC Adv. 4 (2014) 53442-53466. https://doi.org/10.1039/C4RA09105A

[28] D. Conreaux, T. Delaunay, P. Desbordes, N. Monteiro, G. Balme, A convenient access to furo[3,2-c]pyridin-6(5H)-ones by the reaction of 5-iodo-4-methoxy-2-pyridones with terminal alkynes under microwave-enhanced Sonogashira conditions, Tetrahedron Lett. 50 (2009) 3299-3301. https://doi.org/10.1016/j.tetlet.2009.02.073

[29] J.F. Cívicos, D.A. Alonso, C. Nájera, Microwave-promoted copper-free Sonogashira-Hagihara couplings of aryl imidazolylsulfonates in water, Adv. Synth. Catal. 355 (2013) 203-208. https://doi.org/10.1002/adsc.201200629

[30] M.N. Joy, Y.D. Bodke, K.K. Abdul Khader, A.M. Sajith, A rapid approach for the copper, amine, and ligand-free Sonogashira coupling of 4-methyl-7-nonafluorobutylsulfonyloxy

coumarins under microwave irradiation, Tetrahedron Lett. 55 (2014) 2355-2361. https://doi.org/10.1016/j.tetlet.2014.02.094

[31] E. Negishi, A.O. King, N. Okukado, Selective carbon-carbon bond formation via transition metal catalysis. 3. A highly selective synthesis of unsymmetrical biaryls and diarylmethanes by the nickel- or palladium-catalyzed reaction of aryl- and benzylzinc derivatives with aryl halides, J. Org. Chem. 42 (1977) 1821-1823. https://doi.org/10.1021/jo00430a041

[32] E. Negishi, Palladium-or Nickel-Catalyzed Cross-coupling. A new selective method for carbon-carbon bond formation, Acc. Chem. Res. 15 (1982) 340-348. https://doi.org/10.1021/ar00083a001

[33] S. Hua, R. Mroz, J. Carroll, Negishi coupling in the synthesis of advanced electronic, optical, electrochemical, and magnetic materials, Org. Chem. Front. 2 (2015) 416-445. https://doi.org/10.1039/C4QO00322E

[34] P. Walla, C.O. Kappe, Microwave-assisted Negishi and Kumada cross-coupling reactions of aryl chlorides. Chem. Commun. 4 (2004) 564-565. https://doi.org/10.1039/b313887a

[35] I. Mutule, E. Suna, Arylzinc species by microwave-assisted Grignard formation-transmetallation sequence: application in the Negishi coupling, Tetrahedron. 61 (2005) 11168-11176. https://doi.org/10.1016/j.tet.2005.09.022

[36] H.V. Motwani,M. Larhed, Diarylated ethanones from $Mo(Co)_6$-mediated and microwave-assisted palladium-catalysed carbonylative Negishi cross-couplings. Eur. J. Org. Chem. 2013 (2013) 4729-4733. https://doi.org/10.1002/ejoc.201300610

[37] X. Zhang, D. Ye, H. Sun, D. Guo, J. Wang, H. Huang, X. Zhang, H. Jiang, H. Liu, Microwave-assisted synthesis of quinazolinone derivatives by efficient and rapid iron-catalyzed cyclization in water, Green Chem. 11 (2009) 1881-1888. https://doi.org/10.1039/b916124b

[38] T.M.U. Ton, C. Tejo, S. Tania, J.W.W. Chang, P.W.H. Chang, Iron(II)-catalyzed amidation of aldehydes with iminoiodinanes at room temperature and under microwave-assisted conditions, J. Org. Chem. 76 (2011) 4894-4904. https://doi.org/10.1021/jo200284a

[39] Y. Le, Z. Yang, Y. Chen, D. Chen, L. Yan, Z. Wang, G. Ouyang, Microwave-assisted synthesis of 7-azaindoles via iron-catalyzed cyclization of an *o*-haloaromatic amine with terminal alkynes, RSC Advances. 9 (2019) 39684-39688. https://doi.org/10.1039/C9RA08742G

[40] B. Taboada-Seras, C. Santiago, E. Sustatxa, A. Carral-Menoyo, N. Sotomayor, E. Lete, Microwave-assisted cobalt(III)-catalyzed C-H aminocarbonylation reactions with isocyanates for the synthesis of thiophenecarboxamides, Org.Biomol.Chem. 23 (2025) 7283-7293. https://doi.org/10.1039/D5OB00781J

[41] A. Pal, A. Majumdar, S. Sau, A. Thakur, Microwave-assisted cobalt-catalyzed synthesis of quinazolinones via acceptorless dehydrogenative coupling under neat condition. Chem Asian J. 20 (2020) e202401690. https://doi.org/10.1002/asia.202401690

[42] B.H. Lipshutz, B.A. Frieman, C. Lee, A. Lower, D.M. Nihan, B.R. Taft, Microwave-assisted heterogeneous cross-coupling reactions catalyzed by nickel-in-charcoal (Ni/C), Chem. Asian J. 1 (2006) 417-429. https://doi.org/10.1002/asia.200600031

[43] T.S. Chundawat, N. Sharma, P. Kumari, S. Bhagat, Microwave-assisted nickel-catalyzed one-pot synthesis of 2,4,5-trisubstituted imidazoles, Synlett. 27 (2016) 404-408. https://doi.org/10.1055/s-0035-1560825

[44] F. Ke, P. Zhang, Y. Xu, X. Lin, J. Lin, C. Lin, J. Xu, Microwave-assisted nickel-catalyzed synthesis of benzimidazoles: ammonia as a cheap and nontoxic nitrogen source, Synlett. 29 (2018) 2722-2726. https://doi.org/10.1055/s-0037-1610843

[45] Y. Baqi, C.E. Muller, Rapid and efficient microwave-assisted copper(0)-catalyzed Ullmann coupling reaction: general access to anilinoanthraquinone derivatives, Org.Lett. 9 (2007) 1271-1274. https://doi.org/10.1021/ol070102v

[46] L.J. Goossen, F. Manjolinho, B.A. Khan, N. Rodriguez, Microwave-assisted Cu-catalyzed protodecarboxylation of aromatic carboxylic acids, J. Org. Chem. 74 (2009) 2620-2623. https://doi.org/10.1021/jo802628z

[47] J.E.R. do Nascimento, L.C.C. Goncalves, G. Hooyberghs, E.V.V. Eycken, D. Alves, E.J. Lenardao, G. Perin, R.G. Jacob, Synthesis of fused 1,2,3-triazolo-1,3,6-triazonines through copper-catalyzed intramolecular Ullmann cross-coupling reaction, Tetrahedron Lett. 57 (2016) 4885-4889. https://doi.org/10.1016/j.tetlet.2016.09.027

[48] P. Wyrebek, A. Sniady, N. Bewick, Y. Li, A. Mikus, K.A. Wheeler, R. Dembinski, Microwave-assisted zinc chloride-catalyzed synthesis of substituted pyrroles from homopropargyl azides, Tetrahedron. 65 (2009) 1268-1275. https://doi.org/10.1016/j.tet.2008.11.094

[49] V. Srinivasulu, S.McN. Sieburth, R. El-Awady, N.M. Kariem, H. Tarazi, M.J. O'Connor, T.H. Al-Tel, post-Ugi Cascade transformations for accessing diverse chromenopyrrole collections, Org.Lett. 20 (2018) 836-839. https://doi.org/10.1021/acs.orglett.7b03986

Materials Research Foundations 189 (2026) 284-411 https://doi.org/10.21741/9781644904039-12

Chapter 12

Microwave-Assisted Pyrolysis of Biomass for High-Value Products

Denzel C. Makepa[1], Malbenia J Masamvu[2], Charles Rashama[3*]

[1]Department of Fuels and Energy Engineering, Chinhoyi University of Technology, Private Bag 7724, Chinhoyi, Zimbabwe

[2]Department of Mining, Chemical and Metallurgical Engineering; University of Zimbabwe, Faculty of Engineering and The Built Environment, P.O. Box MP 167, Mount Pleasant, Harare, Zimbabwe

[3]Department of Chemical and Processing Engineering; Manicaland State University of Applied Sciences; Fern Hill Campus. P. Bag 7001. Mutare. Zimbabwe

c.rashama@gmail.com

Abstract

Microwave-assisted pyrolysis of biomass was discussed in this chapter in the context of how this technology in comparison to other pyrolysis technologies influences product characteristics, yields and utilization. It has been noted that generally in comparison to other technologies, microwave-assisted pyrolysis ensures more uniform heating and rapid thermal decomposition, resulting in a predictable and consistent product yield and quality. This consistency and predictability facilitate customized product extraction and upgrading processes that are aimed at obtaining value-added end products from biomass microwave-assisted pyrolysis. Biomass microwave-assisted pyrolysis is therefore deemed a sustainable and scalable technology that can meet the growing global demand for clean energy and sustainable chemical production. In this chapter a review of microwave-assisted pyrolysis technology's application towards biomass valorisation to produce high value products is conducted. Aspects that hinder widespread commercialization and therefore require further attention for improving sustainability are also discussed.

Keywords

Microwave-Assisted Pyrolysis. Biochar, Syngas, Bio-Oil, Biochemical, Biofuels

Contents

Materials Research Foundations 189 (2026) 284-411 https://doi.org/10.21741/9781644904039-12

1. Introduction

The urgent global pursuit of sustainable energy systems and climate-resilient industrial practices has intensified interest in renewable resources, particularly biomass, as a cornerstone for clean energy and green chemical production. Biomass, ranging from agricultural residues, forestry byproducts, to organic municipal and industrial wastes, offers a carbon-neutral, abundant, and versatile feedstock for generating heat, power, fuels, and a wide array of value-added chemicals [1,2]. Unlike fossil resources, biomass is part of the contemporary carbon cycle, and its judicious use can contribute significantly to reducing greenhouse gas (GHG) emissions, enhancing rural economies, and promoting energy access in underserved regions [3]. Among various biomass conversion strategies, thermochemical processes, especially pyrolysis, have garnered substantial attention due to their ability to break down complex organic matter into simpler, energy-dense products such as bio-oil, biochar, and non-condensable gases (syngas) [4]. However, conventional pyrolysis techniques face several critical challenges that limit their efficiency, scalability, and product quality. These include non-uniform heat distribution during processing, long heating times, low selectivity, and excessive formation of tar, all of which complicate process control and downstream product utilization [5,6]. Moreover, the reliance on external heat sources, often through conduction or convection, introduces thermal inefficiencies and limits the adaptability of conventional pyrolysis systems, particularly at smaller scales or in decentralized settings. To address these limitations, microwave-assisted pyrolysis has emerged as a promising advanced thermal conversion technology. Microwave-assisted pyrolysis utilizes electromagnetic radiation, typically in the 915 MHz or 2.45 GHz frequency range, to directly heat polar molecules and conductive materials within biomass [7]. This volumetric and selective heating mechanism leads

Materials Research Foundations 189 (2026) 284-411 https://doi.org/10.21741/9781644904039-12

to faster reaction rates, reduced heat losses, and the formation of localized hot spots that can enhance the breakdown of complex organics into more valuable products [7,8]. Microwave-assisted pyrolysis offers several technical advantages over conventional pyrolysis. The most cited advantage of employing microwave-assisted pyrolysis is the rapid and uniform heating, which shortens residence times and improves process efficiency [8]. Microwave-assisted pyrolysis is associated with lower tar and char formation while increasing gas and liquid yields [7]. Another common benefit of microwave-assisted pyrolysis is the enhanced control over product distribution, including higher H_2 and CO concentrations in syngas [9]. There are opportunities for catalytic enhancement offered by microwave-assisted pyrolysis as many microwave-absorbing materials (e.g., biochar, metal oxides) can also function as reforming catalysts [9]. Another notable advantage for microwave-assisted pyrolysis systems is the compact and modular reactor designs, that are suitable for decentralized or mobile applications [9].

In particular, microwave-assisted pyrolysis has demonstrated significant potential in optimizing the yield and quality of all three pyrolysis products (bio-oil, biochar and syngas), making it a highly versatile platform for biomass valorization [7,8]. The syngas produced typically contains elevated levels of hydrogen, carbon monoxide, and light hydrocarbons, enabling its use in a wide range of downstream applications such as Fischer–Tropsch synthesis, methanol and ammonia production, syngas fermentation, and fuel cell technologies [10]. Simultaneously, microwave-assisted pyrolysis can yield bio-oils with improved stability and energy content due to reduced tar and oxygenate content, enhancing their suitability as precursors for renewable fuels and chemicals [7]. Additionally, the biochar generated from microwave-assisted pyrolysis often exhibits enhanced surface area and porosity, along with catalytic or adsorptive properties. These properties make the biochar valuable for soil amendment, carbon sequestration, environmental remediation, or as a functional material in energy systems [7]. This multifaceted product portfolio underlines microwave-assisted pyrolysis's attractiveness for integrated biorefinery applications aimed at achieving both environmental sustainability and economic viability. Despite these advantages, the commercial deployment of microwave-assisted pyrolysis remains limited, and most current activities are at the laboratory or pilot scale [9,11]. Studies have demonstrated microwave-assisted pyrolysis's feasibility across a wide range of feedstocks, including wood chips, rice husk, corn stover, algae, sewage sludge, and food waste. A few notable pilot-scale investigations include a pilot-scale microwave-assisted pyrolysis reactor that was designed to operate continuously at 200-300 °C with a net energy ratio of 72%, that demonstrated the feasibility of small-scale, decentralized biomass conversion systems [12]. Another demonstration involved a continuous microwave-assisted pyrolysis system that was developed to produce syngas from wood pellets at various temperatures. At 800 °C, the system produced a high-quality syngas with a lower heating value of 18.0 MJ/Nm^3 and a syngas content of 67 vol% at a yield of 72.2 wt% [13]. The scale-up of a microwave-assisted pyrolysis reactor using SiC as a microwave absorbent bed was studied for methyl ricinoleate pyrolysis. The reactor was scaled from a lab-scale (1 kW, 126 mm, 0.9 kg/h) to a pilot-scale (7 kW, 580 mm, 7.0 kg/h. The temperature distribution was uniform and stable, achieving selectivity and yield of the target product comparable to lab-scale results [14].

However, challenges such as scalability, non-uniform field distribution in larger reactors, limited penetration depth of microwaves, material compatibility, and high capital costs for microwave generators have previously slowed the transition from pilot to full commercial scale [9,11]. Moreover, the lack of standardized reactor designs and comprehensive techno-economic assessments has made it difficult to benchmark microwave-assisted pyrolysis systems against

Materials Research Foundations 189 (2026) 284-411 https://doi.org/10.21741/9781644904039-12

conventional pyrolysis or gasification alternatives. Nonetheless, recent advances in magnetron technology, reactor design (e.g., rotating beds, hybrid heating), and catalyst development are steadily bridging this gap [15]. Interest in modular, decentralized microwave-assisted pyrolysis systems, especially for rural electrification, waste-to-energy integration, and high-purity hydrogen generation, is driving further R&D, with commercialization likely to follow as these technical barriers are addressed.

This chapter provides an in-depth overview of microwave-assisted pyrolysis for biomass conversion, emphasizing the formation, characteristics, and utilization of its three primary products: syngas, bio-oil, and biochar. It highlights microwave-assisted pyrolysis's advantages over conventional pyrolysis, its integration into circular biorefinery systems, and its role in advancing low-carbon, resource-efficient energy and chemical production.

2. Pyrolysis bio-oil and utilization

2.1 Bio-oil characteristics

Bio-oil is a complex liquid mixture comprising of phenolics, oxygenates, hydrocarbons, aldehydes, organic acids, etc. and its composition vary depending on the feedstock type, pyrolysis conditions (temperature, heating rate, residence time), and catalysts used. Bio-oil is made more useful by separating it into its components [16,17]. Alternatively, the bio-oil can be upgraded into various chemicals without prior separation. Selection of suitable bio-oil valorisation routes depend on understanding the bio-oil's physicochemical properties, its components and effects of these components on the quality and product applications as summarised in Table 1.

Table 1: Bio-oil properties, characteristics, and key compounds

Property	Description	Challenges	Opportunities	References
Composition & Phase Behaviour	Contains 15-40 % water; separates into aqueous and organic phases. Organic phase contains phenols, hydrocarbons, aldehydes, and oxygenates like furans and levoglucosan ($C_6H_{10}O_5$).	Phase separation reduces stability; aqueous phase hard to upgrade; organic phase unstable.	Phenolic compounds valuable for resins, adhesives, and fine chemicals.	[18,19]
Acidity & Corrosiveness	pH is 2-4 due to acids such as acetic acid (CH_3COOH), formic acid ($HCOOH$), and levulinic acid ($C_5H_8O_3$).	Highly corrosive to processing and storage systems; acid-catalyzed polymerization leads to aging.	Organic acids can be recovered and utilized in the chemical and food industries	[19]
High Oxygen Content	30-50 wt% oxygen; includes organic oxygenates: acids, aldehydes, ketones, furans (C_4H_4O), and sugars like levoglucosan.	Reduces calorific value; promotes chemical instability; limits direct fuel applications	Oxygenates serve as precursors for polymers, solvents, and pharmaceuticals.	[20]

Instability & Viscosity	Contains reactive compounds (aldehydes, ketones, phenols) leading to polymerization during storage; viscosity 30 - 120 mPa·s.	Viscosity increases over time, especially at ambient temperatures; clogs nozzles; poor atomization	Stabilization through catalytic hydro-treatment or antioxidant addition improves usability as liquid fuel.	[21]
Ash & Char Content	Contains unconverted solids (0.05 - 0.5 wt%) and alkali metals like Na, K, Ca from biomass ash.	Solids and metals can damage engines, clog filters, and poison catalysts.	Char usable as activated carbon or soil enhancer; metals can be removed by pre-treatment	[17]
Nitrogen & Sulfur Compounds	0 - 0.5 wt% N; trace sulfur present depending on biomass.	Lead to emissions (NOx, SOx), corrosion, and catalyst deactivation.	Selective hydrotreatment can remove N/S; valuable for emission-compliant fuels	[22]

2.2 Bio-oil fractionation to high-value products

Bio-oil can be separated into its distinct chemical fractions, through different unit processes and operations such as distillation, solvent extraction, membrane technologies and adsorption [23]. These separate byproducts can be used for various applications (Table 2). The common bio-oil fractionation processes are briefly described below, highlighting their advantages and disadvantages as well as the target products recovered in each process.

a) Solvent Extraction

The aqueous phase of bio-oil contains predominantly acids, aldehydes, ketones and some water-soluble sugars while the organic phase contain mainly phenols and aromatic hydrocarbons. Solvents enhance separation of the two phases facilitating recovery of the different bio-oil components. The main advantage of this separation technique is that it is simple to implement although it suffers drawbacks from high solvent costs and entrainment of solvent as an impurity in the final products [23]. Advanced techniques, such as the use of supercritical solvents, have been explored to enhance extraction efficiency and enable the separation of a wider range of bio-oil components.

b) Distillation

Bio-oil is separated into distinct product streams using distillation which rely on bio-oil components' varying boiling points [16]. The light fractions will primarily comprise low molecular weight, highly volatile compounds such as aldehydes, furans, alcohols, and certain organic acids. These are ideal for use as fuel additives or precursors for the synthesis of chemicals. Aldehydes, for instance, are useful in various chemical processes and can be utilized in the manufacturing of biodegradable plastics or in the food industry as preservatives. Middle fraction distillates are moderately volatile compounds such as phenols and small oligomers which are particularly important for biofuel production, as they contain bio-oil constituents with a higher energy density and a lower oxygen content than the light fractions. Phenolic compounds, which are prevalent in the middle fractions, can be used for the production of higher-value materials, such as resins, adhesives, and antioxidants. The heavy fractions are high molecular weight compounds, primarily phenolics, lignin-derived oligomers, and biochar. They are typically utilized for biochar production or further chemical upgrading to produce valuable biochemicals. The main disadvantage associated with distillation is the thermal instability of some of the bio-oil components as well as its energy intensity which increases operational costs. This shortcoming is normally addressed by using vacuum distillation at lower operating temperatures.

c) Membrane Technologies

Filtration using membranes of different pore diameters (Ultrafiltration and pervaporation) enables the removal of particulates, water and small oxygenated compounds in bio-oil with minimal thermal degradation as experienced in distillation [24,25]. The resulting bio-oil is of higher stability and quality for compatible use with drop-in fuels, hence more useful. Membrane technologies are considered environmentally friendly and energy-efficient compared to traditional distillation techniques, as they do not require the high heat typically needed for thermal fractionation.

2.3 Bio-oil catalytic upgrading strategies

Upgrading of bio-oil can be implemented without first separating its components. Key strategies for bio-oil upgrading are esterification, catalytic hydrodeoxygenation (HDO) [26], catalytic cracking [27], emulsification [28], supercritical fluids, solvent addition and hot gas filtration (Figure 1). Each of these strategies target to produce stable, energy-dense hydrocarbons that are compatible with conventional drop-in fuels. These processes do not only improve product quality but also increase the economic and environmental value of bio-oil in the economy.

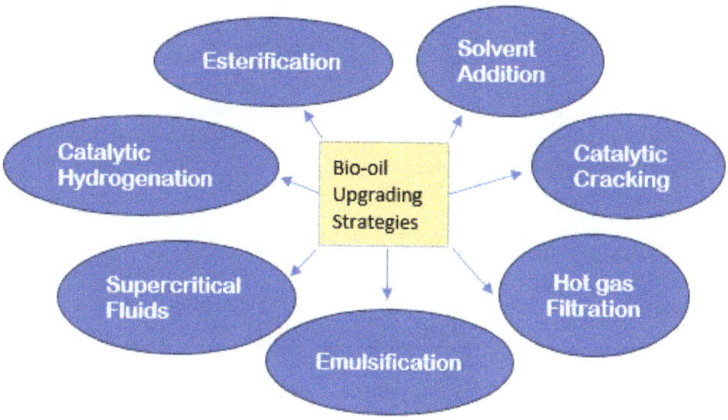

Figure 1: Strategies for bio-oil upgrading. Drawn by Author.

Although HDO produces stable hydrocarbons with reduced oxygen content and higher heating value, the high energy demand for this process is a key drawback towards HDO's widespread use [22]. Zeolite cracking produces aromatic hydrocarbons (BTX) with improved conventional fuel compatibility and low coke formation although the high temperatures during this process pause

Materials Research Foundations 189 (2026) 284-411 https://doi.org/10.21741/9781644904039-12

catalyst deactivation risks. Emulsification confers enhanced flowability, better engine compatibility and direct diesel engine usage to bio-oil although concerns of emulsion instability and water content which may affect combustion have been raised by researchers [29].

Table 2: Platform chemicals that can be extracted from bio-oil

Platform Chemical	Description	Extraction Method	Applications	Reference
Phenols	Aromatic compounds derived from lignin depolymerization, including phenol, guaiacol, and cresols.	Liquid-liquid extraction using solvents like aqueous methanol; advanced methods include supercritical CO_2 and switchable hydrophilicity solvents.	Resins, adhesives, coatings, pharmaceuticals, fuel additives.	[30]
Furfural	Furan compound produced from hemicellulose degradation.	Acid-catalyzed dehydration of pentoses; extraction using biphasic systems with deep eutectic solvents (DES).	Solvents, resins, biofuel intermediates (e.g., butanes, alkenes).	[31]
5-Hydroxymeth-ylfurfural (HMF)	Derived from hexose sugars during pyrolysis.	Solvent extraction; advanced methods involve biphasic systems and DES.	Intermediate for levulinic acid, fuels, plastics, solvents.	[16]
Levoglucosan	Anhydrosugar from cellulose pyrolysis.	Phase separation followed by acid hydrolysis.	Bioethanol production, biodegradable plastics, fermentation feedstock.	[32]
Acetic Acid	Small oxygenate found in the aqueous phase of bio-oil.	Solvent extraction or distillation.	Vinegar, solvent, chemical intermediate.	[33]
Formic Acid	Light carboxylic acid from sugar degradation.	Catalytic oxidation using polyoxometalates (OxFA process).	Fuel cells, preservatives, descaling agents.	[34]
Levulinic Acid	Derived from HMF or cellulose hydrolysis.	Reactive extraction using green solvents like sunflower oil with trioctylamine or tri-n-butyl phosphate.	Plasticizers, fuels, pharmaceuticals.	[35]
Guaiacol	Methoxyphenol from lignin.	Solvent extraction or alkaline extraction.	Antiseptics, flavorings, resin precursor.	[18]

Materials Research Foundations 189 (2026) 284-411 https://doi.org/10.21741/9781644904039-12

3. Biochar characteristics and utilization

3.1 Biochar properties and applications

Biochar produced from microwave-assisted pyrolysis exhibits different properties and characteristics which depend on various factors such as type of biomass processed, feedstock condition and process operating parameters [36]. The important biochar properties include specific surface area, pore volume, stability and calorific value. The biochar characteristics will influence its applications. Some of the key biochar characteristics and use of the biochar are shown in Table 3. Among the different biochar applications are catalysis (guaiacol and propane hydrogenation, esterification, methane reforming, Fischer-Tropsch synthesis), soil amendment, use as a fuel, adsorption (metal cations in water, gases as pollutants, dyes), addition to animal feeds, removing herbicides and other pollutants from water. It also finds use as a carbon sink and in energy (manufacture of electrodes and hydrogen and carbon dioxide storage) including microbial/enzyme immobilizations [37] The biochar applications are enhanced through a process called functionalization which involves the modification of biochar properties to enhance its performance for a specific application. These functionalization techniques broaden the scope of biochar application as an advanced material.

Table 3: Microwave-assisted pyrolysis contribution to biochar properties and applications

Property/Factor	Significance in biochar applications	Microwave-assisted pyrolysis influence	Reference
Fixed carbon (C) and energy yields	High fixed carbon means more energy content hence better fuel use. At least 95%C is required for biochar to be used as coal replacement in metallurgical processes. Fixed C increases with temperature and residence time.	Higher kinetics in Microwave-assisted pyrolysis vs conventional pyrolysis reduce process energy consumption for comparable fixed C levels in biochar.	[38]
Ash content	High ash content in biochar may prevent its use in high-grade industrial applications. The ash content of biochar is largely dependent on the ash content in the parent biomass.	Microwave-assisted pyrolysis has no direct influence on resulting biochar ash content. Whether the biochar is produced through conventional or microwave-assisted pyrolysis route, ash is a concern that needs attention.	[39]
Atomic ratios	Low O/C and H/C ratios are desirable characteristics for better fuels.	Microwave-assisted pyrolysis achieves desirable ratios much faster than conventional pyrolysis	[40]

Materials Research Foundations 189 (2026) 284-411 https://doi.org/10.21741/9781644904039-12

Aromaticity	High aromaticity imparts high thermodynamic stability. Desired for soil amendment or metallurgical purposes.	Higher values of aromaticity were reported at lower temperature in microwave than conventional pyrolysis.	[41]
Persistent free radicals (PFR) and other functional groups	The type, concentration and lifetime of PFRs dictate biochar's pollutant-degrading capabilities. Additionally, functional groups dictate the biochar's alkalinity, which has a bearing on its ability to neutralize acids for soil bioremediation.	microwave-assisted pyrolysis offers improved product quality through inner heating and selectivity to high-polarity components. With Microwave-assisted pyrolysis the surface functional groups are also easier to model and predict.	[42]
pH	This is important for soil amendments as different plant/crops require a narrow-range pH. Higher carbonization increases pH to a maximum of 12.	Process control is much easier with Microwave-assisted pyrolysis than conventional pyrolysis making it easier to customize Microwave-assisted pyrolysis biochar for different crops. Under same pyrolysis conditions Microwave-assisted pyrolysis gives a higher pH biochar.	[36]
Cationic exchange capacity (CEC)	Depends on both functional groups and surface area. The highest CEC are associated with biochar produced at relatively low production temperatures, at which the surface area has significantly increased compared to the feedstock, but sufficient functional groups remain in the structure to provide negative charges.	Microwave-assisted pyrolysis biochar has better electron accepting and electron exchange capacity andenhancement effect than conventional pyrolysis.	[43]
Reactivity	Temperature, residence time, pressure and mineral matter content are key determinants of reactivity as these affect the functional groups on the biochar.	With microwave-assisted pyrolysis the surface functional groups are easier to model and predict than with conventional pyrolysis.	[44]

Materials Research Foundations 189 (2026) 284-411 https://doi.org/10.21741/9781644904039-12

Surface area, porosity, pore volume and distribution	These variables increase with temperature but are more pronounced in the 350°C range during hemicellulose degradation. The percentage distribution of macropores, mesopores and micropores influences gas adsorption cation exchange and plant water uptake capabilities of biochar. The pore volume and distribution are influenced by temperatures. Beyond 800°C decreases have been noted in other studies.	Microwave-assisted pyrolysis generates larger biochar surface areas than conventional pyrolysis due to better heating mechanisms.	[43]
Hydrophobicity (Water-holding capacity)	Higher pyrolysis temperatures result in increased hydrophobicity of the biochar because more polar surface functional groups are removed and the aromaticity is increased. However, contrasting results have been observed and implied that the increasing porosity allowed more water to be held by the char.	Microwave-assisted pyrolysis offers improved product quality through inner heating and selectivity to high-polarity components.	[42]
Mechanical stability	Higher temperatures lower the char's mechanical strength although studies indicate that char could still handle the weight of iron ore in blast furnace operations.	Microwave-assisted pyrolysis offers better product customization through improved control of temperatures.	[45]
Grindability	The change in mechanical stability during carbonization leads to a brittle char with improved grindability compared to raw material.	Microwave-assisted pyrolysis offers better product customization through improved control of temperatures.	[45]
Thermal conductivity and heat capacity	Porosity correlates negatively with thermal conductivity of biochar. Conductivity depends on grain direction and pyrolysis temperature.	Microwave-assisted pyrolysis biochar generally generates larger pores hence poor thermal conductivity than conventional char. This is desirable for construction materials where insulation is key	[46]

Materials Research Foundations 189 (2026) 284-411 https://doi.org/10.21741/9781644904039-12

3.2 Biochar functionalization methods

The most common functionalization methods or purposes are depicted in Figure 2 and briefly described thereafter. These functionalization techniques broaden the scope of biochar application as an advanced material.

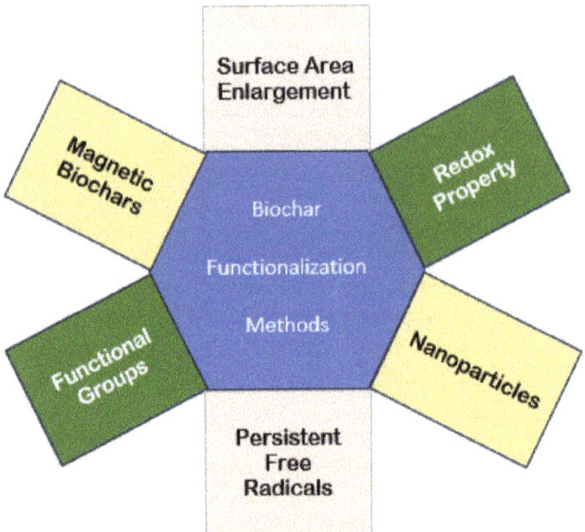

Figure 2: Different biochar functionalization routes. Drawn by Author.

a) Surface area enlargement functionalization

Modification of biochar surface area and pore distribution can be implemented before, during or after pyrolysis depending on the method employed [47]. Common approaches include the biomass pretreatment with alkali, acid or clay before pyrolysis, carbon dioxide activation during pyrolysis and hydrochloric acid washing post pyrolysis.

b) Functionalization through persistent free radical (PFR) manipulation

Four approaches can be used to manipulate the concentration, type, and lifetime of PFRs in biochar. These approaches include varying the pyrolysis temperature and residence time as well as manipulating the metal and metal oxides loading on the biochar surfaces, N-doping and incorporation of functional groups.

c) Magnetization functionalization

Magnetic biochar is produced to facilitate the removal of biochar from slurries after performing its bioremediation duty. Biochar magnetization is achieved through three main techniques which include impregnation followed by pyrolysis, pyrolysis followed by precipitation and hydrothermal method [48]. Additionally, less common practices involve ball-milling of biochar in iron ore, electrodeposition followed by pyrolysis and crosslinking.

Materials Research Foundations 189 (2026) 284-411 https://doi.org/10.21741/9781644904039-12

d) Redox property modification

Biochars have demonstrated their pollution degradation role through influencing the redox potential of contaminated sites. Treating biochar with KMnO4 and the modified Hummer's method functionalises biochar for enhanced electron exchange capacities. Alternatively, H2O2 could also be used to induce redox-active functional groups on the biochar surface.

e) Functional groups modification

Surface functional groups that are known for facilitating metal and organics degradation are promoted on the biochar surface through surface oxidation (using different oxidants after pyrolysis), surface amination (using N-rich materials such as proteins or nitric acid), P–O–P insertion (using phosphoric acid) and surface sulfonation (using sulphuric acid or gaseous sulphur trioxide) [48]. All these modified biochars have specific industrial catalytic or bioremediation applications.

f) Nanoparticle (NP) functionalization

Coating and/or in-situ nanoparticle formation are used to decorate NPs on biochars [47]. This provides a synergy of the biochar and the NP catalytic capabilities which may benefit the bioeremediation efforts.

4. Pyrolytic gaseous products and their utilization

4.1 Overview of gaseous products from biomass pyrolysis

The gaseous products of biomass pyrolysis, while often considered a by-product, have high energy content and significant utility in thermochemical and biorefinery applications [4]. The composition and quality of these gases are highly dependent on the nature of the biomass feedstock and the pyrolysis method employed. The gaseous fraction from biomass pyrolysis is a complex mixture primarily composed of components listed in Table 4. The actual composition of the gas varies with the pyrolysis system, operating parameters, and feedstock characteristics. For instance, hardwoods tend to yield more CO and less H_2 than softwoods, while agricultural residues may generate more CO_2 due to higher hemicellulose content.

4.2 Effect of process parameters on gas yield and composition

The distribution and composition of pyrolysis gases are highly sensitive to operational parameters. Increasing pyrolysis temperature generally increases the gas yield at the expense of bio-oil and biochar. Higher temperatures (≥600°C) enhance secondary cracking of volatiles, leading to more H_2, CO, and light hydrocarbons [53]. The gases CO_2 and CH_4 are more prominent at lower temperatures. Fast pyrolysis, characterized by rapid heating rates (~1000°C/s), typically favors liquid production but also influences the nature of evolved gases by limiting secondary reactions [4]. Slow heating may allow more extensive devolatilization and gas–solid reactions, enhancing gas yields. Prolonged vapor residence time promotes secondary cracking and reforming of volatiles, increasing H_2 and CO yields while reducing CH_4 and tars. An inert atmosphere (e.g., N_2 or Ar) is standard in pyrolysis, but introducing steam or CO_2 can induce reforming reactions (e.g., steam reforming, dry reforming), altering syngas composition [53]. Elevated pressures can enhance gas–solid reactions and affect gas phase equilibrium, favoring CH_4 formation. High lignin content tends to favor CH_4 and phenolic compounds, while high cellulose content favors CO

Materials Research Foundations 189 (2026) 284-411 https://doi.org/10.21741/9781644904039-12

production [9,53]. Moisture content also plays a role as high moisture can reduce peak pyrolysis temperatures and affect gas composition [54]. In microwave-assisted pyrolysis, these parameters are further influenced by microwave-specific factors like dielectric properties of the feedstock, presence of microwave absorbers, and spatial heating uniformity, all of which significantly affect gas yields and composition.

Table 4. Typical gaseous products from biomass pyrolysis

Gas	Characteristics, origins, and utilization potential	Reference
Carbon Monoxide (CO)	A major component, arising from the decomposition of cellulose and hemicellulose and from secondary cracking of volatiles. CO plays a vital role in syngas composition and can be utilized in synthesis reactions like Fischer-Tropsch synthesis and methanol production.	[49]
Carbon Dioxide (CO_2)	Generated mainly from the decarboxylation of hemicellulose and lignin. While CO_2 does not contribute directly to the fuel value, it influences the H_2/CO ratio and overall syngas reactivity.	[50]
Hydrogen (H_2)	Arises from cracking of volatiles, reforming of hydrocarbons, and gas–solid reactions involving char. Its proportion can be enhanced in microwave-assisted pyrolysis due to localized high temperatures and catalytic effects of the biochar.	[51]
Methane (CH_4)	Formed through methanation and decomposition of lignin-derived volatiles. It contributes significantly to the calorific value of the gas mixture.	[52]
Light Hydrocarbons (C_2–C_4)	Includes ethylene, ethane, propylene, propane, and butane. These are produced through thermal cracking of heavier hydrocarbons and contribute to the fuel potential of the gas stream.	[49]
Minor Species	Trace gases such as ammonia (NH_3) from nitrogenous compounds and hydrogen sulfide (H_2S) from sulfur-containing feedstocks are often present. Their presence depends on the nitrogen and sulfur content of the biomass and can influence the need for gas cleanup before utilization.	[49]

4.3 Microwave-assisted pyrolysis influence on syngas composition

Microwave-assisted pyrolysis has emerged as a transformative thermochemical approach for converting biomass into value-added products, particularly syngas with superior quality and tailored composition. Unlike conventional pyrolysis methods that rely on indirect or convective heating, microwave-assisted pyrolysis utilizes microwave radiation to induce dielectric heating within the biomass matrix, resulting in unique thermal behavior that significantly influences gas-phase reactions and the resulting syngas composition [53]. This distinctive mode of energy

Materials Research Foundations 189 (2026) 284-411 https://doi.org/10.21741/9781644904039-12

delivery introduces several key mechanisms that shape the composition and quality of syngas derived from microwave-assisted pyrolysis.

a) Selective heating and hot spots

One of the defining features of microwave-assisted pyrolysis is its capacity for rapid, volumetric, and selective heating. Microwaves penetrate the biomass and generate internal heating directly within polar and conductive components. This leads to the formation of microscale thermal hot spots [55], especially when microwave susceptors like biochar or metal oxides are present. These localized high-temperature zones in microwave-assisted pyrolysis promote enhanced thermal cracking of volatiles, breaking down heavier molecules into smaller gaseous species such as H_2 and CO [53]. They also accelerate endothermic reactions like steam reforming and dry reforming of tars or hydrocarbons, which are typically energy-limited under conventional pyrolysis conditions. Additionally, microwave-assisted pyrolysis improves thermal uniformity in heterogeneous biomass materials, minimizing cold zones that can hinder the complete gasification of volatile intermediates. As a result, syngas produced via microwave-assisted pyrolysis often contains higher concentrations of hydrogen and carbon monoxide compared to that generated through conventional pyrolysis [56].

b) Catalyst synergy with microwave absorbing materials

The use of microwave-absorbing materials such as biochar, silicon carbide (SiC), iron oxide (Fe_2O_3), nickel-based catalysts (e.g., Ni/Al_2O_3), and activated carbon serves a dual function in microwave-assisted pyrolysis: they enhance microwave energy absorption, converting electromagnetic waves into heat more efficiently, and simultaneously act as heterogeneous catalysts that facilitate key reforming and cracking reactions [57]. Under microwave irradiation, these materials generate localized heating at their surfaces and may induce non-thermal effects like increased molecular polarization, both of which contribute to accelerated reaction kinetics [57]. This synergistic behavior leads to improved conversion of heavier intermediates and volatiles into lighter gases, ultimately enhancing the quality of the syngas by increasing its calorific value and selectively enriching it in hydrogen and carbon monoxide.

c) Lower tar formation

Tar formation is a major challenge in thermochemical biomass conversion, often leading to equipment fouling, reduced efficiency, and costly gas cleanup requirements [54]. microwave-assisted pyrolysis offers a significant advantage by substantially reducing tar generation [15]. This reduction is attributed to the rapid heating rates in microwave-assisted pyrolysis, which favor primary devolatilization of biomass before secondary polymerization reactions can produce heavy tars [53]. Additionally, the catalytic properties of biochar and other microwave-absorbing materials help crack long-chain organics into lighter gases, while thermal hot spots created during microwave irradiation facilitate in-situ reforming of condensable volatiles. As a result, more carbon is redirected toward syngas precursors, increasing the overall gas yield and enhancing compatibility with downstream processes such as syngas fermentation or catalytic fuel synthesis [15].

4.4 Hydrogen and methane production potentials

Microwave-assisted pyrolysis holds distinct advantages for the targeted production of H_2 and methane CH_4, two valuable gaseous fuels with broad applicability in clean energy systems and synthetic fuel production.

a) Hydrogen generation

Microwave-assisted pyrolysis inherently favors dehydrogenation and reforming reactions, resulting in enhanced H_2 yields compared to conventional pyrolysis [51]. The rapid and volumetric heating achieved through microwaves promotes thermal cracking of volatiles and facilitates secondary reactions such as steam reforming of hydrocarbons and oxygenates, char–steam gasification, and water–gas shift reaction, especially when water vapor is present [58]. Moreover, microwave-induced plasma effects, observed under specific conditions involving low pressure or high microwave intensity, can further boost H_2 evolution by breaking molecular bonds through non-thermal excitation. The use of transition metal catalysts such as Ni or Fe supported on biochar or metal oxides significantly amplifies hydrogen production. These catalysts, under microwave irradiation, activate reforming pathways and exhibit high local temperatures at active sites, thus improving reaction rates without requiring globally high reactor temperatures [59, 60]. This synergy enables the development of microwave-assisted pyrolysis -based hydrogen-rich gas platforms suitable for integration into fuel cells or green hydrogen supply chains.

b) Methane formation

While microwave-assisted pyrolysis typically results in lower CH_4 yields compared to conventional pyrolysis, owing to the enhanced cracking of volatiles and reduced residence time, there is potential to selectively promote methane generation through tailored process design. It is possible to steer the reaction pathway toward methanation, converting CO and H_2 into CH_4 via the Sabatier reaction by tuning microwave parameters (power level, pulse modulation, irradiation time), and introducing specific catalysts (e.g., Ni/Al_2O_3, Ru-based systems). This opens the door for biomass-to-bio-SNG (synthetic natural gas) conversion under microwave-assisted pyrolysis conditions, especially when coupled with CO_2 or syngas recycling loops [61,62]. Thus, microwave-assisted pyrolysis provides a versatile platform for tuning gaseous product profiles, favoring hydrogen for clean energy or methane for grid-compatible fuel, through controlled energy delivery and catalytic enhancement.

4.5 Integration with other biorefinery processes

Microwave-assisted pyrolysis not only offers a high-efficiency route for converting biomass into valuable gaseous products but also lends itself seamlessly to integration within broader biorefinery frameworks. Its unique operational characteristics such as rapid start-up, tunable outputs, and compact design, make it especially attractive for coupling with other thermochemical, biochemical, and catalytic processes, enabling synergetic energy and material valorization. In this context, Microwave-assisted pyrolysis -derived energy and gas streams can be strategically aligned with downstream and parallel processes to maximize resource utilization and reduce energy losses.

a) Fermentation and anaerobic digestion support

Excess heat and electricity generated during microwave-assisted pyrolysis (particularly when using hybrid heating or conductive susceptors) can be harnessed to maintain optimal conditions for fermentation and anaerobic digestion (AD) processes. This not only improves the energy

Materials Research Foundations 189 (2026) 284-411 https://doi.org/10.21741/9781644904039-12

efficiency of the overall system but supports a cascading valorization of both lignocellulosic and wet biomass fractions.

b) Utilization of hydrogen-rich syngas

The high H_2/CO ratio of microwave-assisted pyrolysis -derived syngas makes it ideal for several high-value downstream processes that include hydrodeoxygenation (HDO) of pyrolysis oils into transport-grade biofuels [63], ammonia synthesis through Haber-Bosch alternatives [63] and fuel cell integration, where clean H_2 supports power generation in off-grid or mobile energy systems [63].

c) Combined heat and power (CHP) generation

Combusting syngas in internal combustion engines, gas turbines, or microturbines provides both electricity and thermal energy, ideal for industrial heat demand, district heating, or on-site energy supply in agro-industrial settings.

d) Carbon utilization in chemical synthesis

The CO content in microwave-assisted pyrolysis syngas enables participation in carbonylation and carboxylation reactions to synthesize fine and bulk chemicals such as acetic acid, formic acid, or methanol, expanding the product portfolio of biorefineries beyond fuels [64].

5. Sustainability aspects of microwave-assisted pyrolysis

5.1 Technical assessments

While microwave-assisted pyrolysis offers many benefits versus traditional pyrolysis systems, the transition from small-scale experimental setups to large-scale industrial applications presents several technical barriers that must be addressed to realize microwave-assisted pyrolysis 's full potential in biomass processing. These challenges include non-uniform energy distribution, limited efficiency of high-power microwave sources, reactor material constraints, and complex process control [65]. Continued interdisciplinary research collaborations between engineers, material scientists, and industry stakeholders will be essential for driving the successful scale-up of microwave-assisted pyrolysis in sustainable biomass valorization. Table 5, highlights major technical barriers in scaling up microwave-assisted pyrolysis, including innovation solutions and limitations. One of the key directions for future research lies in the development of advanced reactor architectures that can ensure uniform microwave energy absorption, better thermal control, and higher throughput. Innovations such as fluidized bed microwave-assisted pyrolysis reactors have demonstrated improved heat and mass transfer characteristics, which are essential for achieving consistent pyrolysis conditions across varying feedstock types [66]. Additionally, continuous-flow microwave-assisted pyrolysis reactors offer a promising route to industrial-scale applications by enabling uninterrupted biomass feeding and product collection, thereby increasing the system's operational efficiency [67]. Multi-zone reactors, designed with segmented heating zones and precise temperature control mechanisms, provide another pathway for process optimization. These systems allow for the sequential application of varying thermal conditions tailored to different stages of pyrolysis (e.g., drying, devolatilization, and cracking), thereby enhancing product selectivity and yield [68]. Furthermore, modular reactor designs could facilitate plug-and-play integration into existing industrial setups, offering flexibility in deployment and scale-up [65].

Materials Research Foundations 189 (2026) 284-411 https://doi.org/10.21741/9781644904039-12

Another transformative avenue is the integration of hybrid systems that couple microwave heating with other thermal, catalytic, or chemical processes. Such hybrid configurations have shown the potential to overcome the limitations of standalone microwave-assisted pyrolysis, particularly in improving bio-oil quality and overall conversion efficiency. For instance, coupling microwave-assisted pyrolysis with catalytic steam reforming has been found to promote hydrogen-rich gas production while reducing tar formation [9]. Similarly, combining microwave-assisted pyrolysis with catalytic cracking, using zeolites or metal-supported catalysts, can enhance the deoxygenation of bio-oil and shift product distribution towards more valuable hydrocarbons [5]. Hybrid systems also extend the feedstock's flexibility of microwave-assisted pyrolysis technologies. By integrating pretreatment steps such as torrefaction, hydrothermal carbonization, or mechanical drying with microwave-assisted pyrolysis, it becomes feasible to process a diverse range of biomass, including wet, fibrous, or mixed municipal solid waste streams. This adaptability is crucial for deploying microwave-assisted pyrolysis systems in decentralized, resource-variable contexts such as rural agro-industrial zones or urban waste recovery centre [69]. Moreover, the inclusion of smart control systems, sensor integration, and AI-based process optimization in microwave-assisted pyrolysis reactors presents a promising frontier. These technologies can facilitate real-time monitoring and dynamic adjustment of key operating parameters such as temperature, microwave power, and residence time, further enhancing process stability and product consistency [70]. Advancements in reactor design and the development of hybrid systems hold the key to unlocking the full potential of microwave-assisted pyrolysis for sustainable and high-value biomass conversion. Future research should focus on the holistic integration of reactor engineering, catalytic science, process control, and feedstock adaptability to create robust, scalable, and commercially viable microwave-assisted pyrolysis technologies.

Table 2: Technical barriers in scaling up microwave-assisted pyrolysis

Barrier	Description	Innovations	Limitation	Reference
Microwave Energy Distribution	Uneven heating due to variation in dielectric properties of biomass in large reactors.	Multi-source reactors, waveguides, CFD/FEA modeling, IR imaging, dielectric sensors.	Thermal runaway, poor yield, localized overheating.	[71]
High-Power Microwave Generators	Industrial-scale systems require 50–100 kW efficient generators, but magnetron sources are outdated.	Solid-state microwave sources offer higher efficiency, precision, and durability.	High capital cost limits industrial adoption.	[72]
Reactor Material Limitations	Traditional materials like steel reflect microwaves and degrade under heat.	Ceramic linings, composites with high thermal/microwave resistance; fluidized/rotating bed designs.	Cost and durability of advanced materials.	[72]
Process Control and Scale-up	Complexities in real-time control due to microwave variability and biomass feed inconsistencies.	AI/ML-integrated feedback systems, LCA, economic assessments, hybrid heating for stability.	Technical complexity, scalability of AI integration.	[11]

Materials Research Foundations 189 (2026) 284-411 https://doi.org/10.21741/9781644904039-12

5.2 The economics of biomass microwave-assisted pyrolysis process

An economic assessment involves evaluating the financial viability of a business model. This is achieved by comparing all the costs associated with constructing and operating the business versus the financial benefits to be accrued from running that business [73]. Different market scenarios can be factored into the evaluation to bring out the business's sensitivity to changing market conditions such as inflation and gender dynamics. Key factors in running a microwave-assisted pyrolysis business include feedstock type, availability and distance from the proposed operational site. The source and price of power play a significant role in determining business profitability. The size of the operation and intended end use of the products also dictate the viability of the microwave-assisted pyrolysis business. A generalization of the microwave-assisted pyrolysis technical and financial viability can be derived from different microwave-assisted pyrolysis projects/studies as reported in Table 5. Findings reported in this Table influence decisions by investors in advancing microwave-assisted pyrolysis technologies for recovering high-value products.

A recent study in Zimbabwe demonstrated that converting pine sawdust into bio-oil via microwave-assisted pyrolysis and subsequently into biodiesel is both economically viable and environmentally beneficial [75]. A comparative economic evaluation of different microwave-assisted pyrolysis plants that employed different biomass feedstocks and operating capacities ranging 267 kg/hr to 100 t/hr was conducted by Mari Selvam et al. (2024) [74]. Profitability in these plants were calculated using biochar sales only. Biochar prices ranged 0.1 USD/kg to 1.909 USD/kg. The payback period across these different plants ranged 1.9 – 20 years. The internal rates of return (IRR) ranged 0.09% to 41% [74,76]. These references [74,76] are informative sources of techno-economic information about biomass microwave-assisted pyrolysis plants may be more economically viable hence favorable in the agricultural sector than the large-scale plants. Incorporating renewable energy, as an input power for operating the microwave-assisted pyrolysis process, improve the reliability, economical, and environmental friendliness of microwave-assisted pyrolysis plants [77]. Check the technology readiness level (TRL) of microwave-assisted pyrolysis is lower than that of conventional pyrolysis [74]. Capital costs do not follow a linear relationship with the plant size and, there is a strong decreasing economy of scale with decreasing plant sizes. The benefit of bigger plant must be weighed against feedstock transportation costs from longer distances.

Materials Research Foundations 189 (2026) 284-411 https://doi.org/10.21741/9781644904039-12

Table 6: Economic assessments of microwave-assisted pyrolysis projects

Project/Study Description	Key findings	Reference
0.75 ton/hr feedstock mobile microwave assisted ex-situ catalytic pyrolysis plant treating Douglas fir in Washington State, USA. Non-condensibles recycled to provide electric power for the facility. 480°C, 700W, 20 min retention time. Operating 8hrs per day and 250 days/year. 10-year lifespan Yields - 30% bio-oil, 40% biochar, 25% syngas.	Acquisition costs for equipment take the biggest share of total capital outlay. The biggest portion of operating costs are from purchasing/moving feedstock and chemicals. Business viability (ROI) is more sensitive to bio-oil yield and selling price. (***viability calculations based on bio-oil sales only***)	[73]
1 ton/batch of feedstock is dried and milled before pyrolysis. Treating sugarcane bagasse. No recirculation of non-condensibles for power generation. Dry at 350°C and pyrolysis at 438°C, 700W, 20 min retention time. Operating 8hrs per day and 300 days/year. 20-year lifespan Yields - 39% biochar, 61% syngas.	Equipment purchases costs contributed the largest portion to total capital outlay. Utilities took the biggest portion of operating costs. Business viability (PBP) is more sensitive to process operational time and biochar selling price. ***(viability calculations based on biochar sales only)***	[74]
2000 ton/day of pine sawdust feedstock is dried, milled and pyrolysed. Recovered bio-oil is converted to biodiesel. No recirculation of non-condensibles for power generation. Dry at 125°C and pyrolysis at 500°C, 25-year lifespan. Yields - 65.8% bio-oil, 8.9% biochar, and 25.3% NCGs. Biodiesel yield was 48% based on raw bio-oil.	Equipment costs contributed the largest portion (37.9%) to the total capital outlay. The price of methanol (catalyst for esterification) takes the biggest share of operating costs. Business viability (MSP)'s sensitivity is due to factors in the order of biodiesel yield and operating costs (methanol, biomass price, activated carbon). ***(viability calculations based on biodiesel sales only)***	[75]

5.3 The environmental impacts assessment of microwave-assisted pyrolysis processes

Environmental sustainability is evaluated by subjecting the microwave-assisted pyrolysis value chain to a life cycle analysis (LCA) to identify and quantify the impacts of running this process from the beginning to the end of its useful life. Although the LCA is traditionally applied to study environmental concerns, it has been recently applied to cover the other two pillars of sustainability: the social and the economic aspects. The standard LCA procedures such as ISO 14040 and 14044 list about seventeen (17) possible environmental indicators although some are more applicable to other processes and irrelevant to other processes. A snapshot of previous LCA studies on microwave-assisted pyrolysis is given in Table 7 to show what could be of importance in ensuring the sustainability of future microwave-assisted pyrolysis projects.

Materials Research Foundations 189 (2026) 284-411 https://doi.org/10.21741/9781644904039-12

Table 7: The environmental sustainability evaluation of microwave-assisted pyrolysis

Project description and LCA approach	Key findings	References
100kW biomass to diesel plant. Emissions from conventional pyrolysis (CFP) were compared to those from microwave-assisted-pyrolysis. Cradle-to-Gate. Functional unit: 1 year production.	Emissions were 917 and 895 tCO_2/yr respectively for CFP and microwave-assisted pyrolysis plants indicating microwave-assisted pyrolysis as more environmentally friendly.	[78]
Microwave-assisted catalytic pyrolysis of forest residues for co-production of upgraded biofuel and high-quality biochar. Cradle-to-Gate. Functional unit: 1MJ of energy consumed.	The carbon intensity ranged -57.3 to 27.4 gCO2-eq/MJ when using either low-carbon electricity or carbon-intense electricity. Compared to petroleum fuels, a 43% to 162% reduction of GHG emissions could be achieved depending on electricity mix and methods to treat biochar coproduct.	[79]
Producing biodiesel from bio-oil derived from the microwave-assisted pyrolysis of pine sawdust. Cradle-to-Gate. Functional unit: 1 kg biodiesel produced.	Global warming potential stood at 1.18 kg CO2 while photochemical oxidant formation was 0.71 kg NMVOC and human toxicity of 2.46 kg 1,4-DCB. A similar study (biofuel from forest biomass) using conventional pyrolysis coupled with different bio-oil upgrading routes resulted in higher GWPs than those of fossil fuel for some of the upgrading routes.	[80,81]

From the ensuing studies, it turns out that microwave-assisted pyrolysis could be a much more favorable route than conventional pyrolysis for better environmental performance. However, variations in the LCA methodology for example choices of functional units, boundaries, site specific variances such as energy costs, biomass variability make comparison of results from different studies inconclusive. More studies to clear out these uncertainties and inconsistencies are required to guide informed comparisons and decisions around the most sustainable choice of pyrolysis technology for each unique circumstance.

6. Concluding Remarks and Future Outlook

To accelerate the adoption of microwave-assisted pyrolysis in biomass processing for value-added chemicals and fuels, research should be directed towards the development of catalysts that enhance product yield and selectivity, particularly those that integrate microwave absorbents to improve heating efficiency and reaction specificity. Integrating microwave-assisted pyrolysis with existing biorefinery processes can optimize resource utilization and align with circular bioeconomy principles, thus enhancing environmental sustainability. Addressing technical barriers highlighted in the technical assessment of microwave-assisted pyrolysis is crucial for scaling up this technology. Innovations in reactor design, including hybrid systems that combine microwave energy with other heating methods, offer potential for improved process efficiency, feedstock flexibility, improved product quality. Furthermore, supportive policy frameworks and regulatory guidelines are essential to overcome commercialization barriers and stimulate investment in microwave-assisted pyrolysis technologies. By focusing on these critical areas, microwave-assisted pyrolysis can significantly contribute to sustainable energy transitions and the production of high-value renewable chemicals. Smart microwave-assisted pyrolysis systems employing real-time sensors, AI algorithms, and digital twins enable adaptive control over process parameters, optimizing performance under variable feedstock conditions. With abundant global biomass resources, microwave-assisted pyrolysis offers a viable and sustainable pathway toward a circular, low-carbon bioeconomy. Crucially, the scalability of microwave-assisted pyrolysis hinges on the resolution of technical barriers such as non-uniform microwave energy distribution, limitations in reactor materials, and the need for high-power, cost-effective microwave sources. Addressing these challenges requires interdisciplinary approaches involving materials science, process engineering, and real-time process control technologies such as AI and smart sensors. Reactor design innovations, including continuous-flow, multi-zone, and fluidized bed systems, will play a central role in enabling industrial-scale deployment. Policy frameworks and supportive regulations are also essential to facilitate the commercial adoption of microwave-assisted pyrolysis technologies. Strategic investment, combined with clear sustainability standards and incentives for bio-based energy, will be key in attracting industry participation and fostering technological development. By addressing these interconnected research opportunities, microwave assisted pyrolysis can evolve from a promising laboratory technique to a cornerstone of next-generation renewable energy and chemical production.

References

[1] E.Y. Shaba, M.J. Jiya, A. Andrew, A.M. Salihu, E. Mamma, S.K. Anyanwu, J.T. Mathew, A. Inobeme, B.N. Yisa, and J.J. Saba, Biomass valorization: A sustainable approach towards carbon neutrality and circular economy, in: R.T. Kapoor, M. Sillanpaa, J. Zdarta, and M. Rafatullah (Eds.), Biomass valorization, Springer, Singapore, 2024, pp. 99-122. https://doi.org/10.1007/978-981-97-8557-5_6

[2] U. Tyagi, N. Anand, A.K. Jain, D. Garg, Forestry biomass as carbon neutral source for the production of biofuels and aromatics, in: A.K. Tiwari and D.B. Pal (Eds.), Sustainable valorization of agriculture and food waste biomass: Application in bioenergy and useful chemicals, Springer, Singapore, 2023, pp. 99-122. https://doi.org/10.1007/978-981-99-0526-3_5

[3] D.C. Makepa, C.H. Chihobo, Sustainable pathways for biomass production and utilization in carbon capture and storage-a review. Biomass Convers. Bior. 8 (2024) 1-23 https://doi.org/10.1007/s13399-024-06010-5

[4] D.C. Makepa, C.H. Chihobo, M. Downmore, Technological advancements in biofuel, bioproducts and bioenergy production from pyrolysis of lignocellulosic biomass, in: B. Gurunathan, G. Venkatesh (Eds.), Biofuels and bioenergy: Biorefinery and circular bioeconomy approaches, Elsevier, Philadelphia, 2025, pp. 85-119. https://doi.org/10.1016/B978-0-443-29254-5.00005-9

[5] G.R. Mong, C.T. Chong, W.W.F. Chong, J.H. Ng, H.C. Ong, V. Ashokkumar, M.V. Tran, S. Karmakar, B.H.H. Goh, M.F. Mohd-Yasin, Progress and challenges in sustainable pyrolysis technology: Reactors, feedstocks and products, Fuel, 324 (2022) 1-20. https://doi.org/10.1016/j.fuel.2022.124777

[6] W.A. Rasaq, M. Golonka, M. Scholz, A. Białowiec, Opportunities and challenges of high-pressure fast pyrolysis of biomass: A review. Energies, 14 (2021) 1-20. https://doi.org/10.3390/en14175426

[7] P.A. Ramírez Cabrera, A.S. Lozano Pérez, C.A. Guerrero Fajardo, Design of a semi-continuous microwave system for pretreatment of microwave-assisted pyrolysis using a theoretical method, Inventions. 10 (2025) 1-25. https://doi.org/10.3390/inventions10020024

[8] J. Ren, X. La, J. Wang, J. Jiang, Beyond microwave susceptors: Exploring 5525 MHz frequency for efficient biomass pyrolysis, Fuel. 372 (2024) 1-11. https://doi.org/10.1016/j.fuel.2024.132220

[9] X. Ren, M. Shanb Ghazani, H. Zhu, W. Ao, H. Zhang, E. Moreside, J. Zhu, P. Yang, N. Zhong, and X. Bi, Challenges and opportunities in microwave-assisted catalytic pyrolysis of biomass: A review, Appl. Energy. 315 (2022) 118970. https://doi.org/10.1016/j.apenergy.2022.118970

[10] F.M. G.Ghazi, M. Abbaspour, R.R. Mohammad, Biofuel production from syngas, in: R.R. Mohammad, A.M. Mohammad, M. Maryam (Eds.), Advances in synthesis gas: methods, technologies and applications, Volume 3: Syngas products and usages, 2023, pp. 271-286. https://doi.org/10.1016/B978-0-323-91878-7.00011-3

[11] R. De La Cruz Iturbides, U. Jauregui Haza, and I. Polaert, Recent technological innovations on continuous microwave-assisted biomass pyrolysis and perspectives for industrial-scale applications, Bioresour. Technol. Reports. 19 (2022) 1-14. https://doi.org/10.1016/j.biteb.2022.101202

[12] H. Luo, Y. Zhang, H. Zhu, X. Zhao, L. Zhu, W. Liu, M. Sun, G. Miao, S. Li, and L. Kong, Microwave-assisted low-temperature biomass pyrolysis: From mechanistic insights to pilot scale, Green Chem. 23 (2021) 821-827. https://doi.org/10.1039/D0GC03348K

[13] N. Zhou, J. Zhou, L. Dai, F. Guo, Y. Wang, H. Li, W. Deng, H. Lei, P. Chen, Y. Liu, and R. Ruan, Syngas production from biomass pyrolysis in a continuous microwave-assisted pyrolysis system, Bioresour. Technol. 314 (2020) 1-8. https://doi.org/10.1016/j.biortech.2020.123756

[14] Y. Duan, P. Yuan, S. Huang, L. Wang, J. Deng, S. Yu, Q. Xie, and Y. Nie, Experimental study on reactor scale-up for microwave-assisted pyrolysis of methyl ricinoleate, Chem. Eng. Process Intensif. 184, (2023) 1-7. https://doi.org/10.1016/j.cep.2023.109293

[15] S. Anju, K. Anamika, S. Suparna, G. Shivani, K. Shakti, K. Anil, N.B. Siddhartha, G. Lalit, and C. M. Hussain, Recent advancement in microwave-assisted pyrolysis for biooil production in: C. M. Hussain, S. Singh, and L. Goswami (Eds.), Waste-to-energy approaches towards zero waste interdisciplinary methods of controlling, Elsevier, Bengaluru, 2022, pp. 197-219. https://doi.org/10.1016/B978-0-323-85387-3.00014-8

[16] X. Yuan, M. Sun, C. Wang, and X. Zhu, Full temperature range study of rice husk bio-oil distillation: Distillation characteristics and product distribution, Sep. Purif. Technol. 263 (2021) 1-9. https://doi.org/10.1016/j.seppur.2021.118382

[17] D. Mohan, C.U. Pittman Jnr, and P.H. Steele, Pyrolysis of wood/biomass for bio-oil: A critical review, Energy and Fuels. 20 (2006) 848-889. https://doi.org/10.1021/ef0502397

[18] A.V. Bridgwater, Review of fast pyrolysis of biomass and product upgrading, Biomass Bioenergy. 38 (2012) 68-94. https://doi.org/10.1016/j.biombioe.2011.01.048

[19] J.S. Kim, Production, separation and applications of phenolic-rich bio-oil: A review, Bioresour. Technol. 178 (2015) 90-98. https://doi.org/10.1016/j.biortech.2014.08.121

[20] S. Czernik and A.V. Bridgwater, Overview of applications of biomass fast pyrolysis oil, Energy and Fuels 18 (2004) 590-598. https://doi.org/10.1021/ef034067u

[21] A. Oasmaa, Bio-oil stabilization in fast pyrolysis of biomass, Adv. Sci. Technol. vol (2017), pp. 138-159. https://doi.org/10.1039/9781788010245-00138

[22] M. Patel, A. Kumar, Production of renewable diesel through the hydroprocessing of lignocellulosic biomass-derived bio-oil: A review. Renew. Sustain. Energy Rev. 58 (2016) 1293-1307. https://doi.org/10.1016/j.rser.2015.12.146

[23] Y.H. Chan, S.K. Loh, B.L.F. Chin, C.L. Yiin, B.S. How, K.W. Cheah, M.K. Wong, A.C. M. Loy, Y.L. Gwee, S.L.Y. Lo, S. Yusup, S.S. Lam, Fractionation and extraction of bio-oil for production of greener fuel and value-added chemicals: Recent advances and future prospects, Chem. Eng. J. 397 (2020) 1-30. https://doi.org/10.1016/j.cej.2020.125406

[24] P. Onsekizoglu, Membrane Distillation: Principle, advances, limitations and future prospects in food industry, in: S. Zereshki (Eds.), Distillation - Advances from modeling to applications, InTech, London (2012) pp 233-266. https://doi.org/10.5772/37625

[25] S. Roy and N.R. Singha, Polymeric nanocomposite membranes for next generation pervaporation process: Strategies, challenges and future prospects, Membranes, 7 (2017) 1-64. https://doi.org/10.3390/membranes7030053

[26] X. Wu, H. Zhang, C. Hu, X. Bao, P. Yuan, Advances in enhancing hydrodeoxygenation selectivity of lignin-derived oxygenates: From synthetic strategies to fundamental techniques, Green Energy Environ. 10 (2024) 292-321. https://doi.org/10.1016/j.gee.2024.05.004

[27] L. Qin, J. Li, S. Zhang, Z. Liu, S. Li, and L. Luo, Catalytic performance of Ni-Co/HZSM-5 catalysts for aromatic compound promotion in simulated bio-oil upgrading, RSC Adv. 13 (2023) 7694-7702. https://doi.org/10.1039/D2RA07706J

[28] A. Farooq, H. Shafaghat, J. Jae, S.C. Jung, and Y.K. Park, Enhanced stability of bio-oil and diesel fuel emulsion using Span 80 and Tween 60 emulsifiers, J. Environ. Manage. 231 (2019) 694-700. https://doi.org/10.1016/j.jenvman.2018.10.098

[29] K. Du, B. Yu, Y. Xiong, L. Jiang, J. Xu, Y. Wang, S. Su, S. Hu, J. Xiang, Hydrodeoxygenation of bio-oil over an enhanced interfacial catalysis of microemulsions stabilized by amphiphilic solid particles, Catalysts 13 (2023) 1-18. https://doi.org/10.3390/catal13030573

[30] D.S. Fardyanti, Megawati, H. Istanto, M.K. Anajib, Prayogo, U. Habibah, Extraction of phenol from bio-oil produced by pyrolysis of coconut shell, J. Phys. Sci. 29 (2018) 195-202. https://doi.org/10.21315/jps2018.29.s2.15

[31] G. Gómez Millán, R.P. Bangalore Ashok, P. Oinas, J. Llorca, H. Sixta, Furfural production from xylose and birch hydrolysate liquor in a biphasic system and techno-economic analysis, Biomass Convers. Biorefinery. 11 (2021) 2095-2106. https://doi.org/10.1007/s13399-020-00702-4

[32] P.R. Patwardhan, J.A. Satrio, R.C. Brown, B.H. Shanks, Product distribution from fast pyrolysis of glucose-based carbohydrates, J. Anal. Appl. Pyrolysis 86 (2009) 323-330. https://doi.org/10.1016/j.jaap.2009.08.007

[33] T. Sarchami, N. Batta, F. Berruti, Production and separation of acetic acid from pyrolysis oil of lignocellulosic biomass: A review, Biofuels, Bioprod. Biorefining 15 (2021) 1912-1937. https://doi.org/10.1002/bbb.2273

[34] J. Reichert, J. Albert, Detailed kinetic investigations on the selective oxidation of biomass to formic acid (OxFA Process) using model substrates and real biomass. ACS Sustain. Chem. Eng. 5 (2017) 7383-7392. https://doi.org/10.1021/acssuschemeng.7b01723

[35] A. Kumar, D.Z. Shende, K.L. Wasewar, Sustainable process development for the recovery of biobased levulinic acid through reactive extraction, J. Chem. Technol. Biotechnol. 98 (2023) 516-525. https://doi.org/10.1002/jctb.7265

[36] C. Nzediegwu, M. Arshad, A. Ulah, M.A. Naeth, and S.X. Chang, Fuel, thermal and surface properties of microwave-pyrolyzed biochars depend on feedstock type and pyrolysis temperature, Bioresour. Technol. 320 (2021) 1-11. https://doi.org/10.1016/j.biortech.2020.124282

[37] R. Potnuri, D. Venkata, C. Sankar, A. Yadav, V. Sridevi, and N. Remya, A review on analysis of biochar produced from microwave-assisted pyrolysis of agricultural waste biomass, J. Anal. Appl. Pyrolysis 173 (2023) 106094. https://doi.org/10.1016/j.jaap.2023.106094

[38] K. Weber and P. Quicker, Properties of biochar, Fuel 217 (2018) 240-261. https://doi.org/10.1016/j.fuel.2017.12.054

[39] A. Atiyyah Ameenah, K.W.A.A.K. Ghani, A. Idris, M.F.Z.Ja'afar, M.A.M. Salleh, N.S. Ishak, Microwave-assisted pyrolysis of EFB-derived biochar as potential renewable solid fuel for power generation: Biochar versus sub-bituminous coal, Renew. Energy 142 (2019) 123-129. https://doi.org/10.1016/j.renene.2019.04.035

[40] X. Li, B. Peng, Q. Liu, and H. Zhang, Microwave pyrolysis coupled with conventional pre-pyrolysis of the stalk for syngas and biochar. Bioresour. Technol. 348 (2022) 1-8. https://doi.org/10.1016/j.biortech.2022.126745

[41] T. Haeldermans, J. Claesen, J. Maggen, R. Carleer, J. Yperman, and P. Adriaensens, Microwave-assisted and conventional pyrolysis of MDF-Characterization of the produced biochars, J. Anal. Appl. Pyrolysis 138 (2019) 218-230. https://doi.org/10.1016/j.jaap.2018.12.027

[42] L. Zhu, H. Lei, L. Wang, G. Yadavalli, X. Zhang, Y. Wei, Y. Liu, D. Yan, S. Chen, and B. Ahring, Biochar of corn stover: Microwave-assisted pyrolysis condition induced changes in surface functional groups and characteristics (2015) 1-40. https://doi.org/10.1016/j.jaap.2015.07.012

[43] C. Long, H. Jian, H. Qian, Z. Jing, Z. Zhi, Z. Nan, Z. Huoqiang, L. Xiangyang, M. Baobin, and W. Fangfang, Resolving the enhancement effect of microwave-assisted pyrolysis on biochar redox properties from the structure-activity relationship, J. Anal. Appl. Pyrolysis 167 (2022) 1-11. https://doi.org/10.1016/j.jaap.2022.105706

[44] F. Qin, C. Zhang, G. Zeng, D. Huang, X. Tan, and A. Duan, Lignocellulosic biomass carbonization for biochar production and characterization of biochar reactivity, Renew. Sustain. Energy Rev. 157 (2022) 1-25. https://doi.org/10.1016/j.rser.2021.112056

[45] X. Wang, W. Morrison, Z. Du, Y. Wan, X. Lin, P. Chen, and R. Ruan, Biomass temperature profile development and its implications under the microwave-assisted pyrolysis condition, Appl. Energy 99 (2012) 386-392. https://doi.org/10.1016/j.apenergy.2012.05.031

[46] M. Boumaaza, A. Belaadi, M. Bourchak, K.A. Juhany, M. Jawaid, M.T. Marvila, and A.R.G. de Azevedo, Optimization of flexural properties and thermal conductivity of Washingtonia plant biomass waste biochar reinforced bio-mortar, J. Mater. Res. Technol. 23 (2023) 3515-3536. https://doi.org/10.1016/j.jmrt.2023.02.009

[47] W.J. Liu, H. Jiang, H.Q. Yu, Development of biochar-based functional materials: Toward a sustainable platform carbon material, Chem. Rev. 115 (2015) 12251-12285. https://doi.org/10.1021/acs.chemrev.5b00195

[48] M. Gęca, A.M. Khalil, M. Tang, A.K. Bhakta, Y. Snoussi, P. Nowicki, M. Wiśniewska, M.M. Chehimi, Surface treatment of biochar-Methods, surface analysis and potential applications: A comprehensive review, Surfaces. 6 (2023) 179-213. https://doi.org/10.3390/surfaces6020013

[49] A. Yahyazadeh, A.K. Dalai, W. Ma, and L. Zhang, Fischer-Tropsch Synthesis for light olefins from syngas: A review of catalyst development, Reactions. 2 (2021) 227-257. https://doi.org/10.3390/reactions2030015

[50] A. Roman, F. Bucura, O.R. Botoran, G.L. Radu, M. Constantinescu, Pyrolysis and gasification of energy crops for phytoremediation in Romania's coal mining region, Int. J. Green Energy. 5075 (2025) 2469139. https://doi.org/10.1080/15435075.2025.2469139

[51] G. Lopez, L. Santamaria, A. Lemonidou, S. Zhang, C. Wu, A.T. Sipra, N. Gao, Hydrogen generation from biomass by pyrolysis, Nat Rev Methods Prim. 2 (2022) 1-13. https://doi.org/10.1038/s43586-022-00097-8

[52] M. Yue, H. Lambert, E. Pahon, R. Roche, S. Jemei, D. Hissel, Hydrogen energy systems: A critical review of technologies, applications, trends and challenges, Renew. Sustain. Energy Rev. 146 (2021) 1-21. https://doi.org/10.1016/j.rser.2021.111180

[53] S. Mutsengerere, C.H. Chihobo, D. Musademba, I. Nhapi, A review of operating parameters affecting bio-oil yield in microwave pyrolysis of lignocellulosic biomass, Renew. Sustain. Energy Rev. 104 (2019) 328-336. https://doi.org/10.1016/j.rser.2019.01.030

[54] Y. Zhang, Y. Liang, S. Li, Y. Yuan, D. Zhang, Y. Wu, H. Xie, K. Brindhadevi, A. Pugazhendhi, C. Xia, A review of biomass pyrolysis gas: Forming mechanisms, influencing parameters, and product application upgrades, Fuel. 347 (2023) 1-10. https://doi.org/10.1016/j.fuel.2023.128461

[55] S. Ethaib, R. Omar, D.R.A. Biak, S.M.M. Kamal, S. L. Zubaidi, Microwave-assisted pyrolysis of biomass waste: A mini Review, Processes. 8 (2020) 1-17. https://doi.org/10.3390/pr8091190

[56] A. Ramos, E. Monteiro, A. Rouboa, Biomass pre-treatment techniques for the production of biofuels using thermal conversion methods - A review, Energy Convers. Manag. 270 (2022) 1-18. https://doi.org/10.1016/j.enconman.2022.116271

[57] M. He, J. Zhao, D. Wang, Q. Liang, T. Wang, S. Zhao, W. Ma, Microwave-assisted catalytic pyrolysis of biomass with biochar materials derived from spent lithium-ion batteries: Microwave absorption and pyrolysis characteristics, J. Environ. Chem. Eng. 12 (2024) 112099. https://doi.org/10.1016/j.jece.2024.112099

[58] J. Chaouki, S. Farag, M. Attia, and J. Doucet, The development of industrial (thermal) processes in the context of sustainability: The case for microwave heating, Can. J. Chem. Eng. 98 (2020) 832-847. https://doi.org/10.1002/cjce.23710

[59] H. Su, S. Wang, W. Liao, R. Gan, Y. Ran, Q. Zhao, L. Fang, Y. Zhang, Synergistic activation of inert iron oxide basal planes through heterostructure formation and doping for efficient hydrogen evolution, Chem. - A Eur. J. 29 (2023) 1-8. https://doi.org/10.1002/chem.202302774

[60] T. Saeki, H. Ohkita, N. Kakuta, T. Mizushima, Effects of metal oxide support and non-noble metal active species on catalytic steam reforming of ethanol for hydrogen production, J. Japan Pet. Inst. 58 (2015) 341-350. https://doi.org/10.1627/jpi.58.341

[61] P. Haro, F. Johnsson, H. Thunman, Improved syngas processing for enhanced Bio-SNG production: A techno-economic assessment, Energy 101 (2016) 380-389. https://doi.org/10.1016/j.energy.2016.02.037

[62] A. Tremel, M. Gaderer, H. Spliethoff, Small-scale production of synthetic natural gas by allothermal biomass gasification, Int. J. Energy Res. 37 (2013) 1318-1330. https://doi.org/10.1002/er.2933

[63] P.K. Ghodke, A.K. Sharma, A. Jayaseelan, K.P. Gopinath, Hydrogen-rich syngas production from the lignocellulosic biomass by catalytic gasification: A state of art review on advance technologies, economic challenges, and future prospectus, Fuel 342 (2023) 1-18. https://doi.org/10.1016/j.fuel.2023.127800

[64] T. Fukuyama, S. Maetani, I. Ryu, Carbonylation and decarbonylation reactions, in P. Knochel (Eds.), Comprehensive organic synthesis, 3 (2014) 1073-1100. https://doi.org/10.1016/B978-0-08-097742-3.00328-1

[65] I.J. Siddique, A.A. Salema, E. Antunes, R. Vinu, Technical challenges in scaling up the microwave technology for biomass processing, Renew. Sustain. Energy Rev. 153 (2022) 1-21. https://doi.org/10.1016/j.rser.2021.111767

[66] B. Zhao, X. Zhang, L. Sun, G. Meng, L. Chen, Y. Xiaolu, Hydrogen production from biomass combining pyrolysis and the secondary decomposition, Int. J. Hydrogen Energy 35 (2010) 2606-2611. https://doi.org/10.1016/j.ijhydene.2009.04.011

[67] J.C.S. Leite, M.J. Suota, L.P. Ramos, M.K. Lenzi, L.F.L. Luz, Development of a microwave-assisted bench reactor for biomass pyrolysis using hybrid heating, ACS Omega 9 (2024) 24987-24997. https://doi.org/10.1021/acsomega.4c02050

[68] J. Robinson, C. Dodds, A. Stavrinides, S. Kingman, J. Katrib, Z. Wu, J. Medrano, R. Overend, Microwave pyrolysis of biomass: Control of process parameters for high pyrolysis oil yields and enhanced oil quality, Energy and Fuels. 29 (2015) 1701-1709 https://doi.org/10.1021/ef502403x

[69] S. Ge, Y. Shi, C. Xia, Z. Huang, M. Manzo, L. Cai, H. Ma, S. Zhang, J. Jiang, C. Sonne, and S.S. Lam, Progress in pyrolysis conversion of waste into value-added liquid pyro-oil, with focus on heating source and machine learning analysis, Energy Convers. Manag. 245 (2021) 1-22. https://doi.org/10.1016/j.enconman.2021.114638

[70] T. Krapivnitckaia, S. Ananicheva, A. Alyeva, A. Denisenko, M. Glyavin, N. Peskov, A. Vikharev, A. Sachkova, S. Zelentsov, and N. Shulaev, Theoretical and experimental demonstration of advantages of microwave peat processing in comparison with thermal exposure during pyrolysis, Processes. 12 (2024) 1-12. https://doi.org/10.3390/pr12010092

[71] E. Antunes, M V. Jacob, G. Brodie, P.A. Schneider, Microwave pyrolysis of sewage biosolids: Dielectric properties, microwave susceptor role and its impact on biochar properties, J. Anal. Appl. Pyrolysis 129 (2018) 93-100. https://doi.org/10.1016/j.jaap.2017.11.023

[72] D. Beneroso, T. Monti, E.T. Kostas, J. Robinson, Microwave pyrolysis of biomass for bio-oil production: Scalable processing concepts, Chem. Eng. J. 316 (2017) 481-498. https://doi.org/10.1016/j.cej.2017.01.130

[73] L. Wang, H. Lei, and R. Ruan, Techno-economic analysis of microwave-assisted pyrolysis for production of biofuels, in: Z. Fang (Eds.), Production of biofuels and chemicals with microwave. 2015, Springer Science, Dordrecht, pp. 251-263. https://doi.org/10.1007/978-94-017-9612-5_12

[74] S. Mari Selvam, P. Balasubramanian, M. Chintala, and L.K.S. Gujjala, Techno-economic analysis of microwave pyrolysis of sugarcane bagasse biochar production, Biomass Convers. Biorefinery 15 (2025) 14229-14239. https://doi.org/10.1007/s13399-024-06232-7

[75] D. C. Makepa, Techno-Economic Analysis and Life-cycle assessment of bio-oil production from microwave-assisted pyrolysis of pine sawdust, MSc Thesis. Chinhoyi University of Technology, 2023. https://doi.org/10.1016/j.heliyon.2023.e22261

[76] S. Ying, Y. Herng, P. Nai, Y. Yek, S. Sow, M. Lock, B. Lai, F. Chin, C. Loong, J.C. Lan, and S. Shiung, Microwave-assisted pyrolysis in biomass and waste valorisation: Insights into the life-cycle assessment (LCA) and techno-economic analysis (TEA). Chem. Eng. J. 491 (2024) 1-11.

[77] A.E.M. Fodah, M.K. Ghosal, D. Behera, Microwave-assisted pyrolysis of agricultural residues: current scenario, challenges, and future direction, Int. J. Environ. Sci. Technol. 19 (2022) 2195-2220. https://doi.org/10.1007/s13762-020-03099-9

[78] P. Hansirisawat, T.R. Srinophakun, Techno-economics of 100 kW power plant from microwave-assisted biodiesel pyrolysis, Int. J. Renew. Energy Res. 10 (2020) 1021-1030.

[79] H. Zhu, J. Saddler, X. Bi, An economic and environmental assessment of biofuel produced via microwave-assisted catalytic pyrolysis of forest residues, Energy Convers. Manag. 263 (2022) 1-13. https://doi.org/10.1016/j.enconman.2022.115723

[80] D.N. Vienescu, J. Wang, A. Le Gresley, J.D. Nixon, A life cycle assessment of options for producing synthetic fuel via pyrolysis, Bioresour. Technol. 249 (2018) 626-634. https://doi.org/10.1016/j.biortech.2017.10.069

[81] D.C. Makepa, H.C. Chihobo, T.T. Manhongo, D. Musademba, Life-cycle assessment of microwave-assisted pyrolysis of pine sawdust as an emerging technology for biodiesel production, Results Eng. 20 (2023) 1-9. https://doi.org/10.1016/j.rineng.2023.101480

Materials Research Foundations 189 (2026) 412-434 https://doi.org/10.21741/9781644904039-13

Chapter 13

Microwave-Assisted Edible Film and Coating Material Development

Abdullah Kurt[1]*

[1]Department of Food Engineering, Akşehir Faculty of Engineering and Architecture, Selçuk University, 42550 Konya, Türkiye

abdullahkurt48@gmail.com

Abstract

Microwave technology provides a fast, energy-efficient approach to improve edible films and coatings for food packaging. Through dipole rotation and ionic conduction, microwaves enable rapid volumetric heating that dries films, homogenizes solutions, and modifies polysaccharide, protein, and composite matrices. Reported outcomes include reduced drying time, stronger networks, and improved barrier, mechanical, optical, and antimicrobial properties. Microwaves also enhance the extraction and activation of biopolymers from agro-industrial by-products. Practical issues such as hot spots and penetration depth are minimized when treating fluid systems or combining microwaves with ultrasound or vacuum, supporting industrial scalability of bio-based edible films.

Keywords

Innovative Packaging, Dielectric Heating, Biopolymers, Modification, Functionalization

Contents

Materials Research Foundations 189 (2026) 412-434 https://doi.org/10.21741/9781644904039-13

1. Introduction

Safe and nutritious food is a fundamental requirement for human health. Food quality is influenced by the nature of raw materials, the processing methods applied, and the stability of products throughout storage and distribution. To maintain safety, nutritional value, and sustainability, the food industry increasingly explores alternative raw materials and innovative processing technologies. Within this context, edible films and coatings have become important tools for extending the shelf life of fresh and processed foods. They also offer an eco-friendly alternative to petroleum-based plastics, since they are biodegradable and safe for both consumers and the environment. By acting as protective layers, these films and coatings help maintain food quality, facilitate storage and transportation, and reduce waste [1].

The food industry has widely adopted plastic as a packaging material because of its low cost, flexibility, chemical stability, and ability to be molded into various forms [2]. In Europe, about 40% of all plastic is currently used for packaging, a trend largely driven by the expansion of retail markets and global food supply chains [3,4]. The most common polymers in food packaging include polypropylene, polyethylene, polyamide, polystyrene, and polyethylene terephthalate. Environmental pollution caused by discarded plastics is often referred to as "white pollution." Most plastic products have a very short service life, typically being used only once. This widespread single use contributes to white pollution and creates major problems in waste management. The persistence of plastics arises from their strong covalent bonds and unique structural properties, which make them biologically non-degradable [5-7].

Edible packaging prepared from natural biopolymers that interact with microwaves has become an environmentally friendly approach to controlling microbial growth, enzymatic activity, and lipid oxidation. Such materials also help reduce moisture loss in highly perishable foods, improve sensory quality, and extend shelf life [8]. Films and coatings create a physical barrier between food and the external environment, limiting the transfer of moisture, oxygen, and soluble substances while protecting the original ingredients. As a thin layer of primary packaging, coatings can even be consumed with the food itself. These packaging materials are produced from isolated biopolymers such as proteins, polysaccharides, lipids, and their composites, as well as from food processing by-products. They often include additives such as plasticizers, cross-linkers, and functional agents. Most of these biopolymers are obtained from plants and are considered sustainable and eco-friendly. Additional sources include biopolymers derived from microorganisms (e.g., polyhydroxyalkanoates) and those synthesized biotechnologically, such as polylactic acid (PLA) and polyglycolic acid. Coatings can be applied to food by immersion, spraying, brushing, or electrospraying, while films are used as solid layers covering the surface of foods [7,9].

Enhancing the functional properties of edible films is essential to broaden their applications in the food industry, particularly by modulating their physical, barrier, mechanical, and chemical features. The use of coatings or films on post-harvest produce and during processing is now common in both research and industrial practice. These coatings help maintain freshness, preserve sensory attributes such as texture, flavor, and appearance, and provide protection against microbial contamination. Consequently, they extend the shelf life of fresh fruits and vegetables [10]. In recent years, both thermal and non-thermal treatments have been widely studied for their effects on biopolymer-based films and coatings. The development of functional films through modification techniques often involves changes in the physiological, conformational, and functional characteristics of film-forming solutions. Such modifications may include the creation of new bonds and cross-linkages or the enhancement of intermolecular electrostatic repulsion [11].

New technologies such as microwaves are applied in many areas, including quality improvement, heat treatment, energy efficiency, protection, texture modification, and analytical or extraction processes. In food packaging, microwaves are also used to extract or isolate bioactive compounds and to synthesize nanoparticles before their incorporation into film formulations for active packaging. In this chapter, however, the focus is on the direct application of microwave treatment to film-forming and coating solutions, either before or after they are applied to a food surface. The underlying mechanisms and their effects on the structural and functional properties of the final films are examined in detail. In addition, when biopolymers are the primary materials used for films and coatings, studies on the microwave-assisted isolation of these polymers are also considered.

2. Microwave

Microwave technology was first developed for radar systems during World War II. Since then, it has been applied in many fields, including ceramics, polymers, and food processing. Today, microwave heating is widely used in food processing operations such as drying, pasteurization, sterilization, thawing, tempering, and baking. Its selective, localized, and easily controllable heating, combined with environmentally friendly characteristics, makes it a strong alternative to conventional thermal technologies in the food industry [12].

Microwave radiation is a non-ionizing electromagnetic energy used as an innovative technique in food processing. It generates internal heating in polar materials by interacting with dipolar molecules under alternating electromagnetic fields. Microwaves operate in the 300 MHz to 300 GHz range, with 915 MHz and 2450 MHz being the most common frequencies in the food industry [13]. Lower frequencies penetrate materials more effectively. At the molecular level, oscillations, bond disruptions, intermolecular friction, and collisions occur, promoting the alignment of reactive groups and facilitating chemical reactions [14]. Microwaves provide a clean, rapid, convenient, and cost-effective heating method. They can also produce non-thermal effects by polarizing macromolecules. Because of their unique dielectric properties, different foods absorb and distribute electromagnetic energy differently, making microwaves useful for processes such as pre-cooking, tempering, and drying [15]. The heat generated inside the material moves outward toward the surface, resulting in fast, uniform, and energy-efficient heating [12].

In general, materials with a higher content of polar molecules respond more effectively to microwave processing. The main mechanisms of microwave pretreatment are dipole rotation and ionic conduction, which promote rapid ion migration in the food matrix and lead to thermal effect [16]. The heating efficiency of a material in a dielectric field is described by the dielectric loss tangent ($\tan \delta$). This parameter is the ratio of the dielectric loss factor (ε''), reflecting the ability to convert electromagnetic energy into heat, to the dielectric constant (ε'). Mathematically, $\tan \delta = \varepsilon''/\varepsilon'$, and higher values indicate stronger microwave absorption and greater heat generation [17].

In recent years, research on modifying and characterizing biopolymers for specific applications has increased [18]. The depolymerization of high-molecular-weight hydrocolloids through non-thermal processes is used to alter intermolecular interactions, thereby enhancing existing functionalities, introducing new properties, and improving economic value. Microwave treatment has emerged as an effective alternative for polymer modification [19]. In film and coating production, it can reduce drying time, modify chemical structures, homogenize composite solutions, and improve the functional properties of the materials (Fig. 1). Non-uniformity in microwave processing results from the field distribution, which creates hot and cold spots, with hot spots occurring in areas of higher electromagnetic intensity. However, because film and coating formulations are treated in fluid form, this limitation is less critical [20]. In practice, microwave drying provides a rapid option after films are spread on plates, while conventional drying can take up to 24 hours [21]. Heating occurs through the interaction between polar molecules and microwave energy, allowing penetration and uniform heating beyond the surface. In addition, the varying dielectric properties of food components cause selective and differential heat absorption [22].

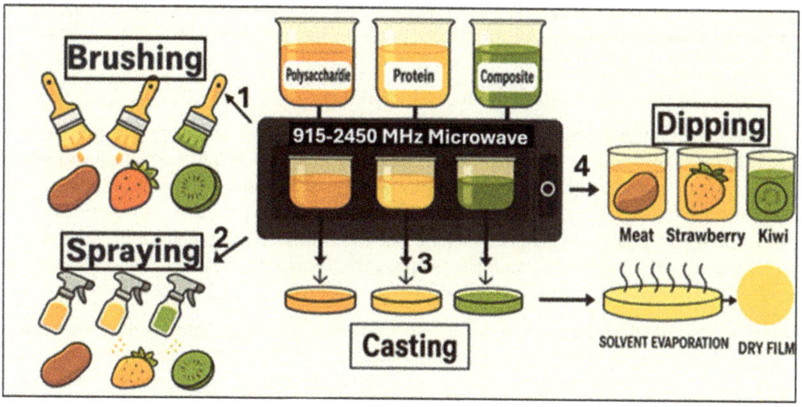

Figure 1. Four application methods of microwave-modified edible film and coating solutions: 1) Brushing, 2) Spraying, 3) Casting, 4) Dipping.

Drying is one of the critical steps in film production and coating applications. Microwave drying, which involves simultaneous heat and mass transfer through volumetric heating, provides an effective alternative to conventional methods. The process generally has two main phases: an initial stage where moisture is removed by evaporation, and a subsequent phase that includes warming, a period of constant moisture loss, and a final stage with a declining drying rate. Compared with conventional drying, microwaves offer several advantages, including lower shrinkage, reduced bulk density, higher rehydration ratio, faster dehydration, and improved energy efficiency [23]. Microwave treatment can also induce the formation of new functional groups and modify polymeric structures, thereby enhancing their functional performance in food applications. As a result, it has been applied in nanoparticle preparation, film formation from food ingredients, and the improvement of emulsion stability [24].

In microwave processing, the drying step can serve two different purposes. In the first approach, the coating is applied to the food and then treated with microwaves, so that the energy acts on both the food substrate and the coating material. This process simultaneously induces thermal modification of the coating matrix and promotes water removal from both the coating and the food

(Fig. 2). In the second approach, the cast film solution is dried directly in a microwave system. This enables rapid moisture removal and, at the same time, drives polymer modification within the forming film (Fig. 3).

Microwave technology also plays an important role in isolating biopolymers and bioactive compounds. It disrupts cell wall structures, allowing solvents to penetrate and release polymers and bioactive molecules. The electric and magnetic fields can also weaken hydrogen bonds, which increases solubility and improves the extraction of valuable substances or the separation of hydrophobic components. Under mild microwave conditions, polar molecules absorb energy rapidly through friction caused by oscillating fields, leading to fast internal heating. When pretreatment intensity increases, high-molecular-weight bioactive compounds may degrade into smaller molecules. This can enhance certain functionalities but may also result in the loss of sensitive compounds. For this reason, optimizing microwave conditions for both the material type and the target compound is essential for selective and efficient extraction [12,24].

3. Polysaccharide-based Edible Films and Coatings

Polysaccharides are natural macromolecular compounds composed of more than ten monosaccharide units linked by glycosidic bonds, and they are found in plants, animals, and microorganisms. These abundant biopolymers are widely used in edible films and coatings because of their physical and chemical properties, film-forming ability, accessibility, biodegradability, and non-toxic nature. Common examples include starch, cellulose, chitin/chitosan, xanthan gum, gellan gum, pullulan, carrageenan, alginate, pectin, galactomannan, glucomannan, and Arabic gum [24].

Polysaccharides contain free hydroxyl groups that promote hydrogen bonding with active agents, producing coatings with excellent oxygen and aroma barrier properties and notable mechanical strength. This performance is partly due to their dense structural organization. However, their inherent hydrophilicity limits packaging integrity under high-humidity conditions. Microwave treatment has become an increasingly popular method to modify polysaccharide functionality by altering structure, enhancing physicochemical characteristics, and improving biological activity for food applications. It can also reduce the hydrophilicity of polysaccharide-based films, thereby improving their stability in moist environment [24,25].

Film-forming solutions prepared with pearl millet starch, vegetable oil, and glycerol, as well as composite mixtures combining native starch with pearl millet starch nanoparticles, were subjected to microwave heating with continuous stirring to ensure uniform gelatinization and film formation. The treatments were applied at 2350 W for 30 to 60 seconds. These methods produced films in which nanoparticle incorporation increased thickness, reduced moisture content, and improved barrier and mechanical properties. However, when nanoparticle levels reached 10%, flexibility and transparency decreased. At moderate concentrations (around 5%), film performance and biodegradability were maintained at optimal levels [26].

Chitosan films have been dried using both conventional convection methods (35 °C for 12 hours) and microwave treatment (2450 MHz, full power, 10 minutes). Microwave drying improved UV–visible light barrier properties without causing major changes in thermal or structural characteristics. In chitosan films containing copper nanoparticles, the particles occupied active binding sites, which enhanced water resistance and promoted faster removal of unbound water during microwave drying [27,28].

Recent research has highlighted the important role of microwave-assisted drying in improving dehydration efficiency and product quality in coated fruits. A study on kiwifruit demonstrated that edible coatings prepared with potato starch, carboxymethyl cellulose (CMC), and pectin, when combined with microwave or microwave–hot air drying, enhanced both efficiency and quality. Microwave treatment at 450 W significantly accelerated drying, improved moisture diffusion, and promoted the retention of bioactive compounds such as vitamin C and total phenolics. In addition, the use of coatings improved the color, texture, and rehydration properties of the dried product [29]. (Fig. 2).

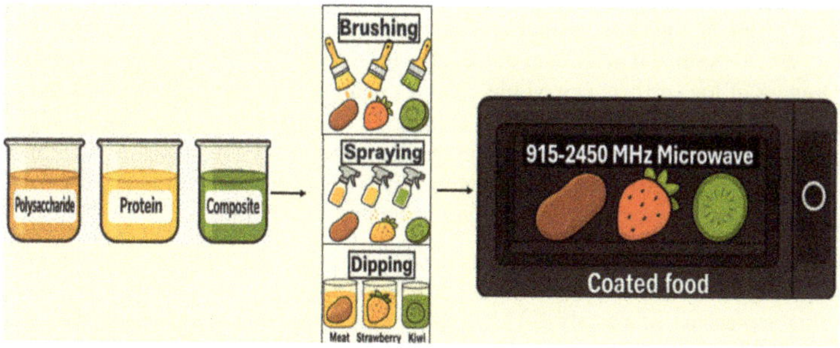

Figure 2. Edible coating applied to food via brushing, spraying, and dipping, followed by microwave treatment of the coated food.

A study investigated the use of edible coatings made of alginate and calcium lactate on strawberries before osmotic dehydration. After coating, the samples were subjected to microwave drying at 1.2 W/g, which resulted in significant reductions in drying time while maintaining selective solute transfer during processing. Microwave treatment improved moisture diffusivity, allowing effective drying to low moisture levels in shorter periods without damaging the coating or degrading the fruit's bioactive compounds [30]. These results indicate that combining microwave energy with functional edible films can improve fruit drying and preservation technologies.

A study developed innovative edible gels that combined sonication with microwave-assisted pasteurization (SMAP) and incorporated citrus peel essential oil to improve the postharvest quality of blackberries (*Rubus fructicosus L.)*. The coating gel formulation included carboxymethyl cellulose (CMC) as the film-forming agent, glycerol as a plasticizer, and Polysorbate 80 (Tween-80) as a non-ionic surfactant to stabilize the emulsion. The SMAP process was carried out at 2450 MHz frequency, 1000 W power, 90 °C, and 60 seconds of microwave treatment. Results indicated that blackberries coated with the SMAP gels showed significantly lower weight loss and decay. They also retained higher levels of flavonoids, phenolics, and antioxidant activity during 21 days of storage at 4 °C. These findings suggest that combining microwave-assisted pasteurization with functional edible coatings is an effective approach for extending fruit shelf life and preserving nutritional quality [31].

Materials Research Foundations 189 (2026) 412-434 https://doi.org/10.21741/9781644904039-13

Microwave radiation used in processing ovens is a type of non-ionizing energy that does not break covalent bonds but interacts with polar molecules such as water, ions, and salts. These molecules continuously adjust their dipole orientation to align with the microwave field. The repeated reorientation generates friction, which contributes to energy dissipation and localized heating within the mixture. In emulsions, water is the main polar component and absorbs microwave energy efficiently. This process disrupts hydrogen bonding and produces heat directly inside the system [31,32].

A recent study investigated the impact of microwave–vacuum drying at 250 W power and reduced pressures on carrots that were pretreated with osmotic enrichment and dehydration using chokeberry juices. The process was combined with edible coatings prepared from sodium alginate and citrus pectin. Applying microwave energy under vacuum significantly improved drying efficiency, resulting in higher dry matter content and lower water activity, particularly at a pressure of 3.5 kPa. When combined with osmotic treatment, microwave drying also increased the total phenolic content of the carrots. The edible coatings had a moderate effect on color attributes and antioxidant activity but did not substantially change the moisture content. Sensory analysis further confirmed that microwave-dried carrots retained desirable texture, color, and aroma. These findings highlight the effectiveness of integrating microwave-assisted vacuum drying with osmotic pretreatment and coating processes to maintain vegetable quality [10].

A recent study demonstrated the effectiveness of microwave irradiation at 300 W for 9 minutes, with stirring every 3 minutes, in a microwave reactor to modify low-methoxy pectin using stearic anhydride in the presence of a catalyst. This technique provided a rapid and efficient way to modify the structure of biopolymers. The microwave-assisted esterification improved grafting efficiency, which in turn enhanced the emulsifying capacity and mechanical strength of the pectin-based films. The incorporation of hydrophobic chains further increased surface hydrophobicity, as confirmed by contact angle measurements. These results show the clear advantages of microwave processing for producing functional edible film materials with enhanced physicochemical properties [33].

Microwave heat treatment is a rapid and eco-friendly physical modification method that alters starch structure by disrupting its molecular arrangement. This process enhances several functional properties of starch, including film formation, mechanical strength, and freeze–thaw stability. Compared with native starch, microwave-modified starch shows reduced hydrophilicity and greater solubility, making it more suitable for food packaging applications [34]. Rising environmental awareness has encouraged the development of biodegradable packaging, such as edible films from breadfruit starch. Microwave treatment significantly improved the physical properties of these films, producing thinner films with lower yellowness (b^*) and higher solubility. Overall, the findings demonstrate that microwave heat treatment is an effective and sustainable approach for enhancing starch-based edible films in eco-friendly packaging applications [35].

Microwave heating has shown strong potential as an alternative drying method for bacterial cellulose (BC) pellicles because of its efficiency and reduced processing time. In a comparative study, BC samples subjected to microwave drying required only 3–5 min, which was about 95% shorter than the time needed for conventional air convection drying. Despite this major reduction in drying time, the structural and mechanical properties of the microwave-dried films remained comparable to those produced by convection. Both retained the cellulose I structure, as confirmed by FTIR and XRD analyses. Microwave-treated films exhibited slightly lower crystallinity and a

Materials Research Foundations 189 (2026) 412-434 https://doi.org/10.21741/9781644904039-13

higher swelling capacity, but their color and tensile strength were similar. These results suggest that microwave drying is an energy-efficient and rapid method for processing BC films without significant loss of quality [36].

In a comparative investigation of drying methods, microwave drying combined with gum Arabic pretreatment was found to be effective in preserving the quality of red beetroot slices. The study showed that freeze-drying resulted in the highest retention of bioactive compounds. However, microwave-dried samples pretreated with gum Arabic also maintained substantial levels of betalain content, antioxidant activity, and sensory properties. The gum Arabic coating reduced the degradation of sensitive compounds and improved the oxidative stability of beetroot extract. These findings suggest that microwave-assisted dehydration, when combined with suitable biopolymer coatings, is a viable method for producing shelf-stable and nutritionally rich dried vegetable products [37].

Microwave-assisted treatment, especially when combined with ultrasonication, improved the functional properties of corn distarch phosphate/corn straw cellulose (CDP/CSC) edible films. The combined ultrasonic and microwave treatment produced a clear increase in tensile strength, elongation at break, and light transmittance. At the same time, it substantially reduced water vapor permeability (WVP). Structural analyses using FTIR, XRD, and SEM confirmed stronger molecular interactions and a more compact, homogeneous film morphology after treatment. These findings show that microwave-based modifications can effectively improve the mechanical and barrier properties of biodegradable films. They also indicate the potential of this approach for advanced food packaging applications [38].

In a study evaluating Na-alginate edible films enriched with traditional Greek alcoholic beverages and 40% ethanol, significant reductions in *Listeria monocytogenes* populations were observed on frankfurters and ham slices during cold storage. However, subsequent microwave reheating of frankfurters did not lead to additional inactivation of the pathogen. This outcome was linked to cross-protection mechanisms developed during storage, including exposure to low pH, indigenous microbiota, and ethanol (~2.4–2.7% w/w). These results emphasize the importance of antimicrobial packaging systems that not only control microbial growth during storage but also prevent the development of increased thermal resistance. Such resistance may reduce the effectiveness of microwave reheating, which is a common consumer practice [39].

A study used microwave irradiation at 640 W for 30 seconds to rapidly gelatinize sago starch, producing a nearly transparent solution suitable for edible film formation. This microwave-assisted process strongly affected the films' water vapor permeability. Permeability decreased as film thickness increased and was further reduced by the incorporation of chitosan microparticles. Films produced by this method and supplemented with EDTA also showed strong antibacterial activity against E. coli. Overall, the results indicate that microwave irradiation is an effective way to develop antimicrobial edible films with improved moisture barrier properties for food packaging applications [40].

In one study, polysaccharides were extracted from loquat (Eriobotrya japonica) seeds using microwave treatment at different power levels (180, 360, and 580 W). The highest yield (48.2 ± 0.53%) was obtained at 580 W. The polysaccharide obtained under this condition showed strong antioxidant activity, including DPPH and hydroxyl radical scavenging capacity as well as ferric-reducing power. It was then incorporated into gelatin-based edible coatings. When applied to rainbow trout (*Oncorhynchus mykiss*) fillets, these coatings significantly delayed microbial

Materials Research Foundations 189 (2026) 412-434 https://doi.org/10.21741/9781644904039-13

spoilage and chemical degradation during 12 days of refrigerated storage, especially at the highest polysaccharide ratio (G1-P1). Improvements were also seen in sensory qualities such as odor, color, and overall acceptability, effectively extending the shelf life of the product. These results suggest that microwave-extracted loquat polysaccharides can be used to develop antioxidant-rich and eco-friendly edible coatings for cold-stored fish products [41].

Microwave-assisted extraction (MAE) efficiently releases polysaccharides by breaking down cell walls through uniform internal heating. The method reduces extraction time while still maintaining yield and bioactivity. In most cases, optimal performance is obtained at 300–900 W for 2–15 minutes. However, using higher power levels can degrade sensitive polysaccharides and lower their functionality [42].

Microwave heating has been recognized as a promising alternative to conventional frying for reheating plant protein-based meat analogs. In a recent study, wheat- and rice-flour-based coatings were applied to MAs, which were first partially fried and then reheated using microwave, infrared, air frying, and deep-fat frying methods. The samples treated with microwave heating showed significantly lower fat content and higher moisture retention compared with deep-fat-fried samples. These findings confirm that microwave heating is an efficient method for maintaining the quality of coated plant-based products during reheating [43].

Microwave-assisted extraction was applied to efficiently obtain pectin and phenolic compounds from the bagasse and pomace of Mexican limes, which are by-products of lime processing. This method caused significant depolymerization of pectin, reducing its molecular weight to 12 kDa compared with 670 kDa under conventional extraction. Such reduction can influence the film-forming properties of pectin. The microwave extracts retained a high degree of esterification (around 70%) and showed bioactive properties related to their phenolic content. They also demonstrated strong antioxidant activities, including Fe^{2+} chelation and inhibition of lipid oxidation. Overall, these results suggest that microwave extraction offers clear benefits for producing functional pectic materials from agricultural by-products for edible film and coating applications [44]. (Table 1).

Table 1. Microwave-modified polysaccharide based edible film and coating applications

Material	Microwave Conditions Applied	Properties Changed	Reference
Pearl millet starch film (with vegetable oil & glycerol) and composite of native starch + PMS nanoparticles	2350 W, 30–60 s	Uniform gelatinization and film formation; increased thickness; reduced moisture content; enhanced barrier and mechanical properties; at 10% nanoparticles: decreased flexibility & transparency; at ~5%: optimal performance & biodegradability	[26]
Chitosan films (with and without Cu nanoparticles)	2450 MHz, full power, 10 min	Improved UV–visible light barrier; no significant change in thermal or structural characteristics; with Cu nanoparticles: enhanced water resistance and faster removal of unbound water during drying	[27, 28]

Edible coatings (potato starch, CMC, pectin) on kiwifruit	450 W microwave and microwave–hot air combination drying	Accelerated drying; improved moisture diffusion; enhanced retention of bioactive compounds (vitamin C, total phenolics); better color, texture, and rehydration characteristics	[29]
SMAP edible gel (CMC, glycerol, Polysorbate 80, citrus peel EO) on blackberries	2450 MHz, 1000 W, 90 °C, 60 s	Reduced weight loss and decay; enhanced retention of flavonoids, phenolics, and antioxidant activity during storage at 4 °C for 21 days	[31]
Carrots pretreated with osmotic enrichment (chokeberry juice) and coated with sodium alginate/citrus pectin	250 W microwave–vacuum drying at 3.5 kPa	Enhanced drying efficiency; higher dry matter content; lower water activity; increased total phenolic content with osmotic pretreatment; coatings moderately influenced color and antioxidant activity; maintained desirable texture, color, aroma	[10]
Low-methoxy pectin modified with stearic anhydride (catalyzed)	300 W, 9 min (stirring every 3 min)	Significantly enhanced grafting efficiency; improved emulsifying properties and mechanical strength; substantial increase in surface hydrophobicity (higher contact angle)	[33]
Bacterial cellulose pellicles	Microwave drying for 3–5 min	≈95% reduction in drying time; retained cellulose I structure; slightly lower crystallinity; higher swelling degree; similar color and tensile properties	[36]
Sago starch films (with and without chitosan microparticles, EDTA)	640 W, 30 s (rapid gelatinization)	Decreased water vapor permeability with increasing thickness; further barrier enhancement by chitosan microparticles; excellent antibacterial activity against E. coli	[40]

4. Protein-based Edible Films and Coatings

The growing concerns about fossil fuel-based plastics make it necessary to shift toward environmentally friendly, biopolymer-based packaging derived from renewable sources. Protein is among the most important and sustainable natural polymers and is found in many different foods. The main sources for producing edible films and coatings include plants such as wheat, soy, corn, mung bean, gluten, and zein, as well as animal-derived proteins such as casein, whey, gelatin, egg albumen, and feather keratin [1]. Protein films can be mechanically strengthened and made less permeable through denaturation, blending with polysaccharides and lipids, or the incorporation of microbial agents, anti-browning agents, probiotics, and antioxidants [45].

Microwave treatment has been reported to improve the water-holding capacity of proteins and to alter their secondary structure, including changes in α-helix and β-sheet content. It also promotes the formation of aggregated intermolecular networks [46]. These structural modifications enhance the film-forming ability and functional properties of proteins when used in edible film and coating production. Microwave processing further affects the physicochemical, secondary, and tertiary structures of protein-based film-forming solutions, thereby strengthening the structure and mechanical performance of the resulting films [47].

Microwave-induced cross-linking, which creates stronger intermolecular covalent bonds in protein networks by reducing polymer mobility, is considered a technological advance [48]. Energy is delivered through the dielectric mechanism of microwave radiation, leading to bulk heating. This allows protein modifications to occur at lower temperatures and shorter durations while facilitating the diffusion of the solid phase [49].

Zein is a by-product of industrial corn starch extraction and is composed of α-, β-, γ-, and δ-zein polypeptides. This non-crystalline biopolymer contains nonpolar hydrophobic amino acids such as leucine, proline, and alanine. Because of these amino acids, zein is insoluble in water, giving edible films a hydrophobic character that limits their applications [50]. Zein-based films are also mechanically fragile and lack the flexibility needed for industrial use, making them less favorable than other protein-derived films.

An investigation showed that, soy protein isolate was blended with microwave-treated zein that had been modified at 60 °C and 350 W for 2 minutes, in varying ratios. The film-forming solutions were prepared in an ethanol–water mixture (20:80, v/v) at pH 12. The influence of microwave-assisted modification on the mechanical behavior and structural properties of the composite films was systematically examined. The microwave-modified blended films showed higher breaking strength and greater fracturing distance, producing more homogeneous films with fewer pores. Improvements were also observed in the surface and cross-section morphologies due to changes in the secondary structure. In addition, the glass transition temperature increased as a result of the modification [18].

Microwave drying, compared with ambient or hot air drying, plays an important role in determining the final properties of edible films. Researchers reported that, microwave drying was applied to soy protein isolate and sunflower oil-based films. This approach reduced the drying time from several days to only a few minutes. Films dried at 800 W showed higher thermal stability, better water barrier properties, greater solubility, and improved color retention. Although some loss in mechanical strength and transparency was noted, microwave treatment still allowed rapid and efficient film production without major compromise in overall quality [51]. (Fig.3).

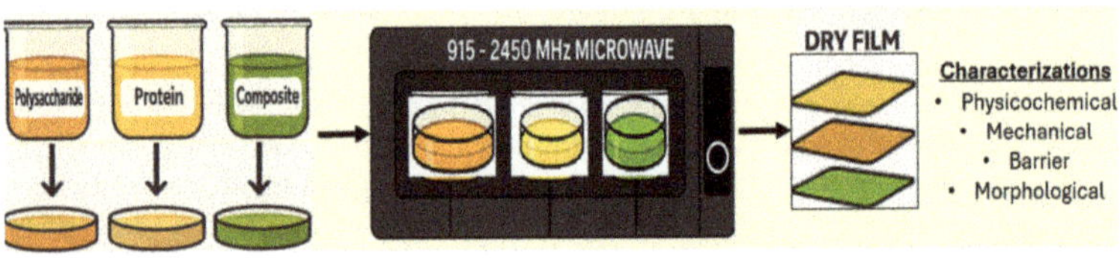

Figure 3. Casting edible film-coating solutions, subjecting them to microwave modification, and characterizing the resulting dried films.

Combining microwave treatment with techniques such as ultrasound has been shown to be more effective for improving the structural integrity and functional characteristics of edible films than using either method alone. In particular, ultrasonic–microwave processing enhanced the functional properties of soy protein isolate edible films embedded with nano-TiO_2. This combined approach reduced both the water vapor transmission rate and the oxygen transmission rate. The improvements were mainly linked to smaller particle size, an increase in free hydroxyl groups, greater particle surface area, and stronger molecular interactions. These changes supported the development of more compact and homogeneous film surfaces. As a result, the films displayed higher tensile strength and density. Optimal outcomes were achieved when the ultrasonic–microwave-assisted treatment time did not exceed 20 minutes, whereas longer treatments damaged the film matrix and reduced mechanical stability [52].

Whey protein isolate (WPI) films containing 8 and 10% (w/w) protein were dried in a microwave oven at 700 W and 2450 MHz for 5 minutes. The microwave-dried films showed improved tensile strength and higher elongation at break. They also exhibited greater haze and gloss compared with films produced by conventional drying. However, microwave treatment did not significantly affect the water vapor permeability of WPI–glycerol-based films [21].

Researchers applied microwave energy as a drying method for edible films prepared from defatted wheat germ protein. The protein was extracted by alkaline solubilization (pH 9.5) followed by isoelectric precipitation (pH 4.0). Microwave drying for only 5 minutes was compared with conventional drying at room temperature, which required 18 hours (Fig. 3). The results showed that microwave treatment greatly reduced the processing time and improved the mechanical performance of the films, producing higher tensile strength and greater elongation at break. Water vapor permeability values were similar for both methods, but microwave-dried films displayed better functional properties, including higher gloss and lower haze. These findings indicate that microwave-assisted drying is an efficient approach for producing protein-based edible films with desirable barrier and mechanical characteristics [53].

To develop a high-performance edible film, whey protein concentrate (WPC) and chitosan (CS) were combined and modified using an ultrasonic/microwave-assisted treatment. Compared with the individual use of ultrasound or microwave applications, the combined method produced a marked improvement in barrier properties. Water vapor permeability (WVP) and oxygen permeability (OP) decreased by 27.9% and 48.1%, respectively, compared with the untreated control. The films also showed improved mechanical strength and higher light transmittance under optimized conditions (WPC:CS = 5.8:6.2, 5 min treatment, pH 5.13). Overall, these results confirm that ultrasonic/microwave synergy can effectively improve the structural integrity and functional performance of biopolymer-based edible films, providing a practical strategy for food packaging applications [54].

Soy protein isolate (SPI)-based edible films were modified using ultrasonic/microwave-assisted treatment to study the effect of different oleic acid (OA) to stearic acid (SA) ratios on barrier properties. Films treated at 20 °C for 15 minutes at 500 W showed the lowest water vapor permeability (0.1×10^{-12} $g \cdot cm^{-1} \cdot s^{-1} \cdot Pa^{-1}$) and the highest contact angle (135°) when the OA:SA ratio was 2:3. The findings indicate that combining lipid addition with microwave-assisted modification can effectively improve the hydrophobicity and moisture resistance of SPI-based films. These effects are likely due to lipid–protein interactions, as evidenced by FTIR spectra (peak at 2920 cm^{-1}) [49].

The applicability of biodegradable films and coatings derived from jellyfish (*Acromitus flagellates*) gelatin, extracted by both conventional hot water and microwave-assisted methods, has been evaluated for the preservation of strawberries during refrigerated storage (4 °C). Microwave-assisted extraction proved especially effective. Gelatin-based coatings significantly reduced weight loss and limited physicochemical changes in the fruit over a 21-day period. Compared with uncoated samples, film-wrapped or coated strawberries maintained better appearance and structural integrity. Gelatin from jellyfish also showed performance comparable

Materials Research Foundations 189 (2026) 412-434 https://doi.org/10.21741/9781644904039-13

to that of cold-water fish and porcine gelatins, indicating its potential as a sustainable option for food packaging applications [55]. (Table 2).

Table 2. Microwave-modified protein based edible film and coating applications

Material	Microwave Conditions Applied	Properties Changed	Reference
Composite film of soy protein isolate and zein	350 W, 60 °C, 2 min	Enhanced breaking strength and fracturing distance; more homogeneous film with fewer pores; improved morphology; increased glass transition temperature	[18]
Soy protein isolate + sunflower oil edible film	800 W, minutes (vs. days by ambient/hot-air drying)	Enhanced thermal stability; improved water barrier; increased solubility; better color retention; slight loss in strength & transparency	[51]
Whey protein isolate (8–10 % w/w) films	700 W, 2450 MHz, 5 min	Improved tensile strength and elongation at break; greater haze and gloss; no change in water vapor permeability	[21]
Defatted wheat germ protein films (alkaline solubilization pH 9.5 → isoelectric precipitation pH 4.0)	Microwave drying, 5 min	Increased tensile strength and elongation; similar water vapor permeability; improved gloss and reduced haze	[53]
Whey protein concentrate + chitosan films (ultrasonic/microwave-assisted)	Ultrasonic/microwave-assisted treatment, 5 min (pH 5.13)	WVP ↓ 27.9 %; OP ↓ 48.1 %; improved mechanical strength; enhanced light transmittance	[54]
SPI films + oleic acid:stearic acid (2:3)	500 W, 20 °C, 15 min (ultrasonic/microwave-assisted)	Lowest WVP (0.1×10^{-12} $g{\cdot}cm^{-1}{\cdot}s^{-1}{\cdot}Pa^{-1}$); highest contact angle (135°); enhanced hydrophobicity and moisture resistance	[49]

5. Polysaccharide- and Protein-Based Composite Edible Films

Combined edible films and coatings made from proteins and polysaccharides offer clear advantages in both functionality and durability for food packaging. Blends such as whey protein isolate and chitosan, when modified with physical treatments like microwave and ultrasound, show significant improvements in barrier properties. These treatments lower water vapor and oxygen permeability, which strengthens protection and extends the shelf life of food products. Microwave-assisted modification of polysaccharides such as konjac glucomannan, chitosan, and citric acid also produces composite films with higher mechanical strength and greater resistance to oxidation. Together, these advances demonstrate that protein–polysaccharide-based films can provide sustainable and effective solutions for food preservation and packaging [47,54,56,57].

Researchers developed microwave-assisted antibacterial nanocomposite edible films (CaO_2@PVP/EA/CMC-Na, or CP/EC) for meat preservation. The films were prepared using waste eggshells as a precursor, with egg albumen and sodium carboxymethylcellulose serving as the matrix, and were produced by the casting method. The 2.5% CP/EC film showed strong bactericidal activity, with inactivation rates of 98.6% against E. coli and 97.2% against S. aureus after 6 minutes of microwave irradiation. Effective meat preservation was achieved through the synergistic action of radical oxygen species (ROS) and the thermal effect of microwaves. This

Materials Research Foundations 189 (2026) 412-434 https://doi.org/10.21741/9781644904039-13

mechanism was linked to disruptions in the metabolic balance of bacterial cells and the breakdown of cell membranes caused by both heat and free radical formation [58].

Striped catfish fillets were immersed in a coating solution for two minutes and then treated with microwaves at different powers (100, 300, and 500 W) for another two minutes. After treatment, the fillets were stored at 4 °C for 15 days. The coating solution was prepared by combining fish gelatin and agar and included lemongrass essential oil. The study assessed how the combination of essential oils and microwaves influenced the quality of the fillets. Microwave treatment decreased protease activity through rapid heating driven by electromagnetic oscillations. It also reduced microbial growth by damaging bacterial proteins and membranes, with the extent of the effect depending on the microwave power. In addition, microwave application lowered the total formation of volatile basic nitrogen, decreased lipid oxidation, and improved the overall quality of the product [8].

In an experimental study on blended films made of whey protein isolate (WPI) and chitosan, a combined ultrasound and microwave treatment was applied to modify film properties. The combined process greatly improved barrier performance, reducing water vapor permeability by 27.9% and oxygen permeability by 48.1%. By contrast, applying ultrasound or microwave individually led to less pronounced improvements. These results show that the combined treatment produces a clear synergistic effect in enhancing the barrier properties of protein-based films [52].

In one investigation, a novel composite film was produced by microwave-assisted modification of konjac glucomannan (KGM), chitosan, and citric acid. Bamboo leaf antioxidants were incorporated to create a functional food packaging material. The film displayed strong oxidation resistance and provided effective UV–Vis light protection. It also showed favorable mechanical properties, with an elongation at break reaching a maximum of 89.12 ± 0.27%. The microwave-modified KGM/CTS/citric acid film exhibited excellent barrier performance, with a water vapor permeability of $(2.562 \pm 0.001) \times 10^{-10}$ g/m·s·Pa. After the addition of 0.2% bamboo leaf antioxidant extract, the WVP decreased further to $(2.494 \pm 0.003) \times 10^{-10}$ g/m·s·Pa, confirming an improvement in moisture barrier properties [56].

In a comparative analysis, microwave treatment at 70 W or 30 W for 1 minute, especially when followed by ultrasound (70M-US, 30M-US), was shown to significantly enhance the functional properties of zein–corn starch-based edible films. The microwave process improved surface homogeneity and mechanical strength, and it also increased the water contact angle, indicating greater hydrophobicity. FTIR analysis revealed more pronounced amide-III bands, suggesting improved biopolymer compatibility due to microwave-induced structural modifications [59].

In one study investigating the drying behavior of apple slices coated with basil, sage, and chitosan, microwave-assisted drying at 90, 180, and 360 W was shown to significantly improve product quality. The results indicated that higher microwave power increased brightness, antioxidant activity, adhesiveness, and cohesiveness. At the same time, it reduced undesirable attributes such as redness, yellowness, browning index, and hardness. These findings demonstrate that microwave application, particularly when combined with functional edible coatings, is an effective strategy for improving the appearance and physicochemical properties of dried fruit products. It may also contribute to shelf-life extension and overall product quality [60].

Materials Research Foundations 189 (2026) 412-434 https://doi.org/10.21741/9781644904039-13

6. Agro-industrial Waste-based Edible Films and Coatings

Microwave technology is an innovative food processing method with strong potential in the valorization of food resources and by-products. This is particularly relevant when biopolymeric components are applied in films and coatings to preserve and enhance nutrients. Food waste and by-products generated across the farm-to-plate supply chain are valuable raw materials for producing high-value compounds. In the EU, the food processing industry generates around 100 million tons of waste and by-products every year, accounting for more than one-third of losses during harvesting and processing. These by-products are rich in polysaccharides, polyphenols, essential oils, pigments, proteins, flavor compounds, enzymes, and dietary fibers, which can be extracted through various methods. Many of these extracted compounds can serve as key components or additives in the development of film and coating materials, contributing to improved functionality and sustainability [61].

Microwave-assisted extraction of pectin from pineapple peel, an abundant agricultural by-product, offers a more efficient alternative to conventional methods and enhances the quality of the extracted pectin. After 30 minutes of microwave irradiation, the pectin showed higher methoxyl content, a greater degree of esterification, and higher galacturonic acid levels compared with traditional extraction. Although the overall yield was relatively low (1.02–2.12%), the extracted pectin displayed promising functional properties. Instead of using the isolated pectin alone, the whole pineapple peel pectin extract solution (PPS) was applied as a natural plasticizer. This improved edible film characteristics, including water vapor barrier capacity and antioxidant activity. These results indicate that microwave-assisted extraction can help valorize pineapple peel waste for the production of functional edible coatings and films in food packaging applications [62].

Microwave-assisted deacetylation has emerged as a promising technique for chitosan production, providing a much faster alternative to conventional methods. In one investigation, chitosan was extracted from Portunus pelagicus L. shells, an underutilized seafood processing waste, using both conventional and microwave-assisted approaches. Microwave deacetylation was performed at 100 °C and 1200 W for 40 minutes. This process achieved a degree of deacetylation (DD) of 77.89%, compared with 76.67% obtained through conventional heating over 120 minutes. These results demonstrate the efficiency of microwave irradiation in improving the deacetylation rate and producing high-quality chitosan within a much shorter processing time [63].

Microwave-assisted synthesis was applied as a green, solvent-reducing approach to prepare iron- and zinc-incorporated soybean oil-based polymers. The process involved condensation polymerization of fatty amidepolyole, sebacic acid, zinc acetate, and ferric chloride. Microwave irradiation not only increased reaction efficiency but also supported eco-friendly processing goals. Structural confirmation of the polymers was obtained by FT-IR and ^{1}H NMR analyses, while their chemical and thermal properties were evaluated using standard methods such as TG/DTA. The resulting polymers displayed notable antifungal activity and were successfully used in film formation, showing potential as protective packaging materials for edible products [64].

Wheat-bran cellulose (WC), microcrystalline wheat-bran cellulose (MWC), and ultrasonic/microwave-modified microcrystalline wheat-bran cellulose (MMWC) were combined with soy protein isolate (SPI) to prepare edible films by casting. The proportion and particle size of these fillers significantly affected the properties of the resulting films. Among them, the SPI/MMWC film displayed superior performance. This was attributed to microwave-assisted modification, which produced MMWC with a finer particle size and more surface hydroxyl groups. These features promoted stronger intermolecular hydrogen bonding with SPI, creating a film with a more compact and homogeneous structure. Structural differences also accounted for the variations in thermal properties observed among the films. In addition, FT-IR analysis confirmed molecular interactions between SPI and the fillers, while SEM imaging showed a smoother and more uniform surface morphology in the SPI/MMWC film [65].

Kafirin, the major prolamin protein of sorghum, is known for its hydrophobic nature and its ability to form biodegradable or edible films with favorable tensile and barrier properties. In one investigation on kafirin-based edible films, microwave heating was applied as a technique to improve film performance. Films cast from glacial acetic acid or aqueous ethanol, with or without plasticizer, were exposed to microwave energy at 80 W for 2 minutes. This treatment produced a notable increase in tensile strength—approximately twofold in non-plasticized films and fourfold in plasticized ones. Microwave heating also raised the glass transition temperature by about 15% and slowed biodegradability. These effects were linked to heat-induced intermolecular disulfide cross-linking, as supported by SDS-PAGE analysis. Scanning electron microscopy further revealed morphological changes in the film matrix, including greater wrinkling in plasticized films and brittleness in non-plasticized ones. Overall, the findings show that microwave treatment can effectively tune the mechanical and structural characteristics of kafirin films, making them more suitable for food packaging applications [66].

Microwave-assisted esterification (800 W, 1 min) has been used as an effective method to modify nanofibrillated cellulose (NFC) and improve its compatibility with ethylcellulose (EC) matrices in fully bio-based nanocomposite films. In this approach, NFC was esterified with different acids of varying carbon chain lengths, including lactic acid (LA), lauric acid (LU), and stearic acid (SA), in solvents such as water, ethanol, and ethyl acetate. The modified NFCs (mNFCs) were successfully characterized and incorporated into EC films, leading to notable improvements in performance. Among the modifications, mNFC-LA provided the strongest reinforcement, increasing tensile strength by up to 130% and reducing the oxygen transmission rate by 98%. These results point to a sustainable and promising alternative to conventional PET films [67].

Industrial oil-processing by-products such as linseed mucilage (LM) have gained attention for their valuable chemical and functional properties, which vary depending on the extraction method. Microwave-assisted extraction has been identified as an effective pretreatment to enhance both the yield and functionality of LM. Research shows that microwave-assisted extraction not only improves extraction efficiency but also helps preserve the bioactive properties of LM. With promising therapeutic effects—including anti-obesity, antidiabetic, and anticancer potential—and its suitability for bioplastic applications, LM obtained through microwave-assisted extraction represents a sustainable material with strong potential for use in food packaging [68].

In one study, microwave treatment was applied to ensure a homogeneous mixture of film-forming solutions prepared with potato peel powder, turmeric powder, lecithin, and whey protein. This process not only facilitated uniform dispersion of the ingredients but also enabled the rapid development of an active coating material designed specifically for Ras cheese. The application of this microwave-assisted coating improved the overall quality of Ras cheese during ripening, enhanced its safety by limiting microbial growth, and extended its shelf life under storage conditions [69].

In one investigation, the potential of agricultural dry wastes such as banana leaves, banana sheets, and coconut leaves for producing biodegradable food plates was explored. Microwave-assisted processing (20 s, 3 mL water, 4% binder) was applied, and the tensile strength of the plates was significantly improved, reaching 0.511 MPa. The microwave treatment proved highly effective in enhancing the mechanical properties while maintaining the biodegradability of the material. Furthermore, applying natural surface coatings improved the water repellency of the plates, making them more resistant to moisture during use [70].

Previous studies have shown that microwave-assisted oxygen plasma treatment is an effective and low-cost method to improve the surface properties of polyvinyl alcohol (PVA) films. The treatment significantly reduces the contact angle, indicating higher hydrophilicity, and introduces polar functional groups such as carbonyls, as confirmed by XPS analysis. In addition, the increase in surface roughness observed through electron and atomic force microscopy enhances adhesion characteristics. Overall, these results demonstrate the potential of microwave plasma treatment in advancing PVA film applications for food packaging and coatings, making it a promising tool for surface modification [71].

7. Future Outlooks and Conclusions

Microwave technology holds strong potential for modifying the physicochemical properties, structural features, and biological activities of polysaccharides, proteins, and composite structures in edible film and coating formulations. It offers key advantages such as reduced drying time, driven by mechanisms like dipole rotation and ion migration. Depending on the power and duration of the microwave process, as well as the type of biopolymer, microwave treatment can alter the microscopic, chemical, and molecular structure of film components. When combined with other non-thermal technologies, such as ultrasound or high-pressure treatment, microwave processing can create synergistic effects and reduce overall energy requirements.

Despite these advantages, wider commercial application still faces challenges. Equipment standardization is lacking, and the precise mechanisms of microwave interactions with biopolymers and film matrices remain incompletely understood. When microwave treatment is applied to food after coating, its specific effects on both the food and the packaging material must also be carefully assessed.

Future research should aim to identify the molecular sites of microwave action to optimize processing outcomes. Approaches such as energetic theory and molecular docking studies may provide deeper insights into the interactions between film and coating components. At the same time, efforts to standardize microwave equipment are crucial for ensuring reproducible performance. The adoption of continuous-flow microwave systems is also expected to support scale-up and industrial implementation, paving the way for broader application in sustainable food packaging.

References

[1] A. Raj, D. Biswas, S. Roy, Recent advances in protein-based edible films and coatings for meat packaging applications, Food Bioprocess Technol. 18 (2025) 8123-8153. https://doi.org/10.1007/s11947-025-03939-x

[2] R. Coles, D. McDowell, M.J. Kirwan, Food and beverage packaging technology, Wiley-Blackwell, Chichester, 2011. https://doi.org/10.1002/9781444392180

[3] S. Gündoğdu, A. Bour, A.R. Köşker, B.A. Walther, D. Napierska, F.-C. Mihai, K. Syberg, S. F. Hansen, T. R. Walker, Review of microplastics and chemical risk posed by plastic packaging on the marine environment to inform the global plastics treaty, Sci. Total Environ. 946 (2024) 174000. https://doi.org/10.1016/j.scitotenv.2024.174000

[4] S. Kumar, A. Mukherjee, J. Dutta, Chitosan based nanocomposite films and coatings: emerging antimicrobial food packaging alternatives, Trends Food Sci. Technol. 97 (2020) 196-209. https://doi.org/10.1016/j.tifs.2020.01.002

[5] A.S. Pottinger, R. Geyer, N. Biyani, C.C. Martinez, N. Nathan, M.R. Morse, C. Liu, S. Hu, M. de Bruyn, C. Boettiger, E. Baker, D.J. McCauley, Pathways to reduce global plastic waste mismanagement and greenhouse gas emissions by 2050, Science 386 (2024) 1168-1173. https://doi.org/10.1126/science.adr3837

[6] A.S. Barone, J.R.V. Matheus, T.S.P. de Souza, R.F.A. Moreira, A.E.C. Fai, Green-based active packaging: Opportunities beyond COVID-19, food applications, and perspectives in circular economy-a brief review, Compr. Rev. Food Sci. Food Saf. 20 (2021) 4881-4905. https://doi.org/10.1111/1541-4337.12812

[7] H. Rostamabadi, I. Demirkesen, R. Colussi, S. Roy, N. Tabassum, J.G. de Oliveira Filho, Y. Bist, Y. Kumar, M. Nowacka, S. Galus, S.R. Falsafi, Recent trends in the application of films and coatings based on starch, cellulose, chitin, chitosan, xanthan, gellan, pullulan, arabic gum, alginate, pectin, and carrageenan in food packaging, Food Front. 5 (2024) 350-391. https://doi.org/10.1002/fft2.342

[8] A.R. Hakim, A. Fauzi, A. Pamungkas, B.B. Sedayu, T.D. Novianto, Enhancing the quality of striped catfish (Pangasius hypophthalmus) fillet: A synergistic effect of microwave treatment and gelatin-lemongrass essential oil coatings, Food Bioprocess Technol. 18 (2025) 4672-4683. https://doi.org/10.1007/s11947-025-03743-7

[9] B. Hassan, S.A.S. Chatha, A.I. Hussain, K.M. Zia, N. Akhtar, Recent advances on polysaccharides, lipids and protein based edible films and coatings: A review, Int. J. Biol. Macromol. 109 (2018) 1095-1107. https://doi.org/10.1016/j.ijbiomac.2017.11.097

[10] A. Ignaczak, Ł. Woźniak, A. Marzec, J. Kowalska, M. Chobot, H. Kowalska, The influence of osmotic treatment, edible coatings application, and reduced pressure on microwave-vacuum-dried carrot properties, Molecules 30 (2025) 1877. https://doi.org/10.3390/molecules30091877

[11] F.T. Saricaoglu, S. Tural, O. Gul, S. Turhan, High pressure homogenization of mechanically deboned chicken meat protein suspensions to improve mechanical and barrier properties of edible films, Food Hydrocoll. 84 (2018) 135-145. https://doi.org/10.1016/j.foodhyd.2018.05.058

[12] F. Chen, B. Xu, W. Cui, Y. Wang, F. Wan, A. Cheng, Effects of microwave treatment on vegetable oil quality and biological activities, Trends Food Sci. Technol. 153 (2024) 104748. https://doi.org/10.1016/j.tifs.2024.104748

[13] L. Román, M.M. Martínez, C.M. Rosell, M. Gómez, Effect of microwave treatment on physicochemical properties of maize flour, Food Bioprocess Technol. 8 (2015) 1330-1335. https://doi.org/10.1007/s11947-015-1493-0

[14] J. Li, Y.-G. Zu, M. Luo, C.-B. Gu, C.-J. Zhao, T. Efferth, Y.-J. Fu, Aqueous enzymatic process assisted by microwave extraction of oil from yellow horn (Xanthoceras sorbifolia Bunge.) seed kernels and its quality evaluation, Food Chem. 138 (2013) 2152-2158. https://doi.org/10.1016/j.foodchem.2012.12.011

[15] J. Tang, F. Hao, M. Lau, Microwave heating in food processing, in: Advances in Bioprocessing Engineering, World Scientific, Singapore (2002) 1-44. https://doi.org/10.1142/9789812706584_0001

[16] Q. Jiang, M. Zhang, A.S. Mujumdar, Application of physical field-assisted freezing and thawing to mitigate damage to frozen food, J. Sci. Food Agric. 103 (2023) 2223-2238. https://doi.org/10.1002/jsfa.12260

[17] D. Dallinger, C.O. Kappe, Microwave-assisted synthesis in water as solvent, Chem. Rev. 107 (2007) 2563-2591. https://doi.org/10.1021/cr0509410

[18] N. Wang, Y.Z. Gao, P. Wang, S. Yang, T.M. Xie, Z.G. Xiao, Effect of microwave modification on mechanical properties and structural characteristics of soy protein isolate and zein blended film, Czech J. Food Sci. 34 (2016) 180-188. https://doi.org/10.17221/442/2015-CJFS

[19] A. Kurt, E. Ozturk, D. Ozmen, M. Yildirim-Yalcin, M. Arici, O.S. Toker, Microwave-modified xanthan gum: Alterations in steady and dynamic rheological behaviors, J. Food Process Eng. 47 (2024) e14525. https://doi.org/10.1111/jfpe.14525

[20] S. Chandrasekaran, S. Ramanathan, T. Basak, Microwave food processing-a review, Food Res. Int. 52 (2013) 243-261. https://doi.org/10.1016/j.foodres.2013.02.033

[21] S. Kaya, A. Kaya, Microwave drying effects on properties of whey protein isolate edible films, J. Food Eng. 43 (2000) 91-96. https://doi.org/10.1016/S0260-8774(99)00136-3

[22] S. Beikzadeh, M. Ghorbani, N. Shahbazi, F. Izadi, Z. Pilevar, A.M. Mortazavian, The effects of novel thermal and nonthermal technologies on the properties of edible food packaging, Food Eng. Rev. 12 (2020) 333-345. https://doi.org/10.1007/s12393-020-09227-y

[23] Q. Guo, D.-W. Sun, J.-H. Cheng, Z. Han, Microwave processing techniques and their recent applications in the food industry, Trends Food Sci. Technol. 67 (2017) 236-247. https://doi.org/10.1016/j.tifs.2017.07.007

[24] X. Chen, J. Yang, M. Shen, Y. Chen, Q. Yu, J. Xie, Structure, function and advance application of microwave-treated polysaccharide: a review, Trends Food Sci. Technol. 123 (2022) 198-209. https://doi.org/10.1016/j.tifs.2022.03.016

[25] L. Kumar, D. Ramakanth, K. Akhila, K. K. Gaikwad, Edible films and coatings for food packaging applications: a review, Environ. Chem. Lett. 20 (2022) 875-900. https://doi.org/10.1007/s10311-021-01339-z

[26] P. Chavan, G.P. Singh, K. Aayush, S. Sharma, P.K. Chauhan, R. Thory, M. Shandilya, D. Kumar, Development and characterization of pearl millet (AHB 1200) starch nanoparticle-

based edible films: A paradigm shift in food packaging, J. Food Process Eng. 47 (2024) e14575. https://doi.org/10.1111/jfpe.14575

[27] G. Cárdenas, J. Díaz, M. Meléndrez, C. Cruzat, Physicochemical properties of edible films from chitosan composites obtained by microwave heating, Polym. Bull. 61 (2008) 737-748. https://doi.org/10.1007/s00289-008-0994-7

[28] G. Cárdenas, J. Díaz, M. Meléndrez, C. Cruzat, A. García Cancino, Colloidal Cu nanoparticles/chitosan composite film obtained by microwave heating for food package applications, Polym. Bull. 62 (2009) 511-524. https://doi.org/10.1007/s00289-008-0031-x

[29] F. Rajabi, S. Karimi, H. Abbasi, N. Layeghinia, Influence of edible coatings pretreatment on the performance of microwave and combined microwave-hot air drying of kiwifruit, Food Bioprod. Process. 153 (2025) 286-297. https://doi.org/10.1016/j.fbp.2025.07.003

[30] J. Gamboa-Santos, L.A. Campañone, Application of osmotic dehydration and microwave drying to strawberries coated with edible films, Dry. Technol. 37 (2019) 1002-1012. https://doi.org/10.1080/07373937.2018.1481426

[31] M. Nadeem, K. Leaks, A. Abdullah, J.S. Adamson Felix, M.A. Shahid, Enhancing postharvest quality of blackberries: impact of sonicated and microwave-assisted pasteurized edible coating gels at different storage temperatures, Gels 11 (2025) 243. https://doi.org/10.3390/gels11040243

[32] R.C. Anantheswaran, H.S. Ramaswamy, Bacterial destruction and enzyme inactivation during microwave heating, in: Handbook of Microwave Technology for Food Application, CRC Press, Boca Raton (2001) 191-214.

[33] P. Wang, P. Fei, C. Zhou, P. Hong, Stearic acid esterified pectin: Preparation, characterization, and application in edible hydrophobic pectin/chitosan composite films, Int. J. Biol. Macromol. 186 (2021) 528-534. https://doi.org/10.1016/j.ijbiomac.2021.06.030

[34] J.N. BeMiller, K.C. Huber, Physical modification of food starch functionalities, Annu. Rev. Food Sci. Technol. 6 (2015) 19-69. https://doi.org/10.1146/annurev-food-022814-015552

[35] H. Arifin, R. Indiarto, D. Ciptaningtiyas, Physical characteristics of edible film from modified breadfruit starch (Artocarpus atilis F.) with glycerol, IOP Conf. Ser. Earth Environ. Sci. 425 (2020) 012028. https://doi.org/10.1088/1755-1315/443/1/012028

[36] I. Indriyati, Y. Irmawati, T. Puspitasari, Comparative study of bacterial cellulose film dried using microwave and air convection heating, J. Eng. Technol. Sci. 51 (2019) 121-132. https://doi.org/10.5614/j.eng.technol.sci.2019.51.1.8

[37] S.M. Kamel, R.M. Alqurashi, A.M. Alshebly, F.A. Saleh, T.M. Alnemr, H.F. Ali, S.A. Ali, S.A. Asiri, M.E. Ibrahim, N.K. Alqahtani, Effect of drying method and coating with gum arabic on physicochemical properties, betalain content, LDL oxidation, and storage stability of dried red beetroot slices, J. Food Process. Preserv. 48 (2024) e14792. https://doi.org/10.1155/2024/8471292

[38] H. Sun, X. Shao, R. Jiang, Z. Ma, H. Wang, Effects of ultrasonic/microwave-assisted treatment on the properties of corn distarch phosphate/corn straw cellulose films and structure characterization, J. Food Sci. Technol. 55 (2018) 1467-1477. https://doi.org/10.1007/s13197-018-3063-8

[39] A.E. Kapetanakou, D. Karyotis, P.N. Skandamis, Control of Listeria monocytogenes by applying ethanol-based antimicrobial edible films on ham slices and microwave-reheated frankfurters, Food Microbiol. 54 (2016) 80-90. https://doi.org/10.1016/j.fm.2015.10.013

[40] S. Bajpai, C. Navin, L. Ruchi, Water vapor permeation and antimicrobial properties of sago starch-based films formed via microwave irradiation, Int. Food Res. J. 18 (2011) 1-6.

[41] M. Nouri, H. Baghaei, M. Kashaninejad, A. Mohammadi Nafchi, Enhanced shelf life of trout fillets using an edible coating containing the polysaccharide extracted from loquat (Eriobotrya japonica) seed, Future Foods 12 (2025) 100693. https://doi.org/10.1016/j.fufo.2025.100693

[42] B.D. Tobar-Espinoza, J.R. Pérez-Correa, Advances in seed mucilage: Chemical characterization, functional properties, extraction, and applications in edible coatings for meat preservation, Appl. Food Res. 5 (2025) 100886. https://doi.org/10.1016/j.afres.2025.100886

[43] M.H.R. Bhuiyan, M.O. Ngadi, Electromagnetic, air and fat frying of plant protein-based batter-coated foods, Foods 12 (2023) 3953. https://doi.org/10.3390/foods12213953

[44] D. Sanchez-Aldana, C.N. Aguilar, G.V. Nevarez-Moorillon, J.C.C. Esquivel, Comparative extraction of pectic and polyphenols from Mexican lime pomace and bagasse, Am. J. Agric. Biol. Sci. 8 (2013) 309-316. https://doi.org/10.3844/ajabssp.2013.309.322

[45] A. Linares-Castañeda, X.M. Sánchez-Chino, Y.d.l.M.G. Gómez, C. Jiménez-Martínez, J. Martínez Herrera, M.S. Cid-Gallegos, L.J. Corzo-Ríos, Cereal and legume protein edible films: A sustainable alternative to conventional food packaging, Int. J. Food Prop. 26 (2023) 3197-3213. https://doi.org/10.1080/10942912.2023.2267785

[46] M.M. Heydari, T. Najib, V. Meda, Investigating starch and protein structure alterations of the processed lentil by microwave-assisted infrared thermal treatment and their correlation with the modified properties, Food Chem. Adv. 1 (2022) 100091. https://doi.org/10.1016/j.focha.2022.100091

[47] A.S. Yashwant, P. Kashyap, G. Goksen, Recent advances in the improvement of protein-based edible films through non-thermal and thermal techniques, Food Biosci. 55 (2023) 103032. https://doi.org/10.1016/j.fbio.2023.103032

[48] S.J. Calva-Estrada, M. Jiménez-Fernández, E. Lugo-Cervantes, Protein-based films: advances in the development of biomaterials applicable to food packaging, Food Eng. Rev. 11 (2019) 78-92. https://doi.org/10.1007/s12393-019-09189-w

[49] Z. Wang, J. Zhou, X.X. Wang, N. Zhang, X.X. Sun, Z.S. Ma, The effects of ultrasonic/microwave-assisted treatment on the water vapor barrier properties of soybean protein isolate-based oleic acid/stearic acid blend edible films, Food Hydrocoll. 35 (2014) 51-58. https://doi.org/10.1016/j.foodhyd.2013.07.006

[50] Y. Chen, R. Ye, X. Li, J. Wang, Preparation and characterization of extruded thermoplastic zein-poly(propylene carbonate) film, Ind. Crops Prod. 49 (2013) 81-87. https://doi.org/10.1016/j.indcrop.2013.04.040

[51] B. G. Erdem, Use of microwave drying for production of protein-oil based edible films, Akademik Gida 21 (2023) 274-283. https://doi.org/10.24323/akademik-gida.1382932

[52] Z. Wang, N. Zhang, H.Y. Wang, S.Y. Sui, X.X. Sun, Z.S. Ma, The effects of ultrasonic/microwave-assisted treatment on the properties of soy protein isolate/titanium dioxide films, LWT 57 (2014) 548-555. https://doi.org/10.1016/j.lwt.2014.01.036

[53] S. Hussien, M. Morgab, Manufacture of edible films from wheat germ protein and study of its mechanical properties, IOP Conf. Ser. Earth Environ. Sci. 1222 (2023) 012167. https://doi.org/10.1088/1755-1315/1252/1/012167

[54] S. Sui, N. Zhang, Y. Wang, Z. Ma, Process optimization of whey protein concentrate-chitosan edible films modified by ultrasonic/microwave assisting treatment, Trans. Chin. Soc. Agric. Eng. 30 (2014) 254-261.

[55] R. Ranasinghe, W. Wijesekara, S. Senanayake, P. Perera, M. Pathmalal, R. Marapana, Preservation of strawberries using jellyfish gelatin-based biodegradable films and coatings under refrigerated storage, Food Sci. Technol. Int. 30 (2024) 10820132241298487. https://doi.org/10.1177/10820132241298487

[56] Q. Wang, Y. Song, J. Sun, G. Jiang, A novel functionalized food packaging film with microwave-modified konjac glucomannan/chitosan/citric acid incorporated with antioxidant of bamboo leaves, LWT 166 (2022) 113780. https://doi.org/10.1016/j.lwt.2022.113780

[57] A.F. Pires, O. Díaz, A. Cobos, C.D. Pereira, A review of recent developments in edible films and coatings-focus on whey-based materials, Foods 13 (2024) 2367. https://doi.org/10.3390/foods13162638

[58] C. Gan, J. Wang, Z. Yuan, M. Cui, S. Sun, M. Alharbi, A.F. Alasmari, W. Du, X. Zhang, D.P. Yang, Polysaccharide- and protein-based edible films combined with microwave technology for meat preservation, Int. J. Biol. Macromol. 270 (2024) 132233. https://doi.org/10.1016/j.ijbiomac.2024.132233

[59] L. Acar, S.O. Keskin, Physical and mechanical characterization of microwave and ultrasound-modified zein-corn starch edible films, Ann. Univ. Dunarea Jos Galati. Fascicle VI-Food Technol. 47 (2023) 154-166. https://doi.org/10.35219/foodtechnology.2023.1.10

[60] S. Karami, R. Farahmandfar, J. Farmani, Z. Raftani Amiri, A. Motevali, Evaluating combined effect of microwave power-edible coating on physicochemical properties of dried apple slices, J. Food Sci. Technol. Iran 20 (2023) 1-16.

[61] M. Marić, A.N. Grassino, Z. Zhu, F.J. Barba, M. Brnčić, S. Rimac Brnčić, An overview of the traditional and innovative approaches for pectin extraction from plant food wastes and by-products: Ultrasound-, microwaves-, and enzyme-assisted extraction, Trends Food Sci. Technol. 76 (2018) 28-37. https://doi.org/10.1016/j.tifs.2018.03.022

[62] P. Rodsamran, R. Sothornvit, Preparation and characterization of pectin fraction from pineapple peel as a natural plasticizer and material for biopolymer film, Food Bioprod. Process. 118 (2019) 198-206. https://doi.org/10.1016/j.fbp.2019.09.010

[63] R. Mahardika, M. Jumnahdi, Y. Widyaningrum, Chitin deacetylation shells of Portunus pelagicus L. using microwave irradiation, IOP Conf. Ser. Earth Environ. Sci. 383 (2019) 012037. https://doi.org/10.1088/1755-1315/353/1/012037

[64] T. Singh, S. Ambreen, S. Shreaz, A.A. Hashmi, Effect of bimetallic soybean oil-based polymer on growth and plasma membrane H+-ATPase activity among fungi, J. Polym. Environ. 21 (2013) 81-87. https://doi.org/10.1007/s10924-012-0454-1

[65] Z. Wang, X.X. Sun, Z.X. Lian, X.X. Wang, J. Zhou, Z.S. Ma, The effects of ultrasonic/microwave-assisted treatment on the properties of soy protein isolate/microcrystalline wheat-bran cellulose film, J. Food Eng. 114 (2013) 183-191. https://doi.org/10.1016/j.jfoodeng.2012.08.004

[66] Y.B. Byaruhanga, C. Erasmus, M.N. Emmambux, J. R. Taylor, Effect of heating cast kafirin films on their functional properties, J. Sci. Food Agric. 87 (2007) 167-175. https://doi.org/10.1002/jsfa.2700

[67] S. Klayya, P. Pripdeevech, E. Bilotti, H. Zhang, N. Soykeabkaew, High barrier bio-nanocomposite films of ethyl cellulose integrated with modified nanocellulose: Effect of nanocellulose's substituted side chains on film performance, RSC Appl. Polym. 3 (2025) 100012. https://doi.org/10.1039/D4LP00356J

[68] M. Chand, R. Chopra, B. Talwar, S. Homroy, P.K. Singh, A. Dhiman, A.W. Payyunni, Unveiling the potential of linseed mucilage, its health benefits, and applications in food packaging, Front. Nutr. 11 (2024) 1334247. https://doi.org/10.3389/fnut.2024.1334247

[69] W.M. Salama, W.A. Ragb, E.S. Abdou, A.S. Dosuky, Enhancing the quality and safety of Ras cheese using an active edible film based on potato peel powder and turmeric, Food Technol. Res. J. 8 (2025) 92-106. https://doi.org/10.21608/ftrj.2025.437558

[70] J.M.R. Tingirikari, S.S. Manthena, S. Ramasamy, Screening and evaluation of agrowaste containing natural polymer as a solid base for biodegradable plate, in: International Conference on Eco-friendly Fibers and Polymeric Materials, Springer, Singapore, 2024, pp. 99-114. https://doi.org/10.1007/978-981-97-7071-7_8

[71] S. Elashry, H. Elsaeed, N.M. El-Siragy, Microwave plasma discharge-assisted surface modification of PVA films: Coatings and food packaging, Eur. Phys. J. Plus 137 (2022) 1252. https://doi.org/10.1140/epjp/s13360-022-03443-7

Materials Research Foundations 189 (2026) 435-467 https://doi.org/10.21741/9781644904039-14

Chapter 14

Microwave-Assisted Synthesis of Photocatalysts for Hydrogen Production

Keerthana Rajaram[1], Balaji Parasuraman[1], Pazhanivel Thangavelu[1*]

[1]Smart Materials laboratory, Department of Physics, Periyar University, Salem -636 011, Tamilnadu, India

pazhanit@gmail.com

Abstract

Hydrogen energy (H_2) is a promising alternative to fossil fuels, with photocatalytic hydrogen production emerging as a sustainable and cost-effective method. However, the development of efficient, eco-friendly photocatalysts remains challenging. Microwave-assisted synthesis offers a better solution by enabling rapid, clean, and energy-efficient fabrication of photocatalytic materials like transition metal oxides, dichalcogenides, and carbon-based compounds. Unlike conventional heating, microwaves generate heat internally within the sample, allowing uniform and precise temperature control. This precise temperature-controlled mechanism makes the microwave-assisted method an attractive technique for large-scale production. In this chapter, we provide a comprehensive overview of microwave-assisted synthesis techniques and their influence in hydrogen production, emphasizing their advantages in photocatalyst fabrication.

Keywords

Photocatalyst, Microwave-Assisted, Hydrogen Production, Synthesis Methods, Rapid Heating

Contents

Materials Research Foundations 189 (2026) 435-467 https://doi.org/10.21741/9781644904039-14

1. Introduction

The need for sustainable and renewable energy sources has intensified due to the growing world's population, coupled with the environmental consequences of fossil fuels. Nearly 84% of the world's energy requirements are fulfilled by the fossil fuels, which simultaneously leads increased emission of greenhouse gases and causing climate change [1]. Among various renewable energy options, solar energy is regarded as the most feasible option due to its inexhaustibility, abundance, and zero operational cost [2,3]. Green hydrogen may be produced in sustainable manner by employing photocatalytic water splitting to convert solar energy into hydrogen energy. It has drawn a lot of attention due it potential in solar-driven hydrogen synthesis since the ground-breaking work of *Fujishima and Honda (1972)* on TiO_2 photoelectromical water splitting [4].

The high energy density (120-143 MJ/kg), zero carbon emission and wide range of applications in the transportation, and industrial sectors making hydrogen (H_2) as the future fuel. However, the economic viability of photocatalytic hydrogen production is constrained due to their low solar-to-hydrogen (STH) conversion efficiencies, slow reaction rate and reliance on expensive cocatalysts [5,6]. Recent advancements in Nanostructuring, elemental doping, defect engineering and plasmonic enhancement have been investigated to improve optical absorption, charge separation and its overall photocatalytic activity. An STH efficiency greater than of 10% and the production cost less than of \$2.0 kg^{-1} are essential for large-scale adoption in order to compete with traditional methods [7,8].

Recently, microwave assisted synthesis shows a great potential in photocatalytic preparation, which enables efficient internal core heating of the material, ensuring uniform temperature distribution by direct coupling of microwave energy to the molecules present in the sample [9,10]. Since its accidental discovery in 1945 by Percy LeBaron Spencer, microwave technology has revolutionized various fields including, material science and synthetic chemistry. This process involved in the conversion of electromagnetic energy into thermal energy by two primary mechanism namely, polarization and conduction, where dipoles attempt to align with an oscillating electromagnetic field leading to a molecular collisions and heat generation on microwave irradiation [11–13]. Unlike high frequency fields that oscillate too rapidly for molecular response, or low frequency fields that allow full alignment without heat generation, microwave method strikes a balance and making it an ideal for efficient heating [14,15].

It's worth noting that photocatalyst prepared via microwave assistance methods show better efficiency than photocatalyst synthesised by using traditional method. For instance, *Yu et al.* synthesised Sn_3O_4/rGO planar heterostructures via microwave-assisted hydrothermal method, which greatly enhanced the light absorption, and interfacial charge transfer kinetics. As a result, the heterostructure demonstrated a hydrogen evolution rate of 20 mmol/g/h. Additionally, microwave assisted method effectively reduced the reaction time from 12 h to 4 h compared to conventional hydrothermal synthesis [16].

Moreover, the localized heating action promotes the formation of nanostructures with tailored morphology and reduced particles. In case of quantum dots like carbon quantum dot (CQDs), microwave irradiation induces carbonization and surface functionalization, which is crucial for achieving desired optical property. This may be effectively used by *Yashwanth et al*, who synthesized nitrogen and phosphorus co-doped CQDs using microwave irradiation, resulting in a 417 $\mu mol\ h^{-1}g^{-1}$ hydrogen production rate under visible light. All of these research works highlight that microwave-assisted techniques accelerate the synthesis of nano photocatalyst with tuned optical properties [17].

In this chapter, we have systematically explored advancements in microwave-assisted synthesis of photocatalysts, focusing on their role in enhancing hydrogen generation efficiency. Additionally, we have reviewed and organized various photocatalytic materials synthesized via different microwave-assisted methods, and their impact in hydrogen production. Finally, we have discussed the challenges and future prospects of the microwave-assisted photocatalyst for large-scale hydrogen production

2. Fundamentals of photocatalytic hydrogen generation

Photocatalytic water splitting involves in the direct conversion of solar energy into hydrogen energy (H_2) through photochemical decomposition of water molecule (H_2O). It utilizes the semiconducting material as a photocatalyst to absorb photon energy, typically from sunlight (UV-Visible range) [18]. The fundamental mechanism shares similarities between photoelectromical (PEC) water splitting, but unlike PEC, where hydrogen and oxygen evolution reaction occur spatially separate electrodes, photocatalytic processes take place on a single photocatalyst surface. The absorbed energy excites electrons from the valence band to the conduction band, generating electron-hole pairs. These charge carriers then separated and migrated to the surface of the semiconductor material, initiating the oxidation and reduction process, leading to the generation of hydrogen H_2 and oxygen O_2 gases [19]. With its ability to efficiently harness solar energy to generate hydrogen fuel without any hazardous byproducts, this technique provides considerable advantages as a clean and sustainable hydrogen production method [20]. The detailed process can be explained as below.

Firstly, the semiconductor absorbs a photon energy of *hv* that is greater than or equal to its bandgap (E_g), exciting an electron (e^-) from the valence band (VB) to the conduction band (CB), leaving a hole (h^+) in the VB. This excitation results in the generation of electron-hole pairs (refer with: Eq. 1):

$$\text{semiconductor} + h\nu \longrightarrow e^-_{CB} + h^+_{VB} \quad (1)$$

For the effective water dissociation process, the photon's energy must be at least of 1.23 eV (ΔG= +237 KJ/mol) to overcome the threshold energy. Secondly, the photogenerated electrons and holes must be effectively separated to prevent recombination. Once the charges are separated and migrate to the surface of photocatalyst, they undergo redox reaction. The holes (h^+) in the VB oxidizes the water molecule, producing proton (H^+) and oxygen gas (O_2) gas (refer with: Eq. 2):

$$2H_2O \rightarrow O_2 + 4H^+ + 4e^- \quad (E = 1.23V \text{ Vs NHE}) \tag{2}$$

Meanwhile, the electrons (e^-) in CB reduce the protons (H^+) to form hydrogen gas (H_2) (refer with: Eq. 3):

$$2H^+ + 2e^- \rightarrow H_2 \quad (E = 0V \text{ Vs NHE}) \tag{3}$$

The overall photocatalytic water splitting reaction (refer with: Fig.1) can be expressed as[21],

$$H_2O \rightarrow H_2 + \frac{1}{2}O_2, \Delta E = 1.23V \text{ Vs NHE} \tag{4}$$

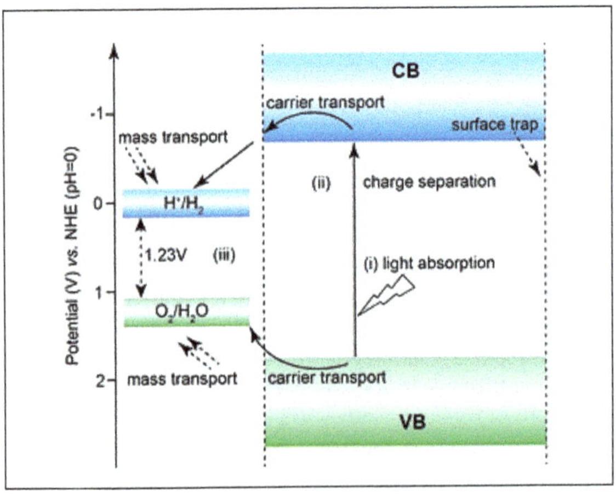

Figure 1. Mechanisms driving photocatalytic hydrogen generation. Reprinted with permission from ref. [21]. Copyright (2020) Elsevier.

Semiconductor's such as metal oxides, sulfides and various carbon-based materials have been extensively studied for photocatalytic hydrogen generation. Among them, metal oxides such as TiO_2, Fe_2O_3, and ZnO were extensively studied due to their availability and stability. In particular, oxide semiconductor with electronic configurations of d^0 (Ti^{4+}, Zr^{4+}) and d^{10} (Sn^{4+}, In^{3+}) are more advantageous when combined with cocatalysts, as d^0 cations possess empty d-orbitals that strengthen metal-oxygen bonding, while d^{10} cations with fully filled d-orbitals provides high stability of the material [22]. Incorporating cocatalysts like Pt and MoS_2 will acts as an electron sink, capturing photogenerated electrons to prevent electron-hole recombination, thereby enhancing hydrogen production by facilitating charge separation [23]. A few layer WS_2/ZnO heterostructure synthesized by Huang et al. demonstrates a 35% increase in H_2 evolution rate of 327 μmol h^{-1} g^{-1}, attributed to the synergistic effect between ZnO and WS_2 [24]. Moreover, wide-bandgap materials like $SrTiO_3$, $KTaO_3$ primarily absorb UV light which is 4% of the solar spectrum, making it necessary to modify these materials for visible-light absorption [25]. For example, a novel $SrTiO_3$ (STO) based heterostructured photocatalyst Ag/Ir/STO developed by Trang and his coworkers, extended its light absorption up to 730 nm and exhibits improved HER efficiency [26]. Additionally, narrow-bandgap materials such as g-C_3N_4 and Ag_3PO_4 have emerged as a potential candidate in visible light driven hydrogen production. Despite the advancements in photocatalytic hydrogen generation, low absorption, rapid charge recombination rate, and poor stability remain as critical challenge in large-scale production. The major factors that determine the photocatalytic activity can be divided into two types: (a) Intrinsic factors such as crystallographic phase, exposed crystal facets, crystallite size, dopants, defects and (b) extrinsic factors like pH, light intensity, and reaction conditions which plays a crucial role in determining photocatalytic performances.

Many of these challenges can be effectively addressed using microwave-assisted synthesis methods, as they enhance crystallinity, regulates the phase and morphology. The rapid and uniform heating provided by the microwave irradiation facilitates the controlled nucleation, better dispersion of dopants, enhances the surface properties and minimizes the charge recombination rates. This chapter will further discuss the basic mechanism of microwave heating and various microwave-assisted methods for synthesizing photocatalyst for hydrogen production.

3. Mechanism of Microwave Heating

A kind of electromagnetic radiation, microwaves operate at frequencies between 300 MHz and 300 GHz within a range of 1mm to 1m of wavelength, shown in Fig. 2. It is classified into three frequency bands with specific characteristic functions, namely, Ultra High Frequency (UHF-300 MHz to 3 GHz), Super High Frequency (SHF-3 GHz to 30 GHz), Extremely High Frequency (EHF-30 GHz to 300 GHz) and are widely used in both commercial & household appliances. Usually operating at 2.45 GHz, purposefully designed to avoid interference with communication and military frequency bands [27]. Depending on their interactions with microwave, materials are categorized as absorbers, reflectors, or transmitters. Absorbers are materials like ferrites, which effectively absorb microwaves and transform them into thermal energy quickly. While the reflectors fail to absorb the microwave, instead they redirect the energy (e.g., metals). Transmitters, which include Teflon and quartz, let microwaves travel through without being absorbed, making it a suitable material as a reaction vessel in microwave systems. Since it travels at the speed of light, they may efficiently transmit energy into the molecules of the material they interact with. This effective energy transfer leads to the rapid internal heating, which forms a foundation of microwave heating [28].

Materials Research Foundations 189 (2026) 435-467 https://doi.org/10.21741/9781644904039-14

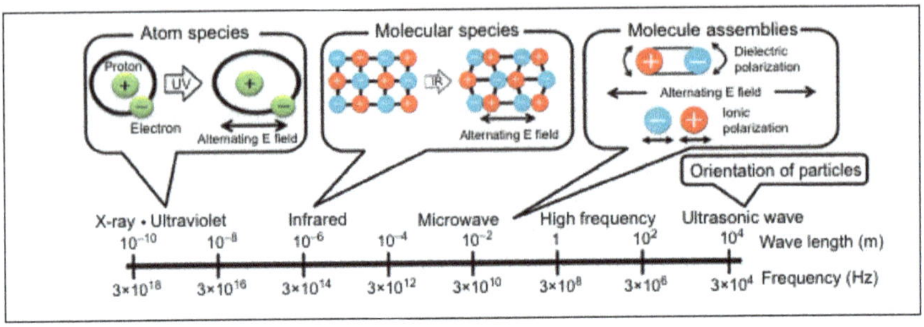

Figure 2. Absorption characteristics of materials across the electromagnetic spectrum. Reprinted with permission from ref. [29]. Copyright (2018) Elsevier.

Microwave heating relies on the interaction of electromagnetic radiation with the molecules in the reaction-medium, utilizing various mechanisms that transforms microwave energy into heat energy [9]. These mechanisms include ion conduction, dipole polarization, nuclear magnetic resonance, and magnetic hysteresis.
Mathematically, microwave heating can be described by the following equation

$$P=\frac{1}{2}\iiint \left(\sigma.E^2+ \omega.\varepsilon_0\varepsilon'' E^2+\omega.\mu_0\mu'' H^2\right)dV \quad (5)$$

Where, P is the power generated (heat energy), σ is the electrical conductivity, ε'' is the dielectric loss, ω is the angular frequency of the oscillating electric field, μ'' represents magnetic loss H and E are the magnetic and electric field. In this equation, first term represents the ohmic heating (from ion conduction), second term represents the dielectric heating (from polarization) and third term represents the heating effect due to magnetic losses [27]. In contrast to conventional techniques that depend on external heat transfer and frequently result in uneven heating, microwaves eliminate temperature gradients by penetrating the reaction medium. This causes heating to occur quickly and steadily, which greatly speeds up reaction kinetics - often up to 1000 times faster than with traditional techniques.

3.1 Microwave-Driven Electric Field Heating

The microwave driven electric field heating mainly depends on two kinds of mechanism, namely, dipole polarisation and ion conduction, as shown in Fig. 3. In the case of ionic conduction, the electric field causes ions to oscillate back and forth within the solution, resulting in sustained agitation due to friction and collisions [30]. This friction leads to an overall increase in temperature, contributing to efficient heating in the reaction medium.

Materials Research Foundations 189 (2026) 435-467 https://doi.org/10.21741/9781644904039-14

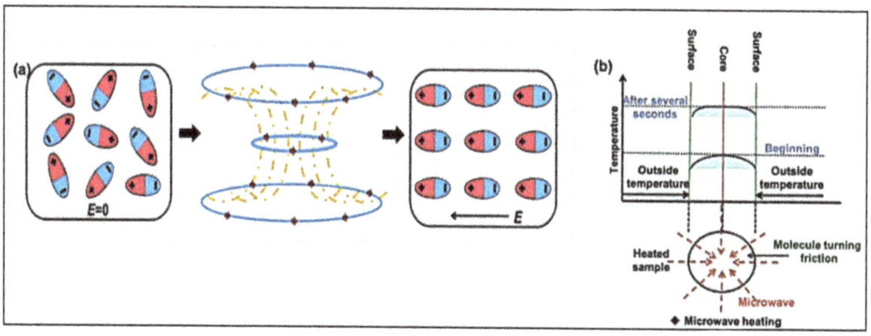

Figure 3. (a) Illustration of molecular polarization and reorientation, (b) Mechanism of microwave heating. Reprinted with permission from ref.[31]. Copyright (2015) Nature.

Dipole polarization occurs when polar molecules are exposed to microwave radiation which excites the molecule due to their inherent dipole characteristics. Simultaneously, the oscillating electromagnetic field induces the reorientation of these molecules and ions relatively in the direction of the electric field. This reorientation process results in molecular rotation tends the molecule to collide with one another causing dielectric losses, which ultimately convert microwave energy into rapid internal heating [32]. Conductive and non-conductive materials experienced heat through different mechanism when exposed to microwave radiation. In non-conductive materials, the alternating electric field causes the molecular rotation of the dipoles, aligning them in the direction of the changing electric field. This leads a friction between the molecules, which generates thermal energy. In contrast, conductive materials experience an additional heating effect due to Joule's law, where the electric field induced by microwave radiation generates an electric current, leading to resistive heating [33]. The efficiency of a material in converting electromagnetic energy into heat when exposed to an alternating electric field (microwave radiation) is given by a dielectric tangent loss factor:

$$\tan\delta_e = \frac{\varepsilon''}{\varepsilon'} \quad (6)$$

Where, ε' is a dielectric constant, which represents the material's ability to store electrical energy, and ε'' is the dielectric, which accounts for the dissipation of electrical energy as heat. In some cases, both the mechanism occurs simultaneously, especially in as system where conductive materials are dispersed within a non-conducting medium [34].

3.2 Microwave-Driven Magnetic Field Heating

The influence of the magnetic field (H-field) on microwave heating has been studied less than that of electric field heating. Nevertheless, its benefits for specific materials like ferrites and ferromagnets are due to their interaction with the magnetic field [22]. The main causes of microwave H-field heating which includes magnetic resonance loss, hysteresis loss, and eddy current loss. When conductive materials are subjected to an alternating magnetic field, they induce a current that generates Joule's heating, which leads to the eddy current loss [35]. It is essential for heating conductors and semiconductors. Hysteresis loss occurs in ferromagnetic materials due to

Materials Research Foundations 189 (2026) 435-467 https://doi.org/10.21741/9781644904039-14

the oscillating magnetic dipoles, producing heat through internal friction. Heating in ferrites and magnetic oxides is caused by magnetic resonance loss, involving electron spin resonance and domain wall resonance. The tangent loss factor for H-field heating is given by:

$$\tan\delta_m = \frac{\mu''}{\mu'} \quad (7)$$

Where, μ' is a real part of permeability (storage component) and μ'' represents a imaginary part of permeability (loss component) [34].

4. Microwave-Assisted synthesis Methods

4.1 Microwave-Assisted Hydrothermal method

Hydrothermal method refers to the synthesis of nanomaterials at high temperature and high vapor pressure to crystalize the material in an aqueous medium (water) [36–39]. It is employed in an autoclave (a sealed vessel) that creates a high-pressure environment for the material synthesis. This method offers several advantages, such as high phase purity, controlled size, morphology, and high crystallinity that are attributed by the controlled environment. Often hydrothermal method is integrated with microwave technique offering several distinct advantages over conventional method due to the combined principle of microwave heating and conventional hydrothermal synthesis [40,41]. Unless like conventional hydrothermal process, the faster and uniform distribution of heat generated due to microwave radiation, typically produces the materials with high crystallinity, well-defined morphology, and uniform particle size (refer with: Fig.4) [42–44].

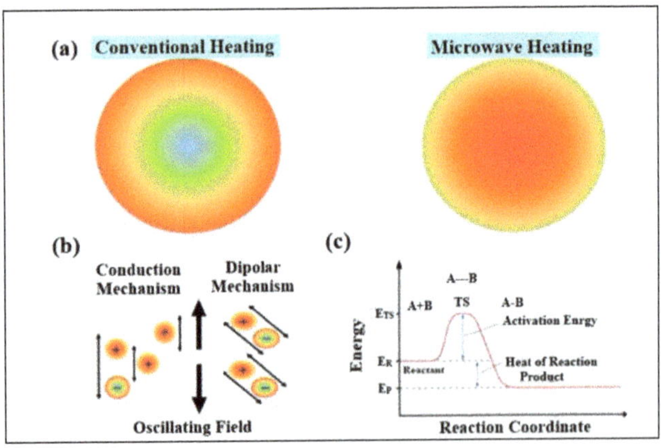

Figure 4. (a) The reaction mixture's temperature profile over time for both conventional heating and microwave treatment, (b) Illustration of the oscillating fields in Ion conduction and Dipolar mechanism, and (c) Activation energy diagram of a chemical process under microwave conditions. Reprinted with permission from ref. [45]. Copyright (2020). Elsevier.

Materials Research Foundations 189 (2026) 435-467 https://doi.org/10.21741/9781644904039-14

For example, Falara et al. synthesized and optimized carbon dot (CDs)/TiO_2 nanohybrids for enhanced photocatalytic hydrogen evolution by using hydrothermal and microwave-assisted methods. The microwave approach produced larger CDs with improved optical properties, leading to better charge transfer efficiency in TiO_2. UV-Vis, and XPS studies confirmed bandgap narrowing (3.21 eV to 3.03eV), increasing the light absorption. Additionally, photocatalytic tests showed that the CDs-2-MW/TiO_2 hybrid achieved the enhanced hydrogen production at a rate of 0.906 $\frac{\mu mol}{min}$ [46].

Recently, Lee et al. effectively showcased the advantages of microwave-assisted hydrothermal synthesis in improving the structural and electrochemical characteristics of MoS_2/graphene (MSG) hybrids for photocatalytic hydrogen evolution. Unlike traditional hydrothermal techniques, microwave heating offered rapid (20 minutes) and uniform heating, ensures high crystallization, and preventing the aggregation of MoS_2 on graphene. Higher defect concentrations and smaller crystallite sizes were shown by XRD analysis, improving the charge transport characteristics. While, PL analysis confirmed reduced charge recombination, attributed to microwave-induced sulfur vacancies that acted as electron traps. After testing a range of reaction time frames (10, 20, 30, and 60 minutes), the study concluded that 20 minutes was the optimum duration since shorter intervals resulted in partial MoS_2 production, while longer times generated particle aggregation, which decreased catalytic surface area [47].

As reported, Khan et al. to developed Yttrium (Y)-doped hydrogen titanate nanorods (HTN) with anatase-rutile heterostructures by employing microwave-assisted hydrothermal synthesis to improve the photocatalytic hydrogen evolution (HER) performance. Microwave heating played a crucial role in the controlled conversion of hydrogen titanate ($H_2Ti_2O_5.H_2O$) into anatase-rutile phases within just 6 hours - a much shorter time than conventional hydrothermal methods. According to XRD analysis, at 1 wt.% Y doping, 46% anatase and 5.7% rutile phases were formed. By facilitating Y^{3+} ion transport, microwave irradiation promoted oxygen vacancies and lattice distortions, as shown by XPS analysis. The TEM images showed that microwave heating resulted in well-dispersed anatase and rutile nanoparticles on HTN rods, ensuring a large surface area and exposure of the active sites. In comparison to pure HTN, the catalyst exhibited 5.5 times boost in H_2 evolution (32.4 μmol in 7 h) under ideal circumstances (1 weight percent Y-doping). This study highlights that microwave-assisted synthesis is an effective and scalable method for controlling phase formation, promoting anatase-rutile heterostructure growth, and enhancing the photocatalytic efficiency of hydrogen titanates for solar-driven hydrogen production [48].

Beyond accelerating synthesis, microwave heating also enhances material homogeneity, reduces the production of secondary phases, and prevents temperature gradients. Unlike conventional hydrothermal methods which involve high temperatures (230–260°C) and extended reaction times (~24 h), microwave assisted method effectively produces highly crystalline $LiNbO_3$ in a just 2-3 h at 260°C. These structurally improved photocatalyst by microwave heating has a directly impact the hydrogen evolution reaction (HER), where 2 h synthesized $LiNbO_3$ exhibited a higher H_2 production rate (10.5 $\mu mol g^{-1} h^{-1}$) compared to the 3 h sample, attributed to its higher surface area (25.7 m^2/g) and better charge separation efficiency (Deon et al.) [49].

Tan and colleagues studied the hydrothermal synthesis of a (2D/3D) $TiO_2/ZnIn_2S_4$ heterostructure with microwave assistance, exhibiting improved photocatalytic hydrogen generation in visible light. $ZnIn_2S_4$ (ZIS) nanostructures were grown in situ on TiO_2 nanosheets with the use of a

Materials Research Foundations 189 (2026) 435-467 https://doi.org/10.21741/9781644904039-14

microwave (150°C, 30 min), which promoted effective charge transfer, homogeneous crystallization, and strong interfacial bonding. FE-SEM and TEM revealed hierarchical nanostructures with flower-like ZIS microspheres anchored to TiO_2 nanosheets. UV-Vis DRS analysis showed a redshift in absorption (~2.28 eV bandgap for ZIS/TiO_2), extending visible light response. EIS measurements showed the lowest charge transfer resistance of $1.11 \times 10^3 \Omega$ and highest photocurrent density (185 μA). The optimized heterostructure catalyst exhibited a hydrogen evolution rate of 1167.38 $mmolg^{-1}h^{-1}$, which is 172.2 times higher than pristine. The catalyst also maintained stable hydrogen production over multiple cycles, demonstrating high durability [50]. These studies demonstrate that microwave-assisted hydrothermal synthesis is a viable method for improving the charge carrier dynamics of the photocatalyst and adjusting nanostructures, providing a feasible path for solar-powered hydrogen production

4.2 Microwave-Assisted Solvothermal method

Unlike the hydrothermal process, the solvothermal method uses non-aqueous/organic solvents like ethanol, glycols, dimethylformamide (DMF), toluene, etc., to crystallize the material at high temperature and pressure [51–53]. This method mainly depends on the choice of the solvent, which influences the solubility of reactants and the rate of nucleation. It operates at high pressure, usually between 1 atm-10,000 atm, and temperatures ranging from 100°C to 1000°C. Using non-aqueous or organic solvents is more benefitted as it provides better crystallinity, controlled morphology and size of the nanomaterial [54–56]. Generally, oxidation, reduction, and hydrolysis are the chemical reactions utilize by the process. This method is widely used to synthesize, transition metal dichalcogenides, quantum dots, etc. The microwave-assisted solvothermal method takes advantage of both the process and provides high-quality material with desired properties [57–59].

For instance, Lee et al. utilized a solvothermal method to synthesize Cu-doped ZnS photocatalysts followed by microwave heating at 450 MHz, highlighting its advantages in rapid reaction kinetics. Structural characterization using FE-SEM and TEM revealed well-defined nanospheres with reduced agglomeration, while XRD analysis confirmed a cubic ZnS phase with no secondary impurities, indicating successful Cu incorporation. While XRD examination verified a cubic ZnS phase free of secondary impurities, suggesting good Cu inclusion, structural characterization using FE-SEM and TEM showed well-defined nanospheres with decreased agglomeration. BET analysis revealed a significant increase in surface area to 91.43 m^2/g Cu-doped ZnS, which contributed to improving charge transfer efficiency. In addition to PL spectra confirming a decrease in charge carrier recombination, which improved photocatalytic performance, UV-Vis DRS findings showed a modest bandgap narrowing, allowing sustained light absorption. Electrochemical studies, including Mott-Schottky analysis, which revealed a decreased charge transfer resistance (Rct = 11.6 Ω) and a negative shift in the conduction band (~ -0.43 eV Vs NHE), facilitating efficient electron transport. Consequently, the improved Cu-doped ZnS maintained consistent catalytic activity across several cycles, achieving an excellent hydrogen evolution rate of 3717.8 $mmolg^{-1}h^{-1}$[60].

As reported by Yue et al. β-ketoenamine-linked covalent organic frameworks (COFs) synthesized via microwave-assisted solvothermal method shows a significant enhancement in the photocatalytic H_2 efficiency in seawater compared to pure water. The synthesized COFs demonstrates a high hydrogen evolution rate of 41.3 $mmolg^{-1}h^{-1}$, which was 1.66 times higher in seawater than in pure water, demonstrating the role of microwave irradiation in boosting catalytic performance in seawater. Fast and energy-efficient COF production was made possible by the microwave-assisted solvothermal approach. Additionally, the adsorption of metal salts allowed for in-situ polarization of the COF framework, lowering the exciton dissociation energy and raising the dielectric constant [61].

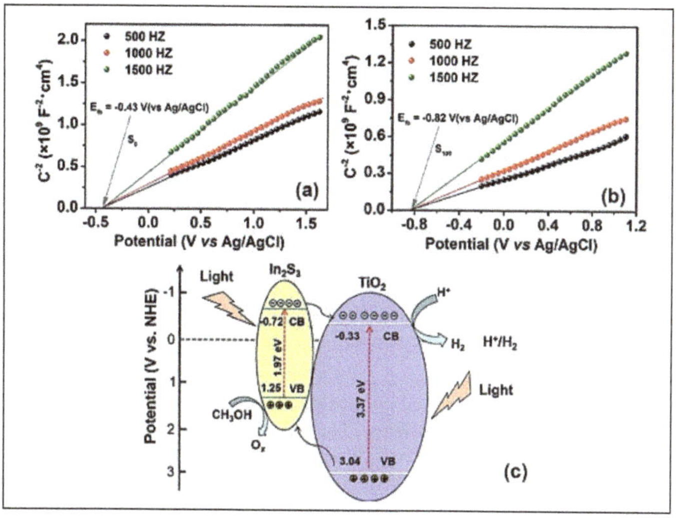

Figure 5. (a) and (b) represents the Mott-Schottky plots of mesoporous TiO_2 and In_2S_3 at 500, 1000, and 1500 Hz, with (c) a schematic illustration of charge transfer mechanism at the interface. Reprinted with permission from ref. [55]. Copyright (2020), Royal Society of Chemistry.

Likewise, Hu et al. highlighted the importance of the microwave-assisted solvothermal method in tailoring material properties for improved photocatalytic hydrogen production. Rapid and even heating is made possible by this method, which results in the controlled formation of In_2S_3/mesoporous TiO_2 nanocomposites with a distinct heterojunction. Ionic liquid-assisted microwave heating serves two purposes: it acts as a morphology-controlling agent to stop particle aggregation and as a microwave absorber to speed up nucleation. As a result, the synthesized nanocomposites exhibit ultrafine In_2S_3 nanoparticles anchored onto a mesoporous TiO_2 matrix, enhances the charge separation (refer with: Fig. 5) and provides abundant active sites for the hydrogen evolution. The optimized In_2S_3/TiO_2 nanocomposite achieved an H_2 evolution rate of 637.9 mmolg^{-1}h^{-1}[55].

4.3 Microwave-Assisted Co-precipitation method

Coprecipitation method involves in precipitation of two or more compounds from a solution, which leads to a supersaturated condition. This condition provides the necessary environment for the nucleation, growth, and agglomeration of the particles in a solution [62]. The primary process of nucleation is responsible for the particle formation, followed by secondary phase called Ostwald ripening which are crucial in determining size and morphology. Reaction rate, temperature, precipitating agent, and stirring speed all have a significant impact on the entire process [63]. The precipitating agent, such sodium hydroxide or ammonium hydroxide, regulates the precipitation process by changing the solution's pH, which influences the metal ions solubility and encourages the production of particles. Mainly, the nucleation is inversely proportionally to the particle size [64]. By carefully controlling the reaction condition, one can tailor the photocatalyst with optimized size, structure and surface area. Traditional coprecipitation method have some drawbacks like long reaction time, inhomogeneity in particle size, poor crystallinity. These issues can be overcome by the microwave assisted coprecipitation method, which promotes uniform heating, and enhance the control over particle size and morphology [65,66]. According to the study by El-Khair et al., microwave-assisted co-precipitation can improve the structural, optical, and electrochemical properties of $CuFe_2O_4$ nanoparticles (NPs) for hydrogen production. $CuFe_2O_4$ was synthesized by precipitating Cu^{2+} and Fe^{3+} with NaOH (pH 12) at 95°C, followed by microwave irradiation (1100 W, 2450 MHz) for 10 minutes and calcination at 600°C. XRD analysis confirmed a highly crystalline $CuFe_2O_4$ spinel structure, while XPS revealed well-defined oxidation states of Cu^{2+} and Fe^{3+}, ensuring effective catalytic activity. UV-Vis spectroscopy demonstrated a bandgap reduction (~2.0 eV for $CuFe_2O_4$) and extension of visible-light absorption, whereas PL spectra confirmed suppressed electron-hole recombination [67].

Likewise, Chen et al. developed a graphene oxide/CdS/single crystal ZnO nanorod (GCZ) hybrid photocatalyst using a microwave-assisted co-precipitation method, which significantly enhanced photocatalytic hydrogen production efficiency. The microwave-assisted synthesis played a crucial role in accelerating nucleation, and promoting uniform dispersion of GO within the ZnO/CdS heterostructure, leading to reduced recombination losses. The influence of graphene oxide (GO) content on photocatalytic performance was systematically analyzed, revealing that the optimized GO loading of 2.91 wt.% resulted in the highest H_2 production rate of 6.511 $\mu mol g^{-1} h^{-1}$. The development of well-integrated heterojunction was achieved by microwave irradiation with enhanced charge carrier dynamics and light absorption [68]. These findings highlight the effectiveness of microwave-assisted synthesis in optimizing crystallinity, and charge transport, making it a potential method for efficient and sustainable hydrogen production

4.4 Microwave-Assisted Sol-gel method

Sol-gel method is of the bottom up technique used for synthesizing nanomaterials, especially metal oxides and ceramics. It occurs by hydrolysing the metal precursors and turning to a sol phase, followed by condensation and polymerization process that transforms into solid gel. The aging of the gel enhances the structural integrity, followed by drying [69–71]. Sol gel derived materials like TiO_2, ZnO and SnO_2 exhibits better photocatalytic performance for hydrogen production. Despite its advantages in controlling stoichiometry of the materials, often it's a time-consuming process which may lead to the aggregation of nanomaterials and in turn affects the surface area. In contrast microwave assisted sol-gel method significantly reduces the reaction time for gelation, drying and heat treatment. This method also minimizes the defect formation resulting in better photocatalytic property [72–74].

For instance, Chen et al. developed a microwave-assisted sol-gel to develop SiO_2 aerogel/ZnCdS composites, aiming to enhance the photocatalytic hydrogen production efficiency. In this approach, SiO_2 aerogel was synthesized via the sol-gel method acting as a carrier to support ZnCdS photocatalysts, and microwave irradiation was employed to facilitate rapid and uniform heating for material synthesis. The results demonstrate that microwave-assisted synthesis improved the dispersion of ZnCdS nanoparticles within the SiO_2 aerogel matrix, leading to the enhanced charge separation and minimized electron-hole recombination. Moreover, the optimized SiO_2/ZnCdS composite exhibited the highest hydrogen evolution rate of 4032.6 $\mu molg^{-1}h^{-1}$, which was 2.4 times higher than that of pristine ZnCdS. This study highlights the synergistic effect of sol-gel derived aerogels and the microwave-assisted synthesis [75].

As reported by Kwon et al., a one-pot microwave-assisted nonaqueous sol-gel method was employed to synthesize Pd-modified TiO_2 nanocrystals, which were subsequently processed into 3D aerogel monoliths. The aerogel structure, obtained through sol-gel processing followed by supercritical drying, retained the Pd-modified TiO_2 nanostructure, ensuring high porosity and a large surface of 422.7 m^2g^{-1}. This approach facilitated the rapid crystallization and simultaneous incorporation of Pd ions into the TiO_2 surface, leading to the enhanced light absorption. As a result, it achieved a remarkable H_2 production rate of 117.5 $\mu molg^{-1}h^{-1}$, maintaining excellent stability over three days of continuous operation. Furthermore, the structural integrity and high surface area achieved through this method contribute to better light absorption and catalytic activity, emphasizing the effectiveness of the microwave-assisted sol-gel method [76].

4.5 Microwave-Assisted Thermolysis method

Thermolysis, also known as thermal decomposition is a chemical process that breaks down materials through an endothermic reaction, requiring external heat supply to break chemical bonds. The decomposition temperature varies depending on the substance [77–80]. Unlike combustion method, thermolysis does not always need oxygen. In contrast, microwave-assisted thermolysis directly interacts with polar molecules or conductive material generating rapid internal heating. While the conventional method often suffers from heat loss and potential overheating, microwave assisted thermolysis provides more controlled and effective heating [81–84].

Zhang et al. addressed the challenge of high charge carrier recombination rates in semiconductor-based photocatalysts, which severely restrict the efficiency of solar-to-fuel conversion. To overcome this limitation, they used a microwave-assisted thermolysis approach with melamine-critic acid aggregates as precursors to developed a π-conjugated in-plane heterostructure between graphitic carbon nitride (g-CN) and carbon rings (Crings), denoted as g-CN/Crings. This well-integrated g-CN/Crings heterostructure was made possible by microwave assisted synthesis, which improved charge carrier utilization during photocatalysis and successfully suppressed excited-state decay. Femtosecond time-resolved transient absorption spectroscopy revealed that Crings incorporation resulted in slower exciton annihilation (τ_1= 7.9 ps) and prolonged shallow electron trapping (τ_2 = 407.1 ps) compared to pristine g-CN (τ_1 = 3.6 ps, τ_2 = 264.1 ps), enabling more photo induced electrons to participate in the photocatalytic reaction [85].

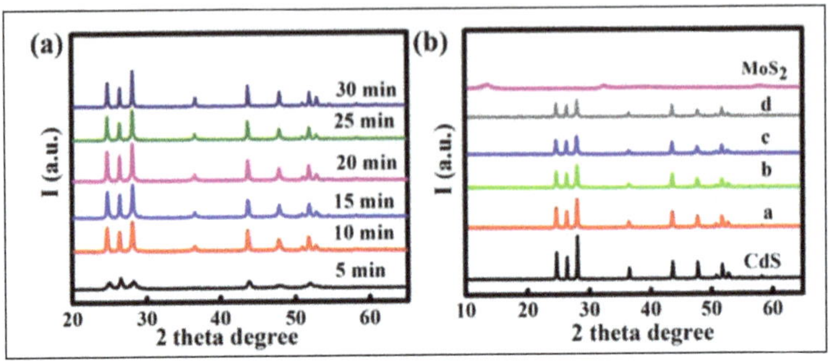

Figure 6. XRD patterns of (a) CdS at different microwave heating times, indicating improved crystallinity up to 20 min and a peak shift at 30 min (b) CdS/MoS_2 after 25 min (optimized duration) of microwave heating, with MoS_2 ratios of 1%, 3%, 5%, and 8%. Reprinted with permission from ref. [86]. Copyright (2018), Royal Society of Chemistry.

Likewise, Jiang et al. demonstrated the effectiveness of microwave-assisted thermolysis in enhancing the photocatalytic activity of MoS_2/CdS composites for hydrogen generation. In this approach simultaneous crystallization (refer with: Fig. 6) and heterojunction formation was achieved by facile single-step microwave process for a chosen duration at 700 W power. With a hydrogen evolution rate of $11.4 mmol g^{-1} h^{-1}$, the optimized MoS_2/CdS hybrid (3 wt. % MoS_2) showed the promise of noble-metal-free photocatalysts. Additionally, the microwave assisted heating method made easier synthesis of ultrafine MoS_2 nanoparticles on CdS, which also increased the surface area available for catalytic reactions and prevented agglomeration [86].

4.6 Microwave-Assisted Pyrolysis method

A high-temperature process that involves the decomposition of organic material in the absence of oxygen is called pyrolysis. During this process, the precursor is thermally decomposed to form a desired product. The temperature and duration of the pyrolysis process significantly influence the crystallinity, morphology, and structure of the resulting product [87,88]. Pyrolysis allows for the synthesis wide range of photocatalyst like TiO_2, Fe_2O_3, and carbon-based hybrid structure. Pyrolysis may risk of particle aggregation at high temperature and the need for precise control over temperature should be ensure to optimal material property. Microwave assisted pyrolysis is beneficial in such that the uniform and rapid heating promote the formation of well-defined crystal structure and improve the charge carrier dynamics. One of the best ways for green synthesis of carbon-based materials [89,90].

Materials Research Foundations 189 (2026) 435-467 https://doi.org/10.21741/9781644904039-14

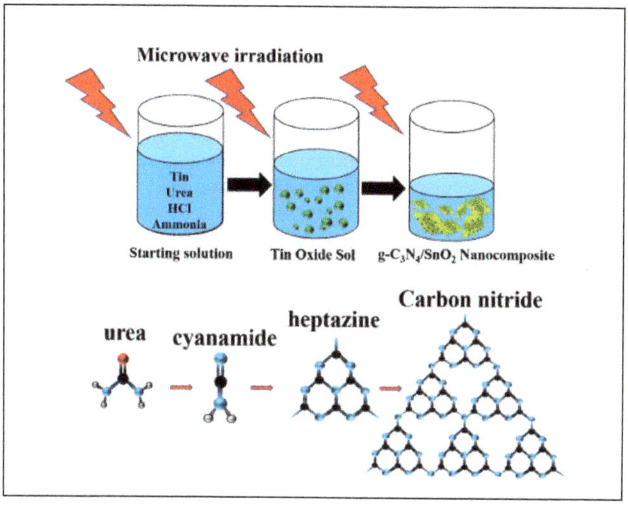

Figure 7. Schematic pattern of carbon nitride synthesized via microwave irradiation and urea pyrolysis with tin derivatives. Reprinted with permission from ref.[91]. Copyright (2018), Elsevier.

Seza et al. synthesized a novel porous $g-C_3N_4/SnO_2$ nanocomposite via a microwave-assisted pyrolysis method for solar water splitting (refer with: Fig. 7). The study explored the effect of SnO_2 content on the nanocomposite's structure and photocatalytic properties. Microwave irradiation of 2.45 GHz and 900 W promoted the in-situ formation of SnO_2 nanoparticles, which in turn induced urea pyrolysis into $g-C_3N_4$, assisted by surface OH groups on SnO_2. The optimized nanocomposite exhibited an enhanced photocurrent density of 33 $\mu A\ cm^{-2}$ under visible light, making it a promising photocatalyst. BET analysis revealed superior porosity of 195 $m^2\ g^{-1}$, indicating enhanced surface activity. Optical analysis (UV-Vis DRS and PL) demonstrated a reduced band gap of 2.98 eV, thereby facilitating efficient charge transfer [91].

5. Photocatalysts from Microwave-Assisted synthesis

As discussed earlier, the photocatalyst synthesised by microwave-assisted methods has drawn a lot of attention due to its quick and energy-efficient production. Among the various classes of material, the photocatalytic hydrogen generation has been primarily explored for, transition metal oxides, chalcogenides, carbon-based materials and MOF-based composites, owing to their photocatalytic properties. Each class offers distinct advantages in terms of light absorption, charge separation, and stability. Microwave-assisted synthesis further enhances these features by promoting better crystallinity and uniform distribution of active sites. Together, these attributes contribute to the development of highly efficient, scalable, and robust photocatalytic systems, as discussed below.

Materials Research Foundations 189 (2026) 435-467 https://doi.org/10.21741/9781644904039-14

5.1 Transition metal oxides

Transition metal oxides are promising material for photocatalytic hydrogen generation because of their flexible electronic structure, less toxicity and high availability. Their bandgap energies approximately of ~2.0-3.2 eV, allows them to absorb visible and UV light, which is necessary for effective water splitting. However, materials like TiO_2 (3.2 eV), Fe_2O_3 (2.1 eV), WO_3 (2.6 eV), and $BiVO_4$ (2.4 eV) requires structural modifications and bandgap adjustment due to their limited light absorption, poor charge separation, and photo-corrosion [92,93]. To improve crystallinity, increase photocatalytic activity, and regulate morphology, microwave-assisted synthesis can often be used. Unlike traditional heating, microwave irradiation generates localized hotspots that speed up the nucleation process, and phase shift by generating quick, homogenous energy transfer through dielectric heating and dipolar polarization [94,95].

For example, Machut et al. used a microwave-assisted technique to synthesize gold-doped TiO_2, which greatly reduced the synthesis time and allowed for fast crystallization at 150°C without the need for extra calcination. Through plasmonic effects, gold nanoparticles were evenly deposited via microwave-assisted reduction, improving charge separation and absorption of visible light. Ethanol served as a hole scavenger in a 20 vol% ethanol-water solution used to assess the photocatalytic hydrogen evolution in a Pyrex reactor under UV LED irradiation. In order to eliminate dissolved oxygen, the catalytic solution was purged with argon gas (420 mL h^{-1}) for 60 minutes. An online micro gas chromatograph with a thermal conductivity detector and separating columns was used to track the H_2 production. This showed excellent photocatalytic hydrogen efficiency and demonstrated the effectiveness of the microwave-assisted method [96].

Spinal oxides like ferrites exhibit excellent thermal and chemical stability. A narrow bandgap of this material makes suitable for visible light absorption, and appropriate energy band levels desirable for efficient water redox reaction. In addition to this, ferrites can be integrated into composites with other materials such as g-C_3N_4, TiO_2, etc. to increase photocatalytic efficiency by reducing electron-hole recombination, enhancing stability, and increasing absorption of visible light [97]. Additionally, spinel ferrites may be effectively reused in photocatalytic systems because to their room temperature magnetic properties will facilitates the recovery and recycling of the material. Yu et al. used a quick and effective microwave-assisted reflux technique in ethylene glycol (EG) without surfactants to create nanostructured M(II)-substituted $M_xFe_2O_4$ (M = Fe, Mn, or Co; x = 0–1) magnetic ferrite spherolites. Under microwave heating, the ferrite nanostructures developed in 30 minutes via an aggregation-based process [98].

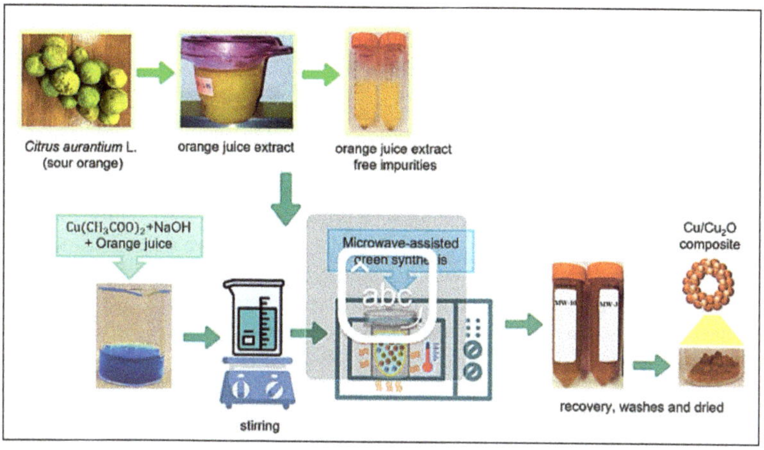

Figure 8. Rapid microwave-assisted green synthesis of Cu/Cu_2O. Reprinted with permission from ref. [99]. Copyright (2023), Elsevier.

As reported, Becerra-Paniagua et al. synthesized Cu/Cu_2O composites using a microwave-assisted green synthesis approach (refer with: Fig.8), employing Citrus aurantium L. as a natural reducing agent to replace conventional toxic chemicals. In comparison to conventional procedures, which need 30 to 60 minutes for reaction time, the microwave heating approach allowed for the production of hollow spherical Cu/Cu_2O structures at a relatively low temperature (80°C) in 3 to 10 minutes. The increased Cu phase content in the composites enhanced visible-light absorption, leading to a significant improvement in photocatalytic hydrogen production. Photocatalytic hydrogen generation experiments demonstrated that the Cu/Cu_2O composite exhibited a linear increase in H_2 evolution, with the MW-10 sample producing 7.06 μmol of H_2 in 60 minutes, achieving a high hydrogen evolution rate of 141 μmol/g, which was double that of the MW-3 sample (54 μmol/g). This enhanced photocatalytic performance is attributed to optimized optical properties, where the MW-10 sample, with a higher Cu phase content and a reduced bandgap (2.0 eV), absorbed more visible light, leading to improved charge separation and transport [99].

Hemandez-Majalca et al. synthesized the GO-TiO_2 nanocomposites via a visible-light-assisted anchoring method, enhancing photocatalytic hydrogen production. They have utilized the microwave pretreatment of graphite for 15s, and synthesized graphene oxide (GO) by using the modified Tour technique. The microwave pretreatment played a crucial role in reducing oxygen-containing functional groups, improving dispersion, and enhancing electronic conductivity. The X-ray diffraction revealed a peak widening, suggesting exfoliation of GO and decomposition of the oxygen group, which improved the electron transport property. FE-SEM confirms a flaking morphology with well-dispersed TiO_2 nanoparticles. A bandgap decreases from 3.1 eV (TiO_2) to 2.6 eV (GO-TiO_2) was shown by UV-Vis spectroscopy, increasing visible-light absorption and improving photocatalytic efficiency. The GO-TiO_2 nanocomposite achieved an impressive H_2 production rate of 6500 mmol H_2/g, significantly surpassing pristine TiO_2 (460 mmol H_2/g). The

Materials Research Foundations 189 (2026) 435-467 https://doi.org/10.21741/9781644904039-14

combination of TiO_2 and microwave-exfoliated GO shows promise as a method for efficient and sustainable photocatalytic hydrogen production [100].

5.2 Transition metal Chalcogenides

Despite tremendous advancement, some problems remain, including large bandgaps in metal oxides, which confine their absorption to the UV range, and poor charge separation due to rapid electron-hole recombination. Furthermore, certain semiconductors are not appropriate for overall water splitting since they prefer either water oxidation or reduction. To address these limitations, transition metal chalcogenides have gained attraction as an effective photocatalyst and cocatalysts in water splitting systems [101]. Among them, transition metal dichalcogenides with generic formula MX_2 (M = Mo, W, Ti, Zr, Hf, Nb; X = S, Se, Te), have been extensively researched. Because of their distinct electrical and structural characteristics, transition metal sulfides are being investigated extensively for photocatalytic hydrogen generation. These materials have high surface-active sites, which promote catalytic processes, and small bandgaps, which enable effective absorption of visible light [102,103]. Additionally, rapid charge carrier separation is made possible by their layered architectures, further decreases the recombination. The great light harvesting properties, adjustable band edges, surface amorphization and outstanding chemical stability, making them an efficient photocatalyst for hydrogen generation [104,105].

$ZnIn_2S_4$, a layered ternary sulfide, is particularly attractive for solar-driven hydrogen production. Its layered crystal structure makes it feasible to produce nanosheets with excellent light absorption efficiency and redox properties. However, it is limited by low electron transport efficiency and rapid recombination rate. For this introducing sulfur or zinc vacancies is seen as a viable option to improve charge separation and carrier mobility. Although the majority of research has been focused on composite making, Chang et al. opted for microwave assisted method to synthesize $ZnIn_2S_4$ nanosheets with sulfur vacancies, which significantly improves the photocatalytic hydrogen evolution. In addition to accelerated the synthesis process to just 1 hour at 140°C, it also regulated sulfur vacancy formation, which lowered the bandgap to 2.82 eV and improving visible-light absorption and electron mobility. Highlighting the benefits of microwave-assisted synthesis in improving catalytic performance, the resultant $ZnIn_2S_4$ nanosheets demonstrated hydrogen generation rates that were 33.3-fold and 16.6-fold greater than those of In_2S_3 and ZnS nanosheets, respectively [106].

A similar study was done by Wang et al. who successfully synthesized ZnCdS (bimetallic sulfides) nanoparticles by using microwave approach owing to its narrow bandgap characteristic and charge separation efficiency. A comprehensive investigation was conducted into the impact of microwave power and irradiation duration on photocatalytic performance. The development of a ZnCdS solid solution with higher H_2 evolution activity was validated by using electron microscopy and X-ray diffraction (XRD) studies. The study revealed that optimal hydrogen production was achieved at a microwave power of 320 W and an irradiation time of 20 minutes, yielding an impressive H_2 evolution rate of 2031 $\mu mol g^{-1} h^{-1}$, significantly outperforming the conventionally synthesized ZnCdS catalyst. Better light absorption and charge carrier mobility were made possible by homogenous elemental distribution, higher crystallinity, and microwave-induced defect engineering, which were all credited with the improved performance [107].

To address the challenges such as poor solubility, limited reactant contact and the toxicity of the traditional precursors like hydrazine hydrate, limiting for large scale application Monga et al. synthesized $MoS_2/MoSe_2$ heterostructures via a microwave-assisted method, demonstrating superior photocatalytic hydrogen production. The mixture was heated for 20 minutes at 180 C in an Anton Paar Monowave 300 microwave synthesizer to create the $MoS_2/MoSe_2$ composites. The microwave-assisted approach lowers the environmental risks by allowing low-temperature processing, quick reaction times, and the creation of well exposed active sites by rapid local heating. Improved charge carrier separation and visible-light absorption resulted from the produced $MoS_2/MoSe_2$ composite's large surface area (60 $m^2\ g^{-1}$), synergistic effect (1.73), and narrow bandgap (~1.89 eV). By facilitating quick synthesis, homogeneous heterostructure development, and well-exposed active sites by the microwave-assisted synthesis greatly promotes charge transfer and decreased recombination rate [108].

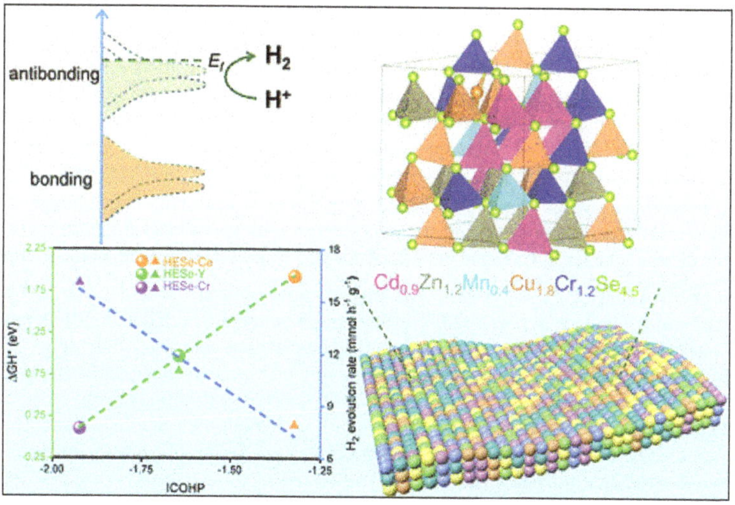

Figure 9. 2D HESes with ICOHP (Integrated Crystal Orbital Hamilton Population) Vs. ΔGH (Hydrogen adsorption free energy correlation plot). Reprinted with permission from ref.[109]. Copyright (2024), American Chemical Society.*

Likewise, Wang et al. explored the potential of high-entropy materials (HEMs) for photocatalytic applications, addressing the challenge of fine-tuning their morphological properties, which has long hindered their practical implementation. To overcome this limitation, they developed a diethylenetriamine (DETA)-assisted microwave hydrothermal method, a rapid and energy-efficient approach for synthesizing two-dimensional high-entropy selenides (HESes). The optimized $Cd_{0.9}Zn_{1.2}Mn_{0.4}Cu_{1.8}Cr_{1.2}Se_{4.6}$ HESes exhibited an impressive hydrogen evolution rate of 16.08 $mmolg^{-1}h^{-1}$ and a quantum efficiency of approximately 30% under 420 nm monochromatic LED irradiation. The enhanced performance was attributed not only to the increased specific surface area and efficient charge carrier separation but also to the microwave-assisted formation of the Cd–H ads bond (refer with: Fig.9), which improved catalytic reaction kinetics [109].

5.3 Carbon-based materials

For solar-driven hydrogen generation, carbon-based nanomaterials and their hybrids, including carbon nitrides, reduced graphene oxides, carbon nanotubes, and carbon dots, were considered as a viable substitution for metal-free photocatalyst owing to its cost-effectiveness and environmentally friendly nature [110]. These materials have appropriate bandgaps (~2 eV) for effective absorption of solar light, as well as sufficient overpotentials and reorganization energies to satisfy the thermodynamic conditions required for photocatalytic water splitting. Furthermore, these materials have a number of structural and elemental flaws that increase the concentration of free carriers on the surface, improving charge separation and removing the need for extra metal-based cocatalysts [111].

For instance, Tan et al. employed a two-step microwave-assisted approach to improve the yield of g-C_2N_3 by five times compared to conventional thermal methods. The microwave pretreatment of a damp urea precursor facilitated the formation of polar functional groups, such as biuret and cyanuric acid, which minimized gas loss during calcination, resulting in a porous structure with a higher surface area (120.7 $m^2\ g^{-1}$) and enhanced light absorption properties. The synthesized g-C_2N_3 exhibited a band gap of 2.94 eV, favoring visible-light-driven photocatalysis. the material exhibited a hydrogen evolution rate of 1.93 $mmolg^{-1}h^{-1}$, outperforming conventional graphitic carbon nitride. Because of their large surface area, stability, adjustable band structure, and capacity to promote charge transfer, g-C_2N_3 are beneficial for photocatalytic hydrogen generation [112].

Graphitic carbon nitride (g-C_3N_4), a well-known photocatalyst for solar-driven hydrogen production, which may achieve up to 80% efficiency in heterostructure based systems and 30% through intrinsic changes. However, its potential for metal-free complete water splitting is still largely untapped, and the majority of investigations rely on hole scavengers or electron-donating molecules to drive the oxidation half-reaction. In order to accomplish total charge separation and accumulation and enable effective overall water splitting without the need for external sacrificial agents, future research should concentrate on the customized design of g-C_3N_4. Wu et al. used a microwave-assisted thermolysis technique to synthesize graphitic carbon nitride (g-C_3N_4) with prominent lone-pair excitation by irradiating with microwaves at 700 W power for 15 min in a domestic microwave oven. The rapid microwave heating process facilitated structural distortion and defect formation, activating the n-p transition* and significantly extending the visible-light absorption range of g-C_3N_4. In comparison to traditionally synthesized g-C_3N_4, which had a maximum H_2 production rate of 3.2 $mmolg^{-1}h^{-1}$, the improved material demonstrated a sevenfold increase in hydrogen evolution, reaching a maximum H_2 generation rate of 21.4 $mmolg^{-1}h^{-1}$. Additionally, the material exhibits a continuous hydrogen evolution at 490 nm, demonstrating how well microwave-assisted lone-pair electron excitation increases photocatalytic activity. Unlike traditional thermal polymerization, which requires prolonged heating, the microwave-assisted method enabled rapid supramolecular polymerization and controlled morphology, leading to enhanced photocatalytic efficiency [113].

Similarly, Genc et al. synthesized Green Carbon Dots (GCDs) from Aloe vera leaves using a microwave-assisted technology, which provides a straightforward, economical, time-efficient, secure, and eco-friendly method. GCDs/TiO_2 composite was obtained by sensitizing TiO_2 with

Materials Research Foundations 189 (2026) 435-467 https://doi.org/10.21741/9781644904039-14

GCDs, enhancing its photocatalytic properties. Using triethanolamine (TEOA) as a sacrificial electron donor, the photocatalytic hydrogen evolution activity of GCDs/TiO_2 was assessed in an aqueous solution under visible-light irradiation ($\lambda > 420$ nm). With a hydrogen production rate of 718 $mmolg^{-1}h^{-1}$, the GCDs/TiO_2 photocatalyst exhibits an enhanced hydrogen evolution [114].

5.4 MOF (Metal-organic Framework) based materials

Metal-organic frameworks (MOFs) have emerged as a promising class of photocatalysts for hydrogen production due to their unique features, including high surface area, structural tunability, and diverse chemical compositions [115]. Their crystalline and extremely porous structure increases photocatalytic performance by facilitating electron-hole separation and exposing more active sites. Pure MOFs usually have poor separation and charge transfer capabilities. To get around this, creating MOF nanocomposites with inorganic semiconductors has been successful in enhancing the rates of hydrogen evolution and photo-stability [116]. Long-wavelength absorbing ligands increase their visible-light responsiveness, whereas MOF's organic linkers with extended conjugation systems improve electron transport. Through heterojunction creation, these alterations enable MOF-based composites to decrease electron-hole recombination in addition to using visible and near-infrared light [117].

For instance, Ayala et al. developed a novel titanium-based MOF, COK-47 (Centrum voor Oppervlaktechemie en Katalyse-47), and optimized it using a microwave-assisted synthesis approach to enhance photocatalytic hydrogen evolution. By using lower temperatures and longer reaction durations, this technique successfully reduced ligand defects that usually result in charge recombination and allowed the creation of isotropic, defect-free particles ($COK\text{-}47_{ISO}$) with a large surface area. With HER rates of 29.5 $\mu mol\ h^{-1}$ (Apparent Quantum Yield (AQY) 2.0%) at 365 nm and 8.6 $\mu mol\ h^{-1}$ (AQY 0.5%) at 405 nm, $COK\text{-}47_{ISO}$ shown exceptional performance under both UV and visible light, making it one of the most effective visible-light-active MOFs to date. Additionally, the material showed durability by maintaining structural stability even after extended exposure to light [118].

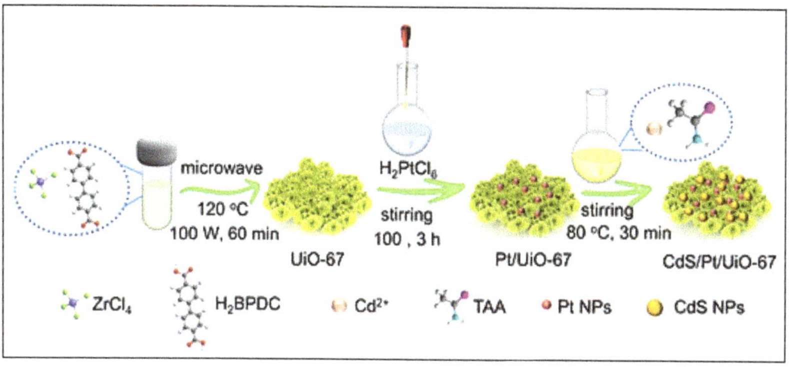

Figure 10. Synthesis of CdS/Pt/UiO-67 Composites. Reprinted with a from permission from ref.[119]. Copyright (2022), Elsevier.

Likewise, Tang et. al. addresses the photo-corrosion and electron-hole recombination issues of CdS by constructing a CdS/Pt/UiO-67 composite. Hierarchically structured Zr-based MOF (UiO-67) was synthesized via microwave-assisted assembly of ultrathin nanosheets and used as support and electron trap (refer with: Fig. 10). This hierarchical structure offers high surface area for uniform Pt and CdS dispersion. CdS nanoparticles were loaded onto Pt-decorated UiO-67 to enhance interfacial contact and charge separation. Among various compositions, CdS/$Pt_{1.5}$/UiO-67(20) exhibited the highest photocatalytic H_2 evolution rate under visible light. This enhancement stems from the strong interfacial contact and dual charge transfer pathways. Mott-Schottky analysis supported thermodynamically favourable electron transfer. The composite also showed good structural stability after repeated use, highlighting the excellence of photocatalyst in H_2 generation [119].

Conclusion

Microwave-assisted synthesis is an advanced technique with broad applications in the environment and energy sectors, particularly in the development of photocatalysts for hydrogen generation. This chapter examines the potential uses of combined microwaves assisted method with conventional methods and gives a general overview of their heating process in preparing photocatalyst for H_2 production. This technique offers several advantages due to its selective, rapid and uniform volumetric heating, resulting in better control over particle size, morphology, and crystallinity. In many cases, microwave irradiation allows direct heating of the reactants, which accelerates homogeneous nucleation and significantly reducing crystallization time. This is particularly beneficial in pyrolysis processes, where it effectively shortens the typically prolonged heating duration. Furthermore, the internal and volumetric heating nature of microwaves has enabled the successful synthesis of highly porous materials with enhanced surface area, which are essential for efficient light absorption and catalytic activity in hydrogen generation. Consequently, photocatalysts produced by this method exhibit improved water-splitting performance under light irradiation with enhanced charge transfer mechanism and demonstrate excellent stability and durability - making them promising candidates for large-scale hydrogen generation. Despite these promising outcomes, several critical challenges must be addressed before microwave assisted synthesis can be fully realized for industrial-scale hydrogen production. For instance, the limited penetration depth of microwaves often leads to localized overheating or non-uniform heating in bulk reaction volumes, which can compromise the structural homogeneity and reproducibility of the photocatalyst. Moreover, the technique remains dependent on the dielectric properties of the precursors and solvents, restricting the versatility of the synthesis and preventing the use of several high-performance materials that poorly absorb microwaves. Some highly efficient photocatalysts cannot be synthesized through microwave-assisted methods due to their poor microwave absorption. With ongoing advancements in reactor design, precursor engineering, and the integration of renewable energy inputs for microwave generation, microwave-assisted synthesis stands out as an ecofriendly, energy-efficient, and scalable approach for producing next-generation photocatalysts for sustainable hydrogen generation.

Reference

[1] T.N.Q. Trang, T.B. Phan, N.D. Nam, V.T.H. Thu, In situ charge transfer at the Ag@ZnO photoelectrochemical interface toward the high photocatalytic performance of H2 Evolution

and RhB degradation, ACS Appl. Mater. Interfaces. 12 (2020) 12195-12206. https://doi.org/10.1021/acsami.9b15578

[2] E. Ha, S. Ruan, D. Li, Y. Zhu, Y. Chen, J. Qiu, Z. Chen, T. Xu, J. Su, L. Wang, J. Hu, Surface disorder engineering in ZnCdS for cocatalyst free visible light driven hydrogen production, Nano Res. 15 (2022) 996-1002. https://doi.org/10.1007/s12274-021-3587-5

[3] T. Banerjee, F. Podjaski, J. Kröger, B.P. Biswal, B.V. Lotsch, Polymer photocatalysts for solar-to-chemical energy conversion, Nat. Rev. Mater. 6 (2020) 168-190. https://doi.org/10.1038/s41578-020-00254-z

[4] W. Xu, X. Li, D. Chen, N. Li, Q. Xu, H. Li, J. Lu, TiO2-Cu7S4 modified with a carbazole-based conjugated porous polymer for adsorption and photocatalytic degradation of bisphenol A, Green Energy Environ. 10 (2025) 598-608. https://doi.org/10.1016/j.gee.2024.05.010

[5] X. Guan, S. Zong, S. Shen, Homojunction photocatalysts for water splitting, Nano Res. 15 (2022) 10171-10184. https://doi.org/10.1007/s12274-022-4704-9

[6] H. Li, H. An, B. Chong, G. Yang, L. Wang, Homogeneous dual-site P lattice doping in CdS quantum rods for visible-light photocatalytic water splitting, Chem. Eng. Sci. 238 (2021) 116594. https://doi.org/10.1016/j.ces.2021.116594

[7] K. Afroz, M. Moniruddin, N. Bakranov, S. Kudaibergenov, N. Nuraje, A heterojunction strategy to improve the visible light sensitive water splitting performance of photocatalytic materials, J. Mater. Chem. A 6 (2018) 21696-21718. https://doi.org/10.1039/C8TA04165B

[8] L. Cheng, Y. Li, G. Cao, G. Sun, J. Cao, Y. Wang, Boosting TEA sensing performance of ZnO porous hollow spheres via in situ construction of ZnS-ZnO heterojunction, Sens. Actuators, B. 364 (2022) 131883. https://doi.org/10.1016/j.snb.2022.131883

[9] R. Krishnan, S.N. Shibu, D. Poelman, A.K. Badyal, A.K. Kunti, H.C. Swart, S.G. Menon, Recent advances in microwave synthesis for photoluminescence and photocatalysis, Mater. Today Commun. 32 (2022) 103890. https://doi.org/10.1016/j.mtcomm.2022.103890

[10] J. Miranda de Carvalho, D. Van der Heggen, L.I.D.J. Martin, P.F. Smet, Microwave-assisted synthesis followed by a reduction step: making persistent phosphors with a large storage capacity, Dalton Trans. 49 (2020) 4518-4527. https://doi.org/10.1039/D0DT00416B

[11] T.H. Chang, Y.C. Chang, C.I. Lee, Y.R. Lin, F.H. Ko, Optimization temperature programming of microwave-assisted synthesis ZnO nanoneedle arrays for optical and surface-enhanced Raman scattering applications, Nanomaterials 12 (2022) 3989. https://doi.org/10.3390/nano12223989

[12] R. He, S. Cao, J. Yu, Y. Yang, Microwave-assisted solvothermal synthesis of Bi4O5I2 hierarchical architectures with high photocatalytic performance, Catal. Today 264 (2016) 221-228. https://doi.org/10.1016/j.cattod.2015.07.029

[13] H. Zhou, L. Wang, H. Shi, H. Zhang, Y. Wang, S. Bai, Y. Yang, Y. Li, T. Zhang, H. Zhang, Highly efficient solar-driven photocatalytic hydrogen evolution by a ternary 3D ZnIn2S4-MoS2 microsphere/1D TiO2 nanobelt heterostructure, New J. Chem. 45 (2021) 14167-14176. https://doi.org/10.1039/D1NJ00608H

[14] Y. Guo, J. Li, Y. Yuan, L. Li, M. Zhang, C. Zhou, Z. Lin, A Rapid microwave-assisted thermolysis route to highly crystalline carbon nitrides for efficient hydrogen generation, Angew. Chem. 128 (2016) 14913-14917. https://doi.org/10.1002/ange.201608453

[15] L. Hadidi, E. Davari, D.G. Ivey, J.G.C. Veinot, Microwave-assisted synthesis and prototype oxygen reduction electrocatalyst application of N-doped carbon-coated Fe3O4 nanorods, Nanotechnology 28 (2017) 095707. https://doi.org/10.1088/1361-6528/aa5716

[16] X. Yu, Z. Zhao, D. Sun, N. Ren, J. Yu, R. Yang, H. Liu, Microwave-assisted hydrothermal synthesis of Sn3O4 nanosheet/rGO planar heterostructure for efficient photocatalytic hydrogen generation, Appl. Catal. B Environ. 227 (2018) 470-476. https://doi.org/10.1016/j.apcatb.2018.01.055

[17] H.J. Yashwanth, S.R. Rondiya, H.I. Eya, N.Z. Dzade, D.M. Phase, S.D. Dhole, K. Hareesh, Synergy between nitrogen, phosphorus co-doped carbon quantum dots and ZnO nanorods for enhanced hydrogen production, J. Alloys Compd. 937 (2023) 168397. https://doi.org/10.1016/j.jallcom.2022.168397

[18] S. Tasleem, M. Tahir, Recent progress in structural development and band engineering of perovskites materials for photocatalytic solar hydrogen production: A review, Int. J. Hydrogen Energy 45 (2020) 19078-19111. https://doi.org/10.1016/j.ijhydene.2020.05.090

[19] H. Nishiyama, T. Yamada, M. Nakabayashi, Y. Maehara, M. Yamaguchi, Y. Kuromiya, Y. Nagatsuma, H. Tokudome, S. Akiyama, T. Watanabe, R. Narushima, S. Okunaka, N. Shibata, T. Takata, T. Hisatomi, K. Domen, Photocatalytic solar hydrogen production from water on a 100-m2 scale, Nature 598 (2021) 304-307. https://doi.org/10.1038/s41586-021-03907-3

[20] S. San Martín, M.J. Rivero, I. Ortiz, Unravelling the mechanisms that drive the performance of photocatalytic hydrogen production, Catalysts 10 (2020) 901. https://doi.org/10.3390/catal10080901

[21] S. Cao, L. Piao, X. Chen, Emerging photocatalysts for hydrogen evolution, Trends Chem. 2 (2020) 57-70. https://doi.org/10.1016/j.trechm.2019.06.009

[22] P. Ganguly, M. Harb, Z. Cao, L. Cavallo, A. Breen, S. Dervin, D.D. Dionysiou, S.C. Pillai, 2D Nanomaterials for photocatalytic hydrogen production, ACS Energy Lett. 4 (2019) 1687-1709. https://doi.org/10.1021/acsenergylett.9b00940

[23] G. Zhao, W. Ma, X. Wang, Y. Xing, S. Hao, X. Xu, Self-water-absorption-type two-dimensional composite photocatalyst with high-efficiency water absorption and overall water-splitting performance, Adv. Powder Mater. 1 (2022) 100008. https://doi.org/10.1016/j.apmate.2021.09.008

[24] Y.J. Huang, L.M. Lyu, C.Y. Lin, G.C. Lee, K.Y. Hsiao, M.Y. Lu, Improved mass-transfer enhances photo-driven dye degradation and H2 Evolution over a few-layer WS2 /ZnO heterostructure, ACS Omega 7 (2022) 2217-2223. https://doi.org/10.1021/acsomega.1c05756

[25] J.E. Ramos-Sanchez, R. Camposeco, S.W. Lee, V. Rodríguez-González, Sustainable synthesis of AgNPs/strontium-titanate-perovskite-like catalysts for the photocatalytic production of hydrogen, Catal. Today 341 (2020) 112-119. https://doi.org/10.1016/j.cattod.2019.08.020

[26] T.N.Q. Trang, M. Tran Van, T.B. Phan, V.T.H. Thu, Spatially controlled photogenerated charge carriers induced by SrTiO3 - architectured heterojunction nanocubes for a photocatalytic hydrogen evolution reaction, ACS Appl. Energy Mater. 4 (2021) 8910-8921. https://doi.org/10.1021/acsaem.1c01163

[27] R.R. Mishra, A.K. Sharma, Microwave-material interaction phenomena: heating mechanisms, challenges and opportunities in material processing, Compos. Part A Appl. Sci. Manuf. 81 (2016) 78-97. https://doi.org/10.1016/j.compositesa.2015.10.035

[28] S. Zhang, Q. Qiu, C. Zeng, K.W. Paik, P. He, S. Zhang, A review on heating mechanism, materials and heating parameters of microwave hybrid heated joining technique, J. Manuf. Process. 116 (2024) 176-191. https://doi.org/10.1016/j.jmapro.2024.02.055

[29] C. Xue, Y. Mao, W. Wang, Z. Song, X. Zhao, J. Sun, Y. Wang, Current status of applying microwave-associated catalysis for the degradation of organics in aqueous phase - A review, J. Environ. Sci. 81 (2019) 119-135. https://doi.org/10.1016/j.jes.2019.01.019

[30] J. Sun, W. Wang, Q. Yue, Review on microwave-matter interaction fundamentals and efficient microwave-associated heating strategies, materials (Basel). 9 (2016) 231. https://doi.org/10.3390/ma9040231

[31] F. Zhang, T. Zhou, Y. Liu, J. Leng, Microwave synthesis and actuation of shape memory polycaprolactone foams with high speed, Sci. Rep. 5 (2015) 11152. https://doi.org/10.1038/srep11152

[32] N. Yoshikawa, Mechanism of microwave heating of matter, in: RF Power Semicond. Gener. Appl. Heat. Energy Util., Springer Singapore, Singapore, 2020, pp. 71-89. https://doi.org/10.1007/978-981-15-3548-2_3

[33] H. Vasudev, G. Singh, A. Bansal, S. Vardhan, L. Thakur, Microwave heating and its applications in surface engineering: a review, Mater. Res. Express 6 (2019) 102001. https://doi.org/10.1088/2053-1591/ab3674

[34] C. Valverde, M.M. Rodríguez-García, E. Rojas, R. Bayón, State of the art of the fundamental aspects in the concept of microwave-assisted heating systems, Int. Commun. Heat Mass Transf. 156 (2024) 107594. https://doi.org/10.1016/j.icheatmasstransfer.2024.107594

[35] Z. Peng, J.Y. Hwang, M. Andriese, Magnetic loss in microwave heating, Appl. Phys. Express 5 (2012) 027304. https://doi.org/10.1143/APEX.5.027304

[36] C. Kampalapura Swamy, A. Hezam, A. Mavinakere Ramesh, D. Habbanakuppe Ramakrishnegowda, D.K. Purushothama, J. Krishnegowda, R. Kanchugarakoppal S., S. Shivanna, Microwave hydrothermal synthesis of copper induced ZnO/gC3N4 heterostructure with efficient photocatalytic degradation through S-scheme mechanism, J. Photochem. Photobiol. A Chem. 418 (2021) 113394. https://doi.org/10.1016/j.jphotochem.2021.113394

[37] Z. Cao, Q. Yin, Y. Zhang, Y. Li, C. Yu, M. Zhang, B. Fan, G. Shao, H. Wang, H. Xu, R. Zhang, H. Lu, Heterostructure composites of TiO2 and CdZnS nanoparticles decorated on Ti3C2Tx nanosheets and their enhanced photocatalytic performance by microwave hydrothermal method, J. Alloys Compd. 918 (2022) 165681. https://doi.org/10.1016/j.jallcom.2022.165681

[38] I. Ahmad, S. Shukrullah, M.Y. Naz, E. Ahmed, M. Ahmad, M.S. Akhtar, S. Ullah, M.U. Farooq, S. Iqbal, M.A. Assiri, M. Alzaid, H. Alrobei, Microwave-assisted one-pot hydrothermal synthesis of V and La co-doped ZnO/CNTs nanocomposite for boosted photocatalytic hydrogen production, Int. J. Hydrogen Energy 47 (2022) 15505-15515. https://doi.org/10.1016/j.ijhydene.2021.09.214

[39] F. Lin, S. Zhou, G. Wang, J. Wang, T. Gao, Y. Su, C.P. Wong, Electrostatic self-assembly combined with microwave hydrothermal strategy: construction of 1D/1D carbon nanofibers/crystalline g-C3N4 heterojunction for boosting photocatalytic hydrogen production, Nano Energy 99 (2022) 107432. https://doi.org/10.1016/j.nanoen.2022.107432

[40] K. Aihara, R. Mizuochi, M. Okazaki, S. Nishioka, S. Yasuda, T. Yokoi, F. Ishiwari, A. Saeki, M. Inada, K. Maeda, Low-temperature microwave-assisted hydrothermal synthesis of Pb2Ti2O5.4F1.2 photocatalyst for improved H2 evolution under visible light, ACS Mater. Lett. 5 (2023) 2355-2360. https://doi.org/10.1021/acsmaterialslett.3c00574

[41] M.A. Mahadik, R.P. Patil, W.S. Chae, H. Hwi Lee, M. Cho, J. Suk Jang, Microwave-assisted rapid synthesis of Cu2S:ZnIn2S4 marigold-like nanoflower heterojunctions and enhanced visible light photocatalytic hydrogen production via Pt sensitization, J. Ind. Eng. Chem. 108 (2022) 203-214. https://doi.org/10.1016/j.jiec.2022.01.001

[42] A. Chachvalvutikul, T. Luangwanta, S. Pattisson, G.J. Hutchings, S. Kaowphong, Enhanced photocatalytic degradation of organic pollutants and hydrogen production by a visible light-responsive Bi2WO6/ZnIn2S4 heterojunction, Appl. Surf. Sci. 544 (2021) 148885. https://doi.org/10.1016/j.apsusc.2020.148885

[43] S. Bao, Q. Tan, S. Wang, J. Guo, K. Lv, S.A.C. Carabineiro, L. Wen, TpBD COF@ZnIn2S4 nanosheets: A novel S-scheme heterojunction with enhanced photoreactivity for hydrogen production, Appl. Catal. B Environ. 330 (2023) 122624. https://doi.org/10.1016/j.apcatb.2023.122624

[44] T. Ma, Z. Li, W. Liu, J. Chen, M. Wu, Z. Wang, Microwave hydrothermal synthesis of WO3(H2O)0.333/CdS nanocomposites for efficient visible-light photocatalytic hydrogen evolution, Front. Mater. Sci. 15 (2021) 589-600. https://doi.org/10.1007/s11706-021-0581-5

[45] A. Kumar, Y. Kuang, Z. Liang, X. Sun, Microwave chemistry, recent advancements, and eco-friendly microwave-assisted synthesis of nanoarchitectures and their applications: a review, Mater. Today Nano 11 (2020) 100076. https://doi.org/10.1016/j.mtnano.2020.100076

[46] P.P. Falara, N. Chatzikonstantinou, A. Zourou, P. Tsipas, E. Sakellis, E. Alexandratou, N.K. Nasikas, K.V. Kordatos, M. Antoniadou, Optimizing carbon dot-TiO2 nanohybrids for enhanced photocatalytic hydrogen evolution, Materials. 18 (2025) 1023. https://doi.org/10.3390/ma18051023

[47] G.J. Lee, Y.H. Hou, C.Y. Chen, C.Y. Tsay, Y.C. Chang, J.H. Chen, T.L. Horng, S. Anandan, J.J. Wu, Enhanced performance for photocatalytic hydrogen evolution using MoS2/graphene hybrids, Int. J. Hydrogen Energy 46 (2021) 5938-5948. https://doi.org/10.1016/j.ijhydene.2020.07.003

[48] S. Khan, H. Ikari, N. Suzuki, K. Nakata, C. Terashima, A. Fujishima, K. Katsumata, V. Rodríguez-González, One-pot synthesis of anatase, rutile-decorated hydrogen titanate

nanorods by Yttrium doping for solar H2 production, ACS Omega 5 (2020) 23081-23089. https://doi.org/10.1021/acsomega.0c02855

[49] V.G. Deon, A. Thesing, L.R. Santana, V.C. Costa, M.O. Vaz, R.M. Silva, M.F. Mesko, N.L. V. Carreno, Synthesis of LiNbO3 nanocrystals by microwave-assisted hydrothermal method: formation mechanism and application to hydrogen evolution reaction, Chem. Pap. 75 (2021) 3807-3815. https://doi.org/10.1007/s11696-021-01616-9

[50] Q. Tan, K. Li, Q. Li, Y. Ding, J. Fan, Z. Xu, K. Lv, Photosensitization of TiO2 nanosheets with ZnIn2S4 for enhanced visible photocatalytic activity toward hydrogen production, Mater. Today Chem. 26 (2022) 101114. https://doi.org/10.1016/j.mtchem.2022.101114

[51] S. Pati, R. Acharya, An overview on g-C3N4 as a robust photocatalyst towards the sustainable generation of H2 energy, Mater. Today Proc. 35 (2021) 175-178. https://doi.org/10.1016/j.matpr.2020.04.178

[52] J. Kundu, D.D. Mal, D. Pradhan, Single-step solvothermal synthesis of highly uniform CdxZn1−xS nanospheres for improved visible light photocatalytic hydrogen generation, Inorg. Chem. Front. 8 (2021) 3055-3065. https://doi.org/10.1039/D0QI00531B

[53] S.H. Chen, Y.S. Jiang, H. Lin, Easy synthesis of BiVO4 for photocatalytic overall water splitting, ACS Omega 5 (2020) 8927-8933. https://doi.org/10.1021/acsomega.0c00699

[54] T. Hu, K. Dai, J. Zhang, G. Zhu, C. Liang, Noble-metal-free Ni2P as cocatalyst decorated rapid microwave solvothermal synthesis of inorganic-organic CdS-DETA hybrids for enhanced photocatalytic hydrogen evolution, Appl. Surf. Sci. 481 (2019) 1385-1393. https://doi.org/10.1016/j.apsusc.2019.03.237

[55] Q. Hu, G. Chen, Y. Wang, J. Jin, M. Hao, J. Li, X. Huang, J. Jiang, Enhancing photocatalytic H2 evolution on In2S3 /mesoporous TiO2 nanocomposites via one-pot microwave-assisted synthesis using an ionic liquid, Nanoscale 12 (2020) 12336-12345. https://doi.org/10.1039/D0NR02958K

[56] G. Li, H. Liang, H. Yu, T. Xu, J. Bai, Interstitial carbon doped of setaria viridis-like Znln2S4 hollow tubes for efficient the performance of photocatalytic hydrogen production, Int. J. Hydrogen Energy 46 (2021) 29951-29959. https://doi.org/10.1016/j.ijhydene.2021.06.150

[57] Z. Zhao, C. Chen, X. Meng, Z. Li, Joule heating assisted solvothermal growth of C3N4 on carbon paper for effective photothermal-photocatalytic water splitting for hydrogen evolution, Sep. Purif. Technol. 354 (2025) 129202. https://doi.org/10.1016/j.seppur.2024.129202

[58] C.J. Chang, M.C. Teng, J. Chen, Y.G. Lin, C.Y. Chen, Microwave solvothermal synthesis of cubic MnS@Ag2S core-shell photocatalysts with improved charge separation and photocatalytic activity, Appl. Surf. Sci. 558 (2021) 149875. https://doi.org/10.1016/j.apsusc.2021.149875

[59] T. Gao, X. Liu, Q. Feng, X. Wu, J. Wang, G. Wang, Microwave assisted rapid synthesis of TiO2 Nfs@COF S-scheme heterojunction photocatalyst for highly efficient photocatalytic hydrogen evolution, J. Colloid Interface Sci, 689 (2025) 138075. https://doi.org/10.1016/j.jcis.2025.138075

[60] G.J. Lee, Y.H. Hou, N. Liu, J.J. Wu, Enhanced photocatalytic hydrogen and methane evolution using chalcogenide with metal ion modification via a microwave-assisted solvothermal method, Catal. Today 355 (2020) 493-501. https://doi.org/10.1016/j.cattod.2019.06.068

[61] Q. Yue, G. Li, P. Fu, B. Meng, F. Ma, Y. Zhou, J. Wang, In-situ polarization of covalent organic frameworks in seawater enables enhanced photocatalytic hydrogen evolution under visible-light irradiation, Nano Res. 16 (2023) 6251-6259. https://doi.org/10.1007/s12274-022-5332-0

[62] J.M. Serra, J.F. Borrás-Morell, B. García-Baños, M. Balaguer, P. PlazaGonzález, J. Santos-Blasco, D. Catalán-Martínez, L. Navarrete, J.M. Catalá-Civera, Hydrogen production via microwave-induced water splitting at low temperature, Nat. Energy 5 (2020) 910-919. https://doi.org/10.1038/s41560-020-00720-6

[63] N.A. Lima, G.C. Mendonça, G.T.S.T. da Silva, B.S. de Lima, E.C. Paris, M.I.B. Bernardi, Influence of the synthesis method on CuWO4 nanoparticles for photocatalytic application, J. Mater. Sci. Mater. Electron. 32 (2021) 1139-1149. https://doi.org/10.1007/s10854-020-04887-2

[64] M.S. Rohokale, D. Dhabliya, T. Sathish, V. Vijayan, N. Senthilkumar, A novel two-step co-precipitation approach of CuS/NiMn2O4 heterostructured nanocatalyst for enhanced visible light driven photocatalytic activity via efficient photo-induced charge separation properties, Phys. B Condens. Matter 610 (2021) 412902. https://doi.org/10.1016/j.physb.2021.412902

[65] R.K. Choudhary, G.N. Kumaraswamy, R. Baitha, M. Kumar, S.R. Shekokar, A. Kumar, M.H. Hussain, P. Kumar, Synthesis of BNiO3 nanocomposites for photocatalytic hydrogen production applications, J. Inst. Eng. (India): D, 2024, pp. 1-11. https://doi.org/10.1007/s40033-024-00725-5

[66] X. Li, J. Xu, X. Zhou, Z. Liu, Y. Li, S. Zhao, L. Li, M. Mao, Amorphous CoS modified nanorod NiMoO4 photocatalysis for hydrogen production, J. Mater. Sci. Mater. Electron. 31 (2020) 182-195. https://doi.org/10.1007/s10854-019-02576-3

[67] M.A.A. El-Khair, A.G. Al-Gamal, K.I. Kabel, W.S. Gado, A.S. Morshedy, Harvesting the synergistic effect of CuFe2O4@Ni-MOF nanomagnetic photocatalyst for enhanced visible light-driven green hydrogen production, Int. J. Hydrogen Energy 101 (2025) 280-294. https://doi.org/10.1016/j.ijhydene.2024.12.449

[68] C. Chen, Y. Huang, S. Huo, Preparation and photocatalytic H2 production property of graphene oxide/CdS/single crystal ZnO nanorod ternary hybrids, Vacuum 205 (2022) 111467. https://doi.org/10.1016/j.vacuum.2022.111467

[69] K. Liu, L. Mi, H. Wang, X. Xiong, K. Zhang, B. Wang, Preparation of Ba1-xSrxTiO3 by the sol-gel assisted solid phase method: Study on its formation mechanism and photocatalytic hydrogen production performance, Ceram. Int. 47 (2021) 22055-22064. https://doi.org/10.1016/j.ceramint.2021.04.226

[70] C. Wang, J. Li, E. Paineau, A. Laachachi, C. Colbeau-Justin, H. Remita, M.N. Ghazzal, A sol-gel biotemplating route with cellulose nanocrystals to design a photocatalyst for improving hydrogen generation, J. Mater. Chem. A 8 (2020) 10779-10786. https://doi.org/10.1039/C9TA12665A

[71] H.M. Gobara, S.A. Hassan, A.M.A. El Naggar, R.S. Mohamed, A.A. Alkahlawy, A.A. Salem, H.M. Salem, Anomalous behavior of H2 storage in photocatalytic splitting of water over Fe nanoparticles loaded on Al-modified SBA-15 composites via microwave or ultrasonic treatment routes, Int. J. Hydrogen Energy 45 (2020) 24710-24725. https://doi.org/10.1016/j.ijhydene.2020.06.223

[72] S.H. Chen, J.H. Su, H. Lin, New modified sol-gel method for the preparation KNb3O8 as a hydrogen evolution photocatalyst in Z-scheme overall water splitting, Top. Catal. 63 (2020) 996-1004. https://doi.org/10.1007/s11244-020-01272-5

[73] F. Abdulaziz, R.R.D. Patel, T.A.M. Taha, Preparation of Co3−xCaxO4 nanostructure catalyst for improved hydrogen production from sodium borohydride, J. Sol-Gel Sci. Technol. 115 (2024) 1208-1220. https://doi.org/10.1007/s10971-024-06343-7

[74] P. Nunocha, M. Kaewpanha, T. Bongkarn, A. Eiad-Ua, T. Suriwong, Effect of Nb doping on the structural, optical, and photocatalytic properties of SrTiO3 nanopowder synthesized by sol-gel auto combustion technique, J. Asian Ceram. Soc. 10 (2022) 583-596. https://doi.org/10.1080/21870764.2022.2094556

[75] C. Chen, Q. Xia, Z. Li, Y. Wang, Microwave synthesis and photocatalytic H2 production performance of SiO2 aerogel/ZnCdS solid solution catalyst, Inorg. Chem. Commun. 168 (2024) 112874. https://doi.org/10.1016/j.inoche.2024.112874

[76] J. Kwon, K. Choi, E. Tervoort, M. Niederberger, One-pot microwave synthesis of Pd modified titanium dioxide nanocrystals for 3D aerogel monoliths with efficient visible-light photocatalytic activity in a heated gas flow reactor, J. Mater. Chem. A 10 (2022) 18383-18395. https://doi.org/10.1039/D2TA04024G

[77] S. Horikoshi, L. Takahashi, K. Sueishi, H. Tanizawa, N. Serpone, Microwave-driven hydrogen production (MDHP) from water and activated carbons (ACs). Application to wastewaters and seawater, RSC Adv. 11 (2021) 31590-31600. https://doi.org/10.1039/D1RA05977G

[78] Z. Saddique, M. Imran, A. Javaid, F. Kanwal, S. Latif, J.C.S. dos Santos, T.H. Kim, G. Boczkaj, Bismuth-based nanomaterials-assisted photocatalytic water splitting for sustainable hydrogen production, Int. J. Hydrogen Energy 52 (2024) 594-611. https://doi.org/10.1016/j.ijhydene.2023.05.047

[79] V.V. Voytovich, S.A. Kurnosenko, O.I. Silyukov, I.A. Rodionov, I.A. Minich, I.A. Zvereva, Study of n-alkylamine intercalated layered perovskite-Like niobates HCa2Nb3O10 as photocatalysts for hydrogen production from an aqueous solution of methanol, Front. Chem. 8 (2020) 300. https://doi.org/10.3389/fchem.2020.00300

[80] J. Wang, Q. Hao, R. Yang, X. Niu, R. Wang, L. Yang, Q. Huang, J. Ye, H. Yang, Y. Wu, A dual S-scheme heterojunction SrTiO3/SrCO3/C-doped TiO2 as H2 production photocatalyst and its charge transfer mechanism, Appl. Catal. B Environ. 356 (2024) 124232. https://doi.org/10.1016/j.apcatb.2024.124232

[81] T.M.F. Marques, R.N. Morais, F.X. Nobre, J.M. Rocha, A. Ghosh, T.A.S. Soares, B.C. Viana, G. Machado, J.C.S. Costa, J.M.E. De Matos, Hydrogen production from aqueous glycerol using titanate nanotubes decorated with Au nanoparticles as photocatalysts, An. Acad. Bras. Cienc. 91 (2019) e20190082. https://doi.org/10.1590/0001-3765201920190082

[82] J.J. Jiang, F.J. Zhang, Y.R. Wang, Review of different series of MOF/g-C3N4 composites for photocatalytic hydrogen production and CO2 reduction, New J. Chem. 47 (2023) 1599-1609. https://doi.org/10.1039/D2NJ05260A

[83] G. Zhang, G. Li, Z. Lan, L. Lin, A. Savateev, T. Heil, S. Zafeiratos, X. Wang, M. Antonietti, Optimizing optical absorption, exciton dissociation, and charge transfer of a polymeric carbon nitride with ultrahigh solar hydrogen production activity, Angew. Chemie 129 (2017) 13630-13634. https://doi.org/10.1002/ange.201706870

[84] S. Hu, D. Jiang, L. Gu, G. Xu, Z. Li, Y. Yuan, Awakening $n \rightarrow \pi^*$ electronic transition by breaking hydrogen bonds in graphitic carbon nitride for increased photocatalytic hydrogen generation, Chem. Eng. J. 399 (2020) 125847. https://doi.org/10.1016/j.cej.2020.125847

[85] Z. Zhang, L. Ren, H. Li, D. Jiang, Y. Fang, H. Du, G. Xu, C. Zhu, H. Li, Z. Lu, Y. Yuan, π-Conjugated in-plane heterostructure enables long-lived shallow trapping in graphitic carbon nitride for increased photocatalytic hydrogen generation, Small 19 (2023) 2207173. https://doi.org/10.1002/smll.202207173

[86] L. Jiang, L. Wang, G. Xu, L. Gu, Y. Yuan, A microwave-assisted thermolysis route to single-step preparation of MoS2 /CdS composite photocatalysts for active hydrogen generation, Sustain. Energy Fuels 2 (2018) 430-435. https://doi.org/10.1039/C7SE00475C

[87] A.A. Ismail, L.A. Al-Hajji, M. Alsaidi, B.N. Nunes, D.W. Bahnemann, Pyrolysis conversion of metal organic frameworks to form uniform codoped C/N-Titania photocatalyst for H2 production through simulated solar light, J. Photochem. Photobiol., A Chem. 407 (2021) 113037. https://doi.org/10.1016/j.jphotochem.2020.113037

[88] M.M. Rizk, M.S. Mahmoud, N.M. Elzahid, I. Ashour, Synthesis of rGO-like materials from local agricultural wastes by microwave treatment followed by pyrolysis, J. Adv. Eng. Trends 43 (2024) 87-99. https://doi.org/10.21608/jaet.2023.161523.1226

[89] Z. Zhang, H. Chen, W. Hu, M. Xie, Y. Pan, B. Niu, D. Duan, L. Ding, D. Long, Y. Zhang, Temperature-dependent NiFeAl catalysts for efficient microwave-assisted catalytic pyrolysis of polyethylene into value-added hydrogen and carbon nanotubes, Fuel 366 (2024) 131390. https://doi.org/10.1016/j.fuel.2024.131390

[90] A. Ekinci, Positive roles of microwave irradiation in hydrogen production from hydrolysis of sodium borohydride using iron oxide catalyst synthesized with watermelon seed peel extract by green method, Int. J. Hydrogen Energy 48 (2023) 29615-29628. https://doi.org/10.1016/j.ijhydene.2023.04.166

[91] A. Seza, F. Soleimani, N. Naseri, M. Soltaninejad, S.M. Montazeri, S.K. Sadrnezhaad, M.R. Mohammadi, H.A. Moghadam, M. Forouzandeh, M.H. Amin, Novel microwave-assisted synthesis of porous g-C3N4/SnO2 nanocomposite for solar water-splitting, Appl. Surf. Sci. 440 (2018) 153-161. https://doi.org/10.1016/j.apsusc.2018.01.133

[92] S. Gholamrezaei, M. Ghiyasiyan-Arani, M. Salavati-Niasari, H. Moayedi, Multidisciplinary methods (co-precipitation, ultrasonic, microwave, reflux and hydrothermal) for synthesis and characterization of CaMn3O6 nanostructures and its photocatalytic water splitting performance, Int. J. Hydrogen Energy 44 (2019) 26373-26386. https://doi.org/10.1016/j.ijhydene.2019.08.141

[93] M.S. Jyothi, V. Nayak, K.R. Reddy, S. Naveen, A.V. Raghu, Non-metal (oxygen, sulphur, nitrogen, boron and phosphorus)-doped metal oxide hybrid nanostructures as highly efficient photocatalysts for water treatment and hydrogen generation, in nanophotocatalysis and environmental applications: 2019: pp. 83-105. https://doi.org/10.1007/978-3-030-10609-6_3

[94] J. Li, Y. Tang, R. Jin, Q. Meng, Y. Chen, X. Long, L. Wang, H. Guo, S. Zhang, Ultrasonic-microwave assisted synthesis of GO/g-C3N4 composites for efficient photocatalytic H2 evolution, Solid State Sci. 97 (2019) 105990. https://doi.org/10.1016/j.solidstatesciences.2019.105990

[95] P. García-Muñoz, J. Ivanez, V.A. de la Peña O'Shea, N. Keller, F. Fresno, Solar hydrogen production from ethanol-water vapours over metal/TiO2 photocatalysts supported on β-SiC alveolar foams, Catal. Today 413 (2023) 113987. https://doi.org/10.1016/j.cattod.2022.12.018

[96] C. Machut, N. Kania, B. Léger, F. Wyrwalski, S. Noël, A. Addad, E. Monflier, A. Ponchel, Fast microwave synthesis of gold-doped TiO2 assisted by modified cyclodextrins for photocatalytic degradation of dye and hydrogen production, Catalysts 10 (2020) 801. https://doi.org/10.3390/catal10070801

[97] B. Liu, Y. Wang, G. Han, L. Zhang, Y. Huang, Facile microwave-assisted synthesis of magnetic ferrite: Rapid interfacial reaction underlying intensifying mechanism, J. Cleaner Prod. 361 (2022) 132181. https://doi.org/10.1016/j.jclepro.2022.132181

[98] S.H. Yu, Q.L. Wang, Y. Chen, Y. Wang, J.H. Wang, Microwave-assisted synthesis of spinel ferrite nanospherolites, Mater. Lett. 278 (2020) 128431. https://doi.org/10.1016/j.matlet.2020.128431

[99] D.K. Becerra-Paniagua, S. Torres-Arellano, C. Martinez-Alonso, E. Luévano-Hipólito, P.J. Sebastian, Facile and green synthesis of Cu/Cu2O composite for photocatalytic H2 generation, Mater. Sci. Semicond. Process. 162 (2023) 107485. https://doi.org/10.1016/j.mssp.2023.107485

[100] B.C. Hernández-Majalca, M.J. Meléndez-Zaragoza, J.M. Salinas-Gutiérrez, A. López-Ortiz, V. Collins-Martínez, Visible-light photo-assisted synthesis of GO-TiO2 composites for the photocatalytic hydrogen production, Int. J. Hydrogen Energy 44 (2019) 12381-12389. https://doi.org/10.1016/j.ijhydene.2018.10.152

[101] J. Zander, R. Marschall, Novel earth-abundant Cu and Fe-based chalcogenide cocatalysts for photocatalytic hydrogen evolution, Sol. RRL 9 (2025) 2500199. https://doi.org/10.1002/solr.202500199

[102] Y. Xu, C. Du, J.D. Steinkruger, C. Zhou, S. Yang, Microwave-assisted synthesis of AuNPs/CdS composite nanorods for enhanced photocatalytic hydrogen evolution, J. Mater. Sci. 54 (2019) 6930-6942. https://doi.org/10.1007/s10853-018-03294-7

[103] R.A. El-Gendy, H.M. El-Bery, M. Farrag, D.M. Fouad, Metal chalcogenides (CuS or MoS2)-modified TiO2 as highly efficient bifunctional photocatalyst nanocomposites for green H2 generation and dye degradation, Sci. Rep. 13 (2023) 7994. https://doi.org/10.1038/s41598-023-34743-2

[104] A. Majumdar, G.M. Karim, P. Dutta, H. Lee, S.K. Deb, A. Sikdar, Y. Oh, U.N. Maiti, Microwave induced rapid surface amorphization of metal oxide nanowire into sulfides shell

for electronically modulated efficient hydrogen evolution catalyst, Catal. Today 423 (2023) 113962. https://doi.org/10.1016/j.cattod.2022.11.023

[105] Y. Wu, Y. Qi, G. Zhou, H. Wang, Z. Yan, Y. Wu, Ternary metal chalcogenides hollow tube derived from MOF sulfurization for photocatalytic hydrogen generation from water and tetracycline wastewater, Int. J. Hydrogen Energy 72 (2024) 179-188. https://doi.org/10.1016/j.ijhydene.2024.05.402

[106] Y.C. Chang, Y.C. Chiao, P.C. Hsu, Rapid microwave-assisted synthesis of ZnIn2S4 nanosheets for highly efficient photocatalytic hydrogen production, Nanomaterials 13 (2023) 1957. https://doi.org/10.3390/nano13131957

[107] Y. Wang, X. Chen, C. Chen, Microwave synthesis of Zn, Cd binary metal sulfides with superior photocatalytic H2 evolution performance, Inorg. Chem. Commun. 134 (2021) 108993. https://doi.org/10.1016/j.inoche.2021.108993

[108] D. Monga, S. Basu, Tuning the photocatalytic/electrocatalytic properties of MoS2 /MoSe2 heterostructures by varying the weight ratios for enhanced wastewater treatment and hydrogen production, RSC Adv. 11 (2021) 22585-22597. https://doi.org/10.1039/D1RA01760H

[109] J. Wang, Z. Wang, J. Zhang, S. Mamatkulov, K. Dai, O. Ruzimuradov, J. Low, Two-dimensional high-entropy selenides for boosting visible-light-driven photocatalytic performance, ACS Nano 18 (2024) 20740-20750. https://doi.org/10.1021/acsnano.4c06954

[110] M.A. Rasool, R. Sattar, A. Anum, S.A. Al-Hussain, S. Ahmad, A. Irfan, M.E.A. Zaki, An insight into carbon nanomaterial-based photocatalytic water splitting for green hydrogen production, Catalysts 13 (2022) 66. https://doi.org/10.3390/catal13010066

[111] H. Jing, M. You, S. Yi, T. Li, H. Ji, Y. Wang, Z. Zhang, R. Zhang, D. Chen, H. Yang, Precursor-engineering coupled microwave molten-salt strategy enhances photocatalytic hydrogen evolution performance of g-C3N4 nanostructures, ChemSusChem 13 (2020) 827-837. https://doi.org/10.1002/cssc.201902730

[112] K.H. Tan, C.Y. Lin, Y. Shih, Microwave-assisted synthesis of hydrangea-Like graphitic carbon nitride as an effective photocatalyst in dye degradation and hydrogen generation, ACS ES&T Water 4 (2024) 5902-5912. https://doi.org/10.1021/acsestwater.4c00879

[113] D. Wu, S. Hu, H. Xue, X. Hou, H. Du, G. Xu, Y. Yuan, Protonation and microwave-assisted heating induced excitation of lone-pair electrons in graphitic carbon nitride for increased photocatalytic hydrogen generation, J. Mater. Chem. A 7 (2019) 20223-20228. https://doi.org/10.1039/C9TA05135J

[114] M.T. Genc, G. Yanalak, I. Aksoy, E. Aslan, I.H. Patır, Green carbon dots (GCDs) for photocatalytic hydrogen evolution and antibacterial applications, ChemistrySelect 6 (2021) 7317-7322. https://doi.org/10.1002/slct.202102135

[115] J. Xiao, R. Li, H. Jiang, Metal-organic framework-based photocatalysis for solar fuel production, Small Methods 7 (2023) 2201258. https://doi.org/10.1002/smtd.202201258

[116] A. Gupta, N. Puri, Recent advances in hydrogen production using metal organic frameworks and their composites, Int. J. Hydrogen Energy 78 (2024) 303-324. https://doi.org/10.1016/j.ijhydene.2024.06.126

[117] T. Yang, H. Zhang, B. Pang, J.W.C. Wong, Recent advances in transition metal-based metal-organic frameworks for hydrogen production, Small Sci. 5 (2025) 2400446. https://doi.org/10.1002/smsc.202400446

[118] P. Ayala, S. Naghdi, S.P. Nandan, S.N. Myakala, J. Rath, H. Saito, P. Guggenberger, L. Lakhanlal, F. Kleitz, M.C. Toroker, A. Cherevan, D. Eder, The emergence of 2D building units in metal-organic frameworks for photocatalytic hydrogen evolution: A case study with COK-47, Adv. Energy Mater. 13 (2023) 2300961. https://doi.org/10.1002/aenm.202300961

[119] J. Tang, X. Ma, J. He, X. Liu, M. Li, Zr (IV) metal-organic framework based cadmium sulfide for enhanced photocatalytic water splitting, J. Environ. Chem. Eng. 10 (2022) 107820. https://doi.org/10.1016/j.jece.2022.107820

Keyword Index

About the Editors

Inamuddin, Ph.D, is an assistant professor at the Department of Applied Chemistry, Zakir Husain College of Engineering and Technology, Faculty of Engineering and Technology, Aligarh Muslim University, Aligarh, India. He has extensive research experience in multidisciplinary fields of analytical chemistry, materials chemistry, electrochemistry, renewable energy, and environmental science. He has worked on different research projects funded by various government agencies and universities and is the recipient of awards, including the Department of Science and Technology, India, Fast-Track Young Scientist Award and Young Researcher of the Year Award 2020 from Aligarh Muslim University. He has published about 216 research articles in various international scientific journals, 18 book chapters, and 230 edited books with multiple well-known publishers. His current research interests include ion exchange materials, a sensor for heavy metal ions, biofuel cells, supercapacitors, and bending actuators.

Dr. Tariq Altalhi is a Professor of Chemistry at Taif University, Saudi Arabia, where he has served as Head of the Chemistry Department and Vice Dean of the College of Science. He earned his Ph.D. from the University of Adelaide, Australia, receiving the Dean's Commendation for Doctoral Thesis Excellence. In 2015, his research was nominated among the top ten entries for the Green Tech Awards, Germany. His research interests include nanomaterial synthesis, membrane technologies, molecular separation, desalination, drug delivery, and biosensing. He published numerous research articles and has co-edited several scientific books and maintains strong collaborations with major industrial partners in Saudi Arabia.

Dr. Arwa Alrooqi is an Assistant Professor in the Department of Chemistry, Faculty of Science, at Al Baha University, Saudi Arabia. She has published numerous research articles in internationally recognized journals. Dr. Alrooqi is actively engaged in multidisciplinary research, focusing on the synthesis and engineering of organic semiconductors and nanomaterials, their characterization, and their applications in smart window technologies. She has published several high-impact research articles and has co-edited numerous books with leading international publishers.

Maha Khan is a Research Scholar in the Department of Applied Chemistry at Aligarh Muslim University (AMU), India. Her research focuses on enzymatic biofuel cells, electrochemical materials, and sustainable energy systems, with particular emphasis on the development of functional electrodes derived from waste plastics. She has published research articles in reputed journals, including *Scientific Reports* and *Electrochimica Acta*. Her innovative work on converting waste-derived ABS plastic into conductive electrode materials was recognized with the Best Paper Award at the International Conference on Sustainable Developments in Air and Waste Management (2024). As a co-editor, she actively contributes to scientific research, academic writing, and the advancement of sustainable electrochemical technologies.

www.ingramcontent.com/pod-product-compliance
Ingram Content Group UK Ltd.
Pitfield, Milton Keynes, MK11 3LW, UK
UKHW020659041026
15590UKWH00001B/364

9 781644 904022